Springer Remote Sensing/Photogrammetry

The Springer Remote Sensing/Photogrammetry series seeks to publish a broad portfolio of scientific books, aiming at researchers, students, and everyone interested in the broad field of geospatial science and technologies. The series includes peer-reviewed monographs, edited volumes, textbooks, and conference proceedings. It covers the entire area of Remote Sensing, including, but not limited to, land, ocean, atmospheric science and meteorology, geophysics and tectonics, hydrology and water resources management, earth resources, geography and land information, image processing and analysis, satellite imagery, global positioning systems, archaeological investigations, and geomorphological surveying.

Onisimo Mutanga · Prem Chandra Pandey ·
Sandipan Das · Uday Chatterjee
Editors

Revealing Ecosystem Services Through Geospatial Technologies

Beyond the Surface

Springer

Editors
Onisimo Mutanga
Discipline of Geography
School of Agricultural, Earth
and Environmental Sciences
University of KwaZulu-Natal
Pietermaritzburg, South Africa

Prem Chandra Pandey
Department of Life Sciences
School of Natural Sciences (SNS)
Shiv Nadar Institution of Eminence
(Deemed to be University)
Greater Noida, Uttar Pradesh, India

Sandipan Das
Symbiosis Institute of Geoinformatics
Symbiosis International (Deemed
University)
Pune, Maharashtra, India

Uday Chatterjee
Department of Geography
Bhatter College, Dantan (Autonomous)
Medinipur, West Bengal, India

ISSN 2198-0721 ISSN 2198-073X (electronic)
Springer Remote Sensing/Photogrammetry
ISBN 978-3-031-98047-3 ISBN 978-3-031-98048-0 (eBook)
https://doi.org/10.1007/978-3-031-98048-0

This work was supported by National Research Foundation (South Africa) Chair in Landuse Planning and Management (84157).

This Springer imprint is published by the registered company Springer Nature Switzerland AG
The registered company address is: Gewerbestrasse 11, 6330 Cham, Switzerland

Dedicated to Young Scholars in the Field of Geography, Environmental Science and Sustainability Science

Foreword

Ecosystem services form the backbone of human survival and well-being. Ecosystems maintain life by offering vital direct resources such as clean water and healthy soil, indirect functions, that include climate regulation and the promotion of cultural and recreational activities. However, as human activities increasingly come into conflict with geographical, ecological, and political boundaries, our understanding and management of these invaluable services must be improved accordingly. The book, *Revealing Ecosystem Services Through Geospatial Technologies: Beyond the Surface,* addresses this critical need by exploring the ecosystems and services they provide for all possible themes.

In an era marked by rapid urbanization, climate change, and increasing biodiversity loss, the need for holistic approaches to monitor, conserve, and manage ecosystems is a priority. This volume explored the advanced methods, such as remote sensing and geospatial tools, with traditional ecological knowledge to present ecosystem service research. It presents a diverse array of case studies and methodologies beyond the conventional methods, demonstrating the universal relevance of ecosystem services across regions.

The chapters in this book collectively showcase the transformative potential of interdisciplinary collaboration. From urban landscapes to coastal mangroves, and from rangelands to wetlands, the contributors explore the complex dynamics that underpin ecosystem services. They employ advanced techniques to analyze, model, and interpret data, providing insight on how ecosystems function, respond to change, and contribute to human well-being. By doing so, they provide actionable insights for policymakers, practitioners, and researchers striving to address global environmental challenges. It provides a comprehensive overview of the current state of the art and highlights the potential for future advancements.

Ecosystem services' investigation beyond boundaries serves as both a call to action and a source of inspiration for those dedicated to fostering sustainable and equitable environmental stewardship. The insights and innovations presented in this volume pave the way for a future where the value of ecosystem services is not only recognized but actively safeguarded for generations to come across the globe.

I commend the independent reviewers, editors, and contributors for their remarkable efforts in bringing this important work to fruition. This book is a significant contribution to the field and a testament to the power of technology in addressing the pressing environmental challenges of our time. I hope that this book will ignite further research, collaboration, and action in the pursuit of a more sustainable world, while maintaining sustainable ecosystem.

Pietermaritzburg, South Africa John Odindi

Preface

Ecosystem services are the foundation of human well-being, providing critical resources, regulating processes, and fostering cultural and supporting functions that sustain life on Earth. In a world increasingly shaped by rapid urbanization, climate change, and land use transitions, understanding and managing these services across boundaries, related to ecological and geographical dimension, are of utmost importance. The book, *Revealing Ecosystem Services Through Geospatial Technologies: Beyond the Surface,* is a compilation of research efforts aimed at addressing the diverse challenges and opportunities in ecosystem services beyond the surface and incorporating advanced remote sensing and geospatial technologies.

The 19 chapters within this volume represent a diverse array of perspectives, methodologies, and case studies, providing insight on different aspects of ecosystem services. These chapters employ traditional knowledge, field inventory, and advanced geospatial techniques to present ecosystems services and provide their mapping, modelling, and risk analysis. The chapters in this volume highlight the critical role of remote sensing, machine learning, and field-based approaches in enhancing our understanding of ecosystem services and their sustainable management. Each chapter offers valuable insights into different kinds of unique research themes, and the last chapter concludes with the challenges and opportunities for further research in diverse ecosystems.

The book is divided into 5 main parts, with the First Part introducing the concepts and summarizing the main findings from various chapters. Second Part delves into the Terrestrial Ecosystem services, exploring the potential of urban city, fallow lands, grasslands, agroecosystems and reforested landscape in providing the services. Third Part demonstrates the advanced geospatial tools and techniques employed in monitoring aquatic ecosystem services, including mangroves, coastal areas and wetland ecosystems. Fourth Part focuses on threats to the ecosystem services, which can impact the life and disrupt smooth functioning of various ecosystems. Finally, the Fifth part unpacks the challenges and opportunities in ecosystem services monitoring for a sustainable future.

We, editors, hope that this book will serve as an invitation to researchers, policymakers, and practitioners to harness the power of technology and interdisciplinary collaboration in safeguarding the ecosystems that sustain us. We extend our gratitude to the contributors and anonymous peer reviewers, whose rigorous research and innovative approaches form the backbone of this book. The peer review process enabled the selection and production of high-quality science products. It is our hope that the book serves as a cornerstone for advancing knowledge and inspiring action in ecosystem services management for generations to come.

Pietermaritzburg, South Africa — Onisimo Mutanga
Greater Noida, India — Prem Chandra Pandey
Pune, India — Sandipan Das
Medinipur, India — Uday Chatterjee

Acknowledgments

We extend our profound gratitude to the contributors, colleagues, publishers, researchers, and readers who had a pivotal role in the preparation and publication of *Revealing Ecosystem Services Through Remote Sensing: Beyond the Surface*. We sincerely thank the authors and researchers whose expertise, dedication, and valuable insights have shaped the content of this volume. Their contributions have enriched this book with diverse perspectives and cutting-edge research in the field of remote sensing and ecosystem services. We express our gratitude to the anonymous reviewers for their helpful critiques, which have enhanced both the quality of the study and the book. We also acknowledge the institutions, universities, and research organizations that have supported this project, providing the necessary resources and encouragement to explore the intersection of remote sensing and ecosystem studies. We extend our heartfelt gratitude to our families, colleagues, and mentors for their constant support, patience, and motivation. Their encouragement has been invaluable in bringing this book to fruition. We express our gratitude to the publishing team at Springer for their unwavering support and professional guidance throughout the publication process. Their meticulous editing, design, and production endeavors have converted the book into a comprehensive and intuitive resource. We hope this book serves as a valuable resource for scholars, practitioners, and policymakers, inspiring further research and innovation in understanding and preserving our planet's ecosystems.

Contents

Editors and Contributors

About the Editors

Onisimo Mutanga is Professor of Remote Sensing and South African Research Chair on Landuse Planning and Management. He received his Ph.D. on Hyperspectral Remote Sensing of Tropical Grass quality and quantity from the ITC, Wageningen University, Netherlands, in 2004. His expertise lies on resource modeling, pattern and condition analysis in the face of global and land use change. This includes the development of innovative remote sensing approaches for improved monitoring and management of various ecosystem traits. Focus areas include vegetation mapping, Ecosystem services quantification, forest fragmentation, disaster risk assessment, rangeland quality and quantity monitoring as well as invasive species modeling, among others. He has published more than 350 articles and has a Scopus H-Index of 58 with over 11,000 citations. He is currently Associate Editor, *ISPRS Photogrammetry and Remote Sensing*, and Editor of the *Journal of Environmental Management* and has been Guest Editor of a number of special issues, including being Editor of the *Journal of Spatial Science*. He is Recipient of the National Science and Technology Forum (NSTF-South 32 Awards) Engineering Research Capacity Development award, Geography Fellowship award, among others.

Dr. Prem Chandra Pandey received his Ph.D. from University of Leicester, United Kingdom. He served as Research Scientist at Tel Aviv University, Israel. Currently, he is working as Assistant Professor at Department of Life Sciences, Shiv Nadar Institution of Eminence (deemed to be University) Greater Noida, India, and also worked with the Center for Environmental Sciences and Engineering, Shiv Nadar University. Previously, he had been associated with Banaras Hindu University India. He received his M.Sc. and M.Tech. in Remote Sensing. He has been Recipient of several awards including The Commonwealth award UK, INSPIRE fellowship, MHRD and UGC fellowships, IT Award, Young Investigator award, and SERB-NPDF from Government of India. He has published more than 90+ peer reviewed journal papers, has edited ten books, several book chapters, and presented his work in national and international conferences. Additionally, he is Life Member of the Indian Society of Geomatics, and the Indian Society of Remote Sensing, as well as IUCN CEM Member (2017–2021; 2022–2025), International Society of Precision Agriculture (ISPA) USA and the Society of Wetland Scientists (2021–2022–2025). He is serving as Editorial Member for the Journal *Geocarto International* (Taylor & Francis), *Scientific Reports* (Nature portfolio) and *Sustainable Development* (Wiley), and associated with *Remote Sensing* (MDPI) as Topic Editor. Dr. Pandey is working in three projects related to Monitoring of Ramsar sites, wetlands/chilika lake mainly focusing on Ramsar sites along with other natural resources based research work funded by NGP and SERB Government of India. Dr. Pandey is also working with science-collaborators for near real-time disaster monitoring at Himalayan region. He has expertise in remote sensing of environment and natural resources for their monitoring, mapping and modeling including forestry, agriculture, and land use land cover.

Dr. Sandipan Das is Assistant Professor at Symbiosis Institute of Geoinformatics, Symbiosis International (Deemed University), Pune, India. Dr. Das pursued his Master's degree in Geoinformatics from Pune University and Ph.D. in Geoinformatics from Symbiosis International (Deemed University). He has 9 years of teaching and 12 years of research experience. His areas of research interest include forest biomass and carbon assessment, water resources management, soil erosion, flood vulnerability, groundwater mapping, drought monitoring, and Geospatial Modelling of Environments. He has worked as Principal Investigator, Co-Principal Investigator, and Team Member on several research projects funded by the Indian Space Research Organisation (ISRO), Department of Science and Technology (DST), and Ministry of Environment and Forest (MoEF). He has published 23 research papers, 4 edited books, 10 book chapters, and 5 conference proceedings. He has organized five capacity building geospatial training programs sponsored by DST and ISRO. Dr. Sandipan Das has completed the Special Issue (S.I.) of *Geophysical, Climatological and Anthropogenic Hazards and Disaster: Vulnerability, Risk Assessment, and Sustainability, Frontiers in Environmental Science Journal* as a Lead Editor. He is Reviewer for the several scientific journals of international reputes. He is Editorial Board Member of Springer *Discover Geoscience Journal*.

Dr. Uday Chatterjee is an Assistant Professor at the Department of Geography, Bhatter College, Dantan, Paschim Medinipur, West Bengal, India and Applied Geographer with a Doctoral Degree in Applied Geography at Ravenshaw University, Cuttack, Odisha, India. His areas of research interest cover Urban Planning, Social and Human geography, Applied Geomorphology, Hazards and Disasters, Environmental Issues. Currently, Dr. Uday Chatterjee has completed the Special Issue (S.I) of *Urbanism, Smart Cities and Modelling*, *Geojournal*, Springer as a Lead Editor, Book Scrics Editor *Development in Environmental Science*, Elsevier.

Contributors

Andrews Amakawa Department of Geography and Resources Development, University of Ghana, Accra, Ghana

Tsitsi Bangira Centre for Transformative Agricultural and Food Systems, School of Agricultural, Earth and Environmental Sciences, University of KwaZulu-Natal (UKZN), Scottsville, Pietermaritzburg, South Africa;
Centre on Climate Change and Planetary Health, London School of Hygiene and Tropical Medicine, London, UK

Mahadev Bera School of Environment and Disaster Management, Ramakrishna Mission Vivekananda Educational and Research Institute, Kolkata, West Bengal, India

Nazuk Bhasin Institute of Environment and Sustainable Development, Banaras Hindu University, Varanasi, India

Naimish Bhatt Department of Civil Engineering, School of Technology, Pandit Deendayal Energy University, Gandhinagar, India

Kuldeep Bora Department of Geography, Assam University, Diphu Campus, Diphu, India

Shobhit Chaturvedi Department of Civil Engineering, School of Technology, Pandit Deendayal Energy University, Gandhinagar, India

Sanjana Chauhan Department of Life Sciences, Sharda School of Basic Sciences and Research, Food Science and Technology, Sharda University, Greater Noida, India

Abel Chemura Department of Natural Resources, Faculty of Geo-Information Science and Earth Observation, University of Twente, Enschede, The Netherlands

Walter Chivasa International Maize and Wheat Improvement Center (CIMMYT), Nairobi, Kenya

Jeganathan Chockalingam Department of Remote Sensing, Birla Institute of Technology, Mesra, Ranchi, India;
Vice Chancellor, Sarala Birla University, Ranchi, India

Sanchali Das Podder Department of Geography, Assam University, Diphu Campus, Diphu, India

Sandipan Das Symbiosis Institute of Geoinformatics, Symbiosis International (Deemed University), Pune, India

Sumanta Das School of Environment and Disaster Management, Ramakrishna Mission Vivekananda Educational and Research Institute, Kolkata, West Bengal, India;
The University of Queensland, St Lucia, QLD, Australia

Timothy Dube Discipline of Earth Sciences, University of the Western Cape, Bellville, South Africa;
Institute for Water Studies, University of the Western Cape, Cape Town, South Africa

Suman Dutta Department of Genetics and Plant Breeding, Ramakrishna Mission Vivekananda Educational and Research Institute, Kolkata, West Bengal, India

Prince Amos Gbatoe Department of Life Sciences, Sharda School of Basic Sciences and Research, Food Science and Technology, Sharda University, Greater Noida, India

Yaqian He Department of Geosciences, University of Arkansas, Fayetteville, AR, USA;
Department of Geography, Indiana University Bloomington, Bloomington, IN, USA

Parag Jyoti Kashyap Department of Geography, Assam University, Diphu Campus, Diphu, India

Rushali Katoch Department of Soil Science, College of Agriculture, Chaudhary Sarwan Kumar Himachal Pradesh Krishi Vishvavidyalaya, Palampur, India

Swadhina Koley Department of Remote Sensing, Birla Institute of Technology, Mesra, Ranchi, India;
Department of Geography, School of Basic and Applied Sciences, Adamas University, Kolkata, India

Siddhi Koli Department of Life Sciences, Sharda School of Basic Sciences and Research, Food Science and Technology, Sharda University, Greater Noida, India

Sudhanshu Kumar Institute of Environment and Sustainable Development, Banaras Hindu University, Varanasi, India

Eungul Lee Department of Geography, Kyung Hee University, Seoul, South Korea;
Department of Climate-Social Science Convergence, Kyung Hee University, Seoul, South Korea

Tafadzwanashe Mabhaudhi Centre for Transformative Agricultural and Food Systems, School of Agricultural, Earth and Environmental Sciences, University of KwaZulu-Natal (UKZN), Scottsville, Pietermaritzburg, South Africa

Rishi Mahajan Department of Microbiology, College of Basic Sciences, Chaudhary Sarwan Kumar Himachal Pradesh Krishi Vishvavidyalaya, Palampur, India

Riya Makwana Department of Civil Engineering, School of Technology, Pandit Deendayal Energy University, Gandhinagar, India

Dadirai Matarira School of Agriculture, Earth and Environmental Science, University of KwaZulu-Natal, Scottsville, Pietermaritzburg, South Africa

Collins Matiza Discipline of Geography, School of Agricultural, Earth and Environmental Sciences, University of KwaZulu-Natal, Scottsville, Pietermaritzburg, South Africa

Trylee Nyasha Matongera School of Biosciences, University of Nottingham Malaysia, Semenyih, Malaysia

Monalisha Mishra School of Tribal Resource Management, KISS Deemed University, Bhubaneswar, Odisha, India

Sfundo Mthiyane Discipline of Geography and Environmental Science, School of Agricultural Earth and Environmental Sciences, University of KwaZulu-Natal, Scottsville, Pietermaritzburg, South Africa

Upenyu Naume Mupfiga Discipline of Geography, School of Agricultural, Earth and Environmental Sciences, University of KwaZulu-Natal, Scottsville, Pietermaritzburg, South Africa;
Department of Geography, Environmental Sustainability and Resilience Building, Midlands State University, Gweru, Zimbabwe

Tatenda Musasa Institute for Water Studies, University of the Western Cape, Cape Town, South Africa;
Department of Geography Environmental Sustainability and Resilience Building, Midlands State University, Gweru, Zimbabwe

Terence D. Mushore Department of Space Science and Applied Physics, Faculty of Science, University of Zimbabwe, Harare, Zimbabwe

Onisimo Mutanga Discipline of Geography and Environmental Science, School of Agricultural Earth and Environmental Sciences, University of KwaZulu-Natal, Scottsville, Pietermaritzburg, South Africa

John Odindi Discipline of Geography and Environmental Science, School of Agricultural Earth and Environmental Sciences, University of KwaZulu-Natal, Scottsville, Pietermaritzburg, South Africa

Anjali Padhi Department of Life Sciences, Sharda School of Basic Sciences and Research, Food Science and Technology, Sharda University, Greater Noida, India

G. K. Panda School of Tribal Resource Management, KISS Deemed University, Bhubaneswar, Odisha, India

Snigdharani Panda School of Tribal Resource Management, KISS Deemed University, Bhubaneswar, Odisha, India

Ashutosh K. Pandey Department of Life Sciences, Sharda School of Basic Sciences and Research, Food Science and Technology, Sharda University, Greater Noida, India

Prem Chandra Pandey Department of Life Sciences, School of Natural Sciences, Shiv Nadar Institution of Eminence (Deemed to Be University), Greater Noida, India

Varun Parmar Department of Soil Science, College of Agriculture, Chaudhary Sarwan Kumar Himachal Pradesh Krishi Vishvavidyalaya, Palampur, India

Prince Patel Department of Civil Engineering, School of Technology, Pandit Deendayal Energy University, Gandhinagar, India

Ankita Paul Department of Environmental Science, Tezpur University, Tezpur, India

Boddupalli M. Prasanna International Maize and Wheat Improvement Center (CIMMYT), Nairobi, Kenya

Ranbir Singh Rana Department of Agronomy, College of Agriculture, Chaudhary Sarwan Kumar Himachal Pradesh Krishi Vishvavidyalaya, Palampur, India

Malini Roy Choudhury School of Environment and Disaster Management, Ramakrishna Mission Vivekananda Educational and Research Institute, Kolkata, West Bengal, India;
The University of Queensland, St Lucia, QLD, Australia

Jayket Sachapara Department of Civil Engineering, School of Technology, Pandit Deendayal Energy University, Gandhinagar, India

Chinmoy Raj Saikia Department of Geography, Bapujee College, Sarukshetri, Assam, India

Narender Kumar Sankhyan Department of Soil Science, College of Agriculture, Chaudhary Sarwan Kumar Himachal Pradesh Krishi Vishvavidyalaya, Palampur, India

Tarun Sharma Centre for Geo-Informatics Research and Training, Chaudhary Sarwan Kumar Himachal Pradesh Krishi Vishvavidyalaya, Palampur, India

Munyaradzi D. Shekede Department of Geography Geospatial Sciences and Earth Observation, University of Zimbabwe, Harare, Zimbabwe;
Department of Geospatial Sciences and Earth Observation, Zimbabwe National Geospatial and Space Agency, Harare, Zimbabwe

Mbulisi Sibanda Department of Geography, Environmental Studies & Tourism, Faculty of Arts, University of the West-Ern Cape, Bellville, Cape Town, South Africa

Gopal Shankar Singh Institute of Environment and Sustainable Development, Banaras Hindu University, Varanasi, India

O. P. Singh Professor, Department of Agricultural Economics, Institute of Agricultural Sciences, Banaras Hindu University, Varanasi, India

Vaishnavi Singh Senior Research Scholar, Department of Agricultural Economics, Institute of Agricultural Sciences, Banaras Hindu University, Varanasi, India

Amit Kumar Tiwari Institute of Environment and Sustainable Development, Banaras Hindu University, Varanasi, India

Benjamin Adjah Torgbor Applied Agricultural Remote Sensing Centre, University of New England, Armidale, Australia;
Forest Commission, Accra, Ghana

Marco Vizzari Department of Agricultural, Food and Environmental Sciences, University of Perugia, Perugia, Italy

Anni Yang Department of Geography and Environmental Sustainability, University of Oklahoma, Norman, OK, USA

Di Yang Department of Geography, University of Florida, Gainesville, FL, USA

Xolile Zuma Discipline of Geography and Environmental Science, School of Agricultural Earth and Environmental Sciences, University of KwaZulu-Natal, Scottsville, Pietermaritzburg, South Africa

General

Introduction to Ecosystem Services and the Role of Geospatial Technologies

Onisimo Mutanga, Prem Chandra Pandey, and Sandipan Das

1 Background

The earth's terrestrial and aquatic systems provide several benefits for human well-being and survival. The Millennium Ecosystem Assessment (2005) defines the benefits that are derived from these systems as Ecosystem services. Ecosystem services, which are categorised into provisioning, regulating, cultural and supporting services have contributed more than 50% of the world's GDP. For example, vegetation provides a number of provisioning (forage production, food), regulatory (climate, temperature), supporting (primary production, nutrient cycling, biomass production) and cultural (educational, recreation) ecosystem services as well as biodiversity maintenance (Masenyama et al. 2022). However, as a result of climate change and anthropogenic impacts, about 60% of the world's ecosystem services have been degraded. This significant reduction of services has serious consequences on food production, climate regulation and net primary production, among other services which directly affect human well-being.

In order to appreciates the benefits of various ecosystems to humanity and monitor their degradation, a systematic quantification of ecosystem services and their change in both time and space is critical. In particular, there is a need to characterize

O. Mutanga (✉)
School of Agricultural, Earth and Environmental Sciences, Discipline of Geography, University of KwaZulu-Natal, Pietermaritzburg, South Africa
e-mail: Mutangao@ukzn.ac.za

P. C. Pandey
Department of Life Sciences, School of Natural Sciences, Shiv Nadar Institution of Eminence (Deemed to Be University), Greater Noida, India

S. Das
Symbiosis Institute of Geoinformatics, Symbiosis International (Deemed University), Pune, India
e-mail: sandipanraj2002@gmai.com

O. Mutanga et al. (eds.), *Revealing Ecosystem Services Through Geospatial Technologies*, Springer Remote Sensing/Photogrammetry,
https://doi.org/10.1007/978-3-031-98048-0_1

the ecological infrastructures, processes and ecosystem functions that underpin the ecosystem services in ways that can be evaluated and tracked. In addition, various intervention measures to restore degraded ecosystems require constant and accurate measurement of the ecosystem characteristics in order to assess the benefits of ecological restoration. Geospatial technologies contribute significantly to mapping, modelling, and quantitative valuation of the ecosystem services in a spatially explicit manner. They provide an opportunity to use standard protocols at various spatial scales as well as wall-to-wall mapping of phenomena through time, which is critical for monitoring on a continuous basis.

This book provides a one stop shop, profiling developments in the geospatial technologies for ecosystem service analysis, with case study demonstration of various application domains. The book compiles traditional as well as latest applications of Remote Sensing and Geographical information systems in quantifying ecosystem services, threats to their sustainability and providing intervention measures. The case study demonstration of various application domains is relevant to ecologists, landuse planners, hydrologists, environmentalists and geographers, among others. The book can be used to better understand the importance of Geospatial technologies in monitoring the changing environment, including assessing the efficacy of various intervention measures. The book demonstrates how single variables (e.g. biomass) measured from remote sensing can be used to predict multiple ecosystem services.

Based on the above statement, the readership of this proposed book will include:

1. Students and academics drawn from Ecology, Geography, Life sciences, Agricultural sciences, Environmental sciences and hydrology
2. Post graduate and senior undergraduate students working on vegetation ecosystems
3. Academic researchers advancing the science and application of remote sensing
4. Management and policy practitioners involved in directing sustainable use of resources

The book is divided into 5 sections, constituting 19 chapters. The first section forms the introductory component and subsequent sections are summarized in the sections below. From the traditional to advanced methods, the authors employed remote sensing time series data to understand diverse services and map their spatial and temporal aspects. Authors employed methods and tools to understand the past, present and future scenarios and linking them to sustainable landuse practices.

2 Terrestrial Ecosystem Services

This section of the book delves into the Terrestrial Ecosystem services, exploring the potential of urban city, fallow lands, grasslands, agroecosystems and reforested landscapes in providing ecosystem services. Chapter 2 highlights recent applications in provisioning, regulating, cultural, and supporting ecosystem services, emphasizing the critical insights gained through remote sensing within an urban landscape (Bhasin et al. 2025). The challenges of data availability, integration, and validation, proposing solutions to overcome these hurdles are also addressed. Chapter 3 by Kolay and Chockalingam (2025) uses remote sensing technology to map and monitor agricultural fallow lands, as supporting ecosystem service for food production which is crucial for regional agro-economic growth. The study showed percentage areas of fallow lands in different locations, thereby facilitating planning for potential crop intensification by utilizing these lands in the seasons apart from the primary growing season. The *Grassland in Focus* chapter 4 by Katoch et al. (2025) explores various methodologies for evaluating and monitoring grassland ecosystem services, emphasizing the ecological, economic and social significance of grasslands. This is followed by the Rice Agroecosystems chapter 5 that provides a methodology for valuing environmental services, emphasizing their relationship with sustainable rice growing techniques (Singh and Singh 2025). The section ends with a detailed profiling of aboveground carbon stocks in reforested landscapes under the past, present and future climate scenarios. The study showed significant above ground carbon dynamics (AGC) under distinct climate scenarios as outlined in CMIP6. The Reforested Landscapes chapter 6, highlights the intersection of urban forestry and climate mitigation, providing insights into carbon sequestration potentials across temporal scales (Matiza and Mutanga 2025).

3 Aquatic Ecosystem Services

The conservation and sustainable management of ecosystems are critical in the context of climate change and increasing anthropogenic pressures. The chapters in this section provide valuable insights into various aspects of ecosystem monitoring, integrating cutting-edge geospatial technologies and field-based methodologies to address ecosystem services along with conservation challenges to support conservation efforts. Few chapters stressed on the coastal and wetland ecosystems, which represent the most dynamic and ecologically sensitive landscapes, subject to complex interactions between natural processes and anthropogenic influences. Advances in remote sensing and geospatial technologies have significantly enhanced the capacity to monitor, analyze, and model these ecosystems (Mohanty and Pandey 2025, Pandey et al. 2025a, 2025b), facilitating high-resolution assessments of vegetation dynamics (Mohanty et al. 2024a, 2024b), chlorophyll content mapping (Mohanty et al. 2024c), biophysical transformations, and hydrological processes, all of which are critical for achieving sustainable development goals (Mohanty et al. 2024d). This section synthesizes a diverse range of studies addressing mangrove and wetland ecosystem dynamics, as well as broader vegetation cover changes, underscoring their ecological significance and the pivotal role of geospatial methodologies in their sustainable management. The protection and sustainable management of aquatic and coastal ecosystems are more important than ever, given the growing challenges posed by climate change and human activities.

Chapter 7 by Kashyap et al. (2025) highlights the significance of mangrove ecosystems in coastal protection, biodiversity conservation, and carbon sequestration. The study emphasizes the need for effective monitoring by integrating geospatial technologies, such as Remote Sensing, GIS, and GNSS, with field inventory methods. This comprehensive approach enhances our understanding of mangrove dynamics and supports conservation planning.

Mishra et al. (2025) in Chap. 8, explores ecosystem disruptions along the Odisha coast, emphasizing the impact of climate change and human activities. Using the Coastal Hazard Wheel (CHW) model and geospatial tools, the study assesses vulnerabilities and proposes management strategies to mitigate risks and promote sustainable coastal management. The findings offer crucial recommendations for preserving biodiversity and ecosystem functions in this region.

Chapter 9 by Dube and Musasa (2025) emphasizes the ecological, economic, and cultural significance of wetlands, which are increasingly threatened by climate change and human activities. The chapter discusses the implications of rising temperatures, sea level rise, and carbon dynamics on wetland ecosystems. It also highlights the role of modern technologies, such as remote sensing, in monitoring and conserving these critical habitats.

In Chap. 10, Chemura et al. (2025) examines vegetation cover dynamics and their impact on water fluxes within the Kafue catchment in Zambia. By integrating remote sensing data and hydrological modeling, the study reveals significant changes in land cover and their implications for water regulation. The findings underscore the importance of considering shrub lands alongside forests in integrated catchment management for sustaining ecosystem services.

Chapter 11 by Bera et al. (2025) focuses on the application of remote sensing technologies for mangrove monitoring. The study reviews advancements in passive and active sensor technologies, including multispectral, hyperspectral, LiDAR, and SAR, to assess mangrove extent, health, and species composition. The integration of machine learning and GIS further enhances the accuracy of mangrove mapping and conservation efforts.

Together, these chapters highlight the role of geospatial technologies, modeling approaches, and field-based assessments in understanding and managing aquatic and coastal ecosystems. The insights presented in this section contribute to ongoing efforts in ecosystem conservation, providing valuable knowledge for researchers, policymakers, and conservation practitioners. As we advance in technology and collaborative research, it is imperative to adopt innovative strategies for the sustainable management of these vital ecosystems. This can lay strong foundation for researchers, policymakers, and conservation practitioners working to protect these vital environments. As environmental challenges intensify, it is essential to adopt forward-thinking strategies that balance technological advancements with sustainable conservation practices.

4 Threats to Ecosystem Services

This section focuses on threats to the ecosystem services, which can impact the life and disrupt smooth functioning of ecosystems. Ecosystem services, the vital benefits that nature provides, face escalating threats from a multitude of human-induced pressures. Habitat destruction and fragmentation, caused by agricultural expansion, urbanization, and resource extraction, significantly reduce ecosystems' ability to provide important functions. Pollution, including air, water, and soil contamination, deteriorates ecosystem health and disrupts natural processes. Climate change acts as an intensifying multiplier, exacerbating existing stressors through altered temperature and precipitation patterns, leading to shifts in species distribution and ecosystem function. Unsustainable farming practices, excessive groundwater extraction, and pollution from industrial and dwelling sources further undermine ecosystem processes, impacting food security, freshwater supply, and air quality. The destruction of natural habitats from infrastructural development and resource extraction reduces cultural and recreational ecosystem services. Geospatial technologies play a crucial role in assessing and mitigating these threats by providing spatially explicit data for monitoring ecosystem health, modeling environmental changes, and informing sustainable management strategies. This section facilitates an understanding of threats to the resilience of systems and to the ecosystem services they offer, and provides possible solutions on how to monitor and assess these threats on an ongoing basis. The threats include pest and disease infestation, Invasive Alien Species (IAP), forest fires, flood risk, temperature extremes and droughts, among others.

Chapter 12 by He et al. (2025) highlights climate regulation as a crucial ecosystem service influenced by land cover and land use change (LCLUC). The study examines how LCLUC affects temperature extremes across China from 1982 to 2013 using remotely sensed data and the Community Earth System Model 2 (CESM2). Findings indicate that LCLUC decreases maximum temperatures in western China but increases them in the east, with similar regional patterns for other temperature indices. Additionally, warm and cold spell durations vary across regions, while warm nights generally decline nationwide.

Chapter 13 by Mupfiga and Mutanga (2025) reviews the use of remote sensing to assess fire impacts on ecosystem services through a systematic literature review. Analyzing 23 studies using the PRISMA protocol, the study highlights increasing research in Asia, with Landsat used for vegetation fire effects and MODIS for fire detection. Findings reveal that fires impact forests, grasslands, wetlands, soils, and the atmosphere through carbon and particulate emissions.

In Chap. 14, Zuma et al. (2025) emphasize the rapid transformation of rangelands due to climate change and human activities, affecting ecosystem services by altering land cover and resource availability. Using a random forest classifier and Landsat data in Google Earth Engine, the study assesses rangeland changes in KwaZulu-Natal from 2000 to 2020. Findings reveal significant fragmentation, with rangeland loss in Vulindlela and fluctuating trends in Inhlazuka communal areas due to forest decline.

Chapter 15 by Chaturvedi et al. (2025) examines the impact of urbanization on coastal infrastructure, environmental health, and ecological balance, emphasizing the importance of land cover analysis for urban planning. Using the Maximum Likelihood Classification (MLC) and Multi-Layer Perceptron—Markov Chain (MLP-MC) model, the study analyzes urban growth in Panaji, Goa, from 1992 to 2042. Findings indicate a 259.24% increase in built-up areas from 1992 to 2022, with further expansion projected by 2042, leading to declines in vegetation, water bodies, and barelands.

In Chap. 16, Chivasa et al. (2025) explore the role of remote sensing in monitoring ecosystem services by assessing the impact of climate change on agricultural productivity and plant disease outbreaks. Remote sensing supports maize disease phenotyping through identification, classification, quantification, and prediction but faces challenges in detecting co-infections. Limited empirical evidence exists on sensor-based phenotyping for mixed infections, as symptom similarities and asymptomatic diseases complicate detection. This review highlights these challenges and explores machine learning and hyperspectral unmixing as potential solutions for improving maize disease phenotyping.

Chapter 17 by Matarira et al. (2025) focuses on assessing human-induced vegetation degradation in the uMngeni Catchment, South Africa, from 2000 to 2023 using spatio-temporal analysis and the RESTREND technique. The study identifies a 25% decline in natural vegetation due to urban and cropland expansion, with major degradation hotspots in Durban Metro and Pietermaritzburg. Approximately 17.1 and 16.2% of the catchment area show severe degradation in rainfall- and moisture-adjusted models, respectively.

Chapter 18 by Mthiyane et al. (2025) explores the impact of woody encroachment on soil organic carbon (SOC) distribution in grasslands, which play a vital role in carbon sequestration and ecosystem services. Using PlanetScope imagery and deep neural networks (DNN), the study quantified SOC variability between pristine and woody encroached grasslands. Results show higher SOC concentrations in encroached areas, with the DNN model achieving an RMSE of 1.90 t/ha and an R^2 of 0.64.

5 Towards a Sustainable Future

Attaining a sustainable future necessitates confronting critical obstacles while capitalizing on the potential offered by geospatial technologies in ecosystem service management. The primary objective of this endeavor is to harmonize human growth with the maintenance of ecological integrity, a challenging undertaking due to the rising world population and consumption trends. Challenges encompass alleviating climate change effects, restoring biodiversity decline, and ensuring equitable resource access. Limited data availability, technological gaps, and socio-economic constraints further hinder effective decision-making. Advancements in remote sensing, artificial intelligence, and big data analytics provide unparalleled prospects to improve environmental monitoring, policy development, and resource conservation. Integrating geospatial technologies with participatory methods enables stakeholders to formulate data-driven, adaptive policies for sustainable land use, biodiversity preservation, and climate resilience. A collaborative and diverse approach is essential for addressing difficulties and maximizing the promise of geospatial tools for a sustainable future.

This book concludes with a chapter by Chauhan et al. (2025) that explores the critical role of ecosystem services in human well-being, including clean water, food security, and climate resilience. The study compares local ecosystem management practices with global strategies to promote sustainability. The research highlights the importance of evidence-based policies and stakeholder engagement in fostering long-term ecosystem resilience and sustainability. Overall, a holistic approach incorporating responsible resource management, sustainable land use, and biodiversity conservation enhances ecosystem vitality while meeting societal demands. Imparting knowledge on the value of ecosystems services to communities through accurate quantification and unpacking the impacts of anthropogenic activities to their existence, is essential for sustainable management of the finite resources.

References

Bera M, Choudhury MR, Das S (2025) Remote sensing for monitoring and assessment of coastal mangrove ecosystems. *In:* Mutanga O, Pandey PC, Das S (eds) Beyond the surface: revealing ecosystem services using geospatial technologies. 199–221

Bhasin N, Kumar S, Tiwari AK, Singh GS (2025) Beyond the Concrete Jungle: Unveiling Urban Ecosystems Through Remote Sensing. *In:* Mutanga O, Pandey PC, Das S (eds) Beyond the surface: revealing ecosystem services using geospatial technologies, 13–39.

Chaturvedi S, Patel P, Makwana R, Sachapara J, Bhati N (2025) Spatiotemporal urban growth assessment in a coastal city from 1992 to 2042 using MLC and MLP-MCA algorithms. In: Mutanga O, Pandey PC, Das S (eds) Beyond the surface: revealing ecosystem services using geospatial technologies, 287–304

Chauhan S, Padhi A, Koli S, Gbatoe PA, Pandey AK (2025) Holistic approaches to ecosystem services: a key to sustainable future. In: Mutanga O, Pandey PC, Das S (eds) Beyond the surface: revealing ecosystem services using geospatial technologies, 373–394.

Chemura A, Shekede MD, Torgbor BA, Mushore TD (2025) Remote sensing vegetation cover change and its impact on water flux parameters at catchment scale: a data-driven approach for the Kafue basin. In: Mutanga O, Pandey PC, Das S (eds) Beyond the surface: revealing ecosystem services using geospatial technologies. 185–198

Chivasa W, Mutanga O, Prasanna BM (2025) Phenotyping complex maize diseases using sensor-based approaches: challenges and prospects. In: Mutanga O, Pandey PC, Das S (eds) Beyond the surface: revealing ecosystem services using geospatial technologies, 305–323.

Dube T, Musasa T (2025) Wetland vegetation species diversity and productivity assessment: implications on ecosystem services. In: Mutanga O, Pandey PC, Das S (eds) Beyond the surface: revealing ecosystem services using geospatial technologies. 171–184

He Y, Yang D, Lee E, Yang A (2025) Land cover and land use change impacts on temperature extremes in China from 1982 to 2013. *In:* Mutanga O, Pandey PC, Das S (eds) Beyond the surface: revealing ecosystem services using geospatial technologies, 225-241

Kashyap PJ, Bora K, Podder SD, Amakawa A, Paul A, Saikia CR (2025) Mangrove monitoring using geospatial technologies incorporating field inventory: a review. In: Mutanga O, Pandey PC, Das S (eds) Beyond the surface: revealing ecosystem services using geospatial technologies. 123–138

Katoch R, Parmar V, Sankhyan NK, Rana RS, Mahajan R, Sharma T (2025) Grasslands in Focus: Evaluating and Monitoring Ecosystem Services. *In:* Mutanga O, Pandey PC, Das S (eds) Beyond the surface: revealing ecosystem services using geospatial technologies, 59–80.

Koley S, Chockalingam J (2025) Extraction of Fallow Lands in the Rabi Season Using Time-Series Landsat 8 Imageries. *In:* Mutanga O, Pandey PC, Das S (eds) Beyond the surface: revealing ecosystem services using geospatial technologies, 41–57.

Masenyama A, Mutanga O, Dube T, Bangira T, Sibanda M, Mabhaudhi T (2022) A systematic review on the use of remote sensing technologies in quantifying grasslands ecosystem services. Gisci Remote Sens 59(1):1000–1025

Matarira D, Mutanga O, Vizzari M (2025) Mapping potential risk areas for ecosystem service loss: an assessment of human induced vegetation degradation in greater uMngeni catchment. In: Mutanga O, Pandey PC, Das S (eds) Beyond the surface: revealing ecosystem services using geospatial technologies. 325–344

Matiza C, Mutanga O (2025) Mapping Aboveground Carbon Stocks Under Past, Present and Future Climate Scenarios in Urban Reforested Landscapes. *In:* Mutanga O, Pandey PC, Das S (eds) Beyond the surface: revealing ecosystem services using geospatial technologies, 97–119.

Millennium Ecosystem Assessment (Program) (ed) (2005) Ecosystems and human well-being: synthesis. Island Press

Mishra M, Panda GK, Panda S (2025) Modelling ecosystem disruption along the Odisha coast of the Bay of Bengal using biophysical parameters. In: Mutanga O, Pandey PC, Das S (eds) Beyond the surface: revealing ecosystem services using geospatial technologies. 139–170

Mohanty S, Srivastava PK, Pandey PC, Singh P, Srivastava S (2024a) Wetland species mapping using advanced technological measurement. Aquat Conser Mar Freshwater Ecosyst 34(12):e70018

Mohanty S, Pandey PC, Srivastava PK (2024b) Aquatic vegetation species identification and mapping using multisource data-review. Aquat Ecosyst Monit 12:166–181

Mohanty S, Pandey PC, Singh P, Dugesar V, Srivastava PK (2024c) Vegetation discrimination based on chlorophyll prediction in Marshy wetland using unmanned aerial vehicles. Aquat Conser Mar Freshwater Ecosyst 34(5):e4170

Mohanty S, Pandey PC, Pandey M, Srivastava PK, Dwivedi CS (2024d) Wetlands contribution and linkage to support SDGs, its indicators and targets-a critical review. Sustain Develop 32(5):5348–5392

Mohanty S, Pandey PC (2025) Spatiotemporal dynamics of Ramsar wetlands and freshwater resources: Technological innovations for ecosystem conservation. Water Environment Research, 97(5), e70072.

Mthiyane S, Matongera TN, Odindi J (2025) Assessment of soil organic carbon across a woody encroached Bisley nature reserve using planetscope data. *In:* Mutanga O, Pandey PC, Das S (eds) Beyond the surface: revealing ecosystem services using geospatial technologies, 345–369

Mupfiga U, Mutanga O (2025) A systematic review of the application of remote sensing in assessing the impact of fire on ecosystem services. In: Mutanga O, Pandey PC, Das S (eds) Beyond the surface: revealing ecosystem services using geospatial technologies, 243–262.

Pandey PC, Srivastava PK, Srivastava SK (eds) (2025a) Aquatic ecosystems monitoring: conventional assessment to advanced remote sensing. CRC Press, New York

Pandey PC, Srivastava PK, Srivastava SK (2025b) Introduction to aquatic ecosystems–editorial message. Aquatic ecosystems monitoring. CRC Press, New York, pp 3–10

Singh V, Singh OP (2025) Comprehensive Valuation of Ecosystem Services in Rice Agroecosystems: Integrating Provisioning, Regulating, Cultural, and Supporting Benefits for Sustainable Practices. *In:* Mutanga O, Pandey PC, Das S (eds) Beyond the surface: revealing ecosystem services using geospatial technologies, 81–96.

Zuma A, Mutanga O, Sibanda M, Dube T, Bangira T, Mabhaudhi T (2025) Land Use Land Cover Changes and Their Implications for Ecosystem Services in a Typical Communal Rangeland Area in Southern Africa, *In:* Mutanga O, Pandey PC, Das S (eds) Beyond the surface: revealing ecosystem services using geospatial technologies, 263–285.

Terrestrial Ecosystem Services

Beyond the Concrete Jungle: Unveiling Urban Ecosystems Through Remote Sensing

Nazuk Bhasin, Sudhanshu Kumar, Amit Kumar Tiwari, and Gopal Shankar Singh

Abstract Urban environments are complex mosaics where natural and built elements coexist and interact, providing a range of ecosystem services crucial for urban sustainability. They play a pivotal role in enhancing the quality of life in cities, providing essential services such as air purification, climate regulation, and recreational spaces. Monitoring and assessing these services are crucial for sustainable urban planning and management. This chapter delves into the innovative use of remote sensing technologies to unveil the hidden dynamics of urban ecosystems. By leveraging images such as optical-hyperspectral-microwave and LiDAR from sensors mounted on satellites, aerial platforms and drones, methodologies for comprehensive data acquisition, processing, and analysis are explored. The chapter highlights recent applications in provisioning, regulating, cultural, and supporting services, emphasizing the critical insights gained through remote sensing. The challenges of data availability, integration, and validation, proposing solutions to overcome these hurdles are also addressed. This chapter underscores the transformative potential of remote sensing in shaping resilient, sustainable urban environments, revealing the intricate interplay between urbanization and natural ecosystems. Through case studies and recent scientific advancements, the path towards a greener, more balanced urban future is illuminated.

Keywords Acoustic · Ecosystem services · LiDAR · Micro-ecosystems · Hyperspectral · Unmanned aerial vehicles

1 Introduction

Urban environments are often considered as static, infrastructure-dominated landscapes, yet they harbour a dynamic interplay of natural and built systems necessary for sustaining urban life. This chapter explores the hidden ecological facets of cities,

N. Bhasin · S. Kumar · A. K. Tiwari · G. S. Singh (✉)
Institute of Environment and Sustainable Development, Banaras Hindu University, Varanasi, India
e-mail: gopalsingh.bhu@gmail.com; gsslab.iesd@yahoo.com

O. Mutanga et al. (eds.), *Revealing Ecosystem Services Through Geospatial Technologies*, Springer Remote Sensing/Photogrammetry,
https://doi.org/10.1007/978-3-031-98048-0_2

their role in resilience and biodiversity conservation, and the transformative potential of remote sensing technologies in uncovering and managing these ecosystems. The following sections explore the overlooked urban ecosystems, innovative remote sensing applications, and the need for integrating ecological perspectives into urban planning.

1.1 Unveiling the Hidden City: Urban Ecosystems Beyond the Concrete

Within the intricate urban fabric, urban ecosystems often remain indiscernible as they are dwarfed by the perceptible built environment (Grey et al. 2023). However, cities conceal a diverse array of life sustaining ecosystems with significant ecological relevance, beneath the overpowering concrete structures. The "hidden" urban ecosystems consist of macro-ecosystems like urban forests, wetlands, rivers, green spaces, as well as of micro-ecosystems prospering in obscure locations such as transportation corridors, vacant spaces, rooftops, abandoned buildings and even cracked pavements (WeForum 2021). These veiled ecosystems intricately amalgamate the natural and artificial components unexpectedly and sustainably and provide critical and fundamental ecosystem services. Yet their contribution is recurrently unnoticed while planning and developing urban paradigms (Oke et al. 2021).

This oversight becomes more apparent as cities continue to expand and transform. Urban environments are characterized by their dynamism and heterogeneity. They are adaptive and are influenced by environmental and anthropogenic forces (Egerer et al. 2021; Alberti 2024). For example, New York's Central Park which functions as an ecological oasis amid a highly urbanized environment (Liu 2023). Similarly, the Meiji Shrine Forests of Tokyo offer major benefits such as biodiversity conservation, carbon sequestration while being located within complex urban setups (Ormsby 2021). Many a times novel ecosystems such as roadside plantations, rooftop gardens, are born out of urban expansion. The vacant spaces left in Berlin after its partition have developed into flourishing meadows and plantations rich in biodiversity (Kowarik 2023). The city's administration has put in concerted efforts to preserve and conserve these unique ecosystems which have emerged and adapted themselves to the constantly altering urban surroundings. Thus, the complexity of urban ecosystems can be understood by their interconnectedness with social, economic, and environmental systems, leading to both challenges and opportunities for sustainable urban development (Kowarik 2023).

This leads to the broader question of the urbanisation paradox: is urbanization a driver of ecological degradation or a catalyst for innovation? Urbanization serves as a driver for ecological deterioration while also potentially spurring ecological innovation. One aspect of urban advancement is frequently associated with resource depletion, habitat fragmentation and ecological deterioration (Wang et al. 2024; Theodorou 2022) while on the contrary it is believed to propose unique prospects for pioneering ecosystem regeneration and management (Lu et al. 2024; Russo and Cirella 2024). Singapore's "City in a Garden" concept (Beatley 2016) is an exemplification of the latter. The massive link of green and biodiversity rich areas around the city is aimed at incorporating natural elements with the human influences. It builds the resilience and sustainability of the urban ecosystem whilst providing essential ecosystem service such as micro-climatic regulation, biodiversity conservation, managing air and heat pollution (Russo and Cirella 2024). Another notable example of urban innovation is Thames River restoration project which was directed at revitalizing the otherwise biologically dead river (Edmonds-Brown 2022). It has contributed significantly to the augmentation of the city's ecosystem services. Thus, cities harbour the potential to conduct ecological experiments with the objective of transforming urban ecosystems into robust, vibrant, multifunctional, resilient ecosystems. Efforts and collaborations between the stakeholders are underway to incorporate the benefits of ecosystem services of natural environments into their urban counterparts (Elmqvist et al. 2015).

1.2 Remote Sensing: A New Lens on Urban Life

The evolution of remote sensing from a simple tool for Earth observation to an advanced method has helped uncover the complexities and intricacies of urban ecosystems (Cracknell 2018). The interactions and associations between the built and natural components of the urban environments have been explored owing to the improved spatio-temporal and spectral resolutions of remote sensing datasets (Metzler et al. 2023). This has given rise to compelling and novel perspectives in the exploration of urbanization and urban life.

At its inception, remote sensing techniques comprised of aerial photography primarily for the temporal mapping of the land use, land cover and forestry (LULCF) changes (Cracknell 2018). In contemporary times expertise in satellite datasets, Light Detection and Ranging (LiDAR), and multispectral and hyperspectral imaging, all contribute to the study and analysis of urban systems (Nguyen et al. 2023). These technologies can now reveal the subtleties of urban landscapes, from vegetation health and soil moisture to pollutant dispersal and habitat fragmentation. For instance, the three-dimensional urban LULCF has been prepared using LiDAR datasets (Chen et al. 2021). One upcoming application of remote sensing for comprehending the complexities of urban ecosystems is the monitoring of urban microclimates, detecting variations in temperatures across different areas of the city, governed by the varying heat absorption and transmission capacities of the land features and materials (Hu et al. 2024; Kousiset al. 2021). Cities like Los Angeles and Tokyo are employing satellite derived datasets to address critical concerns such as urban heat islands and poor air quality (Hulley et al. 2019). Thus, the advancement in remote sensing technologies has contributed to the precise interventions needed to mitigate the deleterious consequences of urbanization. Another notable application is to detect the alterations in the ecosystem composition which are not immediately noticeable to humans. For instance, use of hyperspectral imaging datasets to study the health of vegetation for early signs of stress caused by drought, pollutants or disease (Peng et al. 2022). It has been successfully used to monitor the stress of urban trees to climate change induced drought conditions in Berlin (Kowarik 2023).

Monitoring urban biodiversity using remote sensing capabilities is another promising application. It helps overcome the limitations of time, expense and accessibility posed by traditional biodiversity studies and offers numerous benefits such as improved scales and resolution (Di Martino et al. 2024; Reddy 2021). In Singapore, this approach has been employed to identify key areas for conservation of biodiversity, even in dense urban neighbourhoods. By combining remote sensing data with ecological models, researchers can also track species movement and habitat changes over time, ensuring that conservation efforts remain adaptive and effective (Cavender-Bares et al. 2022; Leitão and Santos 2019). This evolution in remote sensing marks a shift from simple observation to a more sophisticated tool for understanding the hidden and complex layers of urban ecosystems.

The application of remote sensing for urban studies has reconceptualised urban environments as being vibrant, dynamic and multidimensional (Chen et al. 2024; Wellmann et al. 2020). It has brought to fore the variations in the different aspects of urban ecosystems such as LULCF, atmospheric conditions and has changed the outlook of viewing cities as static environments. The real time and temporal monitoring and analysis of these changes, offers insights into the evolution and response of urban ecosystems to various natural and anthropogenic pressures, providing a more nuanced comprehension of the association and interdependencies existing between urban and the surrounding natural environments (Ennouri et al. 2021).

1.3 Scope and Objectives

- The chapter aims to challenge traditional perceptions of urban environments by revealing the unseen ecological dimensions.
- Overview of innovative and disruptive approaches and focus areas.

2 Remote Sensing Technologies for UES Monitoring

2.1 Satellite Platforms: The Unseen Urban Pulse

Satellite platforms have transformed the monitoring and understanding of the complexities of urban environments by delivering a synoptic view of the evolution of the urban landscape (Antrop 2004). These platforms operate as the "unseen pulse" of cities, encapsulating comprehensive information about all aspects of urban life, including land use, environmental changes, and infrastructure developments (Miranda et al. 2017). In Beijing, the use of satellite datasets has assisted in isolating

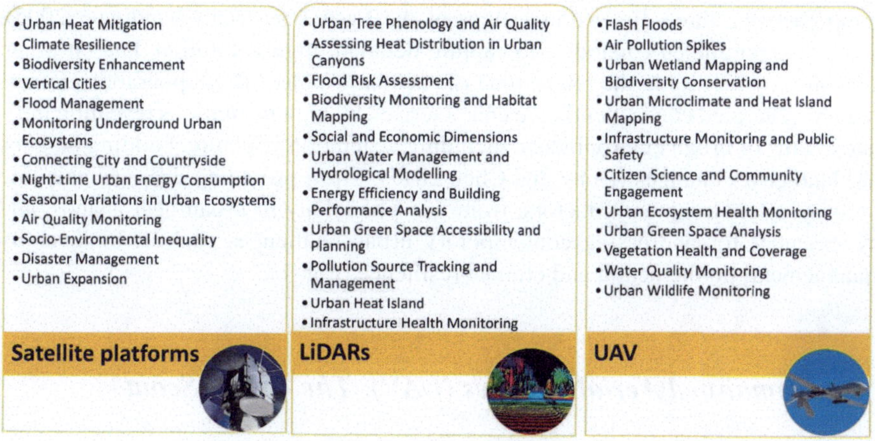

Fig. 1 Applications of remote sensing technologies for monitoring urban ecosystem services

pollution hotspots and evaluating the effectiveness of air quality regulations (Liu et al. 2022). Similarly, in swiftly expanding cities across Africa, satellite imagery facilitates monitoring of deforestation and land-use changes (Masolele et al. 2024). By offering real-time as well as historical data, satellite derived datasets allow policy-makers, urban planners, and environmentalists in informed decision making, aligning with sustainable development goals (SDGs) (Burke et al. 2021). It also ensures more accurate evaluation of urban vulnerabilities and resilience, equipping urban areas to combat climate risks such as floods, droughts, and heatwaves (Zuccaro and Leone 2021).

With the acceleration of urbanization, satellite platforms will see a greater involvement in managing urban growth and ensuring it is carried out within planetary boundaries (Hoornweg et al. 2016) They are quintessential for cultivating data driven urban planning, design and management, based on the principles of protecting and sustaining ecosystems and the fulfilment of human needs (Bibri 2021; Engin et al. 2020). Satellites define the spatio-temporal and spectral aspects of urban environments and offer holistic perceptions regarding the interrelation between anthropogenic activity and urban growth (Fig. 1). The satellite platforms are more than mere observers; they are the architects of insight, enabling cities to understand their indiscernible fluxes—the flows of heat, energy, resources, water, and air that define urban existence (Rios and Ramamurthy 2022; Chrysoulakis et al. 2018).

2.2 *LiDAR: High-Resolution Urban Dynamics*

Light Detection and Ranging (LiDAR) is an airborne platform that performs the remote sensing operation by utilizing laser pulses to quantify distances to construct comprehensive three-dimensional maps of the Earth's surface (Wandinger 2005). It offers unparalleled capability to capture detailed, high-resolution data on urban dynamics by recording the return time of each laser pulse. The topographical maps that are generated consequently are characterised by high resolutions revealing intricate details of urban environments, including vegetation structure, building heights, and land cover changes (Dong and Chen 2017). Thus, providing deep insights into an array of environmental factors, from tree phenology to urban heat islands, and are essential for addressing contemporary urban challenges such as biodiversity management, public health and climate resilience (Fig. 1).

2.3 Unmanned Aerial Vehicles (UAV): The Urban Scout

The extensive and rapid improvement in Unmanned Aerial Vehicles (UAV) technology has extended the purview of urban environmental monitoring (Mohsan et al. 2023). Outfitted with a variety of sensors-thermal, optical, multispectral and LiDAR-UAVs have the ability to capture varied data types for conducting a comprehensive analysis, offering multidimensional insights into transient urban phenomena (Fig. 1). They have emerged as an inexpensive, flexible solution to the demand for concurrent, high-resolution datasets, which are critical for the attainment of urban sustainability. Moreover, UAVs are capable of operating in challenging areas and reach remote areas that are otherwise inaccessible or difficult to access, enhancing data collection in densely built or environmentally sensitive areas (Zhang and Zhu 2023). Their ability to accomplish recurrent surveys also enables the temporal monitoring of changes. For example, thermal cameras can be used for the identification of heat distribution and urban heat islands, whereas multispectral sensors have the ability to monitor vegetation health and land cover changes (Almeida et al. 2021). Integrating data from multiple sensors presents dynamic understanding into urban ecosystems, informing sustainable urban planning and management practices. Thus, helping decision-makers and allowing the multiple stakeholders to address multidimensional environmental issues that may be overlooked otherwise (Wellmann et al. 2020).

2.4 Cutting-Edge Remote Sensing Technologies

The understanding and monitoring of urban ecosystems have been revolutionized by the advancements made in the field of remote sensing and its associated technologies. Integrating and combining multiple datasets and images from various platforms have helped create dynamic and comprehensive models of urban environments (Gallotti et al. 2021). These cutting-edge technologies are introducing new avenues for real-time, multi-dimensional monitoring, advancing insights into microclimates, biodiversity, and infrastructure that were previously obstruse and inaccessible.

2.4.1 From Space to Street: Integrating Satellite, Street-Level, and Citizen Science Data

A multi-scale, thorough analysis of urban ecosystems can be undertaken by amalgamating satellite imagery datasets with street-level data (UAVs and Google Street View). This analysis can improve the understanding of the ecosystem dynamics and processes. For instance, Landsat data and Google Street View imagery was integrated to conduct a detailed study of the urban green canopy in New York City, facilitating tree plantation initiatives (Li 2021). Similarly, for Tokyo city, UAV derived temperature profile was combined with Sentinel-2 datasets to map urban heat distribution, aiding the climate adaptation strategies (Chen et al. 2023a, b). The city of Cape Town, utilised street level camera traps along with satellite images for monitoring urban wildlife (Oladimeji et al. 2024). Citizen science programs with objectives of localized data collection and bridging limitations of traditional methods are aimed at advancing satellite datasets, (Coulson et al. 2021) e.g., Barcelona's Air Quality Egg.

2.4.2 Nano-Satellites, UAV Swarms, and Real-Time Monitoring

Nano-satellites are metamorphosing urban environmental scrutiny with concurrent, high-frequency data. The city of Buenos Aires uses nano-satellites for air quality and vegetation monitoring, supporting swift responses to urban environmental challenges. In Jakarta, nano-satellites deliver real-time flood data, assisting in disaster management (Priyambodoho et al. 2021). UAV swarms also contribute to real-time monitoring. UAV swarms were used to map floodwater levels during Hurricane Harvey, augmenting disaster relief efforts in Houston (Greenwood et al. 2020).

2.4.3 Beyond Light: Multi-Spectral, Hyperspectral, and Acoustic Technologies

Remote sensing technologies have the capability of capturing data beyond the visible spectrum by using multi-spectral, hyperspectral and thermal sensors. The data thus gathered reveals hidden dynamics of urban ecosystems. Multi-spectral imaging has been exploited to detect drought stress in vegetation in the urban areas of Los Angeles (Miller et al. 2022), thus steering water conservation practices. Thermal imaging technology has been deployed to identify areas for retrofitting to promote energy-efficiency after mapping the heat distribution (Deb and Schlueter 2021). Detailed urban tree health assessments have been conducted in Brussels, Belgium using hyperspectral sensors for promoting forest management practices in the urban areas (Degerickx et al. 2018). Acoustic sensors find application in monitoring noise pollution levels (McAuliffe 2016) and its impact on biotic components, e.g., in Seattle.

2.4.4 Immersive Remote Sensing: VR, AR, and AI-Enhanced Visualizations

Integrating immersive advancements such as virtual and augmented reality (VR/AR) along with artificial intelligence (AI) guided analysis of remotely sensed datasets can present dynamic, synergistic and interactional data visualizations for urban planning and design. For analysis of green spaces in Melbourne, AI was used to handle the satellite dataset, the result of this analysis was further visualised using VR to conduct scenario planning for the city (Huerta et al. 2021). AR apps are becoming popular among citizens as they allow citizens to prospect concurrent environmental data along with the prevalent environment, thus, promoting citizen engagement in the management of urban ecosystems e.g., in Amsterdam (Dawkins et al. 2021).

2.5 Emerging Technologies: A Look into the Future

Promising innovations such as quantum computing and sensing are the future and will prove expedient for precision analysis and management of urban environments (Bonab et al. 2023). AI models could be used to forecast heat trends from satellite datasets, facilitating climate adaption, as has been done in Beijing (Liu et al. 2024). Integrating IoT with remote sensing technologies will facilitate the creation of an active digital model of urban areas, promoting informed, concurrent and data driven decision making for urban resilience and sustainability (Gkontzis et al. 2024).

The synthesis of cutting-edge remote sensing advancements—vacillating from satellite and street-level data to VR, AR, and nano-satellites—is revolutionizing the monitoring of urban ecosystems. These innovations empower planners, policymakers, and citizens to better understand and manage urban environments, paving the way for more sustainable and resilient cities (Riaz et al. 2023).

3 Urban Ecosystem Services

Urban ecosystem services (UES) are the resources and benefits provided by urban ecosystems to humans. UES facilitate improvement of quality of life, environmental health, and economic vitality for people dwelling in cities (Lovell and Taylor 2013). Furthermore, they play a crucial role in maintaining ecological homeostasis and promoting sustainability and resilience (Jennings et al. 2016). These services are indispensable for sustenance of urban populations and are inclusive of a wide array of natural processes and resources that urban areas rely on. UES are classically categorized into provisioning, regulating, cultural, and supporting services (Tan et al. 2020). However, in a more contemporary context the former three are considered as primary services while the latter is referred to as a secondary service (Tiwari et al. 2023). Remote sensing technologies have had an effective transformative influence on the mapping, monitoring, and management of these services, delving comprehensively into the dynamic relationship between urban environments and ecosystem functions to derive detailed insights (Reddy 2018; Di Palma et al. 2024).

a. **Provisioning Services**: Provisioning services are the tangible commodities and resources that urban ecosystems provide and supply. In urban settings, the supply of resources like food, water, and raw materials, often derived from urban agriculture, green spaces, and rooftop gardens (Sokolova et al. 2024; Tan et al. 2020). Some examples of provisioning services offered by cities include:

    - Food production: Community gardens and urban farms contributing to food security (Zutter and Stoltz 2023).
    - Water resources: Natural or managed urban wetlands, rivers, and reservoirs provide fresh water for multiple uses (Ganapathi et al. 2024).
    - Raw materials: Urban forests and vegetation provide timber, fibres, and fuel.
    - Medicinal resources: Certain urban plants and green areas may also offer medicinal herbs or serve as reservoirs of genetic diversity with potential pharmaceutical applications (Dutta et al. 2022).

b. **Regulating Services**: Regulating services refer to the maintenance of environmental quality and regulation of natural processes by urban ecosystems. They are fundamental to maintain a steady urban environment, as they mitigate environmental hazards and nourish liveable conditions (Tan et al. 2020; Tiwari et al. 2023). These services are vital for mitigating urban heat, reduce pollution, manage stormwater, and improve air quality and can include:

    - Air quality regulation: Urban trees and green spaces perform filtration of air pollutants, absorb harmful gases and capture particulate matter, thereby improving air quality (Yang et al. 2023).
    - Climate regulation: Vegetation and plantations regulate the micro-climate, minimising the urban heat island effect. Green roofs and urban forests help sequester carbon, contributing to climate change mitigation (Wu et al. 2024).
    - Water regulation and flood control: Urban ecosystems, such as wetlands, green infrastructure, and permeable surfaces, absorb stormwater, reducing runoff and flood risks.
    - Pest and disease control: Natural predators in urban ecosystems help control populations of pests that spread diseases, contributing to public health management.

c. **Cultural Services**: Cultural services are defined as the non-material intangible resources produced by the urban ecosystems. It influences the wellbeing and quality of life of urban populations. These services have profound associations to the recreational, spiritual and aesthetic tenets of urban ecosystems (Tan et al. 2020; Tiwari et al. 2023). These services include but are not limited to:

    - Recreation and tourism: green spaces, parks, rivers, wetlands, and urban forests are deemed to be places for leisure, exercise, and tourism, contributing to physical and mental health of the urban community (Jabbar et al. 2022).

- Aesthetic and spiritual value: Urban green areas often augment the aesthetic allure of the urban area and provide spaces for cultural and spiritual activities. They also promote a sense of identity and belongingness for urban residents.
- Education and knowledge systems: Urban ecosystems serve as living laboratories for environmental education and citizen science, helping urban residents understand ecological processes, biodiversity and sustainability.

d. **Supporting Services**: Supporting services are the secondary underlying ecological processes that enable other ecosystem services to function. Processes such as soil formation, nutrient cycling, and habitat provision, which sustain other ecosystem functions constitute supporting services (Tan et al. 2020; Tiwari et al. 2023). Although these services may not provide instantaneous, explicit benefits, they are central to the sustainability and resilience paradigm of urban ecosystems (McPhearson et al. 2015). They can be exemplified as follows:

- Soil formation and nutrient cycling: Healthy soils in urban areas support plant growth and are critical for sustaining urban agriculture, vegetation, and overall ecosystem productivity (Salomon and Cavagnaro 2022).
- Biodiversity and habitat provision: Urban ecosystems offer a diversity of habitats to several species of birds, insects, and small mammals, contributing to biodiversity conservation within urban areas (Singh et al. 2024).
- Photosynthesis and primary production: Vegetation in urban ecosystems supports photosynthesis, constituting the foundation for producing food, oxygen, and organic matter needed for other services.
- Water cycling: Urban wetlands, rivers, and green infrastructure contribute to the hydrological cycle, by stimulating groundwater recharge and warranting the availability of fresh water.

The four fundamental classes of ecosystem services are imperative for preserving and upholding the economic, social and environmental sustainability of urban ecosystems. They supplement the enhancement of resilience of these ecosystems to environmental challenges like climate change, air pollution, and urban sprawl (McPhearson et al. 2015). Remote sensing technologies perform the ancillary function of provision of data for quantification, visualization, and monitoring UES, enabling cities to make informed decisions for enhancing sustainability and resilience (Table 1). Remote sensing technologies, particularly multispectral and hyperspectral sensors, can be leveraged to assess the health, productivity, and extent of the provisioning services offered by urban ecosystems. For example, UAVs equipped with sensors can monitor urban agriculture practices and optimize resource use by detecting agricultural health and water stress (Rejeb et al. 2022). This information provisions the efficient and effective management of limited urban resources while ensuring food security in urban areas. Satellite imagery can help monitor urban heat islands by pursuing alterations in land surface temperature and vegetation health. LiDAR technology further advances the modelling of stormwater flow by portraying the topography comprehensively, facilitating planned and sustainable drainage system designs and flood mitigation (Trepekli et al. 2022). By incorporating thermal and multispectral data, remote sensing can also support air quality monitoring and carbon sequestration parameters (Vivone 2023). Technologies, such as high-resolution satellite imagery and UAV data, facilitate the mapping of parks and recreational areas, which are elemental for physical and mental well-being of the city dwellers. Integrating this data with GIS can enable visualization of the dissemination, availability and accessibility of cultural ecosystem services across different neighbourhoods, guiding planned and multistakeholder decision making to promote the principle of equitability. Consequently, a greater understanding of cultural services optimisation can be achieved to enhance community awareness, engagement and well-being. Additionally, LiDAR and hyperspectral sensors, allow monitoring of soil health, biodiversity, and habitat connectivity within urban ecosystems. UAVs can also be deployed for assessing the health of wetlands and other critical habitats within urban environments, contributing to the conservation efforts (Mohsan et al. 2023).

4 Challenges in Remote Sensing of Urban Ecosystem Services

Several significant challenges have been posed by the deployment of remote sensing technologies for the monitoring of UES (Dubovik et al. 2021; Luederitz et al. 2015; Kremer et al. 2016). The challenges persist in spatial, spectral and temporal dimensions and are unable to fulfil the requirement of improved visualisations, analysis and for conducting change assessments. These challenges require systematic resolutions to ensure ethical and effective application of these contemporary technologies. The ethical connotations of harmonising environmental monitoring with the potential breach of privacy by performing surveillance is a primary concern (Young et al. 2022). The question of individual rights and liberties is raised when unclear data collection and ownership propositions are put forth. Institutionalising a robust data governance framework is imperative to certify responsible use of remote sensing technologies, providing clear guidelines pertaining to data access, utilization, and sharing, thereby defending personal liberty and rights (Hummel et al. 2021; Sargiotis 2024).

Alternatively, the quantity, volume and complexity of data generated by remote sensing technologies and systems can pose another challenge (Ma et al. 2015). The heterogeneity in the datasets can further overwhelm conventional analytical systems and can complicate the derivation of actionable insights (Sivarajah et al. 2017). The introduction of machine learning (ML) and artificial intelligence (AI) procedures is crucial for enhancement of interpretation capabilities and improvement in processing efficiency (Chen et al. 2023a, b; Himeur et al. 2022). However, the transparency of algorithms used and methodologies employed is a critical parameter for building stakeholder trust, ensuring that the processes used are comprehensible and accessible (Felzmann et al. 2019; Balasubramaniam et al. 2023).

Table 1 Applications of remote sensing technologies to analyse UES

Service	Aspect	Description	Examples
Provisioning services	Urban agriculture	Remote sensing techniques, such as hyperspectral imaging, are employed to monitor vertical farms and rooftop gardens. These methods assess plant health, productivity, and environmental impacts, crucial for optimizing urban agriculture	New York City: hyperspectral imaging is used to monitor plant health and optimize conditions in vertical farms (Qin et al. 2023; Avgoustaki et al. 2022) Singapore: drones equipped with multispectral sensors monitor rooftop gardens for improved yield and sustainability (Muhmad Kamarulzaman et al. 2023)
	Water supply and quality	Remote sensing is used to monitor groundwater levels, track depletion and recharge rates, and connect urban water use with distant ecosystems. This approach enhances urban water management by providing comprehensive data on water resources	Los Angeles: NASA's GRACE satellites monitor groundwater changes (Rodell and Reager 2023) Cape Town: European Space Agency's Sentinel-2 satellites track water quality in the urban water bodies (Nikoo et al. 2024) Sydney: remote sensing is used to monitor water footprints (Vervoort et al. 2022)
	Raw materials and urban mining	The field of urban mining utilizes remote sensing to locate and quantify reusable materials within urban environments. This technology is also applied to map urban biomass, which can be repurposed for energy production, contributing to circular economy initiatives	Tokyo: employs remote sensing to assess metal stockpiles for urban mining (Tanikawa et al. 2015) San Francisco: remote sensing helps convert urban green waste into biogas and compost (Karimi et al. 2023) Barcelona: urban mining techniques are used to recover valuable materials from construction waste (Aldebei and Dombi 2021)

(continued)

Table 1 (continued)

Service	Aspect	Description	Examples
	Urban foraging	Remote sensing facilitates the mapping of edible plants and mushrooms within urban areas. This contributes to food security and community resilience by identifying potential sources of urban food	Berlin: remote sensing utilized to map edible plants and mushrooms, supporting urban foraging initiatives. (Adeyemi and Shackleton 2023) Portland, Oregon: satellite data assists in identifying urban spaces where wild edibles thrive, promoting local foraging (Anderton-Folmer 2013)
	Blue-green infrastructure	The integration of blue (water-related) and green (vegetation-related) infrastructure is monitored through remote sensing. This approach assesses the synergistic effects on urban microclimates and biodiversity, promoting sustainable urban design	Rotterdam and Singapore: manage and monitor blue-green infrastructure projects, enhancing microclimates and biodiversity (Brears 2018) Adelaide: remote sensing is used to optimize the integration of blue and green spaces to improve urban resilience (Nouri et al. 2019)
	Circular economy in urban ecosystems	Remote sensing monitors urban biomass and organic waste, which are crucial for supporting circular economy practices. This includes repurposing biomass for energy production, composting, and construction materials	Copenhagen and Seoul: implement biomass repurposing strategies with remote sensing to monitor and manage resources for energy and composting Paris: remote sensing tracks urban organic waste for recycling and energy production (Riva Sanseverino et al. 2017)
	Green roofs and walls	Remote sensing techniques assess the health and growth of green roofs and living walls. These assessments are used to evaluate their impact on building energy efficiency and urban biodiversity	Toronto: remote sensing is used to evaluate green roofs' effectiveness in reducing energy consumption (Berardi 2016) Hong Kong: remote sensing assesses living walls' impact on urban biodiversity and building insulation (Pan and Chu 2016)

(continued)

Table 1 (continued)

Service	Aspect	Description	Examples
Regulating services	Air quality improvement	Remote sensing is utilized to monitor the dispersion of micro-scale pollutants such as black carbon. Integration with wearable sensors provides personalized air quality information. Additionally, remote sensing detects airborne microplastics, impacting urban health	Delhi: wearable sensors integrated with remote sensing track micro-scale air pollutants (Lambrechts and Sinha 2016) Beijing: remote sensing used to monitor airborne microplastics (Luo et al. 2024) Los Angeles: remote sensing employed to study black carbon dispersion (Kiel et al. 2021)
	Climate regulation	Remote sensing technologies are used to map urban albedo and assess its role in mitigating the urban heat island effect. This includes tracking the effectiveness of green infrastructure and cooling strategies	Los Angeles: Remote sensing evaluates urban albedo and cooling strategies like reflective surfaces (Vahmani and Ban-Weiss 2016) Madrid: remote sensing tracks the effectiveness of green infrastructure in mitigating heat island effects (Fernández and Navarro 2024)
	Flood control	The health and extent of urban wetlands are monitored using remote sensing, which also models the impact of sea-level rise. Real-time data integration with AI-driven flood prediction systems enhances adaptive flood management	Houston and Jakarta: use of remote sensing and AI for flood prediction and management, addressing sea-level rise impacts (Priyambodoho et al. 2021) New Orleans: remote sensing monitors wetland health to support flood mitigation efforts (Li et al. 2016)
	Airborne microplastics detection	Remote sensing technologies detect airborne microplastics, contributing to understanding their dispersion and impact on urban ecosystems and public health	Beijing: remote sensing used to track concentrations of airborne microplastics (Luo et al. 2024)

(continued)

Table 1 (continued)

Service	Aspect	Description	Examples
	Energy efficiency monitoring	Remote sensing tracks urban energy consumption patterns, evaluating the effectiveness of energy-saving measures and urban planning strategies aimed at improving energy efficiency	New York City: remote sensing used to assess the impact of energy-efficient retrofits on building energy use (Lee et al. 2017) Stockholm: remote sensing monitors urban energy consumption patterns to support sustainability initiatives (Bibri and Krogstie 2020)
Cultural services	Urban soundscapes and visual aesthetics	Remote sensing is applied to map urban soundscapes and their correlation with visual aesthetics and mental health, integrating with traditional methods for a comprehensive understanding of urban green spaces' cultural value	Tokyo: integrates acoustic remote sensing with visual data to capture the cultural and aesthetic value of urban green spaces (Zhang et al. 2023) New York City: remote sensing maps soundscapes to assess their impact on urban residents' mental well-being (Sadroleslami 2020)
	Digital preservation of urban heritage	High-resolution remote sensing technologies, such as 3D photogrammetry and LiDAR, create digital twins of cultural heritage sites. This ensures preservation and facilitates monitoring of structural integrity amidst urban expansion	Rome and Istanbul: 3D photogrammetry and LiDAR are used to preserve and monitor the structural integrity of historical buildings (Fregonese et al. 2013; Temizer et al. 2013) Cairo: these technologies are employed to digitize and protect ancient monuments from urban encroachment (Blasco et al. 2017)

(continued)

Table 1 (continued)

Service	Aspect	Description	Examples
	Public health and urban wellness	Remote sensing maps urban heat vulnerability and links green space distribution to public health outcomes. It also monitors "blue spaces" (e.g., rivers, lakes) and their impact on mental well-being	Paris: maps heat vulnerability and green space access (Lemonsu et al. 2015) Vancouver: uses remote sensing to monitor the distribution of blue spaces and their effects on mental health (Rugel et al. 2017) Sydney: remote sensing is used to analyse the impact of urban green spaces on public health (Trinder and Liu 2020)
	Urban wilderness	Remote sensing identifies and protects urban wilderness areas that harbour significant biodiversity, including neglected or rewilded spaces within cities	Detroit: uses remote sensing to map urban wilderness areas, uncovering significant biodiversity hotspots (Rutledge et al. 2021) Philadelphia: remote sensing identifies rewilded spaces that support diverse urban wildlife (Alagona 2022)
	Digital heritage: virtual preservation of cultural sites	Remote sensing and 3D modelling technologies create digital twins of cultural heritage sites, enabling virtual tourism and educational applications	Athens: utilizes digital twins of heritage sites for virtual tourism, supporting cultural preservation and public education (Konstantakis et al. 2023) Istanbul: remote sensing creates detailed digital models of historical sites for preservation and virtual exploration (Büyüksalih et al. 2020)
Supporting services	Urban soil health and ecosystem functions	Soil spectroscopy and remote sensing monitor urban soil health, focusing on carbon content, nutrient cycling, and contamination hotspots. This information informs soil management and remediation efforts	São Paulo and London: use remote sensing to assess urban soil health and contamination, guiding soil management practices (Pons et al. 2007; Lamine et al. 2019) Tokyo: remote sensing aids in monitoring soil carbon content and pollution levels (Inoue 2010)

(continued)

Table 1 (continued)

Service	Aspect	Description	Examples
	Ecosystem productivity and the urban carbon cycle	Remote sensing technologies monitor urban primary productivity and carbon sequestration, contributing to climate action plans and sustainability efforts. This includes quantifying carbon storage in urban green spaces	London and Chicago: utilize remote sensing to assess carbon sequestration in urban green spaces, supporting climate action initiatives (Davies et al. 2011) Toronto: remote sensing is used to monitor urban primary productivity and its role in the carbon cycle (Gonsamo et al. 2013)
	Green space connectivity	Remote sensing evaluates the connectivity between urban green spaces, promoting ecological corridors and enhancing habitat connectivity for urban wildlife	Melbourne: analyses green space connectivity to improve habitat links for urban wildlife (Kirk et al. 2023) Barcelona: remote sensing supports the creation of ecological corridors connecting fragmented green spaces (Basnou et al. 2020)
	Urban microclimate monitoring	Remote sensing tracks microclimate variations within urban areas, providing insights into temperature gradients, humidity levels, and their effects on urban ecosystems	Toronto: uses remote sensing to monitor urban microclimates, focusing on heat islands and moisture levels (Rinner and Hussain 2011) San Francisco: remote sensing maps temperature variations and their impacts on local ecosystems (Huntsman et al. 2024)
	Sustainable urban drainage systems (SUDS)	Remote sensing evaluates the effectiveness of Sustainable Urban Drainage Systems (SUDS) in managing urban runoff and improving water quality, supporting better stormwater management practices	London: uses remote sensing to assess SUDS performance, aiding in effective stormwater management and water quality improvement (Ramovha et al. 2024) Sydney: remote sensing monitors SUDS to enhance urban water management and flood prevention (Ferrans et al. 2022)

Inclusivity is another issue while using remote sensing technologies, as these technologies need to be available, accessible and valuable to every stratum of urban population (Joyce et al. 2022). The marginal and vulnerable communities often lack resources and expertise to engage and utilize these technological advancements (Xiao et al. 2024). Thus, empowering these communities by formulating community-based remote sensing initiatives is necessary to ensure active participation of these communities in monitoring their surrounding urban ecosystems, enhancing local knowledge and fostering environmental stewardship (Svendsen et al. 2016).

Moreover, micro-scale ecological processes are excluded from the purview of existing remote sensing technologies. These micro-scale processes are essential for comprehending ecosystem functions and urban biodiversity. The limitation underscores the urgency of innovating in remote sensing methodology (Tavares et al. 2019). Emerging technologies, such as quantum computing and advanced AI, present opportunities to overcome these limitations, potentially providing extraordinary spatio-temporal resolution in urban ecological assessments (Chen et al. 2021).

Finally, tackling these multifaceted challenges demands transdisciplinary collaboration among diverse fields, including ecology, urban planning, data science, and social sciences (OECD 2020). Such a partnership is crucial for expansively addressing the complexities of urban ecosystems (Zhou et al. 2021). This collaborative approach could expedite the emergence of new domains, for academic and professional pursuits, at the intersection of these disciplines (Beck et al. 2020). Further promotion of holistic strategies for urban ecosystem management and sustainability. By addressing these challenges through collective efforts, stakeholders can leverage remote sensing technologies to enhance the resilience and sustainability of urban environments.

5 Conclusion

The future of urban ecosystem monitoring and analysis requires a paradigm shift towards recognizing and integrating both visible and hidden strata of urban life (Elze and Banzhaf 2022). Remote sensing technologies perform a paramount part in modelling urban sustainability by augmenting resilience and adaptability in response to dynamic contemporary environmental challenges (Rochon et al. 2003). With the evolution of these technologies, integrating them with urban ecosystem management presents immense promises for nurturing and cultivating sustainable and resilient cities (Schewenius et al. 2014). Incessant innovation, interdisciplinary collaboration, and ethical considerations are prerequisites to warrant the effectiveness and social responsibility of remote sensing applications.

The application of remote sensing to urban planning has significant policy implications and can even strengthen environmental management efforts (Netzband et al. 2007). This can be exemplified by the use of remote sensing for global sustainability initiatives such as the New Urban Agenda and the Sustainable Development Goals (SDGs) (Tassopoulou et al. 2019). Emerging innovations such as IoT, AI,

ML presents the opportunities to integrate these technologies with remote sensing datasets and create responsive environmental monitoring systems (Himeur 2022). Progress in the resolution improvements of satellites, thermal LiDAR and UAVs will further improve the applications of remote sensing capabilities for urban environments (Rickens and Tonekaboni 2023). Intensifying remote sensing applications to embrace an extensive array of ecosystem services will promote detailed assessments that will promote sustainable urban development. Ultimately, the ongoing evolution of remote sensing technologies will reinforce evidence-based and informed policy-making, improving the resilience and inclusivity of urban ecosystems in the face of rapid urbanization.

References

Adeyemi O, Shackleton CM (2023) Understanding foraging practices in Lagos metropolis to redesign urban greenspaces in support of human-nature interactions. Urban for Urban Green 79:127805. https://doi.org/10.1016/j.ufug.2022.127805

Alagona PS (2022) The accidental ecosystem: people and wildlife in American cities. Univ of California Press

Alberti M (2024) Cities of the anthropocene: urban sustainability in an eco-evolutionary perspective. Philos Trans Royal Soc B 379:20220264. https://doi.org/10.1098/rstb.2022.0264

Aldebei F, Dombi M (2021) Mining the built environment: telling the story of urban mining. Buildings 11(9):388. https://doi.org/10.3390/buildings11090388

Almeida CRD, Teodoro AC, Gonçalves A (2021) Study of the urban heat island (UHI) using remote sensing data/techniques: a systematic review. Environments 8(10):105. https://doi.org/10.3390/environments8100105

Anderton-Folmer H (2013) The edible desert: an inventory of land suitable for urban agriculture and its economic potential in lower washoe county, Nevada. University of Nevada, Reno ProQuest Dissertations and Theses, 2013. https://www.proquest.com/dissertations-theses/edible-desert-inventory-land-suitable-urban/docview/1446465284/se-2

Antrop M (2004) Landscape change and the urbanization process in Europe. Landscape Urban Plann 67(1–4):9–26. https://doi.org/10.1016/S0169-2046(03)00026-4

Avgoustaki DD, Avgoustakis I, Miralles CC, Sohn J, Xydis G (2022) Autonomous mobile robot with attached multispectral camera to monitor the development of crops and detect nutrient and water deficiencies in vertical farms. Agronomy 12(11):2691. https://doi.org/10.3390/agronomy12112691

Balasubramaniam N, Kauppinen M, Rannisto A, Hiekkanen K, Kujala S (2023) Transparency and explainability of AI systems: From ethical guidelines to requirements. Inform Softw Technol 159:107197. https://doi.org/10.1016/j.infsof.2023.107197

Basnou C, Baró F, Langemeyer J, Castell C, Dalmases C, Pino J (2020) Advancing the green infrastructure approach in the Province of Barcelona: integrating biodiversity, ecosystem functions and services into landscape planning. Urban for Urban Green 55:126797. https://doi.org/10.1016/j.ufug.2020.126797

Beatley T (2016) Singapore city, Singapore: city in a Garden. In: Handbook of biophilic city planning and design. Island Press, Washington, DC

Beck S, Bergenholtz C, Bogers M, Brasseur TM, Conradsen ML, Di Marco D et al (2020) The open innovation in science research field: a collaborative conceptualisation approach. Ind Innov 29(2):136–185. https://doi.org/10.1080/13662716.2020.1792274

Berardi U (2016) The outdoor microclimate benefits and energy saving resulting from green roofs retrofits. Energy Build 121:217–229

Bibri SE (2021) Data-driven smart sustainable cities of the future: urban computing and intelligence for strategic, short-term, and joined-up planning. Comput Urban Sci 1:8. https://doi.org/10.1007/s43762-021-00008-9
Bibri SE, Krogstie J (2020) Environmentally data-driven smart sustainable cities: applied innovative solutions for energy efficiency, pollution reduction, and urban metabolism. Energy Inform 3(1):29
Blasco JMD, Verstraeten G, Hanssen RF (2017) Detecting modern desert to urban transitions from space in the surroundings of the Giza World Heritage site and Greater Cairo. J Cult Herit 23:71–78
Bonab AB, Fedele M, Formisano V, Rudko I (2023) Urban quantum leap: a comprehensive review and analysis of quantum technologies for smart cities. Cities 140:104459. https://doi.org/10.1016/j.cities.2023.104459
Brears RC (2018) Rotterdam becoming a blue-green city. In: Blue and green cities. Palgrave Macmillan, London
Burke M, Driscoll A, Lobell DB, Ermon S (2021) Using satellite imagery to understand and promote sustainable development. Science 371(6535):eabe8628. https://doi.org/10.1126/science.abe8628
Büyüksalih G, Kan T, Özkan GE, Meriç M, Isın L, Kersten TP (2020) Preserving the knowledge of the past through virtual visits: from 3D laser scanning to virtual reality visualisation at the Istanbul Çatalca İnceğiz caves. PFG J Photogr Remote Sens Geoinform Sci 88(2):133–146
Cavender-Bares J, Schneider FD, Santos MJ et al (2022) Integrating remote sensing with ecology and evolution to advance biodiversity conservation. Nat Ecol Evol 6:506–519. https://doi.org/10.1038/s41559-022-01702-5
Chen B, Xu B, Gong P (2021) Mapping essential urban land use categories (EULUC) using geospatial big data: progress, challenges, and opportunities. Big Earth Data 5(3):410–441. https://doi.org/10.1080/20964471.2021.1939243
Chen C, Bagan H, Yoshida T (2023a) Multiscale mapping of local climate zones in Tokyo using airborne LiDAR data, GIS vectors, and Sentinel-2 imagery. Gisci Remote Sens 60(1):970
Chen YN, Fan KC, Chang YL, Moriyama T (2023b) Special issue review: artificial intelligence and machine learning applications in remote sensing. Remote Sens 15(3):569. https://doi.org/10.3390/rs15030569
Chen G, Zhou Y, Voogt JA, Stokes EC (2024) Remote sensing of diverse urban environments: from the single city to multiple cities. Remote Sens Environ 305:114108. https://doi.org/10.1016/j.rse.2024.114108
Chrysoulakis N, Grimmond S, Feigenwinter C, Lindberg F, Gastellu-Etchegorry JP, Marconcini M et al (2018) Urban energy exchanges monitoring from space. Sci Rep 8(1):11498. https://doi.org/10.1038/s41598-018-29873-x
Coulson S, Woods M, Making Sense EU (2021) Citizen sensing: an action-orientated framework for citizen science. Front Commun 6:629700. https://doi.org/10.3389/fcomm.2021.629700
Cracknell AP (2018) The development of remote sensing in the last 40 years. Int J Remote Sens 39(23):8387–8427. https://doi.org/10.1080/01431161.2018.1550919
Davies ZG, Edmondson JL, Heinemeyer A, Leake JR, Gaston KJ (2011) Mapping an urban ecosystem service: quantifying above-ground carbon storage at a city-wide scale. J Appl Ecol 48(5):1125–1134. https://doi.org/10.1111/j.1365-2664.2011.02021.x
Dawkins O, Kitchin R, Young GW, Zawadzki T (2021) City dashboards and 3D geospatial technologies for urban planning and management. In: Rae W (ed) Applied data analysis for urban planning and management. Sage Publications, New York
Deb C, Schlueter A (2021) Review of data-driven energy modelling techniques for building retrofit. Renew Sustain Energy Rev 144:110990. https://doi.org/10.1016/j.rser.2021.110990
Degerickx J, Roberts DA, McFadden JP, Hermy M, Somers B (2018) Urban tree health assessment using airborne hyperspectral and LiDAR imagery. Int J Appl Earth Obser Geoinform 73:26–38. https://doi.org/10.1016/j.jag.2018.05.021

Di Palma M, Rigillo M, Leone MF (2024) Remote sensing technologies for mapping ecosystem services: an analytical approach for urban green infrastructure. Sustainability 16(14):220. https://doi.org/10.3390/su16146220

Dong P, Chen Q (2017) LiDAR remote sensing and applications, 1st edn. CRC Press, New York

Dubovik O, Schuster GL, Xu F, Hu Y, Bösch H, Landgraf J, Li Z (2021) Grand challenges in satellite remote sensing. Front Remote Sens 2:619818. https://doi.org/10.3389/frsen.2021.619818

Dutta T, Nandy S, Dey A (2022) Urban ethnobotany of Kolkata, India: a case study of sustainability, conservation and pluricultural use of medicinal plants in traditional herbal shops. Environ Dev Sustain 24:1207–1240. https://doi.org/10.1007/s10668-021-01493-y

Edmonds-Brown V (2022) From 'biologically dead' to chart-toppingly clean: how the Thames made an extraordinary recovery over 60 years. The Conversation. https://theconversation.com/from-biologically-dead-to-chart-toppingly-clean-how-the-thames-made-an-extraordinary-recovery-over-60-years-180895

Egerer M, Haase D, McPhearson T et al (2021) Urban change as an untapped opportunity for climate adaptation. NPJ Urban Sustain 1:22. https://doi.org/10.1038/s42949-021-00024-y

Elmqvist T, Setälä H, Handel SN, van der Ploeg S, Aronson J, Blignaut JN et al (2015) Benefits of restoring ecosystem services in urban areas. Curr Opin Environ Sustain 14:101–108. https://doi.org/10.1016/j.cosust.2015.05.001

Elze S, Banzhaf E (2022) High-precision monitoring of urban structures to understand changes in multiple ecosystem services. Urban for Urban Green 73:127616. https://doi.org/10.1016/j.ufug.2022.127616

Engin Z, van Dijk J, Lan T, Longley PA, Treleaven P, Batty M, Penn A (2020) Data-driven urban management: mapping the landscape. J Urban Manag 9(2):140–150. https://doi.org/10.1016/j.jum.2019.12.001

Ennouri K, Smaoui S, Triki MA (2021) Detection of urban and environmental changes via remote sensing. Circ Econ Sust 1:1423–1437. https://doi.org/10.1007/s43615-021-00035-y

Felzmann H, Villaronga EF, Lutz C, Tamò-Larrieux A (2019) Transparency you can trust: transparency requirements for artificial intelligence between legal norms and contextual concerns. Big Data Soc 6(1):542. https://doi.org/10.1177/2053951719860542

Fernández CT, Navarro IM (2024) Analysis of the impact of urban green spaces on mitigating the urban heat island effect. Innov Sci Technol 3(3):11–20

Ferrans P, Torres MN, Temprano J, Sánchez JPR (2022) Sustainable urban drainage system (SUDS) modeling supporting decision-making: a systematic quantitative review. Sci Total Environ 806:150447. https://doi.org/10.1016/j.scitotenv.2021.150447

Fregonese L, Barbieri G, Biolzi L, Bocciarelli M, Frigeri A, Taffurelli L (2013) Surveying and monitoring for vulnerability assessment of an ancient building. Sensors 13(8):9747–9773

Gallotti R, Sacco P, De Domenico M (2021) Complex urban systems: challenges and integrated solutions for the sustainability and resilience of cities. Complexity 2021(1):1782354. https://doi.org/10.1155/2021/1782354

Ganapathi H, Awasthi S, Vasudevan P (2024) Wetlands as a nature-based solution for urban water management. In: Stefanakis A, Oral HV, Calheiros C, Carvalho P (eds) Nature-based solutions for circular management of urban water: circular economy and sustainability. Springer, Cham

Gkontzis AF, Kotsiantis S, Feretzakis G, Verykios VS (2024) Enhancing urban resilience: smart city data analyses, forecasts, and digital twin techniques at the neighborhood level. Fut Internet 16(2):47. https://doi.org/10.3390/fi16020047

Gonsamo A, Chen JM, Price DT, Kurz WA, Liu J, Boisvenue C et al (2013) Improved assessment of gross and net primary productivity of Canada's landmass. J Geophys Res Biogeosci 118(4):1546–1560. https://doi.org/10.1002/2013JG002388

Greenwood F, Nelson EL, Greenough PG (2020) Flying into the hurricane: a case study of UAV use in damage assessment during the 2017 hurricanes in Texas and Florida. PLoS ONE 15(2):e0227808. https://doi.org/10.1371/journal.pone.0227808

Grey T, Xidous D, O'Neill D, Collier M (2023) Growing older urbanism: exploring the nexus between ageing, the built environment, and urban ecosystems. Urban Transform 5:8. https://doi.org/10.1186/s42854-023-00053-z

Himeur Y (2022) Using artificial intelligence and data fusion for environmental monitoring: a review and future perspectives. Inform Fus 86:44–75. https://doi.org/10.1016/j.inffus.2022.06.003

Himeur Y, Rimal B, Tiwary A, Amira A (2022) Using artificial intelligence and data fusion for environmental monitoring: a review and future perspectives. Inform Fus 86:44–75. https://doi.org/10.1016/j.inffus.2022.06.003

Hoornweg D, Hosseini M, Kennedy C et al (2016) An urban approach to planetary boundaries. Ambio 45:567–580. https://doi.org/10.1007/s13280-016-0764-y

Hu M, Ghorbany S, Yao S, Wang C (2024) Micro-urban heatmapping: a multi-modal and multi-temporal data collection framework. Buildings 14(9):2751. https://doi.org/10.3390/buildings14092751

Huerta RE, Yépez FD, Lozano-García DF, Guerra Cobian VH, Ferrino Fierro AL, de León Gómez H et al (2021) Mapping urban green spaces at the metropolitan level using very high resolution satellite imagery and deep learning techniques for semantic segmentation. Remote Sens 13(11):2031. https://doi.org/10.3390/rs13112031

Hulley G, Shivers S, Wetherley E, Cudd R (2019) New ECOSTRESS and MODIS land surface temperature data reveal fine-scale heat vulnerability in cities: a case study for Los Angeles County. California. Remote Sens 11(18):2136. https://doi.org/10.3390/rs11182136

Hummel P, Braun M, Dabrock P (2021) Own data? Ethical reflections on data ownership. Philos Technol 34:545–572. https://doi.org/10.1007/s13347-020-00404-9

Huntsman B, Brown LR, Wulff M, Knowles N, Wagner RW, Feyrer F (2024) Climate change scenarios for air and water temperatures in the Upper San Francisco estuary: implications for thermal regimes and delta smelt. San Francisco Estuary Watershed Sci 22(2):22

Inoue Y (2010) Use of remote sensing in assessment of soil and ecosystem carbon status. In: Proceedings of international workshop on evaluation and sustainable management of soil carbon sequestration in Asian countries, Bogor, Indonesia, vol 2010, pp 28–29

Jabbar M, Yusoff MM, Shafie A (2022) Assessing the role of urban green spaces for human well-being: a systematic review. GeoJournal 87:4405–4423. https://doi.org/10.1007/s10708-021-10474-7

Jennings V, Larson L, Yun J (2016) Advancing sustainability through urban green space: cultural ecosystem services, equity, and social determinants of health. Int J Environ Res Public Health 13(2):196. https://doi.org/10.3390/ijerph13020196

Joyce KE, Nakalembe CL, Gómez C, Suresh G, Fickas K, Halabisky M et al (2022) Discovering inclusivity in remote sensing: leaving no one behind. Front Remote Sens 3:869291. https://doi.org/10.3389/frsen.2022.869291

Karimi N, Ng KTW, Mahmud TS, Adusei KK, Kerr S (2023) A systematic review of the latest research trends on the use of satellite imagery in solid waste disposal applications from 2012 to 2021. Environments 10(7):128. https://doi.org/10.3390/environments10070128

Kiel M, Eldering A, Roten DD, Lin JC, Feng S, Lei R et al (2021) Urban-focused satellite CO_2 observations from the orbiting carbon observatory-3: a first look at the Los Angeles megacity. Remote Sens Environ 258:112314

Kirk H, Soanes K, Amati M, Bekessy S, Harrison L, Parris K et al (2023) Ecological connectivity as a planning tool for the conservation of wildlife in cities. MethodsX 10:101989. https://doi.org/10.1016/j.mex.2022.101989

Konstantakis M, Trichopoulos G, Aliprantis J, Michalakis K, Caridakis G, Thanou A et al (2023) An enhanced methodology for creating digital twins within a paleontological museum using photogrammetry and laser scanning techniques. Heritage 6(9):5967–5980

Kousis I, Pigliautile I, Pisello A (2021) Intra-urban microclimate investigation in urban heat island through a novel mobile monitoring system. Sci Rep 11:9732. https://doi.org/10.1038/s41598-021-88344-y

Kowarik I (2023) Urban biodiversity, ecosystems and the city: insights from 50 years of the Berlin School of urban ecology. Landsc Urban Plann 240:104877. https://doi.org/10.1016/j.landurbplan.2023.104877

Kremer P, Hamstead Z, Haase D, McPhearson T, Frantzeskaki N, Andersson E et al (2016) Key insights for the future of urban ecosystem services research. Ecol Soc 21(2):1154

Lambrechts J, Sinha S (2016) Microsensing networks for sustainable cities. Springer, Cham

Lamine S, Petropoulos GP, Brewer PA, Bachari NEI, Srivastava PK, Manevski K et al (2019) Heavy metal soil contamination detection using combined geochemistry and field spectroradiometry in the United Kingdom. Sensors 19(4):762. https://doi.org/10.3390/s19040762

Lee ES, Fernandes LL, Wang T, Selkowitz SE, Mesh S, Frank Y, Yancey R (2017) Demonstration of energy efficient retrofits for lighting and daylighting in New York city office buildings. Lawrence Berkeley National Laboratory, Berkeley, p 208

Leitão PJ, Santos MJ (2019) Improving models of species ecological niches: a remote sensing overview. Front Ecol Evolut 7:9. https://doi.org/10.3389/fevo.2019.00009

Lemonsu A, Viguie V, Daniel M, Masson V (2015) Vulnerability to heat waves: impact of urban expansion scenarios on urban heat island and heat stress in Paris (France). Urban Clim 14:586–605

Li X (2021) Examining the spatial distribution and temporal change of the green view index in New York City using Google street view images and deep learning. Environ Plann B Urban Anal City Sci 48(7):2039–2054. https://doi.org/10.1177/2399808320962511

Li X, Yu L, Xu Y, Yang J, Gong P (2016) Ten years after Hurricane Katrina: monitoring recovery in New Orleans and the surrounding areas using remote sensing. Sci Bull 61:1460–1470

Liu C, Hu Q, Zhang C et al (2022) First Chinese ultraviolet-visible hyperspectral satellite instrument implicating global air quality during the COVID-19 pandemic in early 2020. Light Sci Appl 11:28. https://doi.org/10.1038/s41377-022-00722-x

Liu CC, Hsu K, Peng MS et al (2024) Evaluation of five global AI models for predicting weather in Eastern Asia and Western Pacific. NPJ Clim Atmosp Sci 7:221. https://doi.org/10.1038/s41612-024-00769-0

Liu X (2023) Central park: a green oasis in the heart of New York city. https://storymaps.arcgis.com/stories/47e746779e8e4d648eb67acc3ac1e457

Lovell ST, Taylor JR (2013) Supplying urban ecosystem services through multifunctional green infrastructure in the United States. Landsc Ecol 28:1447–1463. https://doi.org/10.1007/s10980-013-9912-y

Lu J, Cheng Y, Qi X, Chen H, Lin X (2024) Rethinking urban wilderness: status, hotspots, and prospects of ecosystem services. J Environ Manag 364:121366. https://doi.org/10.1016/j.jenvman.2024.121366

Luederitz C, Brink E, Gralla F, Hermelingmeier V, Meyer M, Niven L et al (2015) A review of urban ecosystem services: six key challenges for future research. Ecosyst Serv 14:98–112. https://doi.org/10.1016/j.ecoser.2015.05.001

Luo D, Wang Z, Liao Z, Chen G, Ji X, Sang Y et al (2024) Airborne microplastics in urban, rural and wildland environments on the Tibetan Plateau. J Hazard Mater 465:133177

Ma Y, Wu H, Wang L, Huang B, Ranjan R, Zomaya A, Jie W (2015) Remote sensing big data computing: challenges and opportunities. Fut Gener Comput Syst 51:47–60. https://doi.org/10.1016/j.future.2014.10.029

Di Martino G, Scopece G, Cozzolino S, Riccio D, Ruello G (2024) Remote sensing for monitoring biodiversity in urban environment. In: Proceedings of the 2024 4th URSI Atlantic radio science meeting (AT-RASC) Meloneras, Spain, 2024. IEEE, pp 1–3. https://doi.org/10.46620/URSIATRASC24/LSTE3375

Masolele RN, Marcos D, De Sy V, Abu IO, Verbesselt J, Reiche J, Herold M (2024) Mapping the diversity of land uses following deforestation across Africa. Sci Rep 14(1):1681. https://doi.org/10.1038/s41598-024-52138-9

McAuliffe MP (2016) The ambient soundscape of inland waters in Seattle, Washington: bridge traffic as a source of urban underwater noise pollution?

McPhearson T, Andersson E, Elmqvist T, Frantzeskaki N (2015) Resilience of and through urban ecosystem services. Ecosyst Serv 12:152–156. https://doi.org/10.1016/j.ecoser.2014.07.012

Metzler AB, Nathvani R, Sharmanska V, Bai W, Muller E, Moulds S et al (2023) Phenotyping urban built and natural environments with high-resolution satellite images and unsupervised deep learning. Sci Total Environ 893:164794. https://doi.org/10.1016/j.scitotenv.2023.164794

Miller DL, Wetherley EB, Roberts DA, Tague CL, McFadden JP (2022) Vegetation cover change during a multi-year drought in Los Angeles. Urban Clim 43:101157. https://doi.org/10.1016/j.uclim.2022.101157

Miranda F, Doraiswamy H, Lage M, Zhao K, Gonçalves B, Wilson L, Hsieh M, Silva CT (2017) Urban pulse: capturing the rhythm of cities. IEEE Trans vis Comput Graph 23(1):791–800. https://doi.org/10.1109/TVCG.2016.2598585

Mohsan SAH, Othman NQH, Li Y et al (2023) Unmanned aerial vehicles (UAVs): practical aspects, applications, open challenges, security issues, and future trends. Intell Serv Robot 16:109–137. https://doi.org/10.1007/s11370-022-00452-4

Muhmad Kamarulzaman AM, Wan Mohd Jaafar WS, Mohd Said MN, Saad SNM, Mohan M (2023) UAV implementations in urban planning and related sectors of rapidly developing nations: a review and future perspectives for Malaysia. Remote Sens 15(11):2845. https://doi.org/10.3390/rs15112845

Netzband M, Stefanov WL, Redman CL (2007) Applied remote sensing for urban planning, governance and sustainability. Springer, New York

Nguyen T, Nguyen NP, Savaglio C, Zhang Y, Dumba B (2023) Future of urban remote sensing and new sensors. Eur J Remote Sens 56(1):1073

Nikoo MR, Zamani MG, Zadeh MM et al (2024) Mapping reservoir water quality from Sentinel-2 satellite data based on a new approach of weighted averaging: application of Bayesian maximum entropy. Sci Rep 14:16438. https://doi.org/10.1038/s41598-024-66699-2

Nouri H, Borujeni SC, Hoekstra AY (2019) The blue water footprint of urban green spaces: an example for Adelaide. Australia. Landsc Urban Plann 190:103613

OECD (2020) Addressing societal challenges using transdisciplinary research. In: OECD science, technology and industry policy papers, No. 88. OECD Publishing, Paris. https://doi.org/10.1787/0ca0ca45-en

Oke C, Bekessy SA, Frantzeskaki N et al (2021) Cities should respond to the biodiversity extinction crisis. NPJ Urban Sustain 1:11. https://doi.org/10.1038/s42949-020-00010-w

Oladimeji A, Woodgate Z, O'Riain MJ (2024) Wildlife resilience in an urban landscape: understanding land-use impacts in Cape Town. Urban Ecosyst 27:2517–2530. https://doi.org/10.1007/s11252-024-01606-1

Ormsby AA (2021) Diverse values and benefits of urban sacred natural sites. Trees for People 6:100136. https://doi.org/10.1016/j.tfp.2021.100136

Pan L, Chu LM (2016) Energy saving potential and life cycle environmental impacts of a vertical greenery system in Hong Kong: a case study. Build Environ 96:293–300

Peng Y, Dallas MM, Ascencio-Ibáñez JT et al (2022) Early detection of plant virus infection using multispectral imaging and spatial–spectral machine learning. Sci Rep 12:3113. https://doi.org/10.1038/s41598-022-06372-8

Pons NAD, Pejon OJ, Zuquette LV (2007) Use of geoprocessing in the study of land degradation in urban environments: the case of the city of São Carlos, state of São Paulo, Brazil. Environ Geol 53:727–739. https://doi.org/10.1007/s00254-007-0685-y

Priyambodoho BA, Kure S, Yagi R, Januriyadi NF (2021) Flood inundation simulations based on GSMaP satellite rainfall data in Jakarta. Indonesia. Prog Earth Planet Sci 8(1):34. https://doi.org/10.1186/s40645-021-00425-8

Qin J, Monje O, Nugent MR, Finn JR, O'Rourke AE, Wilson KD et al (2023) A hyperspectral plant health monitoring system for space crop production. Front Plant Sci 14:1133505. https://doi.org/10.3389/fpls.2023.1133505

Ramovha N, Chadyiwa M, Ntuli F et al (2024) The potential of stormwater management strategies and artificial intelligence modeling tools to improve water quality: a review. Water Resour Manag 38:3527–3560. https://doi.org/10.1007/s11269-024-03841-9

Reddy CS (2021) Remote sensing of biodiversity: what to measure and monitor from space to species? Biodivers Conserv 30:2617–2631. https://doi.org/10.1007/s10531-021-02216-5

Reddy GPO (2018) Geospatial technologies in land resources mapping, monitoring, and management: an overview. In: Reddy G, Singh S (eds) Geospatial technologies in land resources mapping, monitoring and management: geotechnologies and the environment, vol 21. Springer, Cham

Rejeb A, Abdollahi A, Rejeb K, Treiblmaier H (2022) Drones in agriculture: a review and bibliometric analysis. Comput Electr Agricult 198:107017. https://doi.org/10.1016/j.compag.2022.107017

Riaz K, McAfee M, Gharbia SS (2023) Management of climate resilience: exploring the potential of digital twin technology, 3D city modelling, and early warning systems. Sensors 23(5):2659. https://doi.org/10.3390/s23052659

Rickens B, Tonekaboni NH (2023) Towards high resolution urban heat analysis: incorporating thermal drones to enhance satellite based urban heatmaps. Companion Proceedings of the ACM Web Conference 2023:707–714. https://doi.org/10.1145/3543873.3587682

Rinner C, Hussain M (2011) Toronto's urban heat island—exploring the relationship between land use and surface temperature. Remote Sens 3(6):1251–1265. https://doi.org/10.3390/rs3061251

Rios G, Ramamurthy P (2022) A novel model to estimate sensible heat fluxes in urban areas using satellite-derived data. Remote Sens Environ 270:112880. https://doi.org/10.1016/j.rse.2021.112880

Riva Sanseverino E, Riva Sanseverino R, Vaccaro V, Macaione I, Anello E (2017) Smart cities: case studies. Smart Cit Atlas West East Intell Commun 12:47–140

Rochon GL, Johannsen CJ, Landgrebe DA, Engel BA, Harbor JM, Majumder S, Biehl L (2003) Remote sensing as a tool for achieving and monitoring progress toward sustainability. Clean Technol Environ Policy 5:310–316. https://doi.org/10.1007/S10098-003-0204-0

Rodell M, Reager JT (2023) Water cycle science enabled by the GRACE and GRACE-FO satellite missions. Nat Water 1:47–59. https://doi.org/10.1038/s44221-022-00005-0

Rugel EJ, Henderson SB, Carpiano RM, Brauer M (2017) Beyond the normalized difference vegetation index (NDVI): developing a natural space index for population-level health research. Environ Res 159:474–483

Russo A, Cirella GT (2024) Urban ecosystem services in a rapidly urbanizing world: scaling up nature's benefits from single trees to thriving urban forests. Land 13:786. https://doi.org/10.3390/land13060786

Rutledge A, Brandt LA, Peters M, Foen F, Harrington A, Rutledge A, Leslie AB (2021) Detroit region vulnerability assessment and synthesis: a report from the urban forestry climate change response framework

Sadroleslami S (2020) Seeing noise: the correlations between urban configuration and spatial perception of noise in NYC

Salomon MJ, Cavagnaro TR (2022) Healthy soils: the backbone of productive, safe and sustainable urban agriculture. J Clean Product 341:130808. https://doi.org/10.1016/j.jclepro.2022.130808

Sargiotis D (2024) Data governance policies and standards: development and implementation. In: Data governance. Springer, Cham

Schewenius M, McPhearson T, Elmqvist T (2014) Opportunities for increasing resilience and sustainability of urban social-ecological systems: insights from the URBES and the cities and biodiversity outlook projects. Ambio 43:434–444. https://doi.org/10.1007/S13280-014-0505-Z

Singh R, Tiwari AK, Kumar S, Singh GS (2024) Avifaunal diversity in managed urban ecosystem: a case study of Banaras Hindu University, Varanasi. Ecol Quest 35(3):1–25

Sivarajah U, Kamal MM, Irani Z, Weerakkody V (2017) Critical analysis of big data challenges and analytical methods. J Bus Res 70:263–286. https://doi.org/10.1016/j.jbusres.2016.08.001

Sokolova MV, Fath BD, Grande U, Buonocore E, Franzese PP (2024) The role of green infrastructure in providing urban ecosystem services: insights from a bibliometric perspective. Land 13(10):1664. https://doi.org/10.3390/land13101664

Svendsen ES, Campbell LK, Fisher DR, Connolly JJ, Johnson ML, Sonti NF et al (2016) Stewardship mapping and assessment project: a framework for understanding community-based environmental stewardship. US Department of Agriculture, Forest Service, Northern Research Station

Tan PY, Zhang J, Masoudi M, Alemu JB, Edwards PJ, Grêt-Regamey A et al (2020) A conceptual framework to untangle the concept of urban ecosystem services. Landsc Urban Plann 200:103837. https://doi.org/10.1016/j.landurbplan.2020.103837

Tanikawa H, Fishman T, Okuoka K, Sugimoto K (2015) The weight of society over time and space: a comprehensive account of the construction material stock of Japan, 1945–2010. J Ind Ecol 19(5):778–791. https://doi.org/10.1111/jiec.12284

Tassopoulou M, Verde N, Mallinis G, Georgiadis H, Kaimaris D, Patias P (2019) Demonstrating the potential of remote sensing to support sustainable development goals implementation: case studies over Greece. In: Proceedings of the seventh international conference on remote sensing and geoinformation of the environment (RSCy2019), p 111740D. https://doi.org/10.1117/12.2533634

Tavares PA, Beltrão N, Guimarães US, Teodoro A, Gonçalves P (2019) Urban ecosystem services quantification through remote sensing approach: a systematic review. Environments 6(5):51. https://doi.org/10.3390/environments6050051

Temizer T, Nemli G, Ekizce E, Ekizce A, Demir S, Bayram B et al (2013) 3D documentation of a historical monument using terrestrial laser scanning case study: Byzantine water Cistern, Istanbul. Int Arch Photogr Remote Sens Spatial Inform Sci 40:623–628

Theodorou P (2022) The effects of urbanisation on ecological interactions. Curr Opin Insect Sci 52:100922. https://doi.org/10.1016/j.cois.2022.100922

Tiwari AK, Singh R, Kumar S, Singh GS (2023) Ecosystem services in the riverine landscapes. In: Advances in water resource planning and sustainability. Springer, Singapore, pp 273–303. https://doi.org/10.1007/978-981-99-3660-1_16

Trepekli K, Balstrøm T, Friborg T et al (2022) UAV-borne, LiDAR-based elevation modelling: a method for improving local-scale urban flood risk assessment. Nat Hazards 113:423–451. https://doi.org/10.1007/s11069-022-05308-9

Trinder J, Liu Q (2020) Assessing environmental impacts of urban growth using remote sensing. Geo-Spatial Inform Sci 23(1):20–39

Vahmani P, Ban-Weiss GA (2016) Impact of remotely sensed albedo and vegetation fraction on simulation of urban climate in WRF-urban canopy model: a case study of the urban heat island in Los Angeles. J Geophys Res Atmosp 121(4):1511–1531

Vervoort RW, Fuentes I, Brombacher J, Degen J, Chambel-Leitão P, Santos F (2022) Progress in developing scale-able approaches to field-scale water accounting based on remote sensing. Sustainability 14(5):2732. https://doi.org/10.3390/su14052732

Vivone G (2023) Multispectral and hyperspectral image fusion in remote sensing: a survey. Inform Fus 89:405–417. https://doi.org/10.1016/j.inffus.2022.08.032

Wandinger U (2005) Introduction to lidar. In: Weitkamp C (eds) Lidar: Springer series in optical sciences, vol 102. Springer, New York

Wang L, Saiz A, Li W (2024) Natural fragmentation increases urban density but impedes transportation and city growth worldwide. Nat Cit 1:642–653. https://doi.org/10.1038/s44284-024-00118-5

WeForum (2021) We must reimagine our cities as nexus of biodiversity, conservation and climate resilience. Here is how. https://www.weforum.org/agenda/2021/06/cities-ecosystems-biodiversity-climate-change/

Wellmann T, Lausch A, Andersson E, Knapp S, Cortinovis C, Jache J et al (2020) Remote sensing in urban planning: contributions towards ecologically sound policies? Landsc Urban Plann 204:103921. https://doi.org/10.1016/j.landurbplan.2020.103921

Wu Z, Zhou Y, Ren Y (2024) Green space-building integration for urban heat Island mitigation: insights from Beijing's fifth ring road district. Sustain Cit Soc 15:105917. https://doi.org/10.1016/j.scs.2024.105917

Xiao A, Xu Z, Skare M et al (2024) Bridging the digital divide: the impact of technological innovation on income inequality and human interactions. Hum Soc Sci Commun 11:809. https://doi.org/10.1057/s41599-024-03307-8

Yang Y, Xu Y, Duan Y, Zhang S, Zhang Y, Xie Y (2023) How can trees protect us from air pollution and urban heat? Associations and pathways at the neighborhood scale. Landsc Urban Plann 236:104779. https://doi.org/10.1016/j.landurbplan.2023.104779

Young N, Roche DG, Lennox RJ, Bennett JR, Cooke SJ (2022) Ethical ecosurveillance: mitigating the potential impacts on humans of widespread environmental monitoring. People Nat 4(4):830–840. https://doi.org/10.1002/pan3.10327

Zhang Z, Zhu L (2023) A review on unmanned aerial vehicle remote sensing: Platforms, sensors, data processing methods, and applications. Drones 7(6):398. https://doi.org/10.3390/drones7060398

Zhang J, Li D, Ning S, Furuya K (2023) Sustainable urban green blue space (UGBS) and public participation: integrating multisensory landscape perception from online reviews. Land 12(7):1360

Zhou W, Pickett STA, McPhearson T (2021) Conceptual frameworks facilitate integration for transdisciplinary urban science. NPJ Urban Sustain 1:1. https://doi.org/10.1038/s42949-020-00011-9

Zuccaro G, Leone MF (2021) Climate services to support disaster risk reduction and climate change adaptation in urban areas: the CLARITY project and the Napoli case study. Front Environ Sci 9:693319. https://doi.org/10.3389/fenvs.2021.693319

Zutter C, Stoltz A (2023) Community gardens and urban agriculture: healthy environment/healthy citizens. Int J Mental Health Nurs 32(6):1452–1461. https://doi.org/10.1111/inm.13149

Extraction of Fallow Lands in the *Rabi* Season Using Time-Series Landsat 8 Imageries

Swadhina Koley and Jeganathan Chockalingam

Abstract Monitoring agricultural fallow lands is crucial for regional agro-economic growth, as it enables potential crop intensification by utilizing these lands in the seasons apart from the primary growing season. This study attempts to map the fallow lands in the *Rabi* season in Jharkhand during 2015–2024 through a spectral index threshold-based approach. Each year's seasonal maximum Normalized Difference Vegetation Index (NDVI) was derived from the Landsat 8 time-series datasets. Sample points collected over the fallow lands in 2015 were used to estimate the NDVI threshold based on the statistical values of the points. The lower and upper thresholds to map fallow lands were derived at 0.214 and 0.312, respectively. The results reveal that the maximum lands were left uncultivated in 2015 (30.86% of the available agricultural area) and 2018 (25.61% of the agricultural area) due to the scanty amount of monsoon and post-monsoon rainfall in the previous growing seasons. The fallow land map of 2015 achieved an 88% accuracy when validated with the reference land use land cover data. Lohardaga and Giridih districts were identified as having a maximum extent of fallow lands, with more than 50% of the agricultural area left uncultivated in these two years. The outcomes of this study would help in understanding the cropping system dynamics of the region and identify the potential lands to grow crops in secondary growing seasons like *Rabi*, further improving crop productivity and ensuring food security.

Keyword Rabi fallow · NDVI threshold · Landsat 8 · Time-series analysis

S. Koley · J. Chockalingam
Department of Remote Sensing, Birla Institute of Technology, Mesra, Ranchi, India
e-mail: jeganathanc@bitmesra.ac.in

S. Koley (✉)
Department of Geography, School of Basic and Applied Sciences, Adamas University, Kolkata, India
e-mail: swadhina.koley1@adamasuniversity.ac.in

J. Chockalingam
Vice Chancellor, Sarala Birla University, Ranchi, India

O. Mutanga et al. (eds.), *Revealing Ecosystem Services Through Geospatial Technologies*, Springer Remote Sensing/Photogrammetry,
https://doi.org/10.1007/978-3-031-98048-0_3

1 Introduction

Monitoring fallow cropland is vital for efficiently using the cultivable land and establishing a sustainable agro-based ecosystem. Leaving the land uncultivated for a longer time could convert the area into abandoned land (Löw et al. 2018), further impacting crop productivity due to the shrinkage in the active croplands. On the other hand, spatial expansion of croplands by forest degradation could harm biodiversity and the environment (Chaplin-Kramer et al. 2015; Zabel et al. 2019). Therefore, identifying and properly monitoring the fallow lands a priori could be a viable solution to increase the cropping intensity and ensure food security. Also, identifying seasonal fallow areas could benefit food grain production (Zhang et al. 2014). Wallace et al. (2017) mentioned that monitoring the fallow or uncultivated land can also provide an idea about the water use in the agricultural sector.

In Jharkhand, most of the cultivable lands are left fallow in the *Rabi* and *Zaid* cropping seasons (an example is shown in Fig. 1). Hence, identifying the seasonal fallow areas would help introduce new cropping patterns and increase the cropping intensity. However, mapping and monitoring fallow lands are challenging, as the reason behind fallow occurrence in a region could be multi-directional. Personal or socio-economic restrictions of the land owner (Alcantara et al. 2012), agricultural practices like shifting cultivation (Das et al. 2021), lack of agricultural infrastructures and facilities (Ghosh et al. 2017), and environmental conditions like drought (Chandna and Mondal 2020) are some of the reasons. These aforementioned factors make fallow lands mapping in conventional methods (such as field surveys) more difficult over a longer period of time, and in larger regions. In contrary, advanced methods like geospatial technology, has proven to be much efficient and less time consuming for similar studies due to its continuous and timely data support over the recent past. This technology is also effective to achieve higher accuracies when integrated with cutting-edge computational methods like automated mapping. By examining different light wavelengths, assessing vegetation health through indices, and observing seasonal shifts, it's possible to pinpoint underused lands during the Rabi season. This technique is beneficial as it provides wide coverage, is cost-efficient, and allows for ongoing monitoring, making it particularly useful for developing sustainable land-use strategies in Jharkhand.

Several studies have been reported before describing several methodologies to map fallow lands using remotely sensed data. Wu et al. (2014) utilized MODIS time-series data for automated fallow land extraction over California. Similarly, Gumma et al. (2016) and Gumma et al. (2018) used the MODIS datasets to monitor fallow lands in South Asia and Myanmar, respectively, with a primary objective of finding lands for short-duration legume cultivation. Wallace et al. (2017) adopted a temporal anomaly-based approach using the MODIS vegetation indices datasets. Apart from the MODIS time-series data, satellite images from various other sources have also been applied to identify fallow lands. Leisz and Rasmussen (2012) implemented a non-parametric classification technique using Landsat for fallow land mapping in Vietnam. Wang and Song (2021) used multi-temporal Landsat images and a random

Fig. 1 Post-Kharif rice fallows in the *Rabi* season (East Singhbhum, Jharkhand)

forest classification approach for fallow identification in China. On the other hand, Chandna and Mondal (2020) used the Landsat images along with Sentinel 1 images for the same purpose in India through a different decision-tree model and Principal Component Analysis (PCA). Some studies extracted fallow lands through land use land cover analysis at various scales (Papadopoulou et al. 2023; Song et al. 2022). Sentinel 2 multispectral imageries were used for the fallow extraction by Ghosh et al. (2017) and Tong et al. (2020).

Apart from the freely available datasets, a few studies have also been reported using datasets obtained from commercially available satellites. Zhang et al. (2014) extracted fallow lands in China using HJ-1 CCD (Huan Jing-1 Charge-Coupled Device) data based on a vegetation indices difference approach. Evstratova and Antoshkin (2021) used images from Quickbird, GeoEye, and WorldView satellites to monitor the growth of shrubs and trees over the fallow lands using wavelet analysis. Rufin et al. (2022) used high-resolution PlanetScope data to map short-term fallow lands in the northern Mozambique region. However, these studies were limited to a minimal spatial extent.

Previous research on fallow-land monitoring focused on two approaches. First, mapping the long-term fallow lands using time-series satellite data (Alcantara et al. 2012; Löw et al. 2018; Wang and Song 2021), and second, mapping the seasonal fallow lands that are left unplanted in other non-primary cropping seasons (Chandna and Mondal 2020; Ghosh et al. 2017; Gumma et al. 2018; Srivastava et al. 2023). Nonetheless, no significant study was reported to estimate the fallow areas in Jharkhand either way. Also, the above-mentioned studies do not discuss about the driving factors behind the fallow lands. Therefore, this study aims to map the fallow lands in Jharkhand using high spatial resolution datasets and tries to evaluate the possible reasons behind these fallow occurrences. This study suggests a more simplified method for feature extraction using NDVI threshold values. Spectral indices are specifically designed to highlight features of interest (e.g., vegetation, water bodies, urban areas). By setting appropriate thresholds, it's possible to clearly differentiate these features from the surrounding background or other types of land cover. Moreover, these threshold-based techniques are easy to use and don't involve complicated tuning or training, which makes them ideal for quick and automated extraction of features.

2 Materials and Methods

2.1 Study Area

The study was carried out in the state of Jharkhand. The state was carved out of another Eastern state, Bihar, on 15 November 2000, with the state capital located at Ranchi. The spatial extent of the study area is 21.97°N–25.34°N and 83.33°E–87.96°E. The total geographical area of the region is about 79,714 km^2, which approximately covers about 2.42% of India. The state shares its administrative boundary with Bihar and a part of Uttar Pradesh in the north, West Bengal in the east, Odisha in the south, and Chhattisgarh in the west. Jharkhand's total agricultural area (TAA) covers about 49% of the total geographical area (TGA). According to (Government of India 2011), the population of Jharkhand is 32.98 Million, of which about 75% belong to the rural area. About 59% of the working population is engaged in the agricultural sector. The average annual rainfall in the study area is about 1400 mm, most experienced in the monsoon season. The area falls under three different agro-climatic sub-zones named Central and North Eastern Plateau (Zone IV), Western Plateau (Zone V), and South-Eastern Plateau (Zone VI) (JICA 2011). Depending on the characteristics of these zones, the climate and physiographic conditions of the state vary over the place. The location of the study area is represented by Fig. 2.

2.2 Materials

Landsat 8 time series data at 30 m spatial resolution (with 16-day revisit time) from 2015 to 2024 was acquired from the United States Geological Survey (USGS) Earth Explorer portal. These multi-spectral datasets were used to generate the NDVI in the *Rabi* season for the specified years. In Jharkhand, the peak growth in the *Rabi* season is usually observed in late January to early February. Hence, Landsat 8 data were obtained for the mentioned period. Though the sowing usually takes place from late October to early December, no data were selected from November and December due to the unavailability of completely cloud-free data. Only the images with <10% cloud cover were selected for the study. High-resolution Google Earth imagery was used to collect the sample points over the fallow areas in the *Rabi* season. About 414 sample points were collected from the seasonal fallow lands using visual interpretation of the standard FCC and the high-resolution Google Earth images. Figure 3 represents the locations of the sample points collected. The land use land cover (LULC) dataset

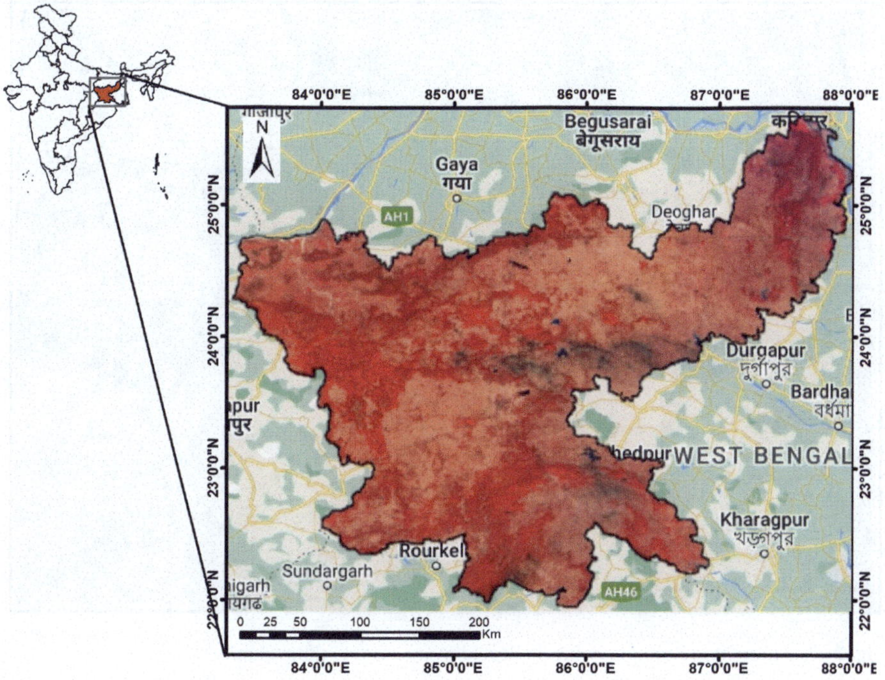

Fig. 2 Location of the study area showing the standard false color composite (FCC) in late January 2015

prepared by the National Remote Sensing Centre (NRSC), Government of India for 2015–2016 was acquired as the reference data for the accuracy assessment.

2.3 *Methods*

2.3.1 Pre-processing of the Landsat 8 Datasets

The Landsat 8 datasets are provided with quantized digital numbers (DNs) representing the different features on the earth's surface. But these numbers are primarily influenced by atmospheric effects like scattering. Hence, before further analysis, the datasets require necessary pre-processing to obtain more accurate information. To convert the DN values to the Top of the Atmosphere (TOA) reflectance, first, the spectral radiance is calculated using Eq. 1. After the spectral radiance was calculated, the TOA spectral reflectance was calculated using Eq. 2.

$$L_\lambda = (M_L \times Q_{Cal}) + A_L \tag{1}$$

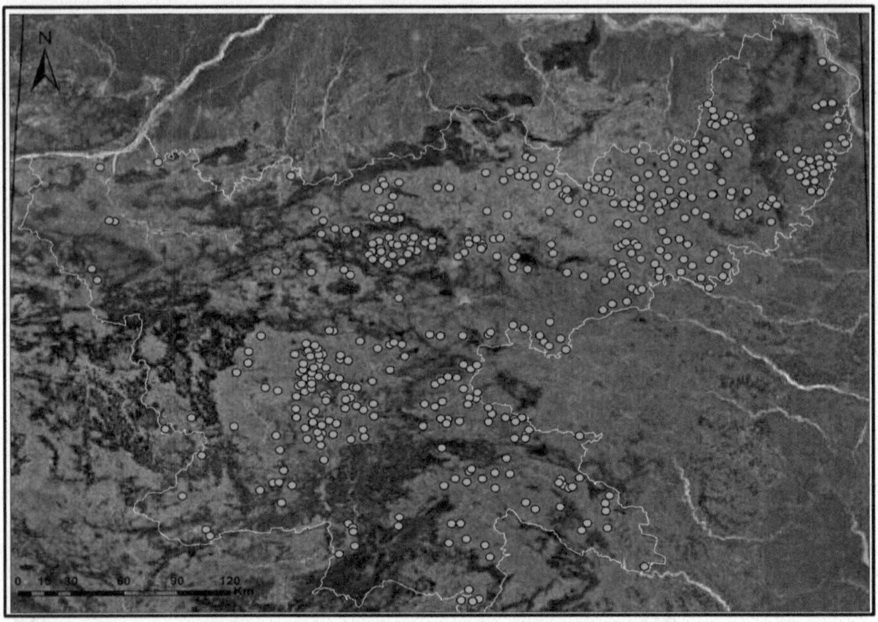

Fig. 3 Location of the sample points collected over seasonal fallow lands

L_λ = Spectral radiance (W m^{-2} sr^{-1} μm^{-1}).
M_L = The radiance multiplicative scaling factor for a particular band.
A_L = The radiance additive scaling factor for a particular band.
Q_{Cal} = The DN value.

$$\rho_{\lambda'} = (M_\rho \times Q_{Cal}) + A_\rho \quad (2)$$

$\rho_\lambda{}'$ = TOA spectral reflectance (unitless).
M_ρ = The reflectance multiplicative scaling factor for a particular band.
A_ρ = The reflectance additive scaling factor for a particular band.
Q_{Cal} = The DN value.

The necessary scaling factors required for the conversion of DN to spectral radiance were obtained from the metadata of the Landsat images. Since only two bands, i.e., NIR and Red bands, were required to calculate the NDVI, these two bands were taken into consideration for necessary corrections.

However, the TOA spectral reflectance estimated using Eq. 2 does not consider the correction for solar angle. Hence, to calculate the accurate TOA reflectance with correction for solar angle, Eq. 3 was used further.

$$\rho_\lambda = \frac{\rho_{\lambda'}}{\sin(\theta_{SE})} \quad (3)$$

ρ_λ = TOA reflectance (with correction for solar angle) (unitless).
θ_{SE} = Sun elevation angle in degrees (obtained from the metadata).

2.3.2 Calculation of Vegetation Index and Determination of the Threshold Values

Seasonal NDVI was used for this study. After acquiring the satellite images in the *Rabi* season from 2015 to 2024, the NDVI (proposed by Rouse et al. 1974) for each image in each year was calculated using Eq. 4. Then the maximum NDVI for the season was estimated for each year. Hence, a total of six NDVI images were prepared for the study. The Global Food-and-Water Security-support Analysis Data (GFSAD) cropland mask at 30 m resolution was used to extract the cropland areas from the NDVI images.

$$\mathrm{NDVI} = \frac{\mathrm{NIR} - \mathrm{Red}}{\mathrm{NIR} + \mathrm{Red}} \tag{4}$$

2015 was considered the base year for estimating the increase or decrease in the fallow areas in the following years. Hence, the 414 sample points were used to determine the threshold values using the 2015 NDVI image. First, the pixel values of the indices were extracted using the corresponding sample points to calculate the threshold values. Next, the mean and standard deviation of the sample points were derived. The threshold was finally decided by taking the twice standard deviation values on either side of the sample mean, as it would statistically cover around 95% of the data values extracted using the sample points (Larson and Farber 2006).

2.3.3 Identification of Fallow Lands, Validation, and Area Estimation

The threshold values were used to differentiate the seasonal fallow from the other classes. The pixels with NDVI values between the achieved thresholds were classified as the seasonal fallow, and the rest were classified as other classes. Two hundred fifty points were generated through stratified random sampling to validate the mapped fallow lands. The LULC data prepared by the NRSC, Government of India for 2015–2016 was fetched through Web Map Services (WMS) to the GIS environment. This LULC layer was considered as the reference map to validate the outcomes. Table 1 represents the structure of the confusion matrix for the accuracy assessment. The overall accuracy was calculated using Eq. 5. The Kappa statistic was calculated using Eq. 6. Finally, district-wise area estimation was done to calculate the area statistics. The methodology of the workflow is represented in Fig. 4.

$$\mathrm{OverallAccuracy(OA)} = \frac{\mathrm{TP} + \mathrm{TN}}{\mathrm{TP} + \mathrm{FP} + \mathrm{FN} + \mathrm{TN}} \times 100 \tag{5}$$

Table 1 Structure of confusion matrix

		Reference	
		Positive	Negative
Classified	Positive	True positive (TP)	False positive (FP)
	Negative	False negative (FN)	True negative (TN)

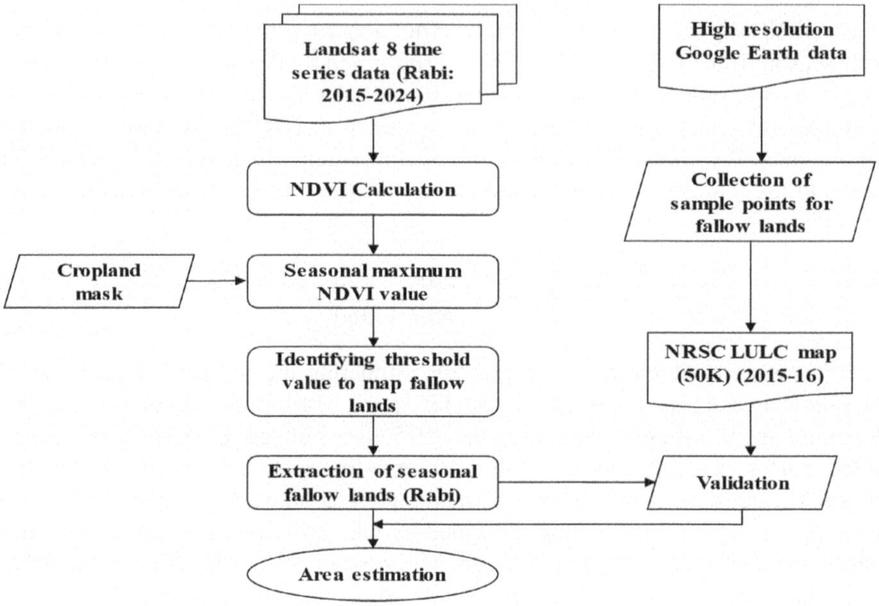

Fig. 4 Methodological diagram to identify fallow lands in the *Rabi* season

$$\text{Kappa}(\kappa) = \frac{P_o - P_e}{1 - P_e} \tag{6}$$

where P_o is the observed overall accuracy, and P_e is the expected accuracy.

3 Results and Discussions

3.1 Results

2015 was considered the base year for mapping the fallow lands in the *Rabi* season. Hence, the threshold values to identify the fallow lands were determined based on the NDVI of 2015. The upper threshold for the fallow land was identified as 0.312, and the lower threshold was identified as 0.214. These estimated threshold values agree with those Chandna and Mondal (2020) obtained for their study to map rice-fallow areas in another Indian region. The pixels with NDVI values within the specified thresholds were classified as the fallow lands, and the rest were considered other classes. However, due to the spectral similarity, other lands like permanent pastures, grasslands, or shrubs might reflect similar NDVI patterns. So, to distinguish between such classes and the crop fallow areas, the GFSAD cropland mask at 30 m resolution was applied to extract the fallow pixels belonging only to the cropland class. The rest of the pixels were categorized as other classes.

To validate the mapped fallow lands, 250 points were generated through stratified random sampling: 175 belonging to the fallow lands and the rest belonging to the other classes. LULC for 2015–2016, prepared by NRSC, was considered as the reference image for accuracy assessment. The mapped fallow lands were 88.4% accurate (Kappa ~0.73) with the reference data. Table 2 represents the confusion matrix and the overall accuracy achieved.

The maximum fallow land area in the *Rabi* season was observed in 2015 (30.86% of the TAA), followed by 2018 (25.61% of the TAA) and 2022 (25.49% of the TAA). The minimum fallow area was observed in 2019, which was 10.45% of the TAA. The yearly distribution of the fallow lands (percentage area) in the *Rabi* season is shown in Fig. 5. The maximum fallow area in 2015 was found in the Lohardaga district (58.95%), followed by the Giridih district (56.87%). These two districts also experienced a high percentage of fallow area in 2018 along with the Khunti district. The fallow area coverage in Lohardaga and Giridih in 2018 was 58.56 and 50.49%, respectively, whereas the area in the Khunti district was 51.98%. The possible reason behind the fallow occurrence in those districts in the years mentioned above could be the insufficient amount of post-monsoon rainfall in the preceding years. *Rabi* cropping mainly depends on surface moisture availability from the monsoon and post-monsoon rainfall during the previous cropping season. In 2014, Lohardaga and Giridih encountered 88 and 56% deficiency in post-monsoon rainfall, respectively (India Meteorological Department 2017). In 2017, the rainfall deficiency in the concerned season was 29% in Giridih, 55% in Khunti, and 22% in Lohardaga (India Meteorological Department 2018). The scanty amount of post-monsoon rainfall in 2014 and 2017 eventually resulted in less cropping in 2015 and 2018, thus leaving many parts of the cultivable croplands as seasonal fallows. Table 3 shows a district-wise distribution of the fallow lands in the *Rabi* season from 2015 to 2024.

Table 2 Confusion matrix for accuracy assessment

		Reference		Total
		Fallow lands	Other	
Classified	Fallow lands	157	18	175
	Other	11	64	75
Total		168	82	250
Overall accuracy				88.4%
Kappa				0.73

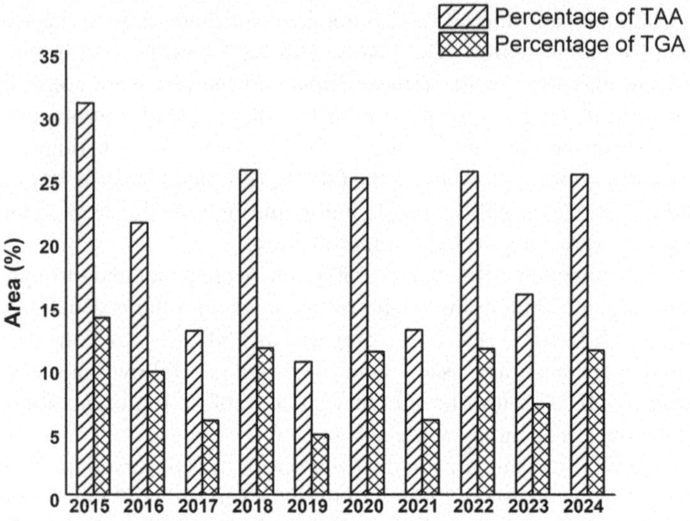

Fig. 5 Yearly distribution of fallow lands in the Rabi season

Figure 6 shows the spatial distribution of fallow lands in the *Rabi* season from 2015 to 2024.

3.2 Discussions

This research suggests a threshold-based method to map the fallow land extent in the *Rabi* season in Jharkhand. The accuracy achieved in mapping the fallow lands (~88%) was more than the benchmark set by Anderson et al. (1976), i.e., 85% overall accuracy and more than the accuracies achieved in previous research (Alcantara et al. 2012; Ghosh et al. 2017; Gumma et al. 2016; Low et al. 2015) ranging between 65 and 83%. This high accuracy proves the effectiveness of the opted method for automated mapping of land features. Previously Hashim et al. (2019) and Idrees et al. (2022) used NDVI-threshold-based techniques to map urban vegetation and urban LULC, respectively. They both achieved an overall accuracy of ~70% for their respective studies. Aryal and Sitaula (2022) used an NDVI threshold of −1.0 to 0.3 to distinguish between non-vegetation and vegetation (NDVI of 0.3 onwards being vegetation) and achieved an overall accuracy of 90%, which closely matches the outcomes of our study. However, it is to be noted that the NDVI is considerably affected by the soil background, atmospheric conditions, solar angle, and the presence of residual vegetation. Hence, the threshold value for NDVI might differ in different regions based on the complexity and heterogeneity of the land surface.

Leaving the lands uncultivated for extended periods can convert the land from seasonal fallow to abandoned agricultural lands. However Gumma et al. (2018)

Table 3 District-wise distribution of fallow lands (% of agriculture area)

District	2015	2016	2017	2018	2019	2020	2021	2022	2023	2024
Bokaro	44.92	17.15	15.67	32.09	3.93	24.98	2.71	19.38	4.47	6.30
Chatra	33.13	17.20	16.65	21.93	4.51	8.43	23.46	29.09	27.89	32.95
Deoghar	39.12	33.11	10.37	35.14	12.18	10.22	12.70	23.94	15.47	28.27
Dhanbad	24.73	10.85	3.80	10.78	3.74	5.22	4.27	23.49	5.52	22.70
Dumka	28.95	26.61	2.71	18.98	5.24	12.38	11.30	30.88	19.47	43.38
East Singbhum	21.14	16.47	5.67	12.07	3.78	17.34	9.45	37.58	9.05	35.84
Garhwa	17.79	5.16	10.89	5.70	2.26	2.32	36.00	41.79	49.99	45.08
Giridih	56.87	38.77	29.43	50.49	22.60	12.36	10.12	12.87	9.45	11.46
Godda	21.63	13.80	2.67	14.96	3.87	5.84	28.67	46.94	28.80	44.91
Gumla	49.33	50.25	21.95	47.27	11.91	9.26	2.78	6.79	3.54	10.12
Hazaribagh	33.52	20.65	19.77	33.42	8.45	11.41	22.26	25.06	22.30	28.29
Jamtara	43.87	29.21	4.87	19.27	8.37	23.37	3.70	21.10	12.83	35.45
Khunti	33.59	17.76	24.55	51.98	9.98	22.87	2.10	12.49	2.83	6.79
Koderma	42.10	24.99	22.50	34.45	11.70	6.90	23.00	23.36	21.28	24.92
Latehar	51.76	36.00	22.87	30.11	8.48	14.17	5.55	9.84	7.21	12.95
Lohardaga	58.95	52.81	29.17	58.56	17.86	17.23	9.92	11.90	9.06	17.30
Pakur	30.52	27.85	4.81	13.29	2.41	19.51	11.15	26.95	20.33	35.55
Palamu	24.09	5.90	11.96	9.83	3.20	3.77	29.91	38.08	32.20	41.66
Ramgarh	44.63	26.98	32.30	46.17	7.18	38.74	2.46	12.35	3.57	2.69
Ranchi	40.80	25.50	25.05	46.32	6.02	32.51	6.81	19.70	8.41	12.41
Sahibganj	13.84	12.84	3.65	13.54	2.33	8.88	25.08	45.66	27.00	42.83
Saraikela-Kharsawan	34.48	20.78	22.71	46.64	3.09	39.23	0.90	14.93	2.13	4.41
Simdega	27.94	39.09	10.23	29.04	8.41	7.41	1.16	8.00	2.32	12.88
West Singbhum	13.91	12.94	11.60	21.29	1.53	27.11	1.22	21.84	3.14	9.15

suggested that it is necessary to understand whether the fallow lands are left to be permanently abandoned or if the emptiness of the land is a part of local crop rotation. Loboda et al. (2017) also agreed with this, stating that leaving the land fallow for a longer time might be necessary for the crop cycle followed in that region. While the coverage of fallow lands decreased by a few per cent in some areas, the extent of uncultivated lands increased significantly in others. The potential reasons behind this increase and decrease in fallow land areas over time might be different factors like socio-economic contribution, improving infrastructure in the society, advanced farming techniques, etc. One of the significant possible reasons could be the declining percentage of workforce engagement in the agricultural sector. About 57.49% of the state's main working population was engaged in agricultural activities in the earlier period (2001), which was cut by 10% in 2011 (Government of India 2011). Districts like Dhanbad, which experienced an increase in fallow land, have active industrial

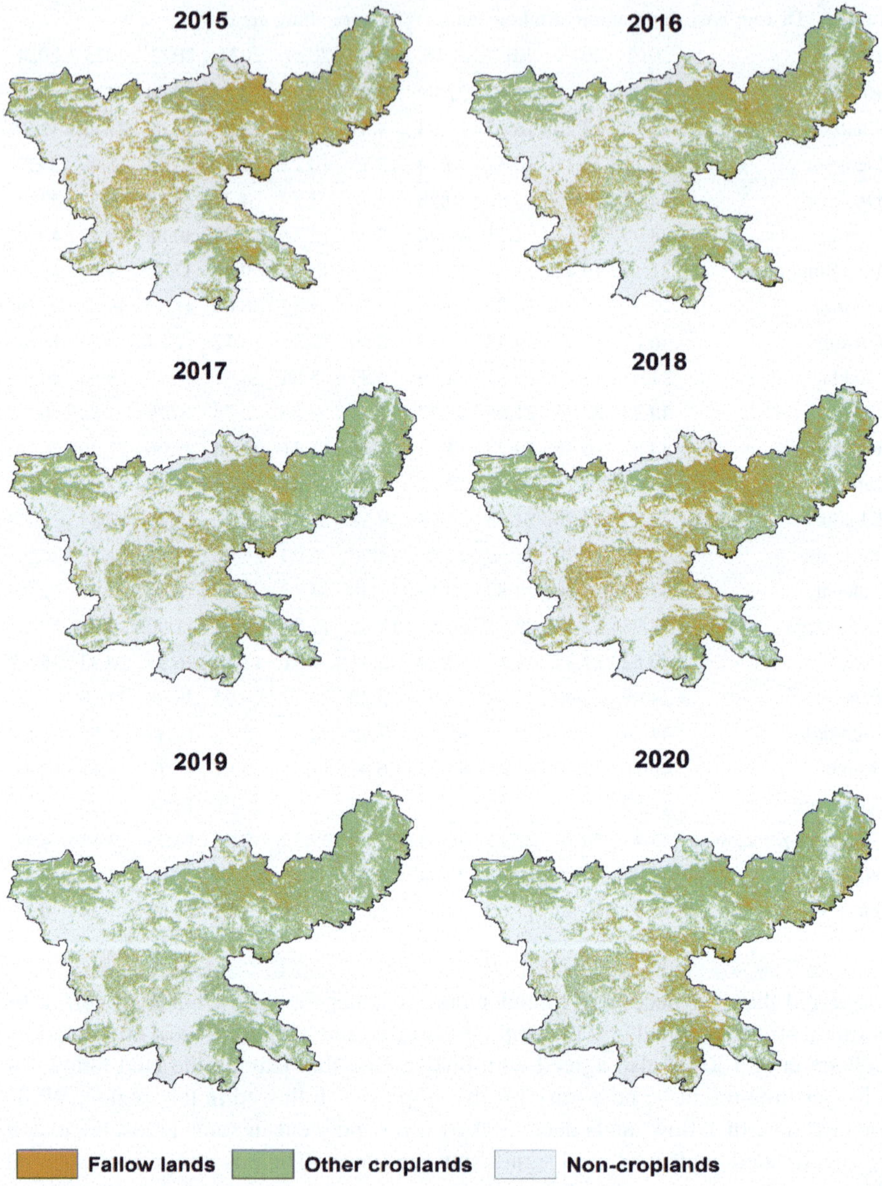

Fig. 6 The spatial distribution of fallow lands in the *Rabi* season

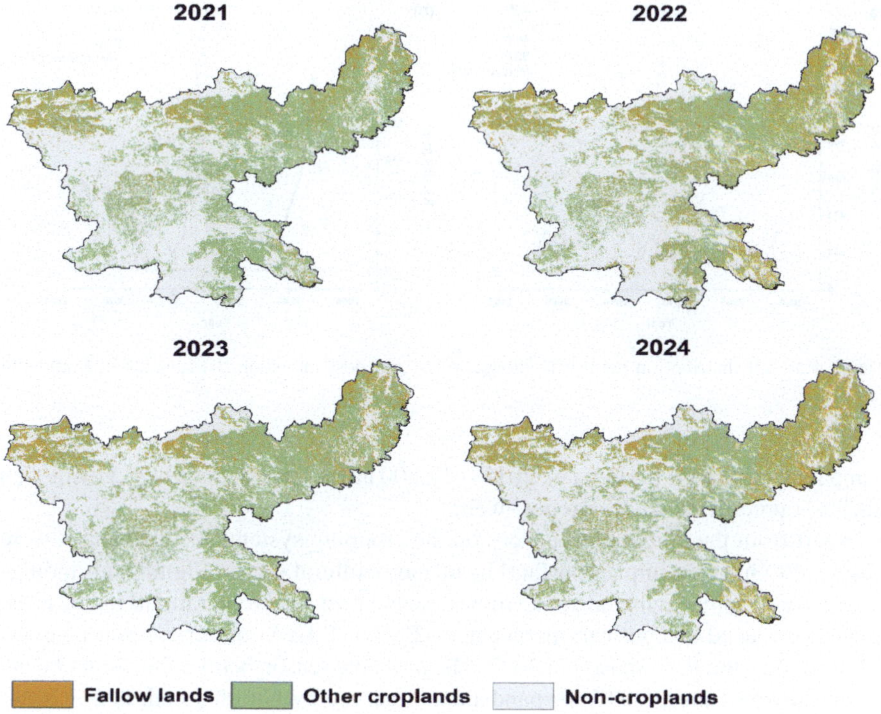

Fig. 6 (continued)

facilities. The need for a sufficient workforce in those industries might have caused a shrinkage in the agricultural population by 11%.

Besides, the existing climatic pattern might have influenced the cropping system. Chatra, Garhwa, and Palamu are the districts that experience the minimum number of rainy days in the monsoon season (Guhathakurta et al. 2020), which could be a reason behind leaving the lands uncultivated. Furthermore, the three different agro-climatic zones in Jharkhand cause uneven monsoon and post-monsoon rainfall distribution. The unreliable rainfall pattern in this region causes inconsistency in the cropping system. The erratic rainfall distribution pattern in the monsoon and post-monsoon seasons in Jharkhand is represented in Fig. 7. Improved farming techniques and facilities provided to the agricultural sector may impact the cropping system, thus reducing the extent of the fallow lands in the *Rabi* season. Increasing irrigation facilities can play a vital role in decreasing the area of fallow lands. Most of the districts showed a fluctuating pattern in the fallow land area over the decade concerned. On the other hand, in Giridih district, a continuously decreasing trend in the fallow area was observed, except in 2015 and 2018, as mentioned earlier. The potential reason behind this reduction in the fallow area could be the increasing irrigation facilities in this district. The gross irrigated area in this district has significantly increased over the past decade, especially in 2020–2023, when it almost tripled (~40,000 ha)

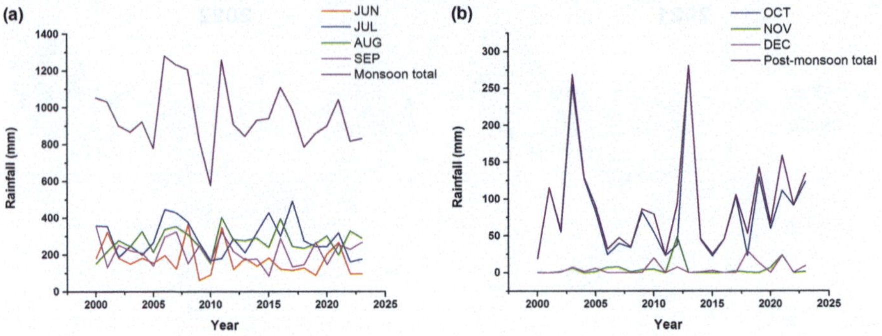

Fig. 7 Rainfall distribution pattern in monsoon (**a**) and post-monsoon (**b**) seasons in Jharkhand (2000–2023)

compared to the base period, i.e., 2015 (~12,800 ha) (https://data.desagri.gov.in/weblus/lus-source-irrigated-area-report-web).

Apart from the climatic influences on the cropping system, social changes were observed to have an impact on Jharkhand's agricultural sector. Significant findings reveal that despite having adequate monsoon and post-monsoon rainfall in 2021, the area experienced a significant percentage (25.5% of TAA) of agricultural land left fallow during the *Rabi* season in 2022. The probable reason behind this might be the second wave of the COVID-19 pandemic. A survey by Ghosh-Jerath et al. (2024) mentioned the adverse impact of the second wave of COVID-19 in the cropping sector of Jharkhand.

The study successfully identified the fallow areas in the *Rabi* season in Jharkhand in the past 10 years based on the NDVI threshold provided. Nonetheless, there remain some uncertainties in the method applied, which allows further research to monitor the fallow lands. Firstly, the fallow land area is random over the years and does not follow any discrete pattern. While some scholars have claimed that if agricultural land is kept uncultivated for three years or even longer, the land is considered abandoned (Wang and Song 2021), some scholars argued that keeping the land fallow for a longer time could be a part of local crop rotation cycle (Alcantara et al. 2012; Loboda et al. 2017; Wallace et al. 2017). No substantial evidence was found in this study indicating permanent agricultural land abandonment. Hence, it is uncertain whether the lands were intentionally uncultivated or if it was natural abandonment. There also lies a scope to investigate various topographic and physiographic factors like the absence of soil nutrients, decreasing soil fertility, lack of adequate groundwater, etc., which could be responsible for temporary or long-term land abandonment.

The outcomes of this research would equip stakeholders with essential insights into neglected farming areas, aiding in thoughtful planning and resource allocation. For those in policymaking roles, detailed maps of fallow lands serve as valuable tools for crafting tailored agricultural policies, potentially encouraging farmers to cultivate these areas through financial assistance, improved irrigation access, or support with farming supplies. Researchers can analyze this information to determine the reasons behind land abandonment, such as insufficient water resources, inadequate soil quality, or socio-economic hurdles, and propose solutions to these challenges. Adopting innovative farming practices, including crop diversification or the use of climate-resilient varieties, can also be explored. By effectively engaging fallow lands, Jharkhand can stimulate agricultural output, enhance food security, and boost farmers' incomes, thereby creating jobs in agriculture and related industries. This approach not only supports economic development but also aligns with the second Sustainable Development Goal of achieving zero hunger, promoting sustainable agricultural practices, and improving livelihoods.

4 Conclusions

The study suggests a threshold-based approach to map seasonal fallow agricultural lands in Jharkhand using time series satellite data. The threshold has been applied to the seasonal maximum NDVI values derived from the data. The seasonal fallow monitoring in the *Rabi* season showed that the rainfall distribution pattern plays a significant role in fallow occurrence. The maximum fallow land in the *Rabi* season was identified in 2015 (30.86%), followed by 2018 (25.51%) due to the lack of post-monsoon rainfall in previous years. Lohardaga and Giridih were two districts identified as having a maximum extent of fallow lands in the *Rabi* season in the abovementioned years. In these years, the two districts had more than 50% of the agricultural area left uncultivated during the *Rabi* season. Conversely, a decreasing trend in the fallow land area in Giridih was observed due to the improved irrigation facilities in this region over time. The study gives further scope to examine different topographic and physiographic parameters to understand their contribution to land abandonment.

Author Contribution Swadhina Koley: Research conceptualization; data curation; primary analysis; writing the original draft. Jeganathan Chockalingam: Editing manuscript; overall supervision.

Funding This research received no external funding from any institution or organization.

Conflict of Interest The authors declare that they have no competing personal or financial interests.

References

Alcantara C, Kuemmerle T, Prishchepov AV, Radeloff VC (2012) Mapping abandoned agriculture with multi-temporal MODIS satellite data. Remote Sens Environ 124:334–347. https://doi.org/10.1016/j.rse.2012.05.019

Anderson JR, Hardy EE, Roach JT, Witmer RE (1976) A land use and land cover classification system for use with remote sensor data. Washington, DC

Aryal J, Sitaula C (2022) NDVI threshold-based urban green space mapping from. Land 11:351. https://doi.org/10.3390/land11030351

Chandna PK, Mondal S (2020) Analyzing multi-year rice-fallow dynamics in Odisha using multi-temporal landsat-8 OLI and sentinel-1 data. Gisci Remote Sens 57(4):431–449. https://doi.org/10.1080/15481603.2020.1731074

Chaplin-kramer R, Sharp RP, Mandle L, Sim S, Johnson J, Butnar I, Milà i Canals L, Eichelberger BA, Ramler I, Mueller C, McLachlan N, Yousefi A, King H, Kareiva PM (2015) Spatial patterns of agricultural expansion determine impacts on biodiversity and carbon storage. Proc Natl Acad Sci 112:7402–7407. https://doi.org/10.1073/pnas.1406485112

Das P, Mudi S, Behera MD, Barik SK, Mishra DR, Roy PS (2021) Automated mapping for long-term analysis of shifting cultivation in northeast India. Remote Sens 13(6):1066. https://doi.org/10.3390/rs13061066

Evstratova L, Antoshkin A (2021) Detection of fallow lands overgrowth from satellite images based on wavelet analysis. E3S Web Confer 333:01003

Ghosh S, Saraf S, Behera M, Biradar C (2017) Estimating agricultural crop types and fallow lands using multi temporal sentinel-2A imageries. Proceed Natl Acad Sci India A Phys Sci 87(4):769–779. https://doi.org/10.1007/s40010-017-0447-5

Ghosh-Jerath S, Dhasmana A, Nair SC, Kapoor R (2024) Impact of the second wave of COVID-19 pandemic on food security among Ho indigenous community of Jharkhand. India. Agricult Food Sec 13(1):1–14. https://doi.org/10.1186/s40066-024-00469-1

Govt. of India (2011) Census of India. https://censusindia.gov.in/2011census/population_enumeration.html. Accessed 8 Aug 2021

Guhathakurta P, Bile L, Khedikar S, Menon P, Prasad AK, Sangwan N, Advani SC (2020) Climate research and services observed rainfall variability and changes over Jharkhand state. IMD Ann Rep 16:27

Gumma MK, Thenkabail PS, Teluguntla P, Rao MN, Mohammed IA, Whitbread AM (2016) Mapping rice-fallow cropland areas for short-season grain legumes intensification in South Asia using MODIS 250 m time-series data. Int J Dig Earth 9(10):981–1003. https://doi.org/10.1080/17538947.2016.1168489

Gumma MK, Thenkabail PS, Deevi KC, Mohammed IA, Teluguntla P, Oliphant A et al (2018) Mapping cropland fallow areas in myanmar to scale up sustainable intensification of pulse crops in the farming system. Gisci Remote Sens 55(6):926–949. https://doi.org/10.1080/15481603.2018.1482855

Hashim H, Latif ZA, Adnan NA (2019) Urban vegetation classification with NDVI threshold value method with very high resolution (VHR) pleiades imagery. Int Arch Photogr Remote Sens Spat Inform Sci 16:237–240

Idrees MO, Omar DM, Babalola A, Ahmadu HA, Yusuf A, Lawal FO (2022) Urban land use land cover mapping in tropical savannah using Landsat-8 derived normalized difference vegetation index (NDVI) threshold. South Afr J Geom 11(1):100–112. https://doi.org/10.4314/sajg.v11i1.8

India Meteorological Department (2017) Rainfall statistics of India—2016 rainfall statistics of India—2016 by (Vol. 01). http://hydro.imd.gov.in/hydrometweb/(S(ltoit4i1zzdzyo55f25yn5eb))/PRODUCTS/Publications/RainfallStatisticsofIndia-2016/RainfallStatisticsofIndia-2016.pdf%0A

India Meteorological Department (2018) Rainfall statistics of India—2017. New Delhi

JICA (2011) Chapter 3: present conditions of Jharkhand state. In: Preparatory survey on initiave for horticulture intensification by micro drip irrigation in Jharkhand. Japan International Cooperation Agency: Nippon Koei Co., Ltd., pp 1–197

Larson R, Farber B (2006) Elementary statistics, 5th edn. Prentice Hall, London

Leisz SJ, Rasmussen MS (2012) Mapping fallow lands in Vietnam's north-central mountains using yearly Landsat imagery and a land-cover succession model. Int J Remote Sens 33(20):6281–6303. https://doi.org/10.1080/01431161.2012.681712

Loboda T, Krankina O, Savin I, Kubanov E, Hall J (2017) Land management and the impact of the 2010 extreme drought event on the agricultural and ecological systems of European Russia.

In: Gutman G, Radeloff V (eds) Land-cover and land-use changes in eastern Europe after the collapse of the soviet union in 1991. Springer, New York, p 20

Low F, Fliemann E, Abdullaev I, Conrad C, Lamers JPA (2015) Mapping abandoned agricultural land in Kyzyl-Orda, Kazakhstan using satellite remote sensing. Appl Geogr 62:377–390. https://doi.org/10.1016/j.apgeog.2015.05.009

Löw F, Prishchepov AV, Waldner F, Dubovyk O, Akramkhanov A, Biradar C, Lamers JPA (2018) Mapping cropland abandonment in the aral sea basin with MODIS time series. Remote Sens 10(2):159. https://doi.org/10.3390/RS10020159

Papadopoulou E, Mallinis G, Siachalou S, Koutsias N, Thanopoulos AC, Tsaklidis G (2023) Agricultural land cover mapping through two deep learning models in the framework of EU's CAP activities using sentinel-2 multitemporal imagery. Remote Sens 15(19):4657. https://doi.org/10.3390/rs15194657

Rouse RWH, Haas JAW, Deering DW (1974) Monitoring vegetation systems in the great plains with ERTS. In: NASA. Goddard space flight center 3d ERTS-1 symposium. United States, pp 309–317. https://ntrs.nasa.gov/search.jsp?R=19740022614. Accessed 19 Nov 2019

Rufin P, Bey A, Picoli M, Meyfroidt P (2022) Large-area mapping of active cropland and short-term fallows in smallholder landscapes using PlanetScope data. Int J Appl Earth Observ Geoinform 112:102937. https://doi.org/10.1016/j.jag.2022.102937

Song W, Prishchepov AV, Song W (2022) Mapping the spatial and temporal patterns of fallow land in mountainous regions of China. Int J Dig Earth 15(1):2148–2167. https://doi.org/10.1080/17538947.2022.2148765

Srivastava AK, Borah SB, Ghosh Dastidar P, Sharma A, Gogoi D, Goswami P et al (2023) Rice-fallow targeting for cropping intensification through geospatial technologies in the rice belt of Northeast India. Agriculture 13(8):1509. https://doi.org/10.3390/agriculture13081509

Tong X, Brandt M, Hiernaux P, Herrmann S, Rasmussen LV, Rasmussen K et al (2020) The forgotten land use class: mapping of fallow fields across the Sahel using Sentinel-2. Remote Sens Environ 239:598. https://doi.org/10.1016/j.rse.2019.111598

Wallace CSA, Thenkabail P, Rodriguez JR, Brown MK (2017) Fallow-land algorithm based on neighborhood and temporal anomalies (FANTA) to map planted versus fallowed croplands using MODIS data to assist in drought studies leading to water and food security assessments. Gisci Remote Sens 54(2):258–282. https://doi.org/10.1080/15481603.2017.1290913

Wang Y, Song W (2021) Mapping abandoned cropland changes in the hilly and gully region of the loess plateau in China. Land 10(12):1341. https://doi.org/10.3390/land10121341

Wu Z, Thenkabail PS, Mueller R, Zakzeski A, Melton F, Johnson L et al (2014) Seasonal cultivated and fallow cropland mapping using MODIS-based automated cropland classification algorithm. J Appl Remote Sens 8(1):083685. https://doi.org/10.1117/1.jrs.8.083685

Zabel F, Delzeit R, Schneider JM, Seppelt R, Mauser W, Václavík T (2019) Global impacts of future cropland expansion and intensification on agricultural markets and biodiversity. Nat Commun 10:1–10. https://doi.org/10.1038/s41467-019-10775-z

Zhang M, Wu B, Meng J, Dong T, You X (2014) Fallow land mapping for better crop monitoring in Huang-Huai-Hai Plain using HJ-1 CCD data. IOP Conf Ser Earth Environ Sci 17(1):12048. https://doi.org/10.1088/1755-1315/17/1/012048

Grasslands in Focus: Evaluating and Monitoring Ecosystem Services

Rushali Katoch, Varun Parmar, Narender Kumar Sankhyan, Ranbir Singh Rana, Rishi Mahajan, and Tarun Sharma

Abstract Grasslands are essential ecosystems that provide a multitude of services vital for environmental health and human well-being. Among these services are carbon sequestration, the support of biodiversity, water regulation, and the enhancement of soil health. This chapter explores various methodologies for evaluating and monitoring these crucial ecosystem services, emphasizing the ecological economic and social significance of grasslands. It examines a variety of assessment frameworks such as field surveys, remote sensing technologies, and ecosystem modelling, that facilitate accurate tracking of grassland health and conditions over time. Case studies from diverse regions worldwide demonstrated how comprehensive monitoring can guide sustainable management practices and inform policy decisions. These practices are essential for ensuring that grasslands continue to provide their critical services amid environmental changes and anthropogenic pressures. Ultimately, the chapter

R. Katoch (✉) · V. Parmar · N. K. Sankhyan
Department of Soil Science, College of Agriculture, Chaudhary Sarwan Kumar Himachal Pradesh Krishi Vishvavidyalaya, Palampur, India
e-mail: rushalikatoch19@gmail.com

V. Parmar
e-mail: v.parmar21505@gmail.com

N. K. Sankhyan
e-mail: nks1998@rediffmail.com

R. S. Rana
Department of Agronomy, College of Agriculture, Chaudhary Sarwan Kumar Himachal Pradesh Krishi Vishvavidyalaya, Palampur, India
e-mail: ranars66@gmail.com

R. Mahajan
Department of Microbiology, College of Basic Sciences, Chaudhary Sarwan Kumar Himachal Pradesh Krishi Vishvavidyalaya, Palampur, India
e-mail: rishisml@gmail.com

T. Sharma
Centre for Geo-Informatics Research and Training, Chaudhary Sarwan Kumar Himachal Pradesh Krishi Vishvavidyalaya, Palampur, India
e-mail: tarunagronomos@gmail.com

O. Mutanga et al. (eds.), *Revealing Ecosystem Services Through Geospatial Technologies*, Springer Remote Sensing/Photogrammetry,
https://doi.org/10.1007/978-3-031-98048-0_4

underscores the importance of adopting integrated scalable methods to enhance the resilience and long-term sustainability of grassland ecosystems globally.

Keywords Biodiversity · Ecosystem · Grasslands · Services · Sustainable

1 Introduction

Grasslands cover nearly 28% of the world's land area, making them the most extensive terrestrial ecosystem on Earth. A majority (63%) of them being concentrated in Russia, Australia, the US, Canada, China, Kazakhstan and Brazil (FAOSTAT 2021). The definition of "grassland" varies depending on the factors or characteristics possessed by them, leading to diverse interpretations. Some definitions focus on vegetation, while others consider climate, land use, or ecological functions (Forage and Grazing Terminology Committee 1991; Barnes et al. 2003). Ecologists define grasslands based on their ecological functions, such as carbon sequestration and water regulation. Grasslands are characterized by a predominance of grasses belonging to the Poaceae family; however, they can also include a smaller presence of plants such as sedges and bushes from Cyperaceae and Juncaceae family, respectively. Different organizations and researchers have defined grasslands, ranging from the Food and Agriculture Organization's (FAO) low tree and shrub cover (<2%) to Pandeya's (1988) high grass cover (>80%) and Dixon et al.'s (2014) minimum vegetation coverage with graminoid or forb dominance and specific tree canopy limits (<10% cover and <5 m height (temperate) or <40% cover and <8 m height (tropical).

While grasslands are broadly classified as tropical and temperate however, a more precise classification by Dixon et al. (2014) and Faber-Langendoen et al. (2016) identified four types viz; tropical, mediterranean, temperate, and semi-desert. A distinct approach was followed by Allen et al. (2011) to delineate two related terms; "pastureland" which refers to land managed for forage production (grazing or cutting), and "rangeland" which describes natural ecosystems with grazable vegetation, encompassing a broader range of habitats like savannas, deserts, shrublands, etc. Figure 1 illustrates the grassland ecosystem diversity as defined by Allen et al. (2011).

Grassland ecosystems provide various benefits, including physical products, resources, and environmental services, that support human survival, livelihood and well-being. They not only conserve biodiversity by providing habitats for a diverse range of wildlife species (Cao et al. 2013) but also offer significant environmental, cultural, and economic benefits to human societies (Bengtsson et al. 2019). However, with the advent of industrialization, grassland ecosystems face escalating threats from climate change and human activities. Rising temperatures, altered precipitation patterns, and extreme weather events induced by climate change disrupt grassland productivity and ecological balance. Additionally, human practices like overgrazing, agricultural expansion, urbanization, and unsustainable land use are exacerbating degradation, leading to soil erosion, biodiversity loss, and reduced ecosystem resilience. The conversion of these natural habitats into farmland, urban

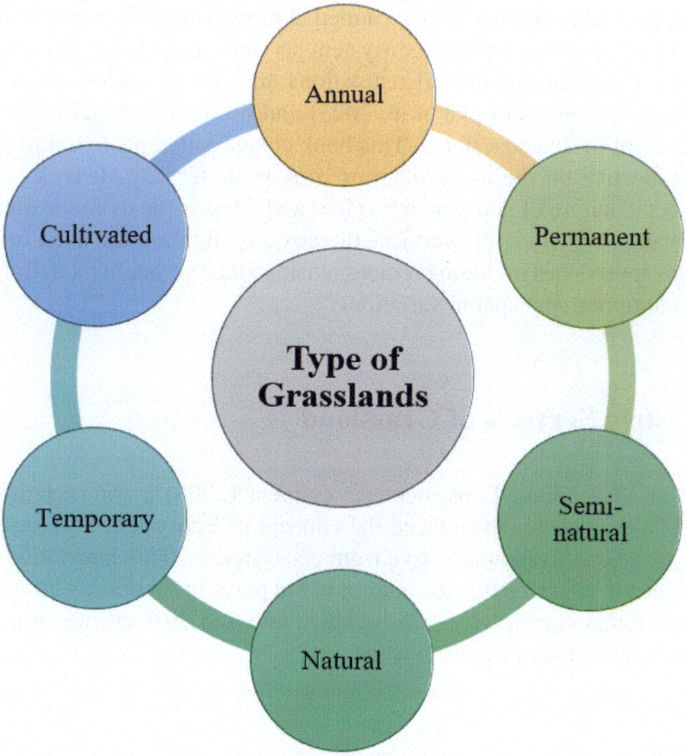

Fig. 1 Type of grasslands

areas, and industrial zones has had a devastating impact on the delicate balance of these ecosystems (Ceballos et al. 2010, Zhu et al. 2023).

Consequently, a thorough understanding and appreciation of these environments is crucial for researchers, policymakers, and the public. This has brought about extensive research on grassland ecosystem services across various areas such as spatial and temporal characteristics (Egoh et al. 2011), influence of climate change and human activities (Byrd et al. 2015; Li et al. 2019) and relationship between biodiversity, ecosystem functioning and human contentment (Wang et al. 2017; Zhao et al. 2020). Comprehensive assessment of the ecosystem services offered by grasslands such as their role in carbon storage (Liu et al. 2023), water filtration (Ren et al. 2023) and biodiversity conservation (Kachler et al. 2023) not only provides valuable insights on the ecosystem processes but also is essential for informed decision-making and effective conservation strategies. While different lab and field methods can assess various indicators of ecosystem services however, economic valuation of these services is also necessary to highlight their immense importance to attract greater investment in conservation and restoration (Egoh et al. 2008; Richter et al. 2021). Furthermore, the recognition of the intrinsic value of these natural landscapes can inspire public engagement and support for initiatives aimed at preserving and restoring grasslands (Cheng and Li 2024). Monitoring ecosystem services also guides policy decisions related to land use, environmental regulations and the allocation of resources to minimize exploitation (Schirpke et al. 2023) and promote long-term conservation and protection of these ecosystems. This book chapter attempts to uniquely present a broad perspective on the assessment of ecosystem services. It seeks to explore the multifaceted nature of ecosystem services and address the dynamic and context-dependent nature of ecosystem services thereby, highlighting the need for adaptive management approaches including remote sensing and ecosystem modelling that can account for temporal and spatial variations.

2 Ecosystem Services of Grassland

According to Millennium Ecosystem Assessment (2005) a comprehensive study of the world's ecosystems introduced the concept of Ecosystem Services, which it defined as "the benefits people derive from ecosystems". This innovative approach acknowledges the vital role that the natural world plays in supporting human welfare and progress. Ecosystem services are classified into four distinct categories based on the specific benefits they provide (Fig. 2).

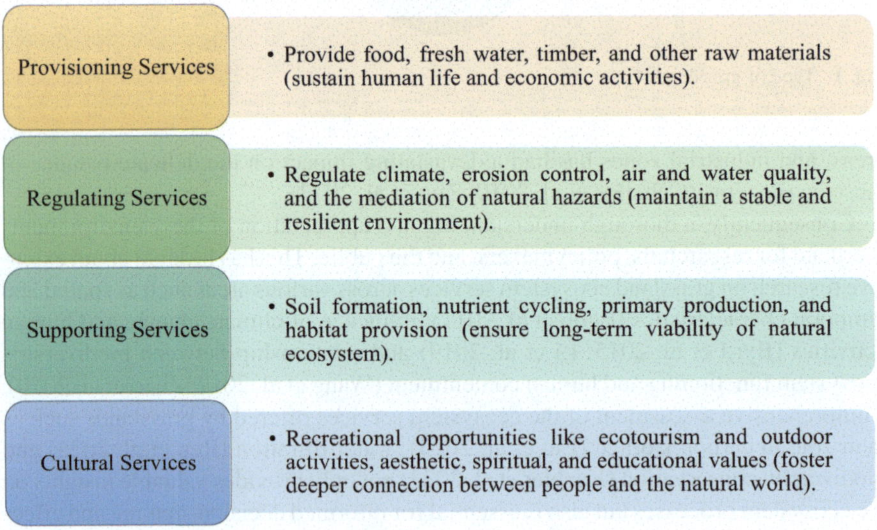

Fig. 2 Categories of grassland ecosystem services

2.1 *Provisioning Services*

Provisioning services encompass the essential resources and materials that ecosystem offer, which are crucial for supporting human existence and prosperity. Grasslands, as integral components of numerous biomes, furnish a diverse range of critical provisioning services that are crucial for maintaining our quality of life.

i. **Provide Food**

The grassland ecosystems supply us with an abundance of food resources, including a variety of edible plants, grains and grazing animals (Brown and MacLeod 2017; Pogue et al. 2018; Bist et al. 2023). Grasslands, with rich soils and abundant plant life, facilitate the growth of staple crops such as wheat, corn, and rice, which form the basis of the global food supply. Additionally, grasslands are home to numerous livestock species, including cattle, sheep, and goats, that provide us with meat, dairy products and other animal-based sustenance.

ii. **Provide Fresh Water**

Alongside food, grasslands also offer a crucial source of fresh water. The deep-rooted grasses and diverse plant life in these ecosystems help to regulate the water cycle, replenishing underground aquifers and maintaining the flow of surface water sources (Huang et al. 2019). This clean freshwater supply is essential for human consumption, agricultural irrigation and various industrial processes.

iii. **Provide Fuel and Timber**

Grasslands are rich source of natural fibre such as wool and cashmere, harvested from grazing animals. These fibres are used to produce textiles, clothing and other essential materials that support human livelihoods and industries (Naik et al. 2023). Additionally, grasslands harbour a wealth of plant species that can be utilized for bioenergy production thus providing a sustainable alternative to fossil fuels (French 2018; Gururani et al. 2022).

Beyond their tangible provisioning services, grasslands also offer intangible benefits. These ecosystems serve as a genetic library, housing a diverse array of plant and animal species that possess unique genetic traits and adaptations. This biodiversity is a valuable resource for scientific research, biotechnology, and the development of new crops and medicines (Dzerefos et al. 2017; Robledo-Abad et al. 2017; Gaurav et al. 2023).

2.2 *Regulating Services*

Grasslands offer a wide range of regulating services crucial for environmental well-being. These services include climate regulation through carbon sequestration, erosion control to safeguard soil integrity, and water regulation to maintain healthy hydrological cycles. Grasslands also play a vital role in air quality improvement, soil formation, pest control, waste treatment and pollination all of which are essential for the overall health and resilience of these natural landscapes.

i. **Climate Regulation and Carbon Sequestration**

Grassland management regimes can significantly impact atmospheric composition by regulating the content of greenhouse gases such as CO_2, CH_4 and N_2O, thereby influencing air temperature (Frolking and Roulet 2007; Offermanns et al. 2023). Grazing ruminants are a crucial source of CH_4 emissions in temperate grasslands (Pogue et al. 2018). Additionally, changes in the surface properties of grasslands can directly alter surface albedo affecting the surface radiation budget and energy balance leading to modifications in weather factors (Cao et al. 2015). Recognizing the complex interactions between grassland management, greenhouse gas emissions and surface energy dynamics is essential for developing comprehensive strategies to mitigate the environmental impacts and maintain the ecological integrity of these vital ecosystems. Carbon sequestration capability is one of the most important regulating ecosystem services that help in mitigating climate change (Chang et al. 2021; Habib et al. 2024). The photosynthetic processes of grasses and other grassland plants capture atmospheric carbon dioxide which is then stored in the biomass and further sequestered in the soil as organic matter. The extensive root systems of grassland vegetation help anchor the soil preventing erosion and enhancing the carbon storage capacity of these ecosystems (Zhang et al. 2024). Grasslands are considered among the most efficient natural carbon sinks, making them a crucial component in the global carbon cycle and a valuable asset in addressing the challenges of climate change.

ii. **Erosion Control**

Grasslands play a role in regulating the environment by providing both wind and water erosion control (Jiang et al. 2025). The deep, fibrous root systems of grassland plants act as anchors, firmly holding the soil in place and preventing it from being swept away by powerful winds. Additionally, the intricate network of roots helps to absorb and retain water, reducing the risk of soil erosion and runoff (Wang et al. 2011). This is particularly important during periods of heavy rainfall or prolonged drought, as the grasslands can help to regulate the flow of water, preventing flooding and mitigating the effects of drought (Liu et al. 2020). Moreover, the dense vegetation of grasslands acts as a natural barrier, diminishing the force of the wind and consequently its capacity to erode the soil. The tightly interwoven network of grasses and other plants obstructs airflow, dissipating the wind's energy and preventing the removal of fertile topsoil. This process helps maintain the land's fertility, allowing grasslands to sustain diverse range of plant and animal life.

iii. **Water and Air Regulation**

Grasslands are instrumental in maintaining both water and air quality. Grassland ecosystem regulates water balance, as their vegetation and water storage capacity affect surface runoff, floods, and aquifer recharge. Compared to farmlands and forests, grasslands have a 20% higher ability to reduce surface runoff, allowing them to store more water (Saha and Kukal 2015). This enhanced water storage capacity helps to mitigate the impact of floods and ensures a more stable water supply for both human and ecological needs (Macleod and Ferrier 2011). In addition to their role in water regulation, grasslands also play a significant part in maintaining water quality. By absorbing and filtering out pollutants and nutrients, grasslands prevent them from reaching waterways and causing harmful algae blooms (Peukert et al. 2014). The deep and extensive root systems of grassland plants act as natural filters, trapping sediments, heavy metals, and other contaminants before they can enter nearby streams, rivers, and lakes. Furthermore, grasslands contribute to the regulation of air quality. The dense vegetation and soil microorganisms in grassland ecosystems have the ability to absorb gaseous pollutants like NO_2, SO_2, and O_3 (Gopalakrishnan et al. 2018). Additionally, grasslands can filter out airborne pollutants, such as particulate matter and volatile organic compounds, thereby improving the overall air quality in surrounding areas.

2.3 Supporting Services

The proper functioning and long-term viability of grassland ecosystems are significantly enhanced by supporting services, which are essential for the conservation of diverse and productive natural environments. Grasslands offer critical supporting services such as soil formation, nutrient cycling, and habitat provisioning.

i. **Soil Formation**

Soil formation is a result of complex interactions among physical, chemical and biological processes and is essential for sustaining the terrestrial ecosystems like grasslands (Weng and Luo 2008). It arises from the combined influence of climate, relief, organisms, parent material and time. Physical weathering of rocks and minerals breaks down larger particles into smaller ones, while chemical weathering alters the molecular structure of these materials. Meanwhile, microorganisms like bacteria and fungi as well as the decomposition of organic matter, further leads to the formation of organic layer over the mineral layer thereby contributing to the soil horizonation (Rizwanuddin et al. 2023). This soil formation process takes place over long timescales, creating the rich, fertile substrate that grassland plants require to thrive.

ii. **Nutrient Cycling**

The cycling of essential nutrients is another key supporting service provided by grassland ecosystems (Radujković et al. 2021; Li et al. 2023). Nutrients like nitrogen, phosphorus and potassium are constantly being exchanged between the physical environment, plants and animals. Grass species take up inorganic forms of these nutrients from the soil incorporating them into their tissues as they grow. When the plants die and decompose, the nutrients are released back into the soil where they can then be absorbed by other living organisms (Kaur et al. 2023). This circular flow of nutrients is what sustains the productivity and biodiversity of the grassland over the long term (Rambaut et al. 2022; Mitra et al. 2022).

iii. **Provisioning of Habitat**

Grassland ecosystems provide diverse habitat resources that support numerous animal and plant species. The varying structures of grasses, forbs and shrubs offer shelter, nesting sites and hiding places, while the abundance of vegetation supplies ample food sources such as nectar, pollen, seeds and foliage. Grasslands are particularly vital for specialist species that have adapted to thrive exclusively in this environment. The mosaic of microhabitats within grasslands enables them to sustain high levels of biodiversity (Villemey et al. 2015; Milić et al. 2024).

2.4 *Cultural Services*

Cultural ecosystem services refer to the non-material benefits that people can derive from their interactions with natural environments (Raymond et al. 2013). These services encompass a profound attachment between communities and their specific environments, which can modify people's physical and mental states leading to psychological and socio-cultural benefits (Ko and Son 2018). These benefits extend beyond the direct use of natural resources and include aesthetic, recreational, spiritual, educational, and creative experiences.

i. **Aesthetic and Recreational Values**

The beauty, tranquility and inspiration that people derive from engaging with natural landscapes, such as enjoying scenic views, participating in outdoor activities or finding solace in natural settings.

ii. **Spiritual and Religious Values**

The sense of connection, meaning and transcendence that people experience in natural environments often associated with cultural or religious traditions.

iii. **Educational and Cognitive Benefits**

The opportunities for learning, exploration and intellectual stimulation that natural environments provide, fostering a deeper understanding and appreciation of ecological processes and the natural world.

iv. **Social and Cultural Identity**

The way in which natural environments contribute to the formation and expression of individual and community identities, shared traditions, and collective memories.

Therefore, recognizing the multifaceted benefits of grassland ecosystem services is essential for developing comprehensive land management strategies that ensure the long-term sustainability and preservation of these valuable natural resources.

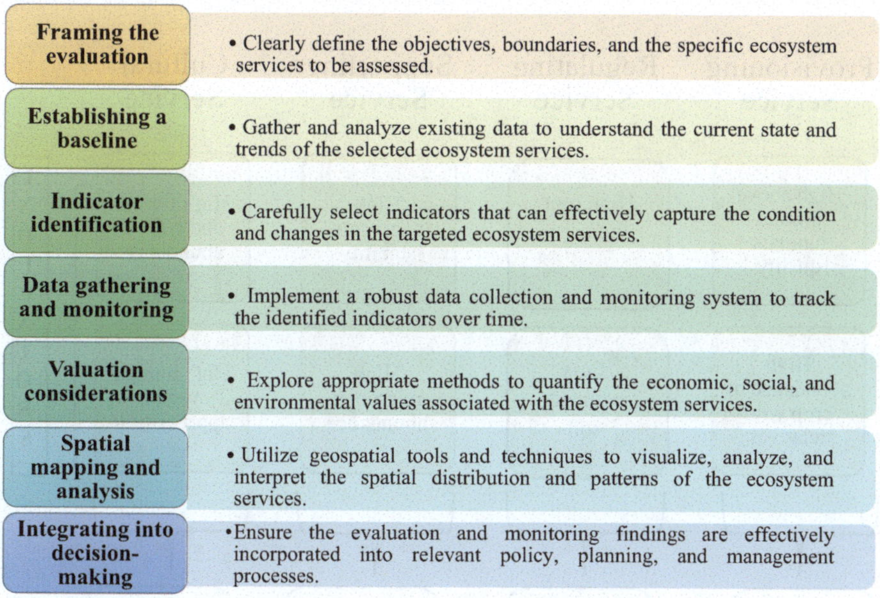

Fig. 3 Steps for evaluating and monitoring grassland ecosystem services

3 Methods for Evaluating and Monitoring Ecosystem Services

Grassland ecosystem services play a vital role in supporting human well-being, making it essential to safeguard their availability amid challenges like climate change, land use shifts and biodiversity loss (Lawler et al. 2014). To facilitate this, evaluating and monitoring these services becomes critical. Although researchers and policy makers have suggested various evaluation criteria, the multi- criteria decision analysis framework is widely adopted method aimed at promoting the long-term viability and resilience of grassland ecosystems. An overview of the fundamental steps for evaluating and monitoring grassland ecosystem services is presented in Fig. 3.

4 Tools for evaluating and monitoring ecosystem services

i. Field Surveys

Field surveys play a crucial role in evaluating the ecosystem services of grasslands. These on-the-ground assessments provide direct firsthand data that can be used to comprehensively analyse the health, functionality and overall benefits offered by these dynamic natural environments. By conducting thorough field surveys, researchers and environmental managers can gather detailed information about the

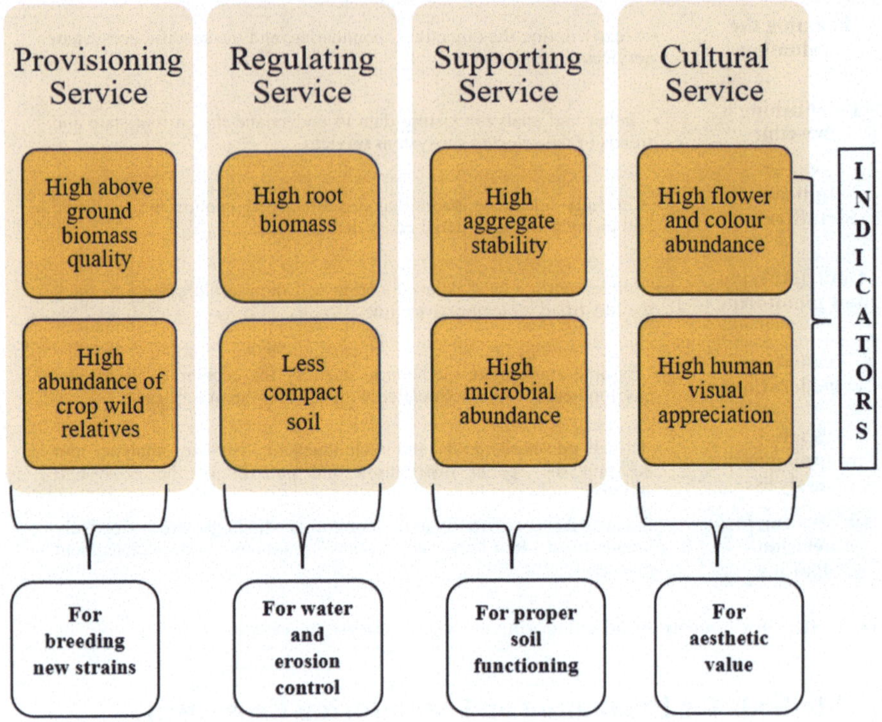

Fig. 4 Ecosystem services illustrating selected indicators and their purpose

composition, structure and productivity of grassland ecosystems (Li et al. 2022). It typically involves the direct measurement of biophysical indicators. These indicators can be assessed using both field and laboratory methods and a single indicator may represent multiple ecosystem services. Few indicators along with the ecosystem services they provide, are illustrated in Fig. 4.

Moreover, these are often interconnected or correlated with each other (Tasser et al. 2020; Haines-Young and Potschin 2018). These indicators can then be used to quantify the specific ecosystem services provided such as carbon sequestration, soil retention, forage production and wildlife habitat. Analysing these on-site measurements allows for a deeper understanding of how grasslands function, the factors that influence their condition and the ways in which they contribute to the overall health and resilience of the broader landscape. Furthermore, field surveys enable the identification of potential threats or disruptions to grassland ecosystems such as invasive species, overgrazing or the impacts of climate change. This information is crucial for developing effective management strategies and conservation plans to protect and enhance the valuable services that grasslands provide. By combining this empirical data with other research methods such as remote sensing and modelling, a comprehensive assessment of grassland ecosystems can be achieved, informing decision-making and guiding sustainable land use practices.

Case Studies

A study was conducted in Germany to assess the spatial distribution of values linked to grasslands and their ecosystem services through field surveys. This fieldwork facilitated a hotspot analysis which identified areas with high (hot spots) and low (cold spots) concentrations of various value types associated with grasslands. The findings indicated that the appreciation for grasslands and their ecosystem services varies across different surveyed locations (Schmitt et al. 2022). In a separate study, a methodological framework was created that focuses on five grass functional types (GFTs) to characterize the ecosystem services derived from grassland diversity and their underlying drivers. This framework involves surveying plant functional diversity at both field and farm levels analysing how it responds to various environmental and management factors and assessing its impact on ecosystem services. The results revealed that FastGFT is positively associated with herbage production but negatively correlated with soil carbon content and species richness at the field level. The diversity of FastGFT within grasslands enables the evaluation of how animal feed requirements align with available resources, as well as the landscape mosaic formed which helps elucidate the relationships between environmental and management drivers and ecosystem services (Duru et al. 2015).

ii. Remote Sensing

Remote sensing is a valuable tool for assessing and monitoring ecosystem services in grasslands, as it delivers large-scale, consistent, and repeatable data that can be challenging to gather through field surveys alone. Traditional ground-based field surveys while valuable, are often time-consuming, labour-intensive and limited in their spatial coverage (Zlinszky et al. 2015). In contrast, remote sensing technology with its ability to capture large-scale data from a bird's-eye view offers a comprehensive and efficient solution for grassland assessment. Satellite and drone-based platforms can provide high-resolution spatially continuous data on the condition, extent and dynamics of grassland ecosystems allowing researchers and policymakers to gain a more holistic understanding of these vital natural resources (Weiss et al. 2020). By leveraging the power of remote sensing, scientists can monitor the health and productivity of grasslands track changes in vegetation cover, identify areas of degradation or encroachment and assess the impacts of various land management practices on ecosystem services (Zhang and Guo 2008). This information is crucial for informing conservation strategies, guiding sustainable grazing and agricultural practices and ensuring the long-term resilience of these vital landscapes. Remote sensing data is acquired via sensors deployed on various platforms including satellites, aircraft and drones (Sozzi et al. 2021; Wu et al. 2023).

Sensors and Platforms

Remote sensing sensors are divided into two main categories: passive and active. Passive sensors detect natural radiation from the Earth's surface or atmosphere, while active sensors emit their own radiation and measure the returning signal (Aqeel et al.

2011). Examples include multispectral and hyperspectral sensors for passive detection and radar and lidar for active sensing. The choice of platform, often satellites, aircraft, or drones, hinges on the required spatial, temporal and spectral resolution, as well as cost and accessibility (Pádua et al. 2017).

Data Gathering, Processing and Feature Analysis

Remote-sensing data is collected as digital images, which can be processed and analysed to extract essential information about the Earth's surface and atmosphere (Richards and Richards 2022). Processing techniques such as radiometric and geometric corrections, image enhancement and classification can be utilized (Sowmya et al. 2017). Image interpretation involves identifying features and patterns through visual analysis or automated algorithms. When combined with Geographic Information System (GIS), remote-sensing data offers insights into ecosystem services and biodiversity conservation. For instance, land cover classification can identify ecosystems and assess their extent while time-series analysis can track ecological changes aiding conservation planning and management (Alleaume et al. 2018).

With technological advancements, the collected data through remote sensing can be integrated with various spatial and non-spatial data—such as climate, topography, and species occurrence records—within GIS to develop species distribution models. These models can predict species distributions under varying environmental conditions. They can aid in conservation efforts by identifying priority areas for habitat protection and restoration as well as assessing the impacts of climate change on biodiversity (De Simone et al. 2021).

Case Studies

Zhang et al. (2023) employed high spatial–temporal resolution multispectral satellite data to evaluate the effectiveness of new-generation sensors including the Sentinel-2 multispectral instrument (MSI) and Tiangong-2 moderate resolution wide-wavelength imager (MWI) for accurately estimating and mapping forage nitrogen (N), phosphorus (P) and potassium (K) contents in alpine grasslands at a regional scale. To predict N, P and K levels, these sensors were combined with various feature selection algorithms and machine learning models. The findings demonstrated that the spectral bands of both Sentinel-2 MSI and Tiangong-2 MWI performed exceptionally well in estimating forage N, P and K contents providing valuable insights for growth monitoring and real-time assessment of forage quality in alpine grasslands. The Sentinel-2 MSI was also employed for the quantitative assessment of grass water content, providing vital pathways for developing a user-friendly robust methodology for the regular monitoring of key rangeland health indicators (Sibanda et al. 2020). This strategy enhanced sustainable conservation and management of the resources while ensured the ongoing provision of ecosystem services. Utilizing remote sensing as an effective tool, González-Dugo et al. (2024) analysed the herbaceous layer's contribution to ecosystem services in olive groves and Mediterranean oak savanna. The study revealed that grasslands in both ecosystems significantly provided provisioning services (such as forage, freshwater, and genetic resources), regulating services (including carbon sequestration, soil conservation and climate and air quality regulation), and cultural services (like aesthetic appreciation and cultural identity). Biomass and above-ground net primary production (ANPP), along with biodiversity were identified as crucial components for integrating ecosystem functioning.

iii. **Environmental Modelling**

Environmental modelling has become an essential tool for understanding and managing the complex interactions between human activities and natural ecosystems. Two widely-used models in this field are Integrated Valuation of Ecosystem Services and Tradeoffs (InVEST) and Artificial Intelligence for Ecosystem Services (ARIES).

InVEST is a suite of models developed by the Natural Capital Project, which is a collaborative effort involving Stanford University, The Nature Conservancy, the World Wildlife Fund and the University of Minnesota. This modelling framework allows researchers and decision-makers to quantify the benefits that people receive from nature such as clean water, flood protection and carbon sequestration. InVEST can be used to assess the current state of ecosystem services as well as to predict how changes in land use, climate or management practices might affect these services in the future.

ARIES, on the other hand, is a web-based platform that uses artificial intelligence and geospatial modelling to map and value ecosystem services. Developed by the University of Vermont, ARIES combines advanced computational techniques with a user-friendly interface making it accessible to a wide range of stakeholders from scientists and policymakers to local communities. The platform can be used to identify the locations and beneficiaries of ecosystem services, as well as to analyse the trade-offs and synergies between different services.

Both InVEST and ARIES have been widely adopted by researchers, conservation organizations and government agencies around the world to support decision making and guide the sustainable management of natural resources. As the importance of ecosystem services becomes increasingly recognized, these and other environmental modelling tools will continue to play a crucial role in shaping our understanding and stewardship of the natural world.

Case Studies

To analyse the spatio-temporal dynamics and factors influencing water conservation in China's Three-River Headstreams Region (TRHR), the InVEST model was employed. During the period from 1980 to 2018, an average water conservation (WC) of 50.36 mm was calculated, amounting to a total of 189.33 billion m^3 (Xie et al. 2024). Moreover, in the above study, precipitation emerged as the primary key driver of WC growth among various climatic variables, while grasslands emerged as the most significant land use type contributing ~84.90–87.78% to the total. In another study, Benra et al. (2021) assessed the performance of the InVEST model against water monitoring records from 224 catchments in southern Chile, covering the years 1998, 2007 and 2013 to capture climatic variations. The results revealed low monthly correlations with significant latitudinal and temporal differences, while annual estimates aligned well with observed data particularly for totals under 1000 mm/year. The model performed better in high rainfall areas and during dry years whereas predictions were less accurate in snow-dominated and arid regions. These findings position the InVEST model as an effective decision support tool for water supply regulation, particularly in data-scarce regions.

The need of integrated, scalable methods becomes crucial for effectively monitoring grassland ecosystem services due to their ability to address the complexity and variability of these ecosystems. Grasslands deliver multiple interconnected services such as carbon storage, water filtration and habitat for biodiversity requiring an integrated approach to assess their overall health. Moreover, scalable techniques like remote sensing and standardized field protocols allow for consistent data collection across diverse regions, accommodating variations in climate, vegetation and soil types. These methods also facilitate adaptability to environmental changes, enabling quick detection and responsive management to maintain ecosystem functions. By optimizing resource use through the combination of remote technologies and ground sampling, they reduce the costs associated with extensive monitoring while ensuring accurate data. Ultimately, these integrated methods provide essential insights for policymakers and land managers supporting data-driven approaches to the sustainable management and conservation of grasslands in the face of future challenges.

5 Global Assessment of Grassland Degradation Using Ecosystem Service Evaluation and Monitoring Tools

Field-based surveys, remote sensing and modelling techniques can also effectively assess the extent and drivers of grassland degradation globally. By integrating these methods, researchers can obtain a comprehensive understanding of degradation patterns, aiding in the development of effective conservation and restoration strategies. However, validating these methodologies robustly and comparably is essential for accurate assessments (Higginbottom and Symeonakis 2014). Field-based assessments can involve vegetation surveys that measure various parameters such as species composition, cover, biomass, and height, to assess the health and productivity of grasslands (Yang et al. 2019). Additionally, analysing soil properties—such as organic matter content, nutrient levels, salinity, and erosion rates—can yield valuable insights into soil condition and potential degradation risks (Kosmas et al. 2014). While field-based assessments, provide valuable ground-truthing and insights into localized degradation processes, their inherent limitations in spatial and temporal coverage restrict their applicability for broad-scale monitoring. For effective large-scale monitoring and early degradation detection, remote sensing and modelling offer significant advantages. Remote sensing provides extensive spatial and temporal data, allowing for the tracking of vegetation dynamics and land cover changes over vast areas. Modelling supported by field data and remote observations, integrates multiple data sources and simulates complex interactions within grassland ecosystems. This combined approach fosters a comprehensive understanding of degradation patterns informing targeted conservation and restoration strategies. Under remote sensing techniques, indices like the Normalized Difference Vegetation Index (NDVI) and Enhanced Vegetation Index (EVI) are widely used to monitor vegetation health and changes over large areas. Multi-temporal analysis of NDVI data can reveal trends in grassland productivity and degradation. The modelling approaches can be used to simulate the impacts of different drivers on grassland ecosystems helping to predict future degradation trends and evaluate the effectiveness of management interventions.

Case Studies

A study by Yan et al. (2023), leveraging Global Land Surface Satellite (GLASS)-NPP long-term data series and the Sen + Mann Kendall (Sen + MK) non-parametric statistical test method (Gilbert 1987), analysed global grassland degradation trends from 2000 to 2020. Their evaluation of climate change's impact on grassland productivity, using moisture index (MI) response relationships, revealed that 1.93 million square kilometres of the world's grasslands have been undergoing degradation since 2000. Among the continents, Africa and South America have shown the highest levels of grassland degradation globally followed by Asia, North America, Europe and Oceania (Fig. 5). In another study, Lyu et al. (2020) utilized Earth Observing One (EO-1)—Hyperion images along with multiple endmember spectral mixture analysis (MESMA) and the fully constrained least squares (FCLS) method to identify common vegetation species and to evaluate the extent of grassland degradation at a regional level, according to established monitoring standards for grassland degradation (Fig. 6).

In Qilian County, located in the northeast of the Qinghai–Tibet Plateau, Hou et al. (2022) analysed the spatial distribution of grassland degradation using the NDVI and its coefficient of variation (CVNDVI). To identify the main climate variables influencing NDVI fluctuations they combined random forest modelling with stepwise regression, enhancing the accuracy of residual trend analysis. Building on this, they developed twelve potential scenarios to examine the driving factors of various levels of degraded grasslands. The study ultimately employed the Hurst index to forecast trends in grassland degradation. The findings indicated that both climate factors and human activities significantly influenced grassland degradation. Areas experiencing slight degradation or recovery (characterized by NDVI slope >0 and CVNDVI slope >0) made up 34.3% of the total grassland area. In contrast, regions of moderate degradation (NDVI slope <0 and CVNDVI slope >0) constituted 3.3%, whereas severely degraded areas (NDVI slope <0 and CVNDVI slope <0) accounted for 1.3%. The forecast suggested that the majority of grasslands in Qilian County are likely to continue degrading, including those previously affected, with areas of ongoing degradation projected to represent 54.78% of the total grassland (Fig. 7).

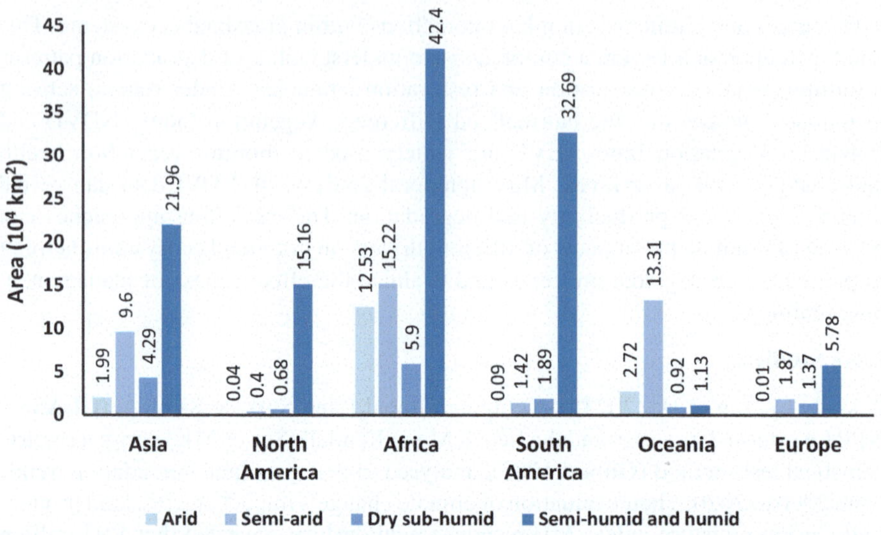

Fig. 5 Global grassland degradation extent in different climatic zones across continents (Yan et al. 2023)

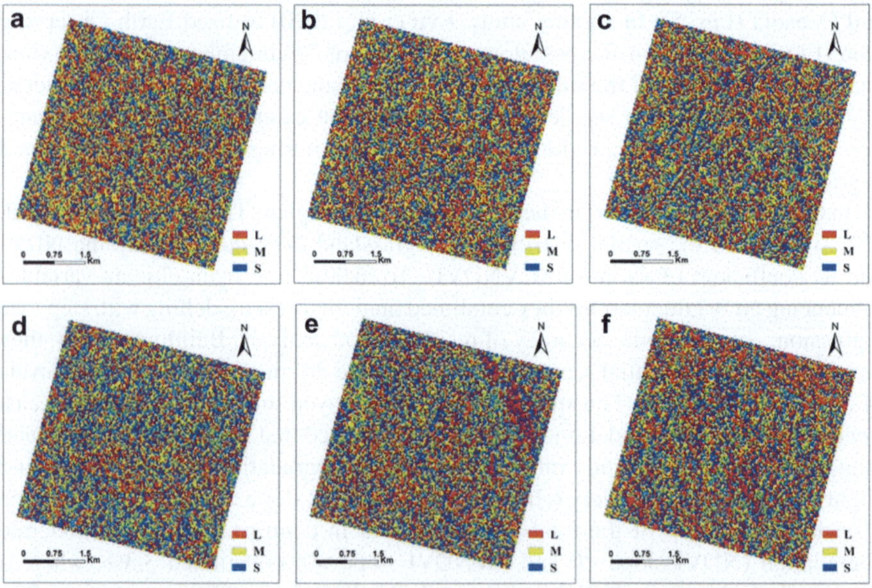

Fig. 6 Distribution of grassland degradation at Inner Mongolia, China. **a**–**f** The different sampling regions. L—Lightly degraded grassland; M—moderately degraded grassland; S—severe degraded grassland (Lyu et al. 2020)

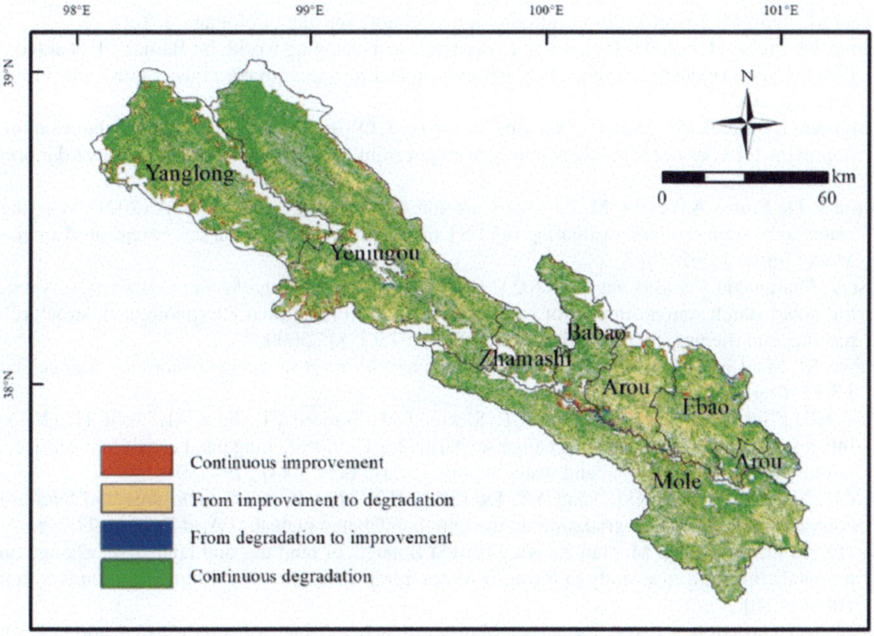

Fig. 7 Spatial distribution of forecasted grassland in Qilian County (Hou et al. 2022)

6 Conclusions

Grasslands are vital for producing ecosystem services, including supporting, provisioning, regulating, and cultural services. To sustain the ecological and economic value of these landscapes, effective monitoring and evaluation of these ecosystem services are essential. Various approaches such as field surveys, remote sensing and modelling approaches are crucial for capturing the complexity and dynamics of grassland ecosystems, facilitating regional and global comparisons and informing sustainable policy and management strategies. Looking ahead, it is imperative to adopt integrated, data-driven frameworks and scalable methods that can adapt to ongoing environmental changes ensuring that grasslands remain resilient and continue to provide essential ecosystem services.

References

Alleaume S, Dusseux P, Thierion V, Commagnac L, Laventure S, Lang M, Luque S (2018) A generic remote sensing approach to derive operational essential biodiversity variables (EBVs) for conservation planning. Methods Ecol Evol 9(8):1822–1836

Allen VG, Batello C, Berretta EJ, Hodgson J, Kothmann M et al (2011) An international terminology for grazing lands and grazing animals. Grass Forage Sci 66(1):2–28

Aqeel M, Jamil M, Yusoff I (2011) Introduction to remote sensing of biomass. InTech

Barnes RF, Nelson CJ (2003) Forages and grasslands in a changing world. In: Barnes RF et al (eds) Forages: an introduction to grassland agriculture and forages, 6th edn. Iowa State Univ. Press, Ames, pp 3–23

Bengtsson J, Bullock JM, Egoh B, Everson C, Everson T, O'Connor T et al (2019) Grasslands–more important for ecosystem services than you might think. Ecosphere 10:e02582. https://doi.org/10.1002/ecs2.2582

Benra F, De Frutos A, Gaglio M, Álvarez-Garretón C, Felipe-Lucia M, Bonn A (2021) Mapping water ecosystem services: evaluating InVEST model predictions in data scarce regions. Environ Model Softw 138:104982

Bist Y, Sharanagat VS, Saxena DC (2023) Guinea grass seed (*Megathyrsus maximus*), a potential novel starch source: impact of isolation methods on functional, morphological, structural, pasting, and rheological properties. Starch-Stärke 75(7–8):2300003

Brown JR, MacLeod ND (2017) An ecosystem services filter for rangeland restoration. Rangeland J 39(5–6):451–459

Byrd KB, Flint LE, Alvarez P, Casey CF, Sleeter BM, Soulard CE, Flint AL, Sohl TL (2015) Integrated climate and land use change scenarios for California rangeland ecosystem services: wildlife habitat, soil carbon, and water supply. Landsc Ecol 30(4):729–750

Cao JJ, Yeh ET, Holden NM, Yang YY, Du GZ (2013) The effects of enclosures and land-use contracts on rangeland degradation on the Qinghai-Tibetan plateau. J Arid Environ 97:3–8

Cao Q, Yu D, Georgescu M, Han Z, Wu J (2015) Impacts of land use and land cover change on regional climate: a case study in the agro-pastoral transitional zone of China. Environ Res Lett 10(12):124025

Ceballos G, Davidson A, List R, Pacheco J, Manzano-Fischer P, Santos-Barrera G, Cruzado J (2010) Rapid decline of a grassland system and its ecological and conservation implications. PLoS ONE 5(1):e8562. https://doi.org/10.1371/journal.pone.0008562

Chang J, Ciais P, Gasser T, Smith P, Herrero M, Havlík P, Obersteiner M, Guenet B, Goll DS, Li W, Naipal V, Shushi Peng S et al (2021) Climate warming from managed grasslands cancels the cooling effect of carbon sinks in sparsely grazed and natural grasslands. Nature Comm 12:118. https://doi.org/10.1038/s41467-020-20406

Cheng C, Li F (2024) Ecosystem restoration and management based on nature-based solutions in China: research progress and representative practices. Nat Based Sol 6:100176. https://doi.org/10.1016/j.nbsj.2024.100176

De Simone W, Allegrezza M, Frattaroli AR, Montecchiari S, Tesei G, Zuccarello V, Di Musciano M (2021) From remote sensing to species distribution modelling: an integrated workflow to monitor spreading species in key grassland habitats. Remote Sens 13:1904

Dixon AP, Faber-Langendoen D, Josse C, Morrison J, Loucks CJ (2014) Distribution mapping of world grassland types. J Biogeo 41:2003–2019

Duru M, Theau JP, Martin G (2015) A methodological framework to facilitate analysis of ecosystem services provided by grassland-based livestock systems. Int J Biodiversity Sci Ecosys Services Manag 32:695. https://doi.org/10.1080/21513732.2015.1030695

Dzerefos CM, Witkowski ETF, Kremer-Köhne S (2017) Aiming for the biodiversity target with the social welfare arrow: medicinal and other useful plants from a critically endangered grassland ecosystem in Limpopo Province, South Africa. Int J Sustain Dev World Ecol 24:52–64. https://doi.org/10.1080/13504509.2016.1174963

Egoh B, Reyers B, Rouget M, Richardson DM, Le Maitre DC, Van Jaarsveld AS (2008) Mapping ecosystem services for planning and management. Agric Ecosyst Environ 127(1–2):135–140. https://doi.org/10.1016/j.agee.2008.03.013

Egoh BN, Reyers B, Rouget M, Richardson DM (2011) Identifying priority areas for ecosystem service management in South African grasslands. J Environ Manag 92(6):1642–1650

Faber-Langendoen D, Keeler-Wolf T, Meidinger D, Josse C, Weakley A, Tart D, Navarro G, Hoagland B, Ponomarenko S, Fults G et al. (2016) Classification and description of world formation types. In: General technical report RMRS-GTR-2016; US Department of Agriculture, Forest Service, Rocky Mountain Research Station: Fort Collins, CO, p 222

FAOSTAT (2021) Data on land cover (grassland). https://www.fao.org/faostat/en/#data/LC. Accessed 10 May 2024

Forage and Grazing Terminology Committee (1991) Terminology for grazing lands and grazing animals. Pocahontas Press, Blacksburg, VA

French KE (2018) Assessing the bioenergy potential of grassland biomass from conservation areas in England. Land Use Pol 82:700–708. https://doi.org/10.1016/j.landusepol.2018.12.001

Frolking S, Roulet N (2007) Holocene radiative forcing impact of northern peatland carbon accumulation and methane emissions. Global Change Biol 13:1079–1088. https://doi.org/10.1111/j.1365-2486.2007.01339.x

Gaurav H, Yadav D, Maurya A, Yadav H, Yadav R, Shukla AC, Palazon J et al (2023) Biodiversity, biochemical profiling, and pharmaco-commercial applications of *Withania somnifera*: a review. Molecules 28(3):1208

Gilbert RO (1987) Statistical methods for environmental pollution monitoring. Wiley, New York

González-Dugo MP, Muñoz-Gómez MJ, Gálvez C, Blázquez-Carrasco A, Dolores Carbonero M, Tortosa FS et al (2024) Remote-sensing based tools to monitor grassland ecosystem services

Gopalakrishnan V, Hirabayashi S, Ziv G, Bakshi BR (2018) Air quality and human health impacts of grasslands and shrublands in the United States. Atmos Environ 182:193–199. https://doi.org/10.1016/j.atmosenv.2018.03.039

Gururani P, Bhatnagar P, Bisht B, Jaiswal KK, Kumar V, Kumar S, Vlaskin MS, Grigorenko AV, Rindin KG (2022) Recent advances and viability in sustainable thermochemical conversion of sludge to bio-fuel production. Fuel 12:351. https://doi.org/10.1016/j.fuel.2022.123351

Habib M, Singh S, Bist Y, Kumar Y, Jan K, Bashir K, Saxena DC et al (2024) Carbon pricing and the food system: implications for sustainability and equity. Trends Food Sci Technol 21:104577

Haines-Young R, Potschin M (2018) CICES V5. 1. Guidance on the application of the revised structure. Fabis Consult 53

Higginbottom TP, Symeonakis E (2014) Assessing land degradation and desertification using vegetation index data: current frameworks and future directions. Remote Sens 6(10):9552–9575. https://doi.org/10.3390/rs6109552

Hou Q, Ji Z, Yang H, Yu X (2022) Impacts of climate change and human activities on different degraded grassland based on NDVI. Sci Rep 12:15918. https://doi.org/10.1038/s41598-022-19943-6

Huang Z, Liu YF, Cui Z, Liu Y, Wang D, Tian FP, Wu GL (2019) Natural grasslands maintain soil water sustainability better than planted grasslands in arid areas. Agric Ecosys Environ 286:106683. https://doi.org/10.1016/j.agee.2019.106683

Jiang H, Guo C, Li X et al (2025) Assessing the contribution of wind and water erosion in the agro-pastoral ecotone of Northern China with 137Cs tracer technology. Sci Rep 15:14699. https://doi.org/10.1038/s41598-025-99457-z

Kachler J, Benra F, Bolliger R, Isaac R, Bonn A, Felipe-Lucia MR (2023) Can we have it all? The role of grassland conservation in supporting forage production and plant diversity. Landscape Ecol 38:4451–4465. https://doi.org/10.1007/s10980-023-01729-4

Kaur H, Hussain SJ, Mir RA, Verma VC, Naik B, Kumar P, Dubey RC (2023) Nanofertilizers–emerging smart fertilizers for modern and sustainable agriculture. Biocat Agricul Biotechnol 54(12):102921. https://doi.org/10.1016/j.bcab.2023.102921

Ko H, Son Y (2018) Perceptions of cultural ecosystem services in urban green spaces: a case study in Gwacheon, Republic of Korea. Ecol Ind 91:299–306. https://doi.org/10.1016/j.ecolind.2018.04.006

Kosmas C, Kairis O, Karavitis C, Ritsema C, Salvati L, Acikalin S, Alcala M, Alfama P et al (2014) Evaluation and selection of indicators for land degradation and desertification monitoring: types of degradation, causes, and implications for management. Environ Manag 54:971–982. https://doi.org/10.1007/s00267-013-0110-0

Lawler JJ, Lewis DJ, Nelson E et al (2014) Projected land-use change impacts on ecosystem services in the United States. Proc Natl Acad Sci USA 111:7492–7497. https://doi.org/10.1073/pnas.1405557111

Li X, Wang Z, Wang D, Wang L, Pan D, Li J, De K, Seastedt TR (2019) Livestock grazing impacts on plateau pika (*Ochotona curzoniae*) vary by species identity. Agric Ecosyst Environ 275:23–31

Li M, Wang X, Chen J (2022) Assessment of grassland ecosystem services and analysis on its driving factors: a case study in hulunbuir grassland. Front Ecol Evol 10:841943. https://doi.org/10.3389/fevo.2022.841943

Li N, Zhang Q, Xu S, Businelli D, Wang Q (2023) Carbon and nitrogen cycling in grassland ecosystems. Front Environ Sci 11:1250061. https://doi.org/10.3389/fenvs.2023.1250061

Liu YF, Liu Y, Shi ZH, López-Vicente M, Wu GL (2020) Effectiveness of re-vegetated forest and grassland on soil erosion control in the semi-arid Loess Plateau. CATENA 195:104787. https://doi.org/10.1016/j.catena.2020.104787

Liu L, Sayer EJ, Denga M, Li P, Liu W, Wang X, Yang S, Huang J, Luo J, Su Y, Grünzweig JM, Jiang L, Hu S (2023) The grassland carbon cycle: mechanisms, responses to global changes, and potential contribution to carbon neutrality. Fundamental Res 3(2):209–218

Lyu X, Li X, Dang D, Dou H, Xuan X, Liu S, Gong J (2020) A new method for grassland degradation monitoring by vegetation species composition using hyperspectral remote sensing. Ecol Indicat 114:106310. https://doi.org/10.1016/j.ecolind.2020.106310

Macleod CJA, Ferrier RC (2011) Temperate grasslands in catchment systems: the role of scale, connectivity and thresholds in the provision and regulation of water quality and quantity. In: Lemaire G, Hodgson J, Chabbi A (eds) Grassland productivity and ecosystem services. CABI International, Oxfordshire, pp 229–238

Milić D, Rat M, Bokić B, Mudri-Stojnić S, Milosević N, Sukur N et al (2024) Exploring the effects of habitat management on grassland biodiversity: a case study from northern Serbia. PLoS ONE 19(3):e0301391. https://doi.org/10.1371/journal.pone.0301391

Millennium Ecosystem Assessment (2005) Ecosystems and human well-being. Island Press

Mitra D, Mondal R, Khoshru B, Senapati A, Radha TK, Mahakur B, Uniyal N et al (2022) Actinobacteria-enhanced plant growth, nutrient acquisition, and crop protection: advances in soil, plant, and microbial multifactorial interactions. Soil Biol Biochem 157:108394. https://doi.org/10.1016/S1002-0160(21)60042-5

Naik B, Kumar V, Gupta AK (2023) Valorization of tender coconut mesocarp for the formulation of ready-to-eat dairy-based dessert (Kheer): utilization of industrial by-product. J Agric Food Res 12:100572

Offermanns L, Tiemeyer B, Dettmann U, Rüffer J, Düvel D, Vogel I, Brümmer C (2023) High greenhouse gas emissions after grassland renewal on bog peat soil. Agric Forest Meteorol 331:109309

Pádua L, Vanko J, Hruška J, Adão T, Sousa JJ, Peres E, Morais R (2017) UAS, sensors, and data processing in agroforestry: a review towards practical applications. Int J Remote Sens 38(8–10):2349–2391

Pandeya SC (1988) Status of Indian rangelands. Range Management Society of India, Jhansi, p 213

Peukert S, Griffith BA, Murray PJ, Macleod CJA, Brazier RE (2014) Intensive management in grasslands causes diffuse water pollution at the farm scale. J Environ Qual 43(6):2009. https://doi.org/10.2134/jeq2014.04.0193

Pogue SJ, Krobel R, Janzen HH, Beauchemin KA, Legesse G, de Souza DM, Iravani M, Selin C, Byrne J, McAllister TA (2018) Beef production and ecosystem services in Canada's prairie provinces: a review. Agric Syst 166:152–172

Radujković D, Verbruggen E, Seabloom EW, Bahn M, Biederman LA, Borer ET, Boughton EH, Catford JA, Campioli M, Donohue I et al (2021) Soil properties as key predictors of global grassland production: have we overlooked micronutrients? Ecol Lett 24:2713–2725

Rambaut LE, Tillard E, Vayssières J, Lecomte P, Salgado P (2022) Trade-off between short and long-term effects of mineral, organic or mixed mineral-organic fertilisation on grass yield of tropical permanent grassland. Eur J Agron 141:126635

Raymond CM, Singh GG, Benessaiah K, Bernhardt JR, Levine J, Nelson H, Turner NJ, Norton B, Tam J, Chan KMA (2013) Ecosystem services and beyond. Bioscience 63(7):536–546. https://doi.org/10.1525/bio.2013.63.7.7

Ren X, Chai X, Qu Y, Xu Y, Khan FU, Wang J, Geming P, Wang W, Zhang Q, Wu Q et al (2023) Restoration of grassland improves soil infiltration capacity in water-wind erosion crisscross region of China's Loess Plateau. Land 12(8):1485. https://doi.org/10.3390/land12081485

Richards JA, Richards JA (2022) Remote sensing digital image analysis, vol 5. Springer, New York

Richter F, Jan P, El Benni N, Lüscher A, Buchmann N, Klaus VH (2021) A guide to assess and value ecosystem services of grasslands. Ecosyst Serv 52:101376. https://doi.org/10.1016/j.ecoser.2021.101376

Rizwanuddin S, Kumar V, Singh P, Naik B, Mishra S, Chauhan M, Verma A, Kumar V (2023) Insight into phytase-producing microorganisms for phytate solubilization and soil sustainability. Front Microbiol 14:1127249. https://doi.org/10.3389/fmicb.2023.1127249

Robledo-Abad C, Althaus HJ, Berndes G, Bolwig S, Corbera E, Creutzig F, Garcia-Ulloa J et al (2017) Bioenergy production and sustainable development: science base for policymaking remains limited. GCB Bioen 9(3):541–556

Saha D, Kukal SS (2015) Soil structural stability and water retention characteristics under different land uses of degraded lower Himalayas of North-west India. Land Degrad Dev 26(3):263–271

Schirpke U, Tasser E, Borsky S, Braun M, Eitzinger J, Gaube V et al (2023) Past and future impacts of land-use changes on ecosystem services in Austria. J Environ Manag 345:118728. https://doi.org/10.1016/j.jenvman.2023.118728

Schmitt TM, Riebl R, Martín-López B, Hänsel M, Koellner T (2022) Plural valuation in space: mapping values of grasslands and their ecosystem services. Ecosyst People 18(1):258–274. https://doi.org/10.1080/26395916.2022.2065361

Sibanda M, Onisimo M, Dube T, Mabhaudhi T (2020) Quantitative assessment of grassland foliar moisture parameters as an inference on rangeland condition in the mesic rangelands of southern Africa. Int J Remote Sens 42(4):1474–1491

Sowmya DR, Deepa Shenoy P, Venugopal KR (2017) Remote sensing satellite image processing techniques for image classification: a comprehensive survey. Int J Comp Appl 161(11):24–37

Sozzi M, Kayad A, Gobbo S, Cogato A, Sartori L, Marinello F (2021) Economic comparison of satellite, plane and UAV-acquired NDVI images for site-specific nitrogen application: observations from Italy. Agron 11(11):2098

Tasser E, Schirpke U, Zoderer BM, Tappeiner U (2020) Towards an integrative assessment of land-use type values from the perspective of ecosystem services. Ecosyst Serv 42:101082. https://doi.org/10.1016/j.ecoser.2020.101082

Villemey A, Van Halder I, Ouin A, Barbaro L, Chenot J, Tessier P et al (2015) Mosaic of grasslands and woodlands is more effective than habitat connectivity to conserve butterflies in French farmland. Biol Conser 191:206–215. https://doi.org/10.1016/j.biocon.2015.06.030

Wang S, Fu BJ, He CS, Sun G, Gao GY (2011) A comparative analysis of forest cover and catchment water yield relationships in northern China. Forest Ecol Manag 262:1189–1198. https://doi.org/10.1016/j.foreco.2011.06.013

Wang B, Tang H, Xu Y (2017) Integrating ecosystem services and human well-being into management practices: insights from a mountain-basin area, China. Ecosyst Serv 27:58–69

Weiss M, Jacob F, Duveiller G (2020) Remote sensing for agricultural applications: a meta-review. Remote Sens Environ 236:111402. https://doi.org/10.1016/j.rse.2019.111402

Weng E, Luo Y (2008) Soil hydrological properties regulate grassland ecosystem responses to multifactor global change: a modelling analysis. J Geophy Res 113:G03003. https://doi.org/10.1029/2007JG000539

Wu J, Li Y, Zhong B, Liu Q, Wu S, Ji C et al (2023) Integrated vegetation cover of typical steppe in China based on mixed decomposing derived from high resolution remote sensing data. Sci Tot Environ 904:166738

Xie X, Peng M, Zhang L, Chen M, Li J, Tuo Y (2024) Assessing the impacts of climate and land use change on water conservation in the three-river headstreams region of China based on the integration of the InVEST model and machine learning. Land 13:352

Yan Z, Gao Z, Sun B, Ding X, Gao T, Li Y (2023) Global degradation trends of grassland and their driving factors since 2000. Int J Digital Earth 16(1):1661–1684. https://doi.org/10.1080/17538947.2023.2207840

Yang Y, Wang J, Chen Y, Cheng F, Liu G, He Z (2019) Remote-sensing monitoring of grassland degradation based on the GDI in Shangri-La. China. Remote Sens 11(24):3030. https://doi.org/10.3390/rs11243030

Zhang C, Guo X (2008) Monitoring northern mixed prairie health using broadband satellite imagery. Int J Remote Sens 29(8):2257–2271. https://doi.org/10.1080/01431160701408378

Zhang Y, Zhao Y, Chen Q, Zhu Y, Liu B, Zhang X, Yin X (2024) Capacity of Forests and Grasslands to Achieve Carbon Neutrality in China. Forests 15(6):1060. https://doi.org/10.3390/f15061060

Zhao Y, Liu Z, Wu J (2020) Grassland ecosystem services: a systematic review of research advances and future directions. Landsc Ecol 35:793–814. https://doi.org/10.1007/s10980-020-00980-3

Zhu Q, Chen H, Peng C, Liu J, Piao S, He JS, Wang S, Zhao X et al (2023) An early warning signal for grassland degradation on the Qinghai-Tibetan Plateau. Nat Commun 14:6406. https://doi.org/10.1038/s41467-023-42099-4

Zlinszky A, Heilmeier H, Balzter H, Czúcz B, Pfeifer N (2015) Remote sensing and GIS for habitat quality monitoring: new approaches and future research. Remote Sens 7:7987–7994

Comprehensive Valuation of Ecosystem Services in Rice Agroecosystems: Integrating Provisioning, Regulating, Cultural, and Supporting Benefits for Sustainable Practices

Vaishnavi Singh and O. P. Singh

Abstract Rice agroecosystems play an important role in global food security while also providing ecosystem services that complement rice production. These services include supplying (food, fodder), regulating (water purification, climate regulation), cultural (aesthetic value, recreational activities), and sustaining (nutrient cycling, soil formation). This chapter provides a thorough methodology for valuing environmental services, emphasising their relationship with sustainable rice growing techniques. The chapter highlights the obstacles and potential in valuing ecosystem services by investigating both ecological and social components. It emphasises the need for integrated techniques that may protect and improve rice agroecosystems while offering incentives to diverse stakeholders. Such a strategy is critical for sustaining the long-term viability of rice production, providing ecosystem services, and supporting the well-being of those who rely on them. The insights offered in this chapter are intended to assist policymakers in developing effective strategies for sustainable rice agroecosystem management.

Keywords Ecosystem services · Rice · Sustainability · Agroecosystems

V. Singh (✉)
Senior Research Scholar, Department of Agricultural Economics, Institute of Agricultural Sciences, Banaras Hindu University, Varanasi 221005, India
e-mail: vagecon@bhu.ac.in

O. P. Singh
Professor, Department of Agricultural Economics, Institute of Agricultural Sciences, Banaras Hindu University, Varanasi 221005, India
e-mail: singhop@bhu.ac.in

O. Mutanga et al. (eds.), *Revealing Ecosystem Services Through Geospatial Technologies*, Springer Remote Sensing/Photogrammetry,
https://doi.org/10.1007/978-3-031-98048-0_5

1 Introduction

Ecosystems are naturally occurring habitats made up of living things and their surroundings. They provide a vast range of services that sustain life on Earth, including those of animals and humans. Agricultural ecosystems, which are mostly under human management, are essential for producing fuel, fibre, and food. These ecosystems are among the most heavily altered to suit human requirements, making up between 24 and 38% of the Earth's surface (Millennium Ecosystem Assessment 2005). Agriculture is unique because it both depends on and provides ecosystem services Among these functions are supplying (such as food and water), regulating (such as flood and pest management), cultural (such as heritage and recreation), and sustaining (such as soil formation and nutrient cycling). However, because of different cropping practices and environmental factors, the intricate relationships between agriculture and ecosystems differ greatly. The significance of appreciating ecological benefits while reducing the adverse effects—or "dis-services"—that agriculture may provide is highlighted by this intricacy. Agricultural productivity depends on important ecosystem services like pollination, nitrogen cycling, soil fertility, and insect control (Power 2010). Improving these services can help in resolving the conflict between the need to boost agricultural output and environmental sustainability (Dyer 2014). But agriculture also has a big influence on natural resources; it causes greenhouse gas emissions, water scarcity, and soil deterioration. These challenges are made worse by population expansion and rising food demand, underscoring the necessity of managing agricultural ecosystems sustainably (Bostian and Lundgren 2022).

Given Earth's finite resources, ensuring food security is a critical worldwide issue. Understanding ecosystem services' contribution to agricultural output and resolving farming-related externalities require valuing them. While non-use values, such protecting ecosystems for future generations, are less prioritized in agricultural contexts, valuation methodologies primarily concentrate on use values—benefits obtained from ecosystem services (Swinton et al. 2015). Finding the right measures and striking a balance between social, ecological, and economic factors are challenges in valuing (Costanza et al. 2017). A major primary crop around the world, rice is essential to maintaining food and nutritional security, particularly in developing nations. It supplies about 19% of dietary energy intake and more than 50% of the world's main food supply. In addition to providing food, rice farming contributes to intricate agroecosystems that provide a range of advantages, such as environmental preservation, cultural heritage preservation, and recreational opportunities (Chiueh 2012). The complex linkages between provisioning, regulatory, cultural, and supporting services make it difficult to value ecosystem services in rice paddies nonetheless their multifunctionality. In order to fully account the variety of benefits that rice agroecosystems offer, recent research supports valuation methodologies that combine biophysical and economic evaluations. Stakeholders can better grasp the value of these services by using techniques like contingent valuation and analytical network processes (Rahman and Baral 2020; Turner et al. 2003). This chapter examines the environmental services that paddy fields provide, how to value them, and the policy implications of acknowledging their many advantages.

Studies emphasise the importance of these ecosystem services in managing our natural capital sustainably (Setälä 2009; Manhães et al. 2018). There are four types of ecosystem services commonly defined as:

- Provisioning services—supply physical things such as food, lumber, and clean water.
- Supporting services—These are the fundamental processes that underpin all other services, including nitrogen cycling, soil formation, and primary production.
- Regulating services—Climate regulation, water purification, and disease management are examples of regulating services that help to keep the environment safe for humans.
- Cultural services—provide intangible advantages including enjoyment, aesthetics, and spiritual experiences (Setälä 2009; Manhães et al. 2018).

These services play an important role in human well-being by affecting fundamental supplies, environmental stability, and cultural demands. Research emphasises the connection between ecosystem health and human well-being, emphasising the significance of sustainable natural resource management (Manhães et al. 2018).

2 Rice Agroecosystems: A Global Perspective

2.1 Global Distribution and Importance of Rice Cultivation

Rice is one of the most frequently cultivated and consumed cereal grains in the world, serving as a primary food source and a source of income for millions of people in a variety of countries and cultures. Over 90% of the world's rice is grown and consumed in Asia, where it is a staple food for more than 60% of the population (Yadav et al. 2020). Rice cropping system covers roughly 161.1 million hectares worldwide, producing over 480.3 million tonnes per year (Yadav et al. 2020). Rice's resilience to a variety of agroclimatic conditions—ranging from tropical to temperate regions—as well as its ability to grow in both irrigated and rainfed systems contribute to its widespread cultivation and significance in the global food chain. Rice is grown under a variety of geographical and environmental conditions, each of which forms its own agricultural practices. These systems range from tropical lowlands in Asia to temperate regions in China and Australia, and from sea-level areas like the Mekong Delta in Vietnam—known as the "Rice Bowl of Vietnam"—to high-altitude regions like Nepal's Jumla District, where rice is grown at the highest elevation (Haefele et al. 2014; Kuenzer and Knauer 2013). While Asia remains the world's greatest rice producer, rice farming has spread to Latin America, the United States, Australia, and Africa. Rice is becoming increasingly important in Africa as a solution to food security issues in low-income and food-insecure areas. Rice cultivation takes place in a wide range of climates, from wet tropics with over 5000 mm of annual rainfall, like Myanmar's Arakan Coast, to desert places with less than 100 mm. The diverse environmental conditions have also influenced the soils in rice-growing locations. For example, the Indo-Gangetic Plain of South Asia, a significant rice producing zone, has issues such as diminishing soil quality caused by continual submergence, uneven fertilisation, and inadequate organic carbon recycling.

Interestingly, balanced fertilisation and soil management approaches have proved that paddy soils may store atmospheric CO_2, increasing soil organic carbon storage and helping to global warming mitigation (Haefele et al. 2014).

2.2 *Ecosystem Services Provided by Rice Agroecosystems*

Rice agroecosystems, which are characterised by complex interactions between agricultural practices, natural ecosystems, and human societies, provide a wide range of ecosystem services in addition to food production. One important service they provide is biodiversity support. Rice production environments, such as submerged paddy fields and adjoining wetlands, provide sanctuaries and resource centres for a diverse range of plant and animal species (DeClerck et al. 2016). This diversification improves ecological balance and resilience because the interactions between different creatures contribute to the stability and functionality of both local and regional ecosystems (Peeters 2009). Additionally, rice agroecosystems play an important role in controlling the hydrological cycle and managing water resources. Rice fields are carefully situated in wetland regions, which helps to manage water flows, decrease flood risks, and replenish groundwater supplies. These functions increase the availability and quality of freshwater resources, which benefits both agriculture and residential use (Saini et al. 2022).

3 Challenges in Valuing Ecosystem Services

While providing a value to rice agroecosystems is necessary to preserve them but it possesses several challenges such as:

1. Non-marketed services: Many services, such as climate regulation or pollination, have no direct market value, making economic valuation difficult. Traditional market-based approaches are lacking, requiring alternatives such as contingent value or choice experiments methods. These strategies, however, are subjective and context-dependent (Xiao et al. 2005). Although valuation methodologies had been developed to calculate and provide monetary values to these non-marketed services but most of them have their own assumptions and limitations, making them difficult for all researchers to adopt easily.

2. Rice agroecosystems are characterised by complex interactions and trade-offs between ecosystem functions. Efforts to improve one service, such as applying more fertiliser to increase yields, may have a negative impact on others, like water quality. Quantifying these trade-offs is normally cumbersome (Zabala et al. 2021).
3. Data limitations: Reliable data on the ecological and socioeconomic features of rice agroecosystems are scarce, especially in underdeveloped countries where these systems are common. This data gap prevents proper assessment and informed decision-making (Nayak et al. 2019). Additionally, many stakeholders in rice farming have limited awareness of the ecosystem service concept, which further hinders data collection directly from the fields.
4. Diverse stakeholder perspectives: Farmers, policymakers, consumers, and local communities may have different ideas on ecological services. Balancing these interests and incorporating them into valuation frameworks necessitates collaborative and inclusive approaches (Chivenge et al. 2020).

Overcoming these problems is essential for developing successful policies that promote the sustainable management of rice agroecosystems while maintaining their long-term potential to provide value.

4 Valuing Provisioning Services of Rice Agroecosystems

4.1 The Economic Value of Rice Production (Grain, Straw, etc.)

Rice farming creates huge economic value, not only from the production of rice grains, but also from secondary by-products such as straw and husk. Rice is a primary meal for billions of people worldwide, with an annual output of almost 600 million tonnes (Abo-El-Enein et al. 2011). China, the main producer, accounts for roughly 30% of world rice supply (Atungulu and Pan 2014).

Aside from its use as a food, rice byproducts provide extensive prospects for economic and environmental benefits. For example, rice husks, which account for around 20% of total harvest, are increasingly being used as a sustainable resource, lowering waste management costs and minimising environmental costs (Abdel-Mohdy et al. 2009). Similarly, rice straw, a significant portion of the crop's biomass, may be recycled as renewable energy or turned into high-value commodities like silica, silicon carbide, and activated carbon (Illankoon et al. 2023; Abo-El-Enein et al. 2011). Research has also shown new possibilities for these byproducts, such as utilising rice husks to create sophisticated materials like silicon-based anodes for lithium-ion batteries. This not only meets the rising demand for renewable energy storage, but also highlights rice leftovers' potential to contribute to a circular economy (Liu et al. 2013). As global food demand rises combined with population expansion, maximising the economic value of rice and its byproducts becomes crucial. A comprehensive strategy to utilising rice byproducts can increase income streams, improve sustainability, and boost the overall economic effect of this important commodity (Abo-El-Enein et al. 2011).

5 Valuing Regulating Services of Rice Agroecosystems

While rice cultivation is well known for its function in food production, studies had highlighted its broader environmental benefits (Usman et al. 2020). Rice fields, which covers a large portion of the Earth's surface, have the ability to act as significant wetland ecosystems, providing key ecosystem services such water purification, flood control, and carbon sequestration (Usman et al. 2020; Eyhorn et al. 2018). Flooding condition of paddy fields produce a unique home for a variety of biological organisms (Usman et al. 2020). These wetland ecosystems have an extraordinary ability to filter and remove impurities, considerably improving water quality—a critical function, especially in areas plagued by water pollution (Usman et al. 2020). Furthermore, rice fields can temporarily hold and gradually release surplus water following heavy rains, aiding in flood control. This role is becoming more significant as climate change increases the frequency and intensity of extreme weather events (Eyhorn et al. 2018). New research emphasises the importance of rice paddies in climate change mitigation through carbon sequestration, expanding its environmental significance (Usman et al. 2020). These varied advantages of rice agriculture emphasise its relevance beyond food production and the necessity to include these ecosystem services into long-term agricultural and development policies.

5.1 *Methodologies for Valuing These Services, Such as Replacement Cost or Avoided Damage*

Valuing ecosystem services is critical for communicating the importance of ecosystem functioning to policymakers, and farmers (Xiao et al. 2005). Ecosystem valuation can help decision-makers understand the non-zero value of ecological services (Lazo 2002). Ecosystem valuation aims to recognise the value of ecosystem services, evaluate policy interventions, understand cost–benefit supply, and create mechanisms to capture ecosystem benefits (Morse-Jones et al. 2011).

The replacement cost technique is a popular method in ecosystem service valuation that assesses the cost of replacing an ecological function with an engineered counterpart. This technique has been used to assess the value of rice paddy ecosystem services such as air regulation and greenhouse gas control (Xiao et al. 2005). Another technique is the averted damage method, which calculates the economic worth of a service by estimating the losses that would result if it were lost. However, recent research on ecosystem service value in agricultural systems has frequently focused on provisioning services, such as food and fibre production. This strategy may neglect the added benefit of controlling and supporting activities, including as gas management, pollination, and nutrient cycling, which are equally important but often underappreciated (Lazo 2002; Xiao et al. 2005).

6 Importance of Rice Agroecosystem Services for Climate Change Mitigation and Adaptation

Aside from their primary role in providing food security, rice agroecosystems provide critical ecosystem services that greatly contribute to climate change mitigation and adaptation. One of these services is water purification, in which rice paddies filter pollutants and enhance water quality downstream (Zabala et al. 2021). This service is especially important in light of climate change, which is causing increased water scarcity and pollution. Rice paddies also serve as natural buffers, contributing significantly to flood prevention. They help decrease flood hazards by absorbing excess rainfall and reducing runoff, both of which are predicted to grow more frequent and intense as climate change proceeds. Perhaps most critically, rice paddies help with carbon sequestration by absorbing atmospheric CO_2 and storing it in soil organic matter (Chivenge et al. 2020). This natural climate mitigation function contributes significantly to reducing greenhouse gas emissions. However, as Langerwisch et al. (2017) suggested land-use changes can have a damaging impact on carbon storage in these agroecosystems, emphasising the significance of protecting rice fields. Furthermore, Nayak et al. (2019) propose using incentives for ecosystem services to encourage sustainable agricultural methods that can improve climate benefits. These services, together with the ability to improve water-use efficiency (Ismail et al. 2020), position rice agroecosystems as key assets in the battle against climate change, providing advantages for both mitigation and adaptation.

7 Valuing Cultural Services of Rice Agroecosystems

Rice agroecosystems are recognised for providing essential cultural functions that go beyond food production. These landscapes provide a variety of recreational, aesthetic, and spiritual advantages that improve human health. Natural landscapes, particularly rice fields, are recognised to improve physical and mental health by providing chances for recreation, relaxation, and inspiration (Berkel et al. 2018). The aesthetic and recreational appeal of these settings is vital, since individuals frequently have an innate psychological desire for nature, whether they realise it or not (Noss et al. 2024). Rice landscapes in developing nations, like as Indonesia, give both agricultural and cultural benefits. For example, the rice fields in Nyuh Kuning Village, Bali, are both culturally significant and popular with visitors, providing a long-term source of revenue for the community when maintained locally. Similarly, rice terraces and ancestral rice varieties in the Philippines' Cordillera area have both historical and aesthetic significance, contributing to tourism and local livelihoods (Bairagi et al. 2021). Numerous studies have found that natural landscapes, especially those with rice production, improve human health, emotional well-being, and intellectual development (Noss et al. 2024). However, the cultural and aesthetic advantages of rice landscapes are under threat from economic expansion, cultural upheavals, and population changes (Maulidi and Wulandari 2017). As a result, protecting these landscapes is critical for both their cultural value and the well-being of local populations.

7.1 Methods for Valuing Cultural Services

The economic value of cultural ecosystem services, such as recreational advantages given by natural regions, has received more attention in recent years (Chen et al. 2004). The travel cost technique and contingent valuation are two popular approaches for assessing the value of these services. The travel cost approach is commonly used to determine the recreational value of natural locations. This strategy posits that the costs people experience while travelling to a destination serve as a proxy for the importance, they have on visiting that area (Chen et al. 2004). By analysing travel-related expenditures and visiting patterns, researchers may evaluate the demand for and economic worth of the site's cultural offerings. The travel cost technique has been used to a variety of scenarios, including the valuing of estuarine and coastal ecosystems. For example, a study on Xiamen Island in China employed this method to determine the site's recreational demand and economic worth (Chen et al. 2004). While it is an effective tool for assessing cultural ecosystem services, it has several drawbacks. For example, it is based on assumptions, such as the belief that travel expenses appropriately represent the value people spend on their visit. Furthermore, it is impossible to account for all necessary trip expenditures. Despite these drawbacks, this technique remains an important tool for analysing the economic significance of cultural activities, giving useful insights for long-term management and development decisions (Chen et al. 2004; Ghermandi et al. 2013).

7.2 Importance of Cultural Heritage and Traditional Knowledge Associated with Rice Farming

In many Asian societies, rice is regarded sacred, representing nutrition and social order. Rice production is intertwined into the fabric of daily life, with rice seen as a source of both physical and spiritual well-being. Traditional rice landraces, passed down through generations, constitute a distinct link between communities and their surroundings, encapsulating millennia of ecological knowledge and cultural behaviours (Wangpan et al. 2019). These types, which are fundamental to many ethnic communities' identity and food security, not only benefit local diets but also help to preserve rich cultural legacy. As modernisation threatens traditional agricultural systems, it is critical to encourage on-farm rice biodiversity conservation in order to protect both food security and cultural identity (Wangpan et al. 2019). Rice variations provide critical nutrients such as carbohydrates and γ-oryzanol, a bioactive molecule that enhances rice's nutritional value, making it a uniquely nutritious diet in Asian civilisations (Wangpan et al. 2019).

8 Valuing Supporting Services of Rice Agroecosystems

Supporting ecosystem services is critical to the long-term production and sustainability of rice agroecosystems. These services, which include nutrient cycling, soil formation, and primary production, lay the groundwork for the supply of other ecosystem services including provisioning, regulation, and cultural benefits (Yuan and Lo 2020). Despite their crucial relevance, the economic value of supporting services is frequently underestimated or overlooked in decision-making processes (Jamouli and Allali 2020; Yuan and Lo 2020). Rice agroecosystems provide a variety of supporting services that are critical to their long-term survival. One important example is nitrogen cycling, which maintains soil fertility and promotes rice fields' long-term productivity. Rice fields also provide home for beneficial creatures such as pollinators and natural pest controllers, which help to maintain the system's general health and resilience.

8.1 Challenges in Valuing the Supporting Services

As we already discussed, providing value to a particular ecosystem service which are generally non-marketed goods is complex. But valuing the ecosystem services provided by agricultural landscapes, such as rice paddy fields, is essential for guiding sustainable management decisions (Marlianingrum et al. 2019; Nayak et al. 2019; Xiao et al. 2005). While the provisioning services related to food and fiber production are well-recognized, the supporting and regulating services offered by these agroecosystems are often underappreciated (Chivenge et al. 2020; Nayak et al. 2019; Xiao et al. 2005). Increasing research has focused on quantifying the economic value of services like gas regulation, nutrient cycling, habitat provision, and the maintenance of genetic diversity in rice paddy systems (Xiao et al. 2005; Nayak et al. 2019). One of the most difficult aspects of this process is valuing supporting services, which generally lack direct market value. Rice agroecosystems rely on services such as soil formation, nitrogen cycling, and primary production to thrive, but their indirect economic benefits can be difficult to quantify. Rice paddies, for example, play an important role in gas control by exchanging greenhouse gases and emitting oxygen. However, turning these biophysical processes into monetary values necessitates complicated models and sophisticated valuation methodologies. Another problem is the temporal and geographical heterogeneity of ecosystem service delivery in rice paddy systems. The services provided by these systems may vary based on crop development stage, management strategies, and environmental factors such as climate and soil condition. These things complicate the proper assessment of ecosystem services, making it impossible to account for their total economic outcome in different conditions.

9 Integrating Ecosystem Service Concept for Sustainable Rice Management

Ecosystem services—the advantages people obtain from nature—have emerged as a key term in environmental management and policy (Yu et al. 2018). Recognising the various ways in which ecosystems contribute to human well-being enables a more systematic assessment of the consequences of land-use decisions (Balzan et al. 2018; Brown 2013).

Several research has looked at the trade-offs and synergies between various ecosystem services (Sharma et al. 2019; Deka et al. 2019). For example, providing services like as food and lumber production may come at the expense of regulating services such as climate regulation and flood management (Deka et al. 2019; Lin et al. 2018). Interventions aimed at improving one service, such as reforestation, can, however, result in several other benefits (Sharma et al. 2019). Understanding these connections is crucial for long-term management, as decisions to maximise one function may unintentionally damage others (Balzan et al. 2018; Yu et al. 2018). Policymakers and land managers must carefully assess how land-use decisions affect the complete range of ecosystem services (Sharma et al. 2019).

Decision-makers can take more comprehensive approaches to land management by measuring the value of ecosystem services, including non-market benefits (Yu et al. 2018; Lin et al. 2018). It is critical to include both economic and social valuation approaches, as price signals alone may underestimate the relevance of ecological services to human well-being (Brown 2013). Additionally, evaluations of ecosystem services should take into consideration their dynamic and geographical character, as the provision and value of services can vary over time and across different landscapes.

9.1 Framework for Integrating Ecosystem Service Values into Decision-Making

Incorporating ecosystem service concept into decision-making is critical to achieving sustainability and effective environmental management. More than a few frameworks have been developed to solve this difficulty, with an emphasis on combining stakeholder viewpoints and recognising nature's diverse values (Raum 2018). A central component of these frameworks is the use of ecosystem services as a "common language" to improve communication and negotiation across stakeholder groups with contrasting interests (Granek et al. 2010). This method facilitates the honest assessment of the trade-offs between various management techniques and creates a common knowledge of how environmental changes may influence human well-being. In spite of the potential benefits, implementing ecosystem service frameworks presents major technical and social hurdles (Daily and Matson 2008). Anticipating how human actions will affect ecosystems and the services they provide, as well as understanding how these changes will affect various social groups, demands interdisciplinary collaboration that brings together knowledge from ecology, economics, and social sciences (Granek et al. 2010). Also, properly integrating this information into institutional frameworks, policies, and decision-making processes poses a significant social and political challenge (Daily & Matson 2008; Bateman et al. 2014). Institutions must be capable of managing, monitoring, and incentivising ecosystem services, as well as developing methods to assure their long-term usage. To overcome these barriers, substantial changes are required to change from conceptual frameworks to practical application, including ecosystem service values into ground-level decision-making processes (Daily and Matson 2008).

9.2 *Real-World Examples of Ecosystem Service Valuation in Rice Agroecosystems*

Valuing ecosystem services is important for measuring the societal benefits of natural settings (Loc et al. 2017; Jiang et al. 2021; Morse-Jones et al. 2011). This is especially important in agricultural settings, where services like water control, soil fertility, and pollination are critical for agricultural output (Loc et al. 2017). One notable example is from Vietnam's Mekong Delta area, where ecosystem service value was used to quantify the advantages of rice paddy development (Loc et al. 2017). The study indicated that while stakeholders acknowledged the significance of ecosystem services in the region, the monetary values given to these services did not adequately capture their true contribution to local incomes, Rice agriculture, the region's primary land use, was shown to create the most monetary value, accounting for around 56% of total value. Surprisingly, stakeholders did not see rice production as the most valued ecosystem function, indicating a mismatch between economic valuation and local views. Similar tendencies have been reported in various wetland and agricultural habitats, where the economic value of ecosystem services is frequently overestimated (Jiang et al. 2021; Morse-Jones et al. 2011). In these circumstances, ecosystem services like as water purification, flood management, and carbon sequestration are usually devalued in conventional economic assessments, despite their importance in ensuring the long-term viability of agricultural landscapes. Incorporating ecosystem service values into land management choices and pricing mechanisms is crucial for sustainable use of natural resources and maintaining important services (Loc et al. 2017; Morse-Jones et al. 2011). However, overcoming challenges such as a lack of clear ecological production functions and difficulties in apprehending trade-offs between various services is required for ecosystem service valuation to fully realise its potential in supporting sustainable agricultural development.

9.3 Valuation of Rice Agroecosystem Services and Policy Management

Rice is indispensable to global food security, accounting for more than 20% of the calories consumed (Lim et al. 2014). As a result, sustainable management of rice agroecosystems is critical not just for food security, but also for reducing environmental impact and improving ecosystem services. Ecosystem service valuation is gradually being used to assist policy and management decisions that try to strike a balance between agricultural production and environmental sustainability. Improving water efficiency in rice farming is an important area of study. Water is a key resource for rice growing, however rice has a far greater water footprint than many other food crops (Tuong 2000). Water-saving devices have been proven in studies to minimise percolation, drainage, and evaporation losses by 15–20% while maintaining output (Bouman et al. 2007). The assessment of water control and conservation services in rice agroecosystems helps to shape policies that support water-saving technology and management techniques. For example, including these values into policy frameworks might encourage the implementation of water-saving measures like alternating wetting and drying (AWD), which decreases water use while preserving rice yield.

However, the possible long-term environmental ramifications of improved aerobic field conditions, including as changes in soil carbon storage and effects on greenhouse gas emissions, must be carefully evaluated. The pricing of climate control services in rice systems might help guide policy to reduce negative environmental consequences like global warming and eutrophication caused by particular water-saving methods (Bouman et al. 2007). Furthermore, in water-scarce areas, new rice varieties paired with targeted management approaches have the potential to increase yields by 50–100% in the next decades (Bouman et al. 2007). The valuation of supporting ecosystem services such as soil fertility and nutrient cycling can provide policymakers with evidence of the need for increased investment in research and extension services, thereby promoting the adoption of sustainable practices that ensure long-term productivity while conserving natural resources. Overall, including ecosystem service value into policy and management choices promotes a more comprehensive knowledge of the trade-offs and synergies between agricultural production and environmental sustainability. By valuing ecosystem services such as water control, soil fertility, and carbon sequestration, policymakers may make more informed decisions that support both food security and ecosystem preservation. The insights gained by ecosystem service value may be used to drive sustainable rice production techniques, reduce negative environmental consequences, and increase climate resilience.

Below are some successful Sustainable Rice Farming Practices that helped in enhancing Ecosystem Services:

- Water-Saving Technologies: Practices like alternating wetting and drying (AWD) cut water usage by 15–20% without hurting yields, save water, and minimise methane emissions (Bouman et al. 2007).
- Improved Rice cultivars: Drought- and salinity-tolerant rice cultivars assist preserve yields in difficult conditions, increasing food security throughout climate change (Zhang et al. 2009).
- Integrated Water Management: Using enhanced rice varieties with effective irrigation increases yields, improves soil quality, and conserves water (Zhang et al. 2009).
- Eco-Friendly Fertiliser Practices: Organic fertilisers and integrated nutrient management decrease chemical consumption, increase soil health, and promote biodiversity.
- Agroecological Approaches: Practices such as agroforestry and crop rotation increase biodiversity, improve soil fertility, and help ecosystem services like pollination.

These strategies increase production while conserving resources and sustaining ecosystem services in rice.

10 Conclusion: Towards a Holistic Approach to Rice Agroecosystem Management

A complete assessment of rice agroecosystem services necessitates an integrated methodology that takes into account the intricate interplay of ecological, economic, and social variables. This method goes beyond evaluating specific ecosystem services and investigates the trade-offs and synergies among them. For example, boosting rice yields may improve food security but may also have unexpected effects such as decreased water quality or biodiversity loss (Xiao et al. 2005). Understanding these interdependencies is critical for making educated crop management decisions, since changes in one service can have a rippling effect on others (Zabala et al. 2021).

Effective rice agroecosystem management necessitates active participation from a wide range of stakeholders, including farmers, consumers, and policymakers. Incorporating multiple views ensures that management techniques are not only ecologically sound, but also socially equitable and economically feasible (Nayak et al. 2019). To assist this, future research should concentrate on building integrated valuation frameworks that incorporate both commercial and non-market advantages of rice agroecosystems, such as cultural and aesthetic qualities. These frameworks should influence policy development, such as the creation of Payment for Ecosystem

Services (PES) programs that provide farmers with financial incentives to adopt sustainable practices. Furthermore, research should look at novel policy tools, such as climate-smart agriculture and water-efficient agricultural approaches, to improve resilience and sustainability.

We can make rice agroecosystems more sustainable and resilient by strengthening holistic valuation methodologies and encouraging policies that recognise the entire spectrum of ecosystem services. This will not only increase long-term agricultural production, but will also protect critical ecosystem services and enhance the livelihoods of those who rely on rice agriculture.

References

Abdel-Mohdy FA, Abdel-Halim ES, Abu-Ayana YM, El-Sawy SM (2009) Rice straw as a new resource for some beneficial uses. Carbohydr Polym 75(1):44–51

Abo-El-Enein SA, Eissa MA, Diafullah AA, Rizk MA, Mohamed FM (2011) Utilization of a low cost agro-residue for production of coagulant aids and their applications. J Hazard Mater 186(2–3):1200–1205

Millennium Ecosystem Assessment (2005) Ecosystems and human well-being: wetlands and water. World Resources Institute

Atungulu GG, Pan Z (2014) Rice industrial processing worldwide and impact on macro-and micronutrient content, stability, and retention. Ann N Y Acad Sci 1324(1):15–28

Bairagi S, Custodio MC, Durand-Morat A, Demont M (2021) Preserving cultural heritage through the valorization of Cordillera heirloom rice in the Philippines. Agric Hum Values 38(1):257–270

Balzan MV, Potschin-Young M, Haines-Young R (2018) Island ecosystem services: insights from a literature review on case-study island ecosystem services and future prospects. Int J Biodivers Sci, Ecosyst Serv Manag 14(1):71–90

Bateman IJ, Harwood AR, Abson DJ, Andrews B, Crowe A, Dugdale S, Termansen M (2014) Economic analysis for the UK national ecosystem assessment: synthesis and scenario valuation of changes in ecosystem services. Environ Resource Econ 57:273–297

Bostian M, Lundgren T (2022) Valuing ecosystem services for agricultural TFP: a review of best practices, challenges, and recommendations. Sustainability 14(5):3035. https://doi.org/10.3390/su14053035

Bouman BA, Humphreys E, Tuong TP, Barker R (2007) Rice and water. Adv Agron 92:187–237

Brown G (2013) The relationship between social values for ecosystem services and global land cover: an empirical analysis. Ecosyst Serv 5:58–68

Chen W, Hong H, Liu Y, Zhang L, Hou X, Raymond M (2004) Recreation demand and economic value: an application of travel cost method for Xiamen Island. China Econ Rev 15(4):398–406. https://doi.org/10.1016/j.chieco.2003.11.001

Chiueh YW (2012) Environmental multifunctionality of paddy fields in Taiwan—a conjunction evaluation method of contingent valuation method and analytic network procedures. Environ Nat Resour Res 2:114–127

Chivenge P, Angeles O, Hadi B, Acuin C, Connor M, Stuart A, Johnson-Beebout S (2020) Ecosystem services in paddy rice systems. In: The role of ecosystem services in sustainable food systems. Academic Press, pp 181–201

Costanza R, De Groot R, Braat L, Kubiszewski I, Fioramonti L, Sutton P, Grasso M (2017) Twenty years of ecosystem services: how far have we come and how far do we still need to go? Ecosyst Serv 28:1–16

Daily GC, Matson PA (2008) Ecosystem services: from theory to implementation. Proc Natl Acad Sci 105(28):9455–9456

DeClerck FA, Jones SK, Attwood S, Bossio D, Girvetz E, Chaplin-Kramer B, Zhang W (2016) Agricultural ecosystems and their services: the vanguard of sustainability? Curr Opin Environ Sustain 23:92–99

Deka S, Tripathi OP, Paul A (2019) Perception-based assessment of ecosystem services of Ghagra Pahar forest of Assam, Northeast India. Geol, Ecol, Landsc 3(3):197–209

Dyer A, Dyer A (2014) Reductionist farming: losing ecosystem services. In: Chasing the Red Queen: the evolutionary race between agricultural pests and poisons, pp 45–60

Eyhorn F, Van den Berg M, Decock C, Maat H, Srivastava A (2018) Does organic farming provide a viable alternative for smallholder rice farmers in India? Sustainability 10(12):4424

Ghermandi A, Nunes PA (2013) A global map of coastal recreation values: results from a spatially explicit meta-analysis. Ecol Econ 86:1–15

Granek EF, Polasky S, Kappel CV, Reed DJ, Stoms DM, Koch EW, Wolanski E (2010) Ecosystem services as a common language for coastal ecosystem-based management. Conserv Biol 24(1):207–216

Haefele SM, Nelson A, Hijmans RJ (2014) Soil quality and constraints in global rice production. Geoderma 235:250–259

Illankoon WAMAN, Milanese C, Collivignarelli MC, Sorlini S (2023) Value chain analysis of rice industry by products in a circular economy context: a review. Waste 1(2):333–369

Ismail H, Kamal MR, bin Abdullah AF, bin Mohd MSF (2020) Climate-smart agro-hydrological model for a large scale rice irrigation scheme in Malaysia. Appl Sci 10(11):3906

Jamouli A, Allali K (2020) Economic valuation of ecosystem services in Africa. E3S Web Conf 183:01002. https://doi.org/10.1051/e3sconf/202018301002

Jiang W, Wu T, Fu B (2021) The value of ecosystem services in China: a systematic review for twenty years. Ecosyst Serv 52:101365. https://doi.org/10.1016/j.ecoser.2021.101365

Kuenzer C, Knauer K (2013) Remote sensing of rice crop areas. Int J Remote Sens 34(6):2101–2139

Langerwisch F, Václavík T, von Bloh W, Vetter T, Thonicke K (2017) Combined effects of climate and land-use change on the provision of ecosystem services in rice agro-ecosystems. Environ Res Lett 13(1):015003

Lazo JK (2002) Economic valuation of ecosystem services: discussion and application. Drug Chem Toxicol 25(4):349–374

Lim JS, Manan ZA, Hashim H, Alwi SRW (2014) Synthesis of a sustainable integrated rice mill complex. J Clean Prod 71:118–127

Lin X, Xu M, Cao C, Singh RP, Chen W, Ju H (2018) Land-use/land-cover changes and their influence on the ecosystem in Chengdu City, China during the period of 1992–2018. Sustainability 10(10):3580

Liu N, Huo K, McDowell MT, Zhao J, Cui Y (2013) Rice husks as a sustainable source of nanostructured silicon for high performance Li-ion battery anodes. Sci Rep 3(1):1–7

Loc HH, Diep NTH, Can NT, Irvine KN, Shimizu Y (2017) Integrated evaluation of ecosystem services in prawn–rice rotational crops, Vietnam. Ecosyst Serv 26: 377–387.

Manhães AP, Loyola R, Mazzochini GG, Ganade G, Oliveira-Filho AT, Carvalho AR (2018) Low-cost strategies for protecting ecosystem services and biodiversity. Biol Cons 217:187–194

Marlianingrum PR, Kusumastanto T, Adrianto L, Fahrudin A (2019) Economic analysis of management option for sustainable mangrove ecosystem in Tangerang District, Banten Province, Indonesia. IOP Conf Ser: Earth Environ Sci 241(1):012026

Maulidi C, Wulandari LD (2017) Changing cultural landscape in post-productivism of rice field in Nyuh Kuning Village Bali. IOP Conf Ser: Earth Environ Sci 70(1):012049

Morizet-Davis J, Marting Vidaurre NA, Reinmuth E, Rezaei-Chiyaneh E, Schlecht V, Schmidt S, von Cossel M (2023) Ecosystem services at the farm level—overview, synergies, trade-offs, and stakeholder analysis. Glob Challenges 7(7):2200225

Morse-Jones S, Luisetti T, Turner RK, Fisher B (2011) Ecosystem valuation: some principles and a partial application. Environmetrics 22(5):675–685

Nayak AK, Shahid M, Nayak AD, Dhal B, Moharana KC, Mondal B, Pathak H (2019) Assessment of ecosystem services of rice farms in eastern India. Ecol Process 8:1–16

Noss R, Aplet G, Comer P, Enquist C, Franklin J, Riley J, Safford H (2024) Natural areas in the twenty-first century. Nat Areas J 44(1):35–40

Peeters A (2009) Production and other ecosystem services from extensive grassland-based systems in Europe: policy options and technical innovations. In: Proceedings of the British Society of Animal Science, vol 2009. Cambridge University Press, pp 247–247

Power AG (2010) Ecosystem services and agriculture: tradeoffs and synergies. Philos Trans R Soc b: Biol Sci 365(1554):2959–2971

Rahman SA, Baral H (2020) Nature-based solution for balancing the food, energy, and environment trilemma: lessons from Indonesia. In: Nature-based solutions for resilient ecosystems and societies, pp 69–82

Raum S (2018) A framework for integrating systematic stakeholder analysis in ecosystem services research: stakeholder mapping for forest ecosystem services in the UK. Ecosyst Serv 29:170–184

Saini A, Manuja S, Kumar S, Shilpa K, Kumari S, Kumar A, Suri D (2022) Effect of different tillage systems and cultivars on yield and yield attributes of rice (Oryza sativa L.). Environ Conserv J 23(1&2):365–369

Setälä H (2009) Ecosystem services and biodiversity in Europe. EASAC Secretariat, The Royal Society

Sharma R, Rimal B, Baral H, Nehren U, Paudyal K, Sharma S, Kandel P (2019) Impact of land cover change on ecosystem services in a tropical forested landscape. Resources 8(1):18

Swinton SM, Jolejole-Foreman CB, Lupi F, Ma S, Zhang W, Chen H (2015) Economic value of ecosystem services from agriculture. In: The ecology of agricultural landscapes: long-term research on the path to sustainability, pp 54–76

Tuong TP (2000) Productive water use in rice production: opportunities and limitations. J Crop Prod 2(2):241–264

Turner RK, Paavola J, Cooper P, Farber S, Jessamy V, Georgiou S (2003) Valuing nature: lessons learned and future research directions. Ecol Econ 46(3):493–510

Usman J, Ogbu JO, Iji SE, Afatar S (2020) Assessment of some wetland potentials in Makurdi supporting rice and sugarcane production. J Hortic Plant Res, p 1

Van Berkel DB, Tabrizian P, Dorning MA, Smart L, Newcomb D, Mehaffey M, Meentemeyer RK (2018) Quantifying the visual-sensory landscape qualities that contribute to cultural ecosystem services using social media and LiDAR. Ecosyst Serv 31:326–335

Wangpan T, Gogoi H, Tapi T, Tangjang S (2019) Bio-cultural diversity and ethnobotanic utility of indigenous rice (*Oryza sativa* L.) in Lakhimpur District, Assam, India. Notulae Scientia Biologicae 11(1):138–144. https://doi.org/10.15835/nsb11110354

Xiao Y, Xie G, Lu C, Ding X, Lu Y (2005) The value of gas exchange as a service by rice paddies in suburban Shanghai, PR China. Agr Ecosyst Environ 109(3–4):273–283

Yadav S, Yadav RB, Chuhan SS, Kumar R, Kumar V (2020) Efficacy of different herbicides and its combination against the weed flora of transplanted rice: a review. Int J Curr Microbiol Appl Sci 9(8):2057–2068

Yu Z, Liu X, Zhang J, Xu D, Cao S (2018) Evaluating the net value of ecosystem services to support ecological engineering: framework and a case study of the Beijing Plains afforestation project. Ecol Eng 112:148–152

Yuan M-H, Lo S-L (2020). Ecosystem services and sustainable development: perspectives from the food-energy-water Nexus. Ecosyst Serv 46:101217. https://doi.org/10.1016/j.ecoser.2020.101217

Zabala JA, Martínez-Paz JM, Alcon F (2021) A comprehensive approach for agroecosystem services and disservices valuation. Sci Total Environ 768:144859

Zhang H, Xue Y, Wang Z, Yang J, Zhang J (2009) An alternate wetting and moderate soil drying regime improves root and shoot growth in rice. Crop Sci 49(6):2246–2260

Mapping Aboveground Carbon Stocks Under Past, Present and Future Climate Scenarios in Urban Reforested Landscapes

Collins Matiza and Onisimo Mutanga

Abstract Understanding reforestation's role is essential for forest ecosystem health, maintaining carbon cycle balance, and addressing climate change. However, the contribution of reforestation to aboveground carbon (AGC) accumulation and its contribution to climate change mitigation in urban landscapes remains poorly understood. This gap limits the development of effective strategies to maximise their carbon accumulation potential. Therefore, this study aims to understand the past, present and future aboveground carbon stocks in reforested urban landscapes using forest inventory observation data, combined with land use, topographic and global climate circulation model data (CMIP6). An artificial neural network regression model was selected to construct the spatial relationship between AGC and the ancillary datasets. Analyzing the spatio-temporal dynamics of aboveground carbon (AGC) from 2012 to 2040, this study employed a resolution of 1 km and a 20-year interval, focusing on three distinct climate scenarios outlined in CMIP6. The cross-validation outcomes substantiate the reliability of AGC predictions in reforested urban landscapes, albeit acknowledging uncertainties in specific study site sections. The mean AGC values recorded over the past three decades were observed at 55.94 and 84.02 t. ha^{-1}. In contrast to historical AGC (2012–2022), future AGC exhibited notable fluctuations under diverse climate scenarios. Specifically, AGC showcased a gradual increasing trend under SSP1-2.6 low emissions, a diminishing trajectory under SSP2-4.5, and a markedly rapid decline under SSP5-8.5. This work enhances the management of urban forest carbon dynamics to support informed climate change mitigation policies and sustainability strategies.

Keywords Aboveground carbon · Reforestation · Climate change · Artificial neural networks · Spatio-temporal evolution

C. Matiza (✉) · O. Mutanga
Discipline of Geography, School of Agricultural, Earth and Environmental Sciences, University of KwaZulu-Natal, Scottsville, Pietermaritzburg, South Africa
e-mail: collinsmatiza3@gmail.com

O. Mutanga et al. (eds.), *Revealing Ecosystem Services Through Geospatial Technologies*, Springer Remote Sensing/Photogrammetry,
https://doi.org/10.1007/978-3-031-98048-0_6

1 Introduction

Over the past two centuries, rapid urbanization and human activity have led to global deforestation, resulting in a significant loss of forest cover, emerging as a major driver of environmental issues like landscape fragmentation, pollution, climate change, biodiversity loss, and thermal stress (He et al. 2014, Laurance 2010, Birdsey and Pan 2015). Urban landscapes account for the majority of global energy consumption and CO_2 emissions, figures projected to rise further by 2030 (United Nations 2023; NRDC 2018; Yue et al. 2023; IEA 2008). Tropical forest loss has also accelerated, reducing carbon storage (Cai and Zhang 2014). Given these challenges, urban reforestation has been proposed as a sustainable climate change mitigation solution by increasing carbon capture through strategies like planting trees and vegetation in cities (Ogunbode and Asifat 2021). Despite the challenges of urbanisation, urban reforestation (UR) provides a sustainable solution through carbon sequestration and storage (Teo et al. 2021; Nave et al. 2019, Anjali et al. 2020). Therefore, assessing the dynamics of AGC stocks over time is critical to quantify the effectiveness of urban greening efforts and guide future investments. Furthermore, projecting climate and land use change influences on future reforested carbon also allows cities to prepare and prioritize interventions that maximize carbon gains.

Climate change and land use change are expected to significantly influence aboveground carbon (AGC) stocks and fluxes in urban reforestation projects. Changing precipitation, temperature, and extreme weather events alter AGC dynamics (Feng et al. 2022). Climatic factors like warming temperatures can suppress plant carbon uptake while accelerating losses through enhanced respiration (Zavaleta et al. 2003; Wan et al. 2005). Topography also modifies carbon density and spatial patterns (Raich et al. 2006). However, land use change poses the greatest threat, contributing substantially to global CO_2 emissions (Hong et al. 2021). Over 1.38 peta-grams of potential AGC loss is projected across the tropics by 2030 due to land use changes (Seto et al. 2012). While reforestation offers climate mitigation potential, substantial knowledge gaps persist regarding precise carbon stock quantification and projection under environmental change (Mitchell et al. 2018; Rodríguez-Veiga et al. 2017).

The advent of remote sensing has significantly advanced carbon mapping, by integrating field data with quantitative methods to map ecosystems and predict tree properties (Xiao et al. 2019; Odebiri et al. 2023; Kowe et al. 2021). Leveraging high resolution satellite imagery, derived vegetation indices and machine learning, enables insights into carbon accumulation patterns (Matiza et al. 2023). For instance, Eckert (2012), utilized WorldView-2 satellite imagery to estimate aboveground biomass density and carbon stocks across landscapes in northeastern Madagascar. Recently Odebiri et al. (2022) combined Sentinel data, climate projections General Circulation Models (GCM) (i.e. CNRM-CM61-1, CanESM5, GFDL-ESM4, ACCESS-ESM1-5, and INM-CM5-0), and soil datasets to model future soil organic carbon across South Africa under land use and climate change. As remote sensing data quality improves, analytical innovations are required to fully exploit the rich details provided.

While statistical models have proven useful for aboveground carbon (AGC) mapping, their predefined forms limit flexibility in capturing intricate multivariate relationships in heterogeneous datasets (Kanevski et al. 2008; Matiza et al. 2023; Koldasbayeva et al. 2023). Machine learning provides distinct advantages, with techniques such as neural networks identifying inherent nonlinear patterns without assumptions, enhancing predictions (López-Serrano et al. 2020; Wang et al. 2022b). Further, complex AGC variability can overwhelm simplistic models. In contrast, ML algorithms such as artificial neural networks (ANNs) excel at extracting robust features invariant to noise through hyperparameter tuning (Pannakkong et al. 2022; Abiodun et al. 2019). This enables more accurate learning from large, noisy datasets. Despite these benefits, applying advanced ML techniques for remote sensing-based AGC simulations remains underexplored, presenting immense research potential.

Therefore, this research leverages remote sensing technology, topo-climatic data and machine learning, specifically artificial neural networks, to predict and analyze dynamics of reforested aboveground carbon (AGC) stocks and distributions in urban areas. The study utilizes a robust multi-source dataset to quantify AGC variability across three key time periods—historical (2012), current (2022) and projected (2040). This temporal analysis of urban forest carbon enables assessment of AGC variability considering evolving land use and climate. Integrating both remotely sensed data, topo-climatic data and machine learning facilitates significant exploration of urban reforestation carbon patterns. Ultimately, these insights can inform sustainable strategies and policies to harness urban forests for enhanced climate change mitigation at local and global scales.

2 Methods and Materials

2.1 Study Area

The Buffelsdraai reforestation site, situated in the eThekwini Municipality of KwaZulu-Natal Province, South Africa, in proximity to the city of Durban, was utilized as the research study area (Fig. 1). The geographic coordinates of the reforestation site fall between 30° 58′ 20.08″ E and 29° 37′ 55.17″ S, encompassing an area of approximately 800–1000 hectares. Established through eThekwini initiatives for the 2010 FIFA World Cup to mitigate greenhouse gas emissions, the reforestation project aimed to create a diverse, functional indigenous forest ecosystem to provide climate regulation services. Specifically, sequestering substantial atmospheric carbon to offset the impacts of the 2010 World Cup and nearby landfill (Mngadi et al. 2022a, Douwes et al. 2015, Mugwedi et al. 2017).

The region exhibits a subtropical climate, characterized by a mean annual temperature ranging between 22 and 27 °C, and a mean annual rainfall varying from 600 to 1000 mm. The study area exhibits an uneven topography supported by a resistant underlying dwyka tillite rock (Mugwedi et al. 2017, Douwes et al. 2015).

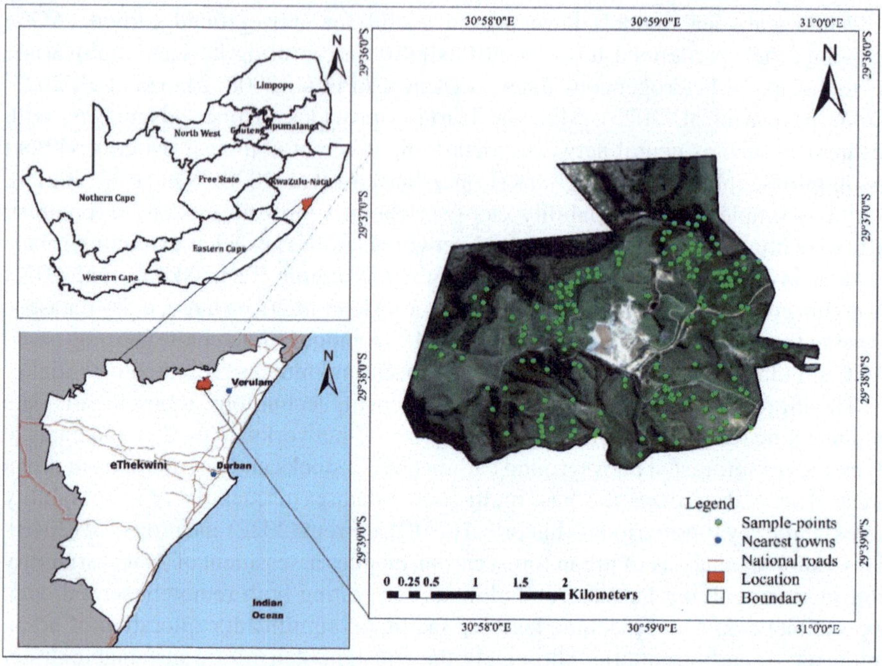

Fig. 1 Spatial context of the Buffelsdraai reforestation site and associated sampling locations within the eThekwini Municipality in KwaZulu-Natal Province, South Africa

Vegetation consists primarily of dense coastal forests dominated by Acacia species (*Caffra and Robusta*), *Syzygium cordatum, Bridelia micrantha* and *Erythrina caffra*. This study analyzes the Buffelsdraai site over a period from reforestation initiation (2012) through 2040 to evaluate historic, current and projected reforestation carbon accumulation potential (Mngadi et al. 2022b).

2.2 Observed Reforested Aboveground Carbon Stock Data Collection

A random field sampling strategy was implemented in the months of February and May 2012 and 2022 across 97 distributed sites spanning the Buffelsdraai reforestation site for data collection. This approach enabled consistent spatial coverage and comparison across the study period. Precise sampling locations were randomly predefined using ArcGIS Pro. At each plot, two 3 m × 3 m quadrats were delineated and situated within larger 10 m × 10 m plots to facilitate multi-scale data collection. These boundary dimensions were chosen based on recommended guidelines for balancing sampling effort with representation of species-level variability (Heiskanen et al. 2013). Every tree within each quadrat was surveyed, recording species, height using a Leica clinometer (accuracy $\pm$ 0.5 m @ 100 m), and diameter at breast height (dbh at 1.3 m) using a tape measure. Additionally, a Trimble global positioning system GPS receiver (accuracy < 50 cm) recorded coordinates. Concurrently, land slope, aspect, and land cover were recorded given their known influence on biomass accumulation (Tang et al. 2016). A total of 97 plots measuring 10 m by 10 m with a minimum of 2 samples per plot giving a total of 194 AGB samples was used.

The collected field measurements served to estimate standing aboveground tree biomass by applying allometric models that mathematically convert structural attributes into biomass estimates. The IPCC (2006) endorses these allometric equations as they enable indirect determination of forest carbon stocks. An extensive review by (Djomo et al. 2016) of allometric models across Africa concluded that generalized species-specific equations provide robust regional estimates in heterogeneous forests. Hence, measured tree height (H) and diameter (D) data were integrated into the pantropical model Eq. 1 to estimate plot-level biomass (Chave et al. 2014).

$$AGB = 0.0673 \times \left(p^{D^2H}\right)^{0.976} \tag{1}$$

where AGB represents aboveground oven-dry biomass (kg), D is diameter-at-breast height (cm), H is tree height (m), and p represents wood density constant (1.2 g/cm^3). Lastly, aboveground biomass values were converted to carbon stock by applying a 0.5 conversion factor, consistent with forest inventory guidelines 0.5 was used to compute the actual carbon stock from above ground biomass (Gara et al. 2016; Solomon et al. 2017, MacDicken 1997).

2.3 *Data Acquisition*

2.3.1 Multispectral Image Collection and Preprocessing

Historical (2012) and current (2022) aboveground carbon (AGC) estimation models were first developed to map the distribution and density of AGC stocks across the reforestation. The analysis integrated multispectral imagery from two sources—Planet Labs' RapidEye and the European Space Agency's (Fassnacht et al.) Sentinel-2 platforms. RapidEye and Sentinel-2 were selected as the best available data sources for the 2012 and 2022 time points. Cloud-free imagery (< 10% cover) was acquired for both sensors during peak productivity months (February to May) in 2012 and 2022. Specifically, the study utilized Sentinel-2 Level 1-C top-of-atmosphere reflectance data, retrieved from ESA's SciHub portal (https://scihub.copernicus.eu). Sentinel-2 provides 10–60-m resolution imagery across 13 spectral bands (443–2190 nm), with a 5-day revisit cycle enabling regular capture (Languille et al. 2015, Mngadi et al. 2021). Preprocessing via ESA's SNAP software included geometric, radiometric and atmospheric corrections. Images were then resampled to a 10-m spatial resolution (Mngadi et al. 2022b). Vegetation indices which combine Sentinel-2 bands to monitor parameters like biomass, leaf area, and chlorophyll content -are all proxies for AGC stocks (Dube and Mutanga 2016; Gara et al. 2016). Consequently, using information derived from Mngadi et al. (2021) and Suardana et al. (2023), vegetation indices including normalized difference vegetation index (NDVI), green normalized vegetation index (GNDVI), enhanced vegetation index (EVI), red-edge normalized vegetation index (RENDVI) and chlorophyll index (red-edge) together with raw sentinel wavebands were used to construct the current AGC estimation 2022 model.

Similarly, Planet Labs' (https://www.planet.com) RapidEye images offer a 5-m pixel size across five bands (440–850 nm), with a daily revisit time. Level 3A products retrieved incorporate radiometric, geometric, and sensor corrections (Anderson et al. 2011). This data covers an orthorectified area of 25 $\times$ 25 km per scene (Jung-Rothenhäusler et al. 2007). As with Sentinel-2 data similar vegetation indices were derived from the RapidEye bands to help explain observed AGC variability, as demonstrated in other studies (Gara et al. 2016; Mngadi et al. 2022a). Similarly, vegetation indices and Rapideye wavebands were integrated to form a model to estimate the past AGC for the year 2012. All remote sensing datasets were clipped and resampled to a standardized 10-m resolution.

2.3.2 Topographic Variables

Topographic covariates were derived from a regional Shuttle Radar Topography Mission (SRTM) digital elevation model (DEM) obtained from the NASA-USGS portal (Cowan and Cooper 2004). Using SAGA GIS software v8.2.3 (https://saga-gis.org), multiple factors were processed: slope, aspect, wetness index, curvature, roughness and elevation. The topographic data were resampled to 10 m for standardized analysis with other input predictors.

2.3.3 Bioclimatic Variables and Future Climate Scenarios

Gridded historical and future climate data were obtained from the WorldClim portal (WorldClim) at approximately 1 square km resolution (Fick and Hijmans 2017). Variables include monthly precipitation, temperatures, and derived bioclimatic factors representing annual/seasonal trends. Bioclimatic descriptors condense complex climatic factors (e.g. seasonality) into localized metrics useful for ecological modeling. Future climate projections were accessed from three CMIP6 general circulation models (GCMs)—ACCESS-ESM1-5, MIROC6, and EC-EARTH2—under three different greenhouse gas emissions scenarios. Using multiple climate models captures a range of sensitivity and better constrains uncertainty (Wang et al. 2022a). Representative concentration pathway shared socio-economic scenarios (RCP-SSP) were selected to represent low (SSP1-26), medium (SSP2-4.5), and high (SSP5-8.5) emission trajectories (Gao et al. 2013). These three RCPs were chosen to capture a range of plausible futures. The raw climate model outputs for these scenarios were then statistically downscaled using statistical downscaling model (SDSM) to generate higher resolution projections better aligned with the spatial and temporal scale of the other predictors in this case study.

2.3.4 Projected Landcover

This analysis integrated multi-temporal LULC maps obtained from the ESRI landcover classifications based on Sentinel 2 archive (2017–2021) encompassing 12 cover class types at 10 m resolution (> 70% accuracy)(Venter et al. 2022). The South African Landcover map (accessed through the specified portal (https://egis.environment.gov.za/gis_data_downloads/) for 2012 served as the basis for the 2012 land cover classification (SANLC 2020). Maps were reclassified into three major categories based on vegetation descriptions.

An artificial neural network approach within the MOLUSCE QGIS plugin was implemented to project LULC forward based on historical changes and relationships with slope, aspect and other spatial factors. Projected maps for 2022 and 2040 were generated and validated against 2021 reference data, achieving 81% correct match and 0.774 kappa statistic. All LULC and environmental data outputs were resampled to a 60 m grid for consistent integration. Table 1 summarizes all the input predictor variables used in the models.

2.3.5 Spatial Prediction Model

Artificial neural networks (ANNs) provide flexible machine learning frameworks adept at capturing complex multivariate relationships. Their multi-layer structure enables both linear and non-linear feature transformations, making ANNs well-suited for the intricate connections between aboveground carbon stocks, climate factors, topography, land use, vegetation indices and other covariates (Pannakkong et al. 2022; Abiodun et al. 2019). The core architecture of an Artificial Neural Network (ANN) model commonly utilizes a Multi-Layer Perceptron (MLP) structure. This structure features an input layer, one or more hidden layers, and an output layer. Interconnected neurons within each layer transmit information unidirectionally from input to output, enabling the model to capture both linear and non-linear relationships between input and output variables (Abiodun et al. 2019; Ghorbanian et al. 2022). Overfitting, a concern across modeling approaches, also affects neural networks with inaccurate predictions on new data. Mitigation requires careful hyperparameter tuning and training techniques.

The initial 2012 ANN model was used to predict 2022 carbon stocks without incorporation of actual 2022 field data. The 2022 predictions were validated against in situ measurements, serving to evaluate model accuracy. Adjustments were made based on this validation to improve performance. With the model refined and validated on 2022 data, it was then leveraged in a two-step process to predict 2040 carbon stocks. In stage 1, the ANN utilizes embedded regression relationships to estimate

Table 1 Covariates used for past (2012), current (2021) and future (2040) AGC simulation

Predictor variable	Description/formulae	Source
Sentinel-2 bands 2, 3, 4, 5, 6, 7, 8, 8A, 11, and 12	Wavebands of sentinel ranging from 490–2190 nm	(Drusch et al. 2012)
RapidEye bands 1, 2, 3, 4, and 5	Waveband s of RapidEye ranging from 440–850 nm	(Stoll et al. 2012)
Normalized difference vegetation index (NDVI)	$\frac{NIR-RED}{NIR+RED}$	(Pettorelli 2013)
Green normalized difference vegetation index (GNDVI)	$\frac{NIR-Green}{NIR+Green}$	(Jiang et al. 2007)
Enhanced vegetation index (EVI)	$2.5 \times \frac{NIR-RED}{(NIR+6\times RED-7.5\times BLUE+1)}$	(Jiang et al. 2007)
Red-edge normal difference vegetation index (RENDVI)	$\frac{NIR-RedEdge}{NIR+RedEdge}$	(Evangelides and Nobajas 2020)
Chlorophyll index red-edge (CIRE) Sentinel-2A	$\frac{Rededge(763nm)}{Rededge(705nm)}^{(-1)}$	(Evangelides and Nobajas 2020)
Chlorophyll index red-edge (CIRE) RapidEye	$\frac{NearInfrared}{Rededge}^{(-1)}$	(Clevers and Gitelson 2012)
Elevation	Height above sea level in meters	(Jacobsen 2005)
Topographic Wetness index (TWI)	Relative wetness due to terrain factors within catchments a measure of position on the slope	(Languille et al. 2015)
Surface Roughness	The irregularities and variations in the elevation of a land surface	(Lang et al. 2013)
Aspect	Slope direction approximation of the amount of radiation received by a site	(Lang et al. 2013)
Slope	The gradient of a specific site	(Lang et al. 2013)
Landcover classification map 2012	South Africa National Land use Land cover map (2012)	(SANLC 2020)
Landcover classification map 2022	Projected Land use Land cover map (2022)	(Venter et al. 2022)
Landcover classification map 2040	Projected Land use Land cover map (2040)	
Bioclimatic variable 12	Mean Annual Precipitation	(Fick and Hijmans 2017)
Bioclimatic variable 1	Mean annual temperature	(Fick and Hijmans 2017)
Bioclimatic variable 4		(Fick and Hijmans 2017)
Bioclimatic variable 5	Maximum temperature of the warmest month	(Fick and Hijmans 2017)
Bioclimatic variable 13	Precipitation of wettest month	(Fick and Hijmans 2017)
Average of 3 general circulation models (GCM)	Mean annual precipitation/ Temperature	(McBride et al. 2021)

2040 carbon levels based on climate and land use projections. Stage 2 further refines these projections through additional training on error residuals.

The same model architecture and covariates were maintained from 2012 to 2040 projections, with only the climate, terrain, and land cover input data updated to represent projected 2040 conditions. Three shared socioeconomic pathway (SSP) scenarios were simulated to evaluate a range of outcomes. The ANN flexibility enables re-training if additional future data becomes available, further improving accuracy. Validating the 2022 predictions demonstrates the model's capability for projecting future carbon stocks.

2.3.6 Model Evaluation

We evaluated model performance using complementary goodness-of-fit metrics, widely used in remote sensing studies (Odebiri et al. 2022; Matiza et al. 2023). The coefficient of determination (R^2) was leveraged to quantify the proportion of variance explained by the model. Additionally, we considered the mean absolute error (MAE) as a crucial indicator of the average magnitude of discrepancies between observed and predicted values. Lin's concordance correlation coefficient (LCCC) provided insight into both precision and accuracy, offering a comprehensive assessment of agreement. Furthermore, the root mean square error (RMSE) was employed to gauge the square root of the mean of squared errors, providing a measure of the model's overall accuracy. A higher R^2 and LCCC along with lower MAE and RMSE indicate better model fit (Morley et al. 2018).

To evaluate model performance, we adopted a rigorous validation approach. The ANN model was developed on a training set comprising 70% of the 2012 data (135 samples). The remaining 30% (59 samples) were reserved for testing model predictions. We developed an initial model on just the 2012 data, then projected the 2022 carbon stocks and validated its predictions on the actual 2022 data not used for training. Comparing these 2022 predictions to actual observations would assess generalization ability before finalizing the model. This phased validation strategy would further verify accurate generalization to new time periods, enhancing reliability evaluation, especially for long-term forecasts. Our use of training/testing splits across multiple years, and prospective testing on future data provides robust validation of model performance over time.

SHapley Additive exPlanations (SHAP) is a model interpretation method grounded in cooperative game theory that explains the output of machine learning models by quantifying the contribution of each feature to an individual prediction (Liben-Nowell et al. 2012). The algorithm is derived from Shapley values, a solution concept in coalitional game theory that distributes payouts among players according to their marginal contribution. In the SHAP methodology, each feature is conceptualized as a "player" in the model (Liben-Nowell et al. 2012). The Shapley value φi for feature is computed by comparing the model output to the expected output over all possible feature coalitions. The SHAP method has also been shown to enhance model transparency and interpretability in remote sensing applications (Temenos

Table 2 Descriptive statistics of measured aboveground carbon stock (in t. ha^{-1})

Year	Mean	Standard deviation	Minimum	Maximum
2012	57.82	26.75	1.05	116.33
2022	87.45	32.09	17.43	177.84

et al. 2023). Huang et al. (2022) applied SHAP to a random forest model for grassland aboveground biomass prediction in China, identifying precipitation, temperature, and slope as key predictors. Due to its documented strengths in providing granular insights into feature importance and individual predictions, the SHAP method was chosen as the preferred tool for interpreting the model's performance (Movsessian et al. 2021).

3 Results

3.1 Descriptive Statistics of AGC Stock Measurements

The descriptive analysis of the data across (Table 2) various periods indicates that observed AGC exhibited a moderate range and variation throughout the study period. The distributions were found to be asymmetrical, with skewness values reaching 1.05. In terms of variance homogeneity, Levene's test produced p-values below 0.05 for all pairwise comparisons of periods, suggesting statistically significant differences between samples across different time frames.

3.2 Model Performance

Table 3 presents validation results for the artificial neural network (ANN) model developed to estimate aboveground carbon (AGC) stock changes over time. The ANN regression approach explained 78% of the observed variation in AGC within the validation dataset, assessed using a tenfold cross-validation. Accuracy was evaluated using root mean square error (RMSE), mean absolute error (MAE), and the Lin Concordance Correlation Coefficient (LCCC). Analysis of both spatial and temporal prediction accuracy revealed 2012 had the lowest precision (R^2 0.68, RMSE 27.55 t. ha^{-1}, MAE 21.70 t. ha^1 and LCCC 0.78) based on these metrics, with accuracy improving in 2022 (R^2 0.80, RMSE 25.68 t. ha^{-1}, MAE 22.16 t. ha^1 and LCCC 0.83). The model achieved an overall mean carbon stock of 55.94 t. ha^{-1}, slightly lower than the reference data's mean carbon stock of 57.75 t. ha^{-1}, indicating acceptable AGC predictions. Assessment by year shows the model captured interannual AGC fluctuations. The relationship between observed and predicted stocks (Fig. 2) strengthened over time as biomass accumulated within the reforestation area as supported by the

Table 3 Model validation results based on the tenfold cross validation for ANN regression model

Year	R^2	RMSE (t. ha^{-1})	MAE (t. ha^{-1})	LCCC	Mean carbon (t. ha^{-1})
2012	0.68	27.5546	21.70	0.78	55.94
2022	0.80	25.6899	22.1698	0.83	84.02

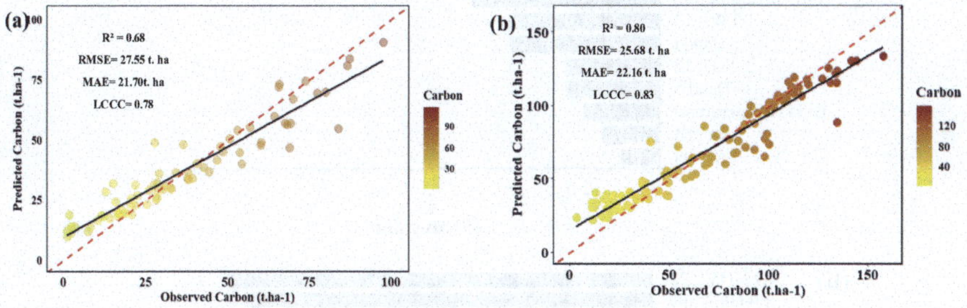

Fig. 2 Correlation between the observed and predicted aboveground carbon stocks **a** validation scatterplot for 2012, **b** validation plot for 2022

model's 2022 mean carbon stock of 87.45 t. ha^{-1} for the reference data and 84.02 t. ha^{-1} achieved by the model. By robustly validating both the spatial and temporal performance of the ANN approach, this study demonstrates reliable landscape-scale mapping of carbon stock accumulation over decades within the regenerating forest.

The SHAPley method identified 10 key predictors with high importance (SHAP value $\geq$ 3.0) for modeling aboveground carbon (AGC) in urban reforested landscapes. Precipitation was the most influential variable, followed by NDVI, band 8, land use type, temperature, EVI, RENDVI, elevation, topographic wetness index (TWI), and band 4 (Fig. 3). Notably, while elevation and TWI ranked near the bottom of the 10 most important variables at the plot level, together with other topographic factors like slope and aspect, they still contributed significantly to landscape-level AGC stock predictions. This suggests complex interactions between climate, vegetation and terrain in determining carbon storage across this heterogeneous urban ecosystems (Timilsina et al. 2014).

3.3 *Mapping of AGC Stocks*

The landscape-level maps of AGC stocks were generated using tenfold cross-validation for each time interval in Python 3.9.1. Spatially, the maps revealed an increase in mean AGC stocks from 55.94 t. ha^{-1} in 2012 to 84.02 t. ha^{-1} in 2022. Additionally, the total AGC stocks were calculated. Notably, low AGC values were concentrated in the central sections of the study area, corresponding to the current land use (2022 map) of a landfill. The lack of vegetation in the central section explains

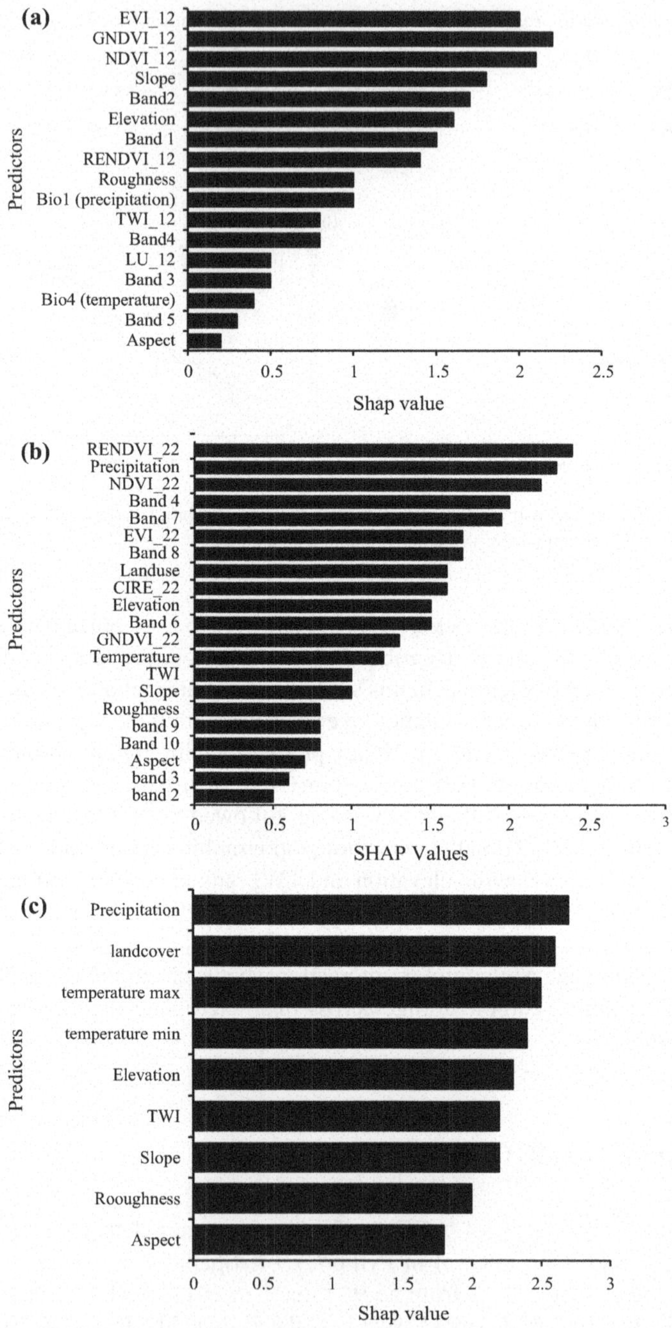

Fig. 3 Shapley Importance ranking of predictors used for the estimation and simulation of past (**a**), current (**b**), and projected (**c**) future aboveground carbon stock

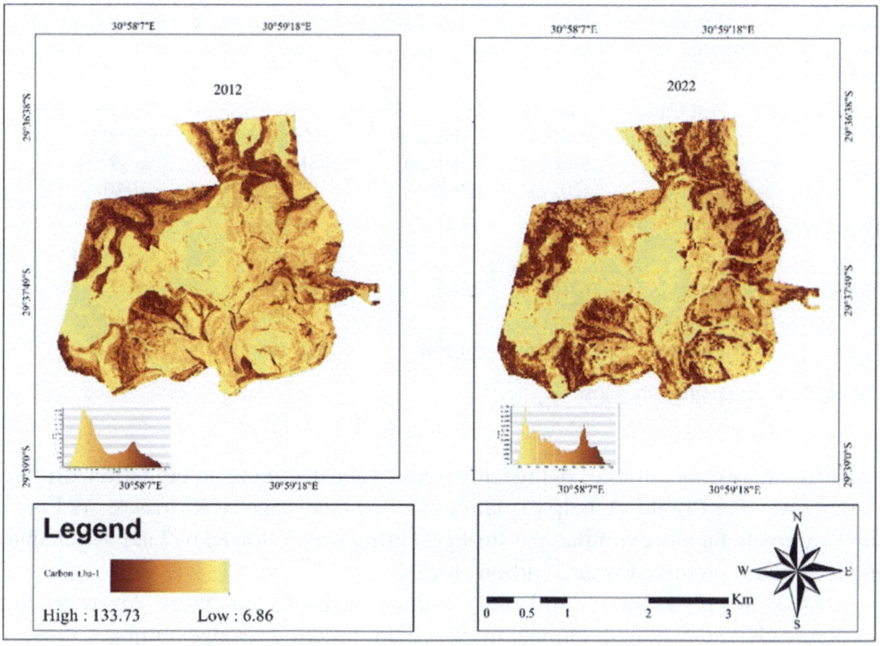

Fig. 4 Spatial distribution of past (2012) and current (2022) *AGC* stocks estimated using ANN

the low aboveground carbon reserves (Fig. 4). The maps show a gradual increase in AGC from 2012 to 2022 with an approximate total increase of 13 557t of carbon over the last decade. This is evident on the generated maps with a gradual increase in the northern, southern and western sections of the map. Aboveground carbon density also significantly rose over time.

The predicted future AGC stocks differed amongst the three shared socio-economic pathways including SSP 126, SSP245. And SSP585. The lowest emission shared socio-economic pathway (SSP126) produced the highest accuracy with the following metrics ($R^2 = 0.63$, RMSE = 27.44 t. ha^{-1}, MAE = 25.54 t. ha^{-1} and LCCC = 0.75). SSP 245 (medium to high emissions) followed with metrics of ($R^2 = 0.61$, RMSE = 28.44 t. ha^{-1}, MAE = 26.54t. ha^{-1} and LCCC = 0.74) and lastly SSP 585 (high emissions) with the following metrics ($R^2 = 0.60$, RMSE = 29.23 t. ha^{-1}, MAE = 27.54 t. ha^{-1} and LCCC = 0.70). Table 4 shows the decrease of AGC stocks across the different SSPs, with an approximate total aboveground carbon stocks estimated at 39206 t (SSP126), 26400 t (SSP245), and 25200 t (SSP585) respectively. A notable diminishing trend in AGC is evident across the different SSP scenarios, with the lowest accumulation of AGC under SSP585, which coincides with the high emissions scenario.

Figure 5 maps this diminishing spatial pattern in stocks over time as climate change intensifies, especially in northern and southern sections. The central landfill area persists as lowest AGC due to lack of vegetation limiting carbon accumulation.

Table 4 Changes between past (2012), current (2022) and future (2040) aboveground carbon stocks expressed in tonnes the numbers in parenthesis show gain or losses in carbon stocks across the two landcover classes

Land use type	Past Carbon Stock (t) (2012)	Current Carbon Stock (t) (2022)	Future SSP126 Carbon Stock (t) (2040)	Future SSP245 Carbon Stock (t) (2040)	Future SSP585 Carbon Stock (t) (2040)
Forest cover	9290	19,257	31 956	20,400 (− 900)	19,100 (− 1300)
Non-forest	3200	6700	7340	6000 (− 1340)	6100 (− 100)
Total	**12,400**	**25,957**	**39,206**	**26,400**	**25,200**

The negative sign represents a loss

CMIP6 climate projections of rising temperatures and declining precipitation, driving extreme weather (Table 4), help explain modelled landscape AGC losses. As Fig. 3 shows, climate factors exhibited the strongest influence, followed by land use change and terrain, on predicted future carbon stocks.

Most of the predictors significantly influenced projected future Aboveground Carbon (AGC) stocks, with climate models exhibiting the strongest impact, closely followed by land-use change and terrain variables (Fig. 3).

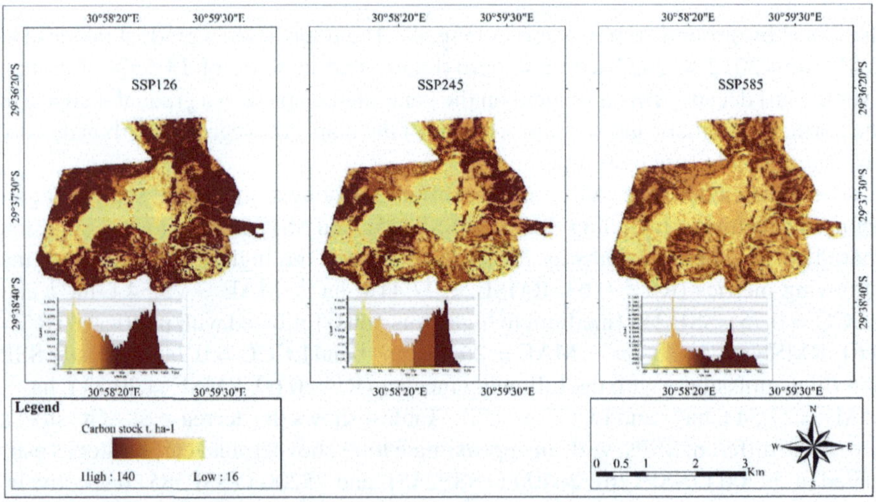

Fig. 5 Spatial distribution of projected (2040) AGC stocks estimated using ANN under the three share socio-economic pathways

3.4 Evaluation of Past, Current and Future AGC Stocks

Table 4 shows changes in forest cover and non-forest cover in hectares (ha) over the past, present, and future projected time periods. Currently, non-forest land cover dominates the landscape at 697.19 ha compared to 350.19 ha of forest cover. However, forests are projected to increase substantially from 350.19 ha in 2022 to 526.30 ha by 2040. This expanded forest cover will occur concurrently with declines in non-forest land area over time (Table 5).

Tables 4 and 6 present the estimated total aboveground carbon stocks (tonnes) and mean aboveground carbon stock density (tonnes per hectare) for the two studied land cover types under the three shared socio-economic pathways, projected over time. There is a notable increase in AGC stocks over time from 12,400 tonnes in 2012 to 39,206 tonnes after two decades under the low emission SSP 126 pathway. This aligns with substantial projected forest cover gains (Table 3.4). Non-forest cover type maintained lower density (4.06–14.08 t. ha^{-1}) and stocks throughout from 2012 to 2040. Future variations in AGC vary by emission scenario and period. For instance, Table 3.5 Forest carbon density varied widely, with the 2040 SSP126 projection doubling the 2012 baseline (60.22 vs. 35.66 t. ha^{-1}). Negligible changes occurred in stocks/density under high-emission SSP585, in contrast to SSP126.The mechanisms underlying these changes need further investigation in the light of internal processes, human activities and other external forces thereof.

In general, modeled carbon stocks demonstrated continued carbon uptake across the landscape over time, with sink strength magnified under sustainable pathways

Table 5 Projected Forest cover and non-forest cover change amongst the three time periods expressed in hectares (ha) and % as well as change (Δ) with negative sign showing a decline in land area

Land use type	2012 (ha)	2022 (ha)	2040 (ha)	Δ (ha)	2012 (%)	2022(%)	2040 (%)	Δ (%)
Forest cover	260.45	350.19	526.30	266.85	24	33.43	50.27	32.27
Non-forest	787.93	697.19 (– 90.74)	521.08 (– 176.11)	– 266.85	76	66.67	49.73	– 32.27

Table 6 Mean aboveground carbon stock density t. ha^{-1} for past, current and future predictions, the value in parenthesis shows differences in mean stock density

Land use type	Past (2012)	Current (2022)	Future (SSP 126)	Future (SSP 245)	Future (SSP 585)
Forest cover	35.66 t. ha^{-1}	54.99	60.72	56.34	55..66
Non-forest cover	4.06 t. ha^{-1}	9.61	14.08	11.51	11.23

allowing forest recovery. Tables 4 and 6 quantify these gradient effects on density and total stocks under contrasting scenarios.

4 Discussion

4.1 Model Performance

The results demonstrate future climate and land use changes substantially impact urban reforestation AGC stocks, with differing trajectories across emission scenarios. Integrating remote sensing and machine learning gives valuable insights to inform land management and climate mitigation strategies, highlighting the carbon sequestration potential of urban reforestation. The results of the ANN model performance and visual fit indicate that the model performed well in estimating the dynamic changes in AGC in 2022 compared to 2012 and 2040 (Table 2 and Fig. 2) with high metrics $R^2 = 0.80$ on the validation dataset. Our estimates showed less bias than previous landscape models Verly et al. (2023), and an R^2 comparable to recent similar studies Datta et al. (2023) The strength of ANN lies in handling complex relationships between numerous environmental factors, often outperforming traditional methods. By combining historical data, remote sensing, and future projections, ANNs provide valuable insights into how urban development and climate change might impact carbon storage in urban forests. This is especially relevant in the context of eThekwini Municipality, where rapid urbanization and climate variability present ongoing challenges (Gharaibeh et al. 2020, Douwes et al. 2015). The performance of ANNs in this study aligns with previous research (Dou and Yang 2018; Zhang et al. 2022) that has underscored the efficacy of machine learning approaches in environmental forecasting (Gharaibeh et al. 2020).

The future AGC simulations, however, displayed a reduction in the accuracies (from $R^2 = 0.60$–0.63, across all three shared socio-economic pathways (SSP126, SSP245, and SSP585) developed from the average of the three CMIP6 models including—ACCESS-ESM1-5, MIROC6, and EC-EARTH2(van Noije et al. 2021; Hajima et al. 2020). Furthermore, the absence of Sentinel-2 data and derived indices hindered accurate modeling of future aboveground carbon stocks. Literature highlights the importance of spectral data for AGC mapping, utilizing specific spectral bands and combinations (Mngadi et al. 2022a; Odebiri et al. 2023; Gara et al. 2016). The NIR band reveals canopy structure and density, the red band assesses photosynthetic activity, and the SWIR band enables vegetation differentiation. Indices such as NDVI and EVI integrate NIR and red bands for robust vegetation density measures (Baldocchi et al. 2020). References Dube and Mutanga (2016), Gara et al. (2016). The red-edge band, sensitive to chlorophyll, provides insights into plant stress and biomass (Gara et al. 2016). In this study, Sentinel-2 and Rapid-eye spectral metrics, along with derived vegetation indices, including red band 4, red-edge bands 7 and 6, near-infrared band 8, NDVI, RENDVI, and GNDVI, were identified as critical variables for AGC estimation in both 2012 and 2022 (Fig. 3). However, the reduced accuracy of the future projections is understandable given the data limitations, but these initial simulations still offer valuable insights into potential reforestation contributions and aboveground carbon accumulation moving forward.

4.2 *Mapping of AGC Stocks*

Key findings show pronounced northern, eastern and southern AGC density increases over the past decade, with various patches seeing gains exceeding 60 t/ha—contrasting with more modest accruals averaging 20–30 t/ha in central zones. These accumulation patterns likely stem from recent concentrated reforestation efforts alongside supportive localized climate and soil conditions enabling higher survival rates and carbon sequestration capacities in thriving young forests (Matiza et al. 2024).

The AGC projections under SSP 245 and 585 pathways reveal minimal variances, with density estimates ranging from 55.66–56.34 t/ha and total stocks of 25,200–26,400 tonnes between scenarios. This is because shared socioeconomic pathways may incorporate comparable narratives around aspects like population growth, economic development, technological innovations, and environmental policies between scenarios. This could lead to limited divergence in model outputs. These small AGC differences align with other studies applying similar socio-economic pathways (Gao et al. 2023; Ter-Mikaelian et al. 2021; Wang et al. 2022c). However, intermittent rainfall and temperature extremes from intensifying climate change could transform recovering forests into neutral or negative carbon sinks over time. Adopting a low-emissions pathway appears to support balanced, eco-friendly urban ecosystems where forests significantly thrive. For example, the SSP 126 scenario showed a 102.71 million Mg carbon stock increase by 2050, reflecting substantial sequestration potential given proper conditions (Gao et al. 2023). Overall, quantifying spatial variability in historic AGC densities and projecting future stocks under varying pathways provides insights into targeted reforestation and climate mitigation needs.

4.3 *Past, Current and Future Changes in AGC Stocks*

Future projections reveal a divergence in reforestation success and carbon storage potential under varying emissions scenarios. The SSP126 low-emission pathway enables forest cover to expand to 50% by 2040 as trees thrive under reduced climate stress (Velasco et al. 2016). Improved growth conditions reinforce carbon uptake, with reforested areas continuing to effectively sequester $CO2$ per SSP126 goals. However, under SSP245 and SSP585, forest cover declines to only 35% by 2040, with carbon accumulation plateauing or diminishing over time. Rising emissions severely threaten reforestation efforts and carbon capture viability (Yan et al. 2023). Contrasting outcomes highlight the detrimental impacts of unrestrained emissions compared to an SSP126 trajectory where changes expanding reforested areas and biodiversity conservation unlock vegetation carbon storage and resilience (Lewis et al. 2019). Aligned climate and land use policies therefore hold the key to sustainable carbon management via urban reforestation.

4.4 Ecological Services and Implications

The carbon sequestration potential revealed in reforested urban areas underscores the need to integrate such efforts into city planning and development strategies. Prioritizing reforestation initiatives can maximize carbon uptake while enhancing urban ecosystem resilience to intensifying climate change. Moreover, the study highlights the importance of adaptive forest management attuned to environmental influences on carbon stock dynamics. This includes emphasizing sustainable land use and biodiversity conservation to support reforestation success. From a policy perspective, the findings provide actionable insights by quantifying the role urban forests can play in meeting global carbon goals, such as those outlined in the Paris Agreement. Urban reforestation can serve as a nature-based solution to offset emissions and contribute to national and local climate targets. However, realizing this potential requires integrating these insights across urban governance and planning frameworks. This includes developing policies that incentivize green infrastructure investments, fostering cross-sectoral collaboration, and engaging local communities in reforestation efforts. Such measures are key to nurturing more sustainable and climate-resilient cities. However, uncertainties remain in the initial carbon mapping and future projections which rely on modeled nonlinear responses. Abnormal precipitation, increased wildfires, and land use changes could all impact actual stocks differently than predicted. Ongoing monitoring and model refinement is therefore needed to track the complex dynamics of reforested carbon in the urban environment.

4.5 Conclusion

The study concludes that:

- Urban reforestation's contribution to increasing AGC stocks diminishes in the future, especially under high emission scenarios.
- Future climate change and land-use changes significantly impact urban reforested AGC stocks.
- Remote sensing and machine learning are complementary tools that can improve forest monitoring and inform policy and planning.

These projections provide valuable data to inform land use management and climate mitigation strategies by stakeholders and city authorities. However, fully unlocking the carbon storage capacity of cities requires integrated policy making and land management strategies. Overall, the analysis not only improves understanding of reforestation in urban zones but also demonstrates the need for cross-sectoral coordination on climate action plans and adaptive ecosystem governance. Targeted monitoring, sustainable development policies, and public–private collaborations will be essential to overcoming uncertainties and steering cities toward more sustainable, climate-resilient trajectories.

Ultimately, this work underscores the importance of leveraging technological advancements, such as remote sensing and machine learning, to support evidence-based decision-making in urban environmental management. By fostering interdisciplinary collaboration and prioritizing long-term sustainability, cities can harness the full potential of urban reforestation as a nature-based solution to climate change.

Acknowledgements We would like to thank the editors for the invitation to contribute to this book volume.

References

Abiodun OI, Kiru MU, Jantan AB, Omolara AE, Dada KV, Umar A, Linus OU, Arshad H, Kazaure AA, Gana UM (2019) Comprehensive review of artificial neural network applications to pattern recognition. IEEE Access 7:158820–158846

Anderson, C., Naughton, D., Brunn, A. & Thiele, M. Radiometric correction of RapidEye imagery using the on-orbit side-slither method. Remote Sensing, 2011

ANJALI, K., KHUMAN, Y. & SOKHI, J. 2020. A Review of the interrelations of terrestrial carbon sequestration and urban forests: Terrestrial carbon sequestration and urban forests. *AIMS Environmental Science,* 7.

BALDOCCHI, D. D., RYU, Y., DECHANT, B., EICHELMANN, E., HEMES, K. S., MA, S., SANCHEZ, C. R., SHORTT, R., SZUTU, D. J., VALACH, A. C., VERFAILLIE, J. G., BADGLEY, G., ZENG, Y. & BERRY, J. A. 2020. Outgoing Near-Infrared Radiation From Vegetation Scales With Canopy Photosynthesis Across a Spectrum of Function, Structure, Physiological Capacity, and Weather. *Journal of Geophysical Research: Biogeosciences,* 125.

Birdsey R, Pan Y (2015) Trends in management of the world's forests and impacts on carbon stocks. For Ecol Manage 355:83–90

Cai B, Zhang L (2014) Urban CO_2 emissions in China: spatial boundary and performance comparison. Energy Policy 66:557–567

CHAVE, J., RéJOU-MéCHAIN, M., BúRQUEZ, A., CHIDUMAYO, E., COLGAN, M. S., DELITTI, W. B., DUQUE, A., EID, T., FEARNSIDE, P. M. & GOODMAN, R. C. 2014. Improved allometric models to estimate the aboveground biomass of tropical trees. *Global change biology,* 20**,** 3177–3190

CLEVERS, J. & GITELSON, A. Using the red-edge bands on Sentinel-2 for retrieving canopy chlorophyll and nitrogen content. 2012.

Cowan DR, Cooper GRJ (2004) The Shuttle Radar Topography Mission–a new source of near-global digital elevation data. ASEG Extended Abstracts 2004:1–4

Djomo AN, Picard N, Fayolle A, Henry M, Ngomanda A, Ploton P, McLellan J, Saborowski J, Adamou I, Lejeune P (2016) Tree allometry for estimation of carbon stocks in African tropical forests. Forestry 89:446–455

Dou X, Yang Y (2018) Estimating forest carbon fluxes using four different data-driven techniques based on long-term eddy covariance measurements: Model comparison and evaluation. The Science of the Total Environment 627:78–94

DOUWES, E., ROUGET, M., DIEDERICHS, N., O'DONOGHUE, S., ROY, K. & ROBERTS, D. Buffelsdraai landfill site community reforestation project. XIV World Forestry Congress, 2015.

Drusch M, del Bello U, Carlier S, Colin O, Fernandez V, Gascon F, Hoersch B, Isola C, Laberinti P, Martimort P (2012) Sentinel-2: ESA's optical high-resolution mission for GMES operational services. Remote Sens Environ 120:25–36

Dube T, Mutanga O (2016) The impact of integrating WorldView-2 sensor and environmental variables in estimating plantation forest species aboveground biomass and carbon stocks in uMgeni Catchment, South Africa. ISPRS J Photogramm Remote Sens 119:415–425

Eckert S (2012) Improved forest biomass and carbon estimations using texture measures from WorldView-2 satellite data. Remote Sens 4:810–829

Evangelides C, Nobajas A (2020) Red-Edge Normalised Difference Vegetation Index (NDVI705) from Sentinel-2 imagery to assess post-fire regeneration. Remote Sensing Applications: Society and Environment 17:100283

Fassnacht KS, Gower ST, Mackenzie MD, Nordheim EV, Lillesand TM (1997) Estimating the leaf area index of North Central Wisconsin forests using the landsat thematic mapper. Remote Sens Environ 61:229–245

Feng Y, Zeng Z, Searchinger TD, Ziegler AD, Wu J, Wang D, He X, Elsen PR, Ciais P, Xu R (2022) Doubling of annual forest carbon loss over the tropics during the early twenty-first century. Nature Sustainability 5:444–451

Fick SE, Hijmans RJ (2017) WorldClim 2: new 1-km spatial resolution climate surfaces for global land areas. Int J Climatol 37:4302–4315

Gao F, Xin X, Song J, Li X, Zhang L, Zhang Y, Liu J (2023) Simulation of LUCC Dynamics and Estimation of Carbon Stock under Different SSP-RCP Scenarios in Heilongjiang Province. Land 12:1665

Gao Y, Fu JS, Drake JB, Lamarque JF, Liu Y (2013) The impact of emission and climate change on ozone in the United States under representative concentration pathways (RCPs). Atmos Chem Phys 13(18):9607–9621

Gara TW, Murwira A, Ndaimani H (2016) Predicting forest carbon stocks from high resolution satellite data in dry forests of Zimbabwe: exploring the effect of the red-edge band in forest carbon stocks estimation. Geocarto Int 31:176–192

GHARAIBEH, A. A., SHAAMALA, A., OBEIDAT, R. & AL-KOFAHI, S. D. 2020. Improving land-use change modeling by integrating ANN with Cellular Automata-Markov Chain model. *Heliyon,* 6.

Ghorbanian A, Ahmadi SA, Amani M, Mohammadzadeh A, Jamali S (2022) Application of Artificial Neural Networks for Mangrove Mapping Using Multi-Temporal and Multi-Source Remote Sensing Imagery. Water 14:244

HAJIMA, T., WATANABE, M., YAMAMOTO, A., TATEBE, H., NOGUCHI, M. A., ABE, M., OHGAITO, R., ITO, A., YAMAZAKI, D., OKAJIMA, H., ITO, A., TAKATA, K., OGOCHI, K., WATANABE, S. & KAWAMIYA, M. 2020. Development of the MIROC-ES2L Earth system model and the evaluation of biogeochemical processes and feedbacks. *Geoscientific Model Development.*

He C, Liu Z, Tian J, Ma Q (2014) Urban expansion dynamics and natural habitat loss in China: A multiscale landscape perspective. Glob Change Biol 20:2886–2902

HEISKANEN, J., PELLIKKA, P., BETEMARIAM, E. A. & PACKALEN, P. 2013. Field measurement guidelines for aboveground biomass and fuel wood stocks. *Building Biocarbon and Rural Development in West Africa (BIODEV); World Agroforestry Centre: Nairobi, Kenya.*

Hong C, Burney JA, Pongratz J, Nabel JE, Mueller ND, Jackson RB, Davis SJ (2021) Global and regional drivers of land-use emissions in 1961–2017. Nature 589:554–561

Huang W, Li W, Xu J, Ma X, Li C, Liu C (2022) Hyperspectral Monitoring Driven by Machine Learning Methods for Grassland Above-Ground Biomass. Remote Sensing 14:2086

IEA 2008. World energy outlook. Washington dc, usa IEA.

IPCC 2006. 2006 IPCC guidelines for national greenhouse gas inventories. *Institute for Global Environmental Strategies, Hayama, Kanagawa, Japan.*

JACOBSEN, K. Analysis of SRTM elevation models. EARSeL 3D-Remote Sensing Workshop, Porto, 2005.

JIANG, Z., HUETE, A. R., KIM, Y. & DIDAN, K. 2-band enhanced vegetation index without a blue band and its application to AVHRR data. Remote Sensing and Modeling of Ecosystems for Sustainability IV, 2007. SPIE, 45–53.

JUNG-ROTHENHäUSLER, F., WEICHELT, H. & PACH, M. Rapideye-A novel approach to space borne geo-information solutions. ISPRS Hannover Workshop, 2007. 4–7.

KANEVSKI, M. F., POZDNUKHOV, A. & TIMONIN, V. Machine Learning Algorithms for GeoSpatial Data. Applications and Software Tools. 2008.

KOLDASBAYEVA, D., TREGUBOVA, P., GASANOV, M., ZAYTSEV, A., PETROVSKAIA, A. & BURNAEV, E. V. 2023. Challenges in data-based geospatial modeling for environmental research and practice. *ArXiv,* abs/2311.11057.

Kowe P, Mutanga O, Dube T (2021) Advancements in the remote sensing of landscape pattern of urban green spaces and vegetation fragmentation. Int J Remote Sens 42:3797–3832

Lang M, McCarty G, Oesterling R, Yeo I-Y (2013) Topographic metrics for improved mapping of forested wetlands. Wetlands 33:141–155

LANGUILLE, F., DECHOZ, C., GAUDEL, A., GRESLOU, D., DE LUSSY, F., TRéMAS, T. & POULAIN, V. Sentinel-2 geometric image quality commissioning: first results. SPIE Remote Sensing, 2015.

LAURANCE, W. F. 2010. Habitat destruction: death by a thousand cuts. *Conservation biology for all,* 1:73–88

Lewis T, Verstraten L, Hogg B, Wehr BJ, Swift S, Tindale N, Menzies NW, Dalal RC, Bryant P, Francis B (2019) Reforestation of agricultural land in the tropics: The relative contribution of soil, living biomass and debris pools to carbon sequestration. Sci Total Environ 649:1502–1513

LIBEN-NOWELL, D., SHARP, A., WEXLER, T. & WOODS, K. M. Computing Shapley Value in Supermodular Coalitional Games. International Computing and Combinatorics Conference, 2012.

LóPEZ-SERRANO, P. M., CáRDENAS DOMíNGUEZ, J. L., CORRAL-RIVAS, J. J., JIMéNEZ, E., LóPEZ-SáNCHEZ, C. A. & VEGA-NIEVA, D. J. 2020. Modeling of aboveground biomass with Landsat 8 OLI and machine learning in temperate forests. *Forests,* 11:11

MACDICKEN, K. G. 1997. A guide to monitoring carbon storage in forestry and agroforestry projects.

Matiza C, Mutanga O, Odindi J, Mngadi M (2024) The utility of Planetscope spectral data in quantifying above-ground carbon stock in an urban reforested landscape. Eco Inform 80:102472

Matiza C, Mutanga O, Peerbhay K, Odindi J, Lottering R (2023) A systematic review of remote sensing and machine learning approaches for accurate carbon storage estimation in natural forests. Southern Forests 85:123–141

McBride LA, Hope AP, Canty TP, Bennett BF, Tribett WR, Salawitch RJ (2021) Comparison of CMIP6 historical climate simulations and future projected warming to an empirical model of global climate. Earth System Dynamics 12:545–579

Mitchell MGE, Johansen KL, Maron M, McAlpine CA, Wu D, Rhodes JR (2018) Identification of fine scale and landscape scale drivers of urban aboveground carbon stocks using high-resolution modeling and mapping. The Science of the Total Environment 622–623:57–70

Mngadi M, Odindi J, Mutanga O (2021) The utility of Sentinel-2 spectral data in quantifying above-ground carbon stock in an urban reforested landscape. Remote Sensing 13:4281

MNGADI, M., ODINDI, J. & MUTANGA, O. 2022a. Quantifying Carbon Stock Variability of Species Within a Reforested Urban Landscape Using Texture Measures Derived from Remotely Sensed Imagery. *Advances in Remote Sensing for Forest Monitoring*, 150–166.

Mngadi M, Odindi J, Mutanga O, Sibanda M (2022b) Estimating aboveground net primary productivity of reforested trees in an urban landscape using biophysical variables and remotely sensed data. Sci Total Environ 802:149958

Morley SK, Brito T, Welling DT (2018) Measures of Model Performance Based On the Log Accuracy Ratio. Space Weather 16:69–88

MOVSESSIAN, A., CAVA, D. G. & TCHERNIAK, D. 2021. Interpretable machine learning in damage detection using Shapley Additive Explanations. *ASCE-ASME J Risk and Uncert in Engrg Sys Part B Mech Engrg.*

MUGWEDI, L., ROUGET, M., EGOH, B. N., SERSHEN, RAMDHANI, S., SLOTOW, R. & RENTERIA, J. L. 2017. An Assessment of a Community-Based, Forest Restoration Programme in Durban (eThekwini), South Africa. *Forests,* 8:255

Nave LE, Walters BF, Hofmeister K, Perry CH, Mishra U, Domke GM, Swanston C (2019) The role of reforestation in carbon sequestration. New for 50:115–137

NRDC. 2018. *Fossil Fuels: The Dirty Facts* [Online]. online: NRDC. Available: https://www.nrdc.org/stories/fossil-fuels-dirty-facts#sec-whatis [Accessed 29 december 2023].

Odebiri O, Mutanga O, Odindi J (2022) Deep learning-based national scale soil organic carbon mapping with Sentinel-3 data. Geoderma 411:115695

Odebiri O, Mutanga O, Odindi J, Naicker R, Slotow R, Mngadi M (2023) Evaluation of projected soil organic carbon stocks under future climate and land cover changes in South Africa using a deep learning approach. J Environ Manage 330:117127

Ogunbode TO, Asifat JT (2021) Sustainability and challenges of climate change mitigation through urban reforestation-a review. Journal of Forest and Environmental Science 37:1–13

PANNAKKONG, W., THIWA-ANONT, K., SINGTHONG, K., PARTHANADEE, P. & BUDDHAKULSOMSIRI, J. 2022. Hyperparameter Tuning of Machine Learning Algorithms Using Response Surface Methodology: A Case Study of ANN, SVM, and DBN. *Mathematical Problems in Engineering.*

PETTORELLI, N. 2013. *The normalized difference vegetation index*, Oxford University Press.

Raich JW, Russell AE, Kitayama K, Parton WJ, Vitousek PM (2006) Temperature influences carbon accumulation in moist tropical forests. Ecology 87:76–87

RODRíGUEZ-VEIGA, P., WHEELER, J., LOUIS, V., TANSEY, K. & BALZTER, H. 2017. Quantifying Forest Biomass Carbon Stocks From Space. *Current Forestry Reports,* 3:1–18

SANLC 2020. South African National Landcover 2020 Accuracy assessment report. Pretoria, South Africa Department of Environment, forestry and fisheries

SETO, K. C., GüNERALP, B. & HUTYRA, L. R. 2012. Global forecasts of urban expansion to 2030 and direct impacts on biodiversity and carbon pools. *Proceedings of the National Academy of Sciences,* 109:16083–16088

Solomon N, Birhane E, Tadesse T, Treydte AC, Meles K (2017) Carbon stocks and sequestration potential of dry forests under community management in Tigray, Ethiopia. Ecol Process 6:1–11

STOLL, E., KONSTANSKI, H., ANDERSON, C., DOUGLASS, K. & OXFORT, M. The RapidEye constellation and its data products. 2012 IEEE Aerospace Conference, 2012. IEEE, 1–9.

Suardana AMAP, Anggraini N, Nandika MR, Aziz K, As-Syakur AR, Ulfa A, Wijaya AD, Prasetio W, Winarso G, Dewanti R (2023) Estimation and Mapping Above-Ground Mangrove Carbon Stock Using Sentinel-2 Data Derived Vegetation Indices in Benoa Bay of Bali Province, Indonesia. Forest and Society 7:116–134

Tang XL, Perez-Cruzado C, Vor T, Fehrmann L, Alvarez-Gonzalez JG, Kleinn C (2016) Development of stand density management diagrams for Chinese fir plantations. Forestry 89:36–45

Temenos A, Temenos N, Kaselimi M, Doulamis A, Doulamis N (2023) Interpretable deep learning framework for land use and land cover classification in remote sensing using SHAP. IEEE Geosci Remote Sens Lett 20:1–5

Teo HC, Zeng Y, Sarira TV, Fung TK, Zheng Q, Song XP, Chong KY, Koh LP (2021) Global urban reforestation can be an important natural climate solution. Environ Res Lett 16:034059

Ter-Mikaelian MT, Gonsamo A, Chen JM, Mo G, Chen J (2021) Historical and future carbon stocks in forests of northern Ontario, Canada. Carbon Balance Manage 16:1–18

Timilsina N, Escobedo FJ, Staudhammer CL, Brandeis TJ (2014) Analyzing the causal factors of carbon stores in a subtropical urban forest. Ecol Complex 20:23–32

UNITED NATIONS. 2023. *Causes and Effects of Climate Change.* [Online]. online United Nations Available: https://www.un.org/en/climatechange/science/causes-effects-climate-change [Accessed 29 December 2023].

VAN NOIJE, T., BERGMAN, T., LE SAGER, P., O'DONNELL, D., MAKKONEN, R., GONçALVES-AGEITOS, M., DöSCHER, R., FLADRICH, U., VON HARDENBERG, J., KESKINEN, J.-P., KORHONEN, H., LAAKSO, A., MYRIOKEFALITAKIS, S., OLLINAHO, P., GARCíA-PANDO, C. P., REERINK, T. J., SCHRöDNER, R., WYSER, K. & YANG, S. 2021. EC-Earth3-AerChem: a global climate model with interactive aerosols and atmospheric chemistry participating in CMIP6. *Geoscientific Model Development.*

Velasco E, Roth M, Norford L, Molina LT (2016) Does urban vegetation enhance carbon sequestration? Landsc Urban Plan 148:99–107

Venter ZS, Barton DN, Chakraborty T, Simensen T, Singh G (2022) Global 10 m Land Use Land Cover Datasets: A Comparison of Dynamic World, World Cover and Esri Land Cover. Remote Sensing 14:4101

WAN, S., HUI, D., WALLACE, L. & LUO, Y. 2005. Direct and indirect effects of experimental warming on ecosystem carbon processes in a tallgrass prairie. *Global Biogeochemical Cycles,* 19.

Wang B, Gray JM, Waters CM, Anwar MR, Orgill SE, Cowie AL, Feng P, LI LIU, D. (2022a) Modelling and mapping soil organic carbon stocks under future climate change in south-eastern Australia. Geoderma 405:115442

Wang Y, Qin R, Cheng H, Liang T, Zhang K, Chai N, Gao J, Feng Q, Hou M, Liu J, Liu C, Zhang W, Fang Y, Huang J, Zhang F (2022b) Can Machine Learning Algorithms Successfully Predict Grassland Aboveground Biomass? Remote Sens 14:3843

Wang Z, Li X, Mao Y, Li L, Wang X, Lin Q (2022c) Dynamic simulation of land use change and assessment of carbon storage based on climate change scenarios at the city level: A case study of Bortala, China. Ecol Ind 134:108499

XIAO, J., CHEVALLIER, F. D. R., GOMEZ, C., GUANTER, L., HICKE, J. A., HUETE, A. R., ICHII, K., NI, W., PANG, Y., RAHMAN, A. F., SUN, G., YUAN, W., ZHANG, L. & ZHANG, X. 2019. Remote sensing of the terrestrial carbon cycle: A review of advances over 50 years. *Remote Sensing of Environment.*

Yan L, Meng S, Yang F, Dai X, Wang H (2023) Changes in Forest Vegetation Carbon Storage and Its Driving Forces in Subtropical Red Soil Hilly Region over the Past 34 Years: A Case Study of Taihe County, China. Forests 14:602

Yue S, Ji G, Chen W, Huang J, Guo Y, Cheng M (2023) Spatial and Temporal Variability Characteristics of Future Carbon Stocks in Anhui Province under Different SSP Scenarios Based on PLUS and InVEST Models. Land 12:1668

Zavaleta ES, Thomas BD, Chiariello NR, Asner GP, Shaw MR, Field CB (2003) Plants reverse warming effect on ecosystem water balance. Proc Natl Acad Sci 100:9892–9893

Zhang F, Tian X, Zhang H, Jiang M (2022) Estimation of Aboveground Carbon Density of Forests Using Deep Learning and Multisource Remote Sensing. Remote Sens 14:3022

Wetlands and Aquatic Ecosystem Services

Mangrove Monitoring Using Geospatial Technologies Incorporating Field Inventory: A Review

Parag Jyoti Kashyap, Kuldeep Bora, Sanchali Das Podder, Andrews Amakawa, Ankita Paul, and Chinmoy Raj Saikia

Abstract Mangrove ecosystems are vital for coastal protection, biodiversity, and carbon sequestration, yet face increasing threats from climate change and human activities. Effective monitoring, integrating diverse data sources, is crucial for their conservation and management. This study examines the integration of geospatial technologies (Remote Sensing, Geographic Information Systems, and Global Navigation Satellite Systems) with field inventory methods for comprehensive mangrove monitoring. Satellite-based Remote Sensing provides broad-scale data for tracking changes in mangrove cover and health, while Light Detection and Ranging enables precise measurement of canopy height and forest structure. Unmanned Aerial Vehicles deliver high-resolution imagery for detailed assessments. Field inventories, incorporating ground-truthing, vegetation surveys, and biomass measurements, are essential for validating remote sensing models and ensuring data accuracy. This review highlights the synergistic value of combining geospatial technologies and field data for effective mangrove monitoring, addressing key challenges and knowledge gaps to support improved coastal zone management and promote sustainable conservation efforts.

Keywords Mangrove · Geospatial technology · Ecosystem · Biodiversity · Management · Synergy

P. J. Kashyap (✉) · K. Bora · S. Das Podder
Department of Geography, Assam University, Diphu Campus, Diphu 782462, India
e-mail: kashyapparagjyoti@gmail.com

A. Amakawa
Department of Geography and Resources Development, University of Ghana, Accra, Ghana

A. Paul
Department of Environmental Science, Tezpur University, Tezpur, India

C. R. Saikia
Department of Geography, Bapujee College, Sarukshetri, Assam, India

O. Mutanga et al. (eds.), *Revealing Ecosystem Services Through Geospatial Technologies*, Springer Remote Sensing/Photogrammetry,
https://doi.org/10.1007/978-3-031-98048-0_7

1 Introduction

Mangrove forests, the guardians of our coastlines, are disappearing at an alarming rate, with an estimated 35% already lost globally (Cebrian 2002). These vital ecosystems face unprecedented threats from rising sea levels, coastal development, and unsustainable practices, jeopardizing their ability to protect shorelines, harbor biodiversity, and sequester carbon. To combat this crisis, effective and accurate monitoring is paramount, demanding innovative approaches that can provide timely and spatially explicit data. Geospatial technologies offer a powerful solution, enabling researchers and policymakers to assess mangrove health, distribution, and changes over time with unprecedented efficiency.

Mangrove forests are especially suited ecosystems that are found in tropical and subtropical locations all over the world. Numerous plant and animal species, many of which are highly specialized and dependent on the unique mangrove environment, can be found in these coastal ecosystems. Mangroves are acknowledged as one of the most abundant repositories of biological and genetic diversity on Earth, since about 90% of marine creatures pass through this ecosystem during some stage of their life cycle (Sandilyan and Kathiresan 2012). Many different types of wildlife, such as crabs, amphibious fish, and birds, can find a complicated and thriving home in the intricate network of mangrove roots, branches, and canopy. Mangrove ecosystems are dynamic due to the special adaptations of certain species, like vivipary (live birth), salt tolerance in trees, air breathing in crabs, and amphibious activities in fish (Guebas et al. 2020). Mangrove ecosystems are highly resilient and remarkably adapted to their surroundings, which make them vital resources in defending coastal areas against the catastrophic effects of natural disasters (Prakash and Chauhan 2017). In light of rising sea levels and stronger storm events linked to climate change, their extensive root systems protect shorelines, lessen wave energy, and minimize coastal erosion—all of which are becoming more and more important (Setturu 2014). Additionally, by functioning as a natural barrier that absorbs excess water and keeps inland areas safe from flooding, these ecosystems also act as a buffer against storm surges, emphasize their crucial role in disaster risk reduction and climate adaption measures in coastal towns (Giri 2016). Mangroves are also essential for removing particles from the water that are high in carbon. By encouraging sediment accretion and carbon burial, this mechanism results in the storage of carbon in sediments. Mangrove forests therefore function as substantial carbon sinks, assisting in the mitigation of climate change. Climate change is exacerbated by the loss of mangroves, which releases stored carbon dioxide emissions back into the atmosphere (Ginalyn, and Ocay, 2024). Mangroves are among the most productive ecosystems; their above- and below-ground sections store more carbon than those of terrestrial forests. Sequestering carbon is essential for reducing the effects of climate change (Kumari 2021).

Beyond their ecological value, mangroves provide essential ecosystem services, including storm surge protection, fisheries support, and water filtration (Giri 2016; Rizal 2018). Their complex root systems dissipate wave energy, protecting coastal communities from erosion and flooding, a service that is increasingly critical in the face of rising sea levels and intensified storms (Setturu, 2014). Mangrove forests also serve as vital nurseries for a wide array of marine species, supporting local fisheries and contributing to food security (Sandilyan & Kathiresan, 2012). Moreover, they act as natural filters, trapping sediments and pollutants, thereby maintaining water quality and protecting sensitive coastal habitats (Tam & Wong, 1999).

However, effective mangrove monitoring faces significant challenges. Their remote location, often in difficult-to-access coastal areas, makes traditional field surveys time-consuming and costly (Kumar et al. 2013). The complex structure of mangrove forests, with their dense canopies and intricate root systems, complicates accurate assessment of forest health and species composition (Wang and Gu 2021). Furthermore, distinguishing between different mangrove species based on spectral signatures alone can be challenging, particularly in areas with high species diversity (Faridah-Hanum et al. 2021).

Geospatial technologies offer powerful solutions to these challenges, providing efficient and cost-effective tools for comprehensive mangrove monitoring. Remote sensing, Geographic Information Systems (GIS), and Global Navigation Satellite Systems (GNSS) enable researchers to overcome the limitations of traditional methods, offering broad-scale data acquisition, detailed spatial analysis, and accurate mapping capabilities (Hussin et al. 2007; Wang et al. 2004). Integrating these technologies with field inventory methods enhances assessment accuracy and supports informed coastal management policies. As shown in Fig. 1, mangroves play a crucial role in coastal protection, buffering the coastline against erosion and storm surges, especially in vulnerable regions like the Sundarbans in India.

This review examines the application of various geospatial technologies—Satellite-based Remote Sensing (RS), Unmanned Aerial Vehicles (UAVs), LiDAR, GIS, and GNSS—to mangrove ecosystem monitoring, highlighting their specific uses, strengths, and limitations. Also to identify and limit the gaps in the current knowledge repository.

The Main Objectives of This Review Are to:

- Assess the effectiveness of integrating geospatial technologies (Remote Sensing, Geographic Information System (GIS) and Global Navigation Satellite System (GNSS)) with field inventory methods for mangrove monitoring.
- Examine the specific applications of Satellite Imagery, Aerial Photography, and Drones in mangrove monitoring through Remote Sensing.
- Describe the role of GIS in Data integration and Spatial Analysis for mangrove ecosystems.
- Evaluate Data collection techniques used in Field Inventories.
- Identify and limit the gaps in the current knowledge repository.

Fig. 1 Mangrove in coastal protection in Sundarbans (India). *Photo credit* Dr. Raju Gogoi

2 Remote Sensing Based Studies of Mangrove Ecosystems

2.1 Satellite-Based Remote Sensing (RS): Balancing Resolution and Coverage

Satellite-based Remote Sensing (RS) is essential for monitoring mangrove ecosystems, providing high-resolution, temporal data. Landsat satellites, with their long-term data archive, are widely used for tracking changes in mangrove cover and health over time. Studies have utilized these satellites to monitor mangrove degradation and regeneration (Giri et al. 2010). Similarly, Sentinel-2 imagery has been used for global mangrove mapping, with its high-resolution data allowing for the detection of changes over time (Navarro et al. 2019). In a study on the Sundarbans, satellite data was used to monitor changes caused by sea-level rise and sedimentation (Rahman et al. 2011). Satellite multi-spectral and hyper-spectral data have also proven valuable for tracking mangrove deforestation and recovery (Chaube et al. 2019).

The strength of satellite RS lies in its broad spatial coverage, temporal consistency allowing for change detection over long periods, and relative cost-effectiveness compared to aerial surveys. However, it has weaknesses including susceptibility to cloud cover especially in tropical mangrove regions and relatively coarse spatial resolution compared to UAVs or LiDAR, limiting detailed species identification and structural analysis. Spectral confusion with other vegetation types can also be a limitation. Specific applications include broad-scale mangrove mapping, monitoring deforestation rates, and assessing mangrove response to large-scale disturbances like cyclones. Landsat's 30 m resolution is suitable for detecting significant changes in mangrove extent, but not for detailed canopy structure or species-level discrimination. Sentinel-2's 10 m resolution offers improvement, but cloud cover remains a challenge. Selecting the appropriate sensor requires balancing the need for temporal frequency, spatial detail, and budget constraints.

2.2 SAR versus LiDAR: Penetrating the Canopy

When considering mangrove monitoring, both Synthetic Aperture Radar (SAR) and Light Detection and Ranging (LiDAR) technologies offer unique advantages and disadvantages. Table 1 summarizes these features:

SAR's strengths include its all-weather capability, sensitivity to vegetation structure, and ability to penetrate the canopy to some extent, providing information on biomass and inundation patterns. LiDAR's strengths include high accuracy in measuring canopy height and vertical structure, detailed information for biomass estimation and carbon stock assessment, and the ability to derive Digital Terrain Models (DTMs) even under dense canopies. LiDAR is ideal for detailed structural analysis and biomass estimation in clear conditions. SAR, conversely, offers an advantage in cloudy regions and provides some structural information, making it suitable for regions with persistent cloud cover. The choice depends on the specific research question and environmental conditions.

Table 1 Comparison of SAR and LiDAR technologies for mangrove monitoring

Feature	SAR	LiDAR
Strengths	All-weather capability; sensitivity to vegetation structure; ability to penetrate canopy to some extent; provides information on biomass and inundation patterns	High accuracy in measuring canopy height and vertical structure; detailed information for biomass estimation and carbon stock assessment; ability to derive DTMs even under dense canopies
Weaknesses	Complex data interpretation; signal saturation in dense mangrove forests can limit biomass estimation accuracy; less effective for species-level identification	Cannot penetrate water; expensive data acquisition; computationally intensive processing; limited spectral information for species identification; affected by heavy rain
Specific Applications	Mapping mangrove extent in persistently cloudy regions; estimating above-ground biomass; monitoring inundation patterns; detecting disturbances like oil spills	Creating high-resolution DTMs of mangrove ecosystems; estimating above-ground biomass and carbon sequestration; analyzing mangrove forest structure and complexity; generating elevation models

2.3 *Unmanned Aerial Vehicles (UAVs): High-Resolution Detail and Flexibility*

Aerial photography, historically a key method for environmental monitoring, continues to play a crucial role in mangrove studies. Recent advancements in drone technology have transformed mangrove monitoring. High-resolution aerial images offer detailed views of mangrove canopies and are particularly useful in small or inaccessible areas. Advancements in camera technology have improved the quality of aerial photographs, enhancing their utility in vegetation mapping and health assessment Unmanned Aerial Vehicles (UAVs) provide near-real-time, high-resolution images essential for detailed monitoring, including health assessments and spatial distribution mapping. Drones have been particularly useful in regions where traditional satellite imagery may be limited and mangrove species identification (Zimudzi et al. 2021). UAVs' ability to access remote areas adds flexibility to the monitoring process, making it a critical tool for conservation, making it a valuable tool for comprehensive mangrove monitoring. (Hsu et al. 2020).

The strength of UAVs lies in ultra-high spatial resolution, flexible deployment, near-real-time data acquisition, and cost-effectiveness for small areas. However, they have limited spatial coverage, battery life constraints, regulatory restrictions on airspace, are weather-dependent, and require skilled operators. Specific applications include mapping mangrove species distribution, assessing mangrove health, monitoring small-scale changes like replanting efforts and erosion, and generating orthomosaics and 3D models. UAVs provide invaluable detailed information but are best suited for targeted monitoring in smaller areas, complementing satellite data by providing validation data and detailed information in specific locations.

2.4 *LiDAR: Mapping the Forest Structure*

LiDAR technology is crucial for precise elevation data and understanding mangrove forest structure. LiDAR provides critical insights into biomass estimation and canopy height mapping. Studies in Guarás Island have demonstrated its effectiveness in analysing mangrove forest structure and carbon sequestration potential (Salum et al. 2020). The use of LiDAR has enhanced the ability to conduct detailed ecological studies, offering insights into mangrove management (Feliciano et al. 2017).

LiDAR offers high-resolution 3D data, accurate measurement of canopy height and vertical structure, the ability to penetrate the canopy and provide information about the understory, and usefulness for estimating biomass and carbon stocks. Drawbacks include the high cost of data acquisition, complex data processing, the potential for being affected by cloud cover for airborne LiDAR, and limited spectral information for species identification. Specific applications include mapping mangrove canopy height and structure, estimating above-ground biomass, creating digital elevation models (DEMs), and assessing the impact of sea-level rise.

3 Overview of Field Validation Techniques

The combination of field inventory methods and geospatial technologies in mangrove monitoring is effective in assessing forest health, structure, and carbon stock at various scales. Field inventories offer detailed data but lack spatial coverage, while geospatial technologies offer efficient monitoring but lack precision in species identification and ecological conditions. A study on the west coast of India emphasizes the importance of field inventory methods in calibrating and validating remote sensing models. They use open-source geospatial software to process field data and map mangrove extents, enhancing the accuracy of satellite and UAV-derived models (Bharath Setturu 2014).

3.1 Data Collection Techniques

Ground-truthing methods are crucial for geospatial analysis, particularly in mangrove monitoring, as they ensure accuracy by validating remote sensing models with actual on-the-ground observations, bridging the gap between satellite imagery and field inventory methods.

3.1.1 Ground-Truthing

Ground-truthing in geospatial analysis involves collecting field measurements to verify and calibrate with remote sensing data. It involves integrating field inventory data with remote sensing imagery to ensure accuracy and reliability of the remotely sensed data. For mangrove monitoring, ground-truthing plays a critical role in several areas:

Field-based identification of mangrove species helps validate classifications made by remote sensing technologies, ensuring that satellite or drone imagery correctly identifies different species or forest zones. Field data is crucial for accurate remote sensing technologies used in mangrove mapping and health monitoring. Santos and (Santos and Bitencourt 2016) emphasize the importance of field data for precise mapping of mangrove structures, while (Nayak and Bahuguna 2001) highlight the need for ground validation in remote sensing techniques for mangroves in India, as classification accuracy predictions rely on ground-validated maps. Field validation is crucial for ensuring the reliability of drone-derived measurements, as even high-resolution drone imagery can lead to misclassifications, especially when distinguishing between mangrove species or health statuses, despite drones reducing the need for extensive fieldwork (Dronova et al. 2021).

Field measurements, such as diameter at breast height (DBH) and tree height, are essential for calibrating LIDAR and satellite-derived forest structure data. Field validation plays a crucial role in integrating remote sensing data with ground-based observations, enhancing structural characterization, and ensuring reliable mapping and ecological monitoring of mangrove forests (Lucas et al. 2020).

Remote sensing tools, including indices like the Normalized Difference Vegetation Index (NDVI), can detect vegetation health. However, ground-truthing is necessary to determine stress factors such as salinity or disease. The combination of drone technology and field-based data collection improves the detection and assessment of mangrove health, enabling early identification of stress, disease, or pollution. This real-time data acquisition is particularly valuable for monitoring ecosystem health and responding to environmental threats, offering a higher temporal resolution compared to traditional satellite-based methods (Ruwaimana et al. 2018).

Field validation is also essential for assessing mangrove health and post-disturbance recovery. Martínez et al. (2022) demonstrated the value of multisource remote sensing data combined with field surveys in evaluating hurricane damage and recovery in mangrove ecosystems. Their findings highlight how field validation enhances the model's ability to detect early signs of stress and damage.

3.1.2 Vegetation Surveys

One of the popular techniques for evaluating the vegetation in mangrove ecosystems is to conduct plot-based field surveys. These surveys divide the study area into plots, ranging from 10 × 10 m to 20 × 20 m, based on the density and complexity of the forest. Information is gathered on the species composition, tree height, diameter at breast height, tree density, and canopy cover. This information is essential for verifying the classification of remotely sense satellite imagery and calibrating geographic models to have fewer misclassification and errors in the result. For instance, Pandey et al. (2019) combined field plot data with hyperspectral imagery to map mangrove species and biomass more accurately, emphasizing that ground-based plot data is critical for reducing misclassification errors in remote sensing data.

Transect sampling involves setting up straight lines (transects) through the mangrove forest, along which measurements are taken at regular intervals. This method is particularly useful in large mangrove areas where it is challenging to implement plot-based surveys. Data collected along transects, including species counts, tree measurements, and environmental conditions, is used to assess forest structure and composition. Otero et al. (2018) used transect sampling in combination with UAV imagery in the Matang Mangrove Forest Reserve, demonstrating that transects provide detailed and spatially consistent data to validate geospatial models (Otero et al. 2018).

Allometric equations are used in the field to estimate above-ground biomass and carbon stock by measuring tree parameters such as DBH and tree height. These measurements are critical for estimating the carbon sequestration potential of mangrove forests. The equations vary based on species, region, and forest type, requiring region-specific calibration. Nguyen & Nguyen, (2021) highlighted that integrating allometric measurements with LiDAR data from UAVs significantly improves biomass estimates in mangrove ecosystems (Nguyen & Nguyen, 2021).

3.1.3 Biomass and Carbon Stock Measurement

Allometric equations are used to calculate biomass in mangroves based on tree features like diameter at breast height (DBH) and height. These metrics are critical for creating region-specific models. The combination of field data and remote sensing enables large-scale biomass estimate. For example, Nguyen and Nguyen (2021) employed allometric equations and UAV-LiDAR data to estimate mangrove biomass in Southeast Asia, increasing the accuracy of biomass and carbon stock estimations. Aslan et al. (2016) employed remote sensing techniques, including multispectral and hyperspectral imagery, to estimate biomass and carbon stock in Indonesian mangrove forests. LiDAR technology gives precise estimations of above-ground biomass, such as canopy height and tree structure (Maeda et al. 2016). Because of its capacity to penetrate the forest canopy and gather structural data even under foggy situations, Synthetic Aperture Radar (SAR) technology is useful for biomass assessment. Pham et al. (2019) used SAR data to better identify and map carbon stocks in Southeast Asia's mangrove environment. Machine learning algorithms are utilised to automate biomass and carbon stock estimates utilising big datasets from remote sensing platforms, outperforming traditional methods. Jones et al. (2020) showed that merging machine learning with UAV data yields very accurate and scalable biomass assessments.

3.2 Challenges and Solutions

Accessibility, data quality, and labour intensity are all obstacles that might impede efficiency and precision in the field inventory for mangrove monitoring. Geospatial technologies, on the other hand, appear to provide viable solutions to these difficulties.

3.2.1 Accessibility Challenges

Mangrove forests, often in remote coastal areas with challenging terrain, pose challenges for researchers to access and measure. To overcome these challenges, researchers are using Unmanned Aerial Vehicles (UAVs) or drones equipped with multispectral and LiDAR sensors to provide high-resolution imagery without direct field access. UAVs can quickly cover extensive areas, capturing real-time data on forest structure and species composition. Otero et al. (2018) found that this strategy is successful for mapping mangrove forests in inaccessible areas, reducing the need for manual data collecting.

3.2.2 Data Accuracy

The complexity of mangrove ecosystems makes it difficult to accurately estimate forest biomass, species diversity, and structure. Field data collection is time-consuming and often fails to capture full variability, leading to under- or overestimation of biomass or species counts. Combining field-based measurements with LiDAR and satellite remote sensing can improve data accuracy. LiDAR provides detailed canopy height and biomass estimation, while multispectral and hyperspectral imagery maps species composition across large areas. Maeda et al. (2016) used LiDAR data for highly accurate biomass estimates.

3.2.3 Labor Intensity

Field inventory methods such as plot-based surveys and transect sampling are labour-intensive and time-consuming, necessitating large human resources and additional expenditures. Machine learning algorithms can automate data collection and analysis, predicting species composition and biomass across vast regions. This minimises the need for large field surveys. Pham et al. (2019) demonstrated the application of machine learning to automate the categorisation of mangrove species and biomass, considerably lowering the need for manual labour.

3.2.4 Temporal and Spatial Monitoring Limitations

Traditional field methods for monitoring mangroves over time can lead to data gaps and hinder long-term efforts. Satellite-based Remote Sensing and Time-Series Analysis can provide continuous, large-scale monitoring of mangrove forests. Free-access satellites like Sentinel-2 offer frequent imagery that can be analyzed to track changes in forest structure and health over time. Giri (2016) used multi-temporal satellite imagery to monitor changes in mangrove forests globally, providing accurate monitoring of forest dynamics and insights into long-term ecological trends.

Geospatial technologies such as UAVs, LiDAR, machine learning, and satellite-based remote sensing enhance mangrove monitoring and conservation efforts by increasing accessibility, data accuracy, reducing labour intensity, and enabling continuous monitoring, thereby overcoming traditional inventory challenges.

4 Discussion of Key Findings: Challenges and Opportunities

Integrating geospatial technologies with field data has proven to be a transformative approach for enhancing mangrove monitoring. This combination enables large-scale assessments while ensuring accuracy through field verification (Maurya et al. 2020). Conventional approaches to mapping mangroves, such as field surveys, tend to be complex and time-consuming, complicating the assessment of various elements within the mangrove ecosystem (Faridah-Hanum et al. 2021).

Recent advancements in satellite sensor technology have significantly improved the classification and species differentiation capabilities in mangrove studies compared to traditional methods. Synthetic Aperture Radar (SAR) has been identified as an effective alternative for characterizing mangrove ecosystems, particularly in regions with persistent cloud cover (Hanum et al. 2021). The combination of SAR with hyperspectral imagery has been shown to enhance mapping precision (Kuenzer et al. 2011). For instance, Held et al. (2003) demonstrated that the integration of AIRSAR data with hyperspectral imagery in North Queensland, Australia, improved classification accuracy by approximately 3% in identifying species communities. Similarly, Wang et al. (2004) highlighted the effectiveness of IKONOS imagery in providing detailed spectral reflectance, aiding in precise mangrove differentiation.

The effectiveness of integrating remote sensing data with field surveys has also been demonstrated in various case studies. Mesta et al. (2014) successfully combined Landsat and IRS multispectral imagery with field surveys to monitor mangrove changes in the Aghnashini estuary, India. Their analysis revealed an increase in mangrove cover from 40.79% in 1989 to 44.92% in 2010 due to reforestation efforts. Likewise, Giri (2010) utilized Global Land Cover (GLC) data and Landsat imagery to create the first global mangrove map, validated using ground truth data, high-resolution satellite imagery, and expert input.

One of the major strengths of remote sensing in mangrove monitoring is its ability to provide large-scale and long-term observations with high temporal frequency. Satellite imagery enables the detection of spatial and temporal changes in mangrove cover, allowing researchers to track deforestation, degradation, and restoration trends efficiently (Giri 2016). Additionally, remote sensing techniques, particularly those using SAR and hyperspectral sensors, enhance species differentiation and habitat characterization, even in challenging environments (Kuenzer et al. 2011). However, remote sensing has limitations, including challenges in distinguishing between mangrove species due to spectral similarities, especially in low-resolution imagery. The presence of cloud cover can obstruct optical sensors, limiting their effectiveness in tropical and subtropical regions (Hanum et al. 2021). While SAR overcomes this limitation, its interpretation requires specialized knowledge and extensive field validation.

Field surveys provide critical ground-truth data necessary for validating remote sensing classifications and ensuring the accuracy of geospatial analyses. They enable direct species identification, health assessments, and ecological studies that remote sensing alone cannot achieve (Sandilyan and Kathiresan 2012). Moreover, fieldwork allows for community engagement and participatory research, fostering local conservation efforts. Despite these advantages, field surveys are time-consuming, labor-intensive, and often restricted to specific areas due to logistical and financial constraints. The accessibility of certain mangrove ecosystems, particularly in remote or densely vegetated regions, further limits fieldwork feasibility. Consequently, integrating remote sensing with selective field validation is the most effective approach for comprehensive mangrove monitoring (Wang and Gu 2021).

Despite the advances of utilizing geospatial techniques, there is still a need for studies to:

- **Improve species identification algorithms.** Enhanced algorithms utilizing hyperspectral data and machine learning can better differentiate mangrove species, even in mixed forests.
- **Develop better methods for mapping mangrove health under stress.** Integration of thermal imagery and physiological data can aid in early detection of stress from pollution, salinity changes, and disease.
- **Improve biomass estimation in complex mangrove forests.** Combining LiDAR data with allometric equations and machine learning can provide more accurate biomass estimates, accounting for forest structural complexity.
- **Standardize protocols for integrating remote sensing and field data.** Developing standardized protocols to facilitate data sharing and comparison across different studies and regions, is crucial for global monitoring efforts.

Future research should focus on improving the integration of remote sensing and field-based techniques to enhance mangrove classification and monitoring. The use of machine learning and artificial intelligence (AI) in processing high-resolution satellite imagery and UAV (Unmanned Aerial Vehicle) data presents significant opportunities for automating species classification and health assessments. Additionally, exploring multi-temporal analysis through advanced sensors, such as LiDAR, can provide more accurate insights into biomass estimation and structural dynamics of mangrove forests (Nagelkerken et al. 2008). Furthermore, expanding community-based monitoring programs by training local stakeholders in the use of geospatial technologies can improve conservation efforts. Promoting collaborations between governmental agencies, research institutions, and local communities will facilitate sustainable mangrove management while addressing socio-economic and environmental challenges.

5 Conclusion

Mangroves, known for their resilience and ecological diversity, are vital for maintaining the Earth's natural balance and conserving biodiversity. They protect against natural disasters, as evidenced during the Asian Tsunami (Danielsen 2005), and play a key role in filtering carbon-rich particles from water, aiding in carbon sequestration and climate change mitigation (Kumari 2021). However, these ecosystems face threats from climate change, coastal development, and overexploitation. Geospatial technologies help monitor mangrove health and distribution, supporting conservation and sustainable management (Sandilyan and Kathiresan 2012).

Mangroves span 123 countries, covering approximately 150,000 km^2 (Rafique 2018). This review examines recent geospatial methodologies for monitoring mangroves, highlighting advancements in Satellite-based Remote Sensing (RS) for high-resolution data, as well as the roles of aerial photography and drone technology. LiDAR has improved ecological studies, aiding mangrove management (Feliciano et al. 2017). The integration of Geographic Information System (GIS) data enhances assessment accuracy, supporting informed decision-making. Cloud computing platforms like Google Earth Engine facilitate the processing of large satellite datasets, revolutionizing mangrove monitoring (Ditian et al. 2021). Combining field inventory techniques with geospatial technologies effectively evaluates forest health, structure, and carbon stock, while ground truth data bolsters the reliability of remote sensing for assessing species composition and biomass. Vegetation surveys are crucial, utilizing methods like plot-based surveys, transect sampling, and allometric measurements for biomass and carbon assessment. The integration of geospatial technologies with field data has enhanced mangrove monitoring, enabling extensive assessments with accurate field verification (Maurya et al. 2020). In Pambala, Sri Lanka, Guebas et al. (2005) utilized IKONOS imagery to evaluate the function of mangroves in providing tsunami protection. The classified images were then analyzed in conjunction with on-screen digitized images and ground truth data, demonstrating that this methodology successfully differentiated between Rhizophora apiculata and R. mucronate.

The continued integration of geospatial technologies and field-based methods is essential for effective mangrove conservation. By embracing these advanced tools and fostering collaboration among researchers, policymakers, and local communities, we can ensure the long-term health and resilience of these vital coastal ecosystems, safeguarding their invaluable contributions to biodiversity, carbon sequestration, and coastal protection.

References

Ajai, Chauhan HB (2017). Mangrove inventory, monitoring, and health assessment. In: Finkl C, Makowski C (eds) Coastal wetlands: alteration and remediation. Coastal Research Library, vol 21. Springer, Cham. https://doi.org/10.1007/978-3-319-56179-0_19

Kumari A, Rathore MS (2021) Roles of mangroves in combating the climate change, pp 225–255. https://doi.org/10.1007/978-981-16-2494-0_10

Aslan A, Rahman AF, Warren MW, Robeson SM (2016) Mapping spatial distribution and biomass of coastal wetland vegetation in Indonesian Papua by combining active and passive remotely sensed data. Remote Sens Environ 183:65–81. https://doi.org/10.1016/j.rse.2016.04.026

Avtar R, Kumar P, Oono A, Saraswat C, Dorji S, Hlaing Z (2016) Potential application of remote sensing in monitoring ecosystem services of forests, mangroves and urban areas. Geocarto Int. https://doi.org/10.1080/10106049.2016.1206974

Bharath Setturu PNM (2014) Inventorying, mapping and monitoring of mangroves towards sustainable management of West Coast, India. J Geophys Remote Sens 03(03). https://doi.org/10.4172/2169-0049.1000130

Chaube NR, Lele N, Misra A, Murthy TVR, Manna S, Hazra S, Samal RN (2019) Mangrove species discrimination and health assessment using AVIRIS-NG hyperspectral data. Curr Sci 116(7):1136–1142

Chen R, Zhang R, Zhao C, Wang Z, Jia M (2023) High-resolution mapping of mangrove species height in Fujian Zhangjiangkou National Mangrove Nature Reserve combined GF-2, GF-3, and UAV-LiDAR. Remote Sens 15(24):5645. https://doi.org/10.3390/rs15245645

Cummings AR, Shah M (2018) Mangroves in the global climate and environmental mix. Geogr Compass 12:e12353. https://doi.org/10.1111/gec3.12353

Dahdouh-Guebas F, Ajonina GN, Amir AA, Andradi-Brown DA, Aziz I, Balke T, Barbier EB, Cannicci S, Cragg SM, Cunha-Lignon M, Curnick DJ, Duarte CM, Duke NC, Endsor C, Fratini S, Feller IC, Fromard F, Hugé J, Huxham M, Kairo JG, Kajita T, Kathiresan K, Koedam N, Lee SY, Lin H-J, Mackenzie JR, Mangora MM, Marchand C, Meziane T, Minchinton TE, Pettorelli N, Polanía J, Polgar G, Poti M, Primavera J, Quarto A, Rog SM, Satyanarayana B, Schaeffer-Novelli Y, Spalding M, Van der Stocken T, Wodehouse D, Yong JWH, Zimmer M, Friess DA (2020) Public perceptions of mangrove forests matter for their conservation. Front Mar Sci 7:603651. https://doi.org/10.3389/fmars.2020.603651

Danielsen F (2005) The Asian tsunami: a protective role for coastal vegetation. Science 310(5748):643–643. https://doi.org/10.1126/science.1118387

Ditian D, Meganatha RM, Prasetyo WE, Hartanto S, Hartono R (2021) Mangrove extent and change mapping of Muaragembong from 1990 to 2020 using Google Earth Engine (GEE). In: 2021 IEEE ocean engineering technology and innovation conference: ocean observation, technology and innovation in support of ocean decade of science (OETIC), 35–38. https://doi.org/10.1109/OETIC53770.2021.9733736

Dronova I, Kislik C, Dinh Z, Kelly M (2021) A review of unoccupied aerial vehicle use in wetland applications: emerging opportunities in approach, technology, and data. Drones 5(2). https://doi.org/10.3390/drones5020045

Faridah-Hanum I et al (eds) (2014) Mangrove ecosystems of Asia, © Springer Science+Business Media, New York. https://doi.org/10.1007/978-1-4614-8582-7_14

Feliciano EA, Wdowinski S, Potts MD, Lee SK, Fatoyinbo TE (2017) Estimating mangrove canopy height and above-ground biomass in the Everglades National Park with airborne LiDAR and TanDEM-X data. Remote Sens 9(7):702. https://doi.org/10.3390/rs9070702

Flores-de-Santiago F, Flores-Verdugo F (2024) A comparison of forest structural methods of semi-arid mangrove species using a field-based approach. Cien Marinas 50. https://doi.org/10.7773/cm.y2024.3432

Garg JK (2013) Wetland assessment, monitoring and management in India using geospatial techniques. J Environ Manage. https://doi.org/10.1016/j.jenvman.2013.12.018

Cuenca-Ocay G (2024) Mangrove ecosystems' role in climate change mitigation. https://doi.org/10.59120/drj.v12i2.168

Giri C, Ochieng E, Tieszen LL, Zhu Z, Singh A, Loveland T, Duke N (2010) Status and distribution of mangrove forests of the world using earth observation satellite data. Glob Ecol Biogeogr 20(1):154–159. https://doi.org/10.1111/j.1466-8238.2010.00584.x

Giri C (2016) Observation and monitoring of mangrove forests using remote sensing: opportunities and challenges. Remote Sens 8(9):783. https://doi.org/10.3390/rs8090783

Hamzah ML, Amir AA, Maulud KNA, Sharma SAHADEV, Mohd FA, Selamat SN, Begum RA (2020) Assessment of the mangrove forest changes along the Pahang coast using remote sensing and GIS technology. J Sustain Sci Manag 15(5):43–58

Hidayah Z, Andriyani L (2019) IOP Conf Ser: Earth Environ Sci 276:012034

Hsu AJ, Kumagai J, Favoretto F, Dorian J, Guerrero Martinez B, Aburto-Oropeza O (2020) Driven by drones: improving mangrove extent maps using high-resolution remote sensing. Remote Sens 12(23):3986. https://doi.org/10.3390/rs12233986

Hussin YA, Zuhair MM, Weir M (2007) Monitoring mangrove forests using remote sensing and GIS. gisdevelopment.net

Jones AR, Raja Segaran R, Clarke KD, Waycott M, Goh WSH, Gillanders BM (2020) Estimating mangrove tree biomass and carbon content: a comparison of forest inventory techniques and drone imagery. Front Mar Sci 6(January):1–13. https://doi.org/10.3389/fmars.2019.00784

Jusoff K (2008) Geospatial information technology for conservation of coastal forest and mangroves environment in Malaysia. Comput Inf Sci 1(2):129–134

Lucas R, Van De Kerchove R, Otero V, Lagomasino D, Fatoyinbo L, Omar H, Satyanarayana B, Dahdouh-Guebas F (2020) Structural characterisation of mangrove forests achieved through combining multiple sources of remote sensing data. Remote Sens Environ 237(December 2019):111543. https://doi.org/10.1016/j.rse.2019.111543

Maeda Y, Fukushima A, Imai Y, Tanahashi Y, Nakama E, Ohta S, Kawazoe K, Akune N (2016) Estimating carbon stock changes of mangrove forests using satellite imagery and airborne Lidar data in the South Sumatra State, Indonesia. In: The international archives of the photogrammetry, remote sensing and spatial information sciences, XLI-B8, pp 705–709. https://doi.org/10.5194/isprs-archives-XLI-B8-705-2016

Maurya K, Mahajan S, Chaube N (2021) Remote sensing techniques: mapping and monitoring of mangrove ecosystem—a review. 7:2797–2818. https://doi.org/10.1007/s40747-021-00457-z

Mesta P, Setturu B, Chandran MD, Rajan K, Ramachandra TV (2014) Inventorying, mapping and monitoring of mangroves towards sustainable management of West Coast, India. J Geophys Remote Sens 3:130. https://doi.org/10.4172/2169-0049.1000130

Nagelkerken I, Blaber SJM, Bouillon S, Green P, Haywood M, Kirton LG, Meynecke J-O, Pawlik J, Penrose HM, Sasekumar A, Somerfield PJ (2008) The habitat function of mangroves for terrestrial and marine fauna: a review. Aquat Bot 89(2):155–185, ISSN 0304-3770. https://doi.org/10.1016/j.aquabot.2007.12.007

Navarro JA, Algeet N, Fernández-Landa A, Esteban J, Rodríguez-Noriega P, Guillén-Climent ML (2019) Integration of UAV, Sentinel-1, and Sentinel-2 data for mangrove plantation aboveground biomass monitoring in Senegal. Remote Sens 11(1):77. https://doi.org/10.3390/rs11010077

Nayak S, Bahuguna A (2001) Application of remote sensing data to monitor mangroves and other coastal vegetation of India. Indian J Mar Sci 30(4):195–213

Nguyen H-H, Nguyen TTH (2021) Above-ground biomass estimation models of mangrove forests based on remote sensing and field-surveyed data: implications for C-PFES implementation in Quang Ninh Province, Vietnam. Reg Stud Mar Sci 48:101985. https://doi.org/10.1016/j.rsma.2021.101985

Otero V, Van De Kerchove R, Satyanarayana B, Martínez-Espinosa C, Fisol MA Bin Ibrahim MR, Bin, Sulong I, Mohd-Lokman H, Lucas R, Dahdouh-Guebas F (2018) Managing mangrove forests from the sky: forest inventory using field data and Unmanned Aerial Vehicle (UAV) imagery in the Matang Mangrove Forest Reserve, peninsular Malaysia. For Ecol Manag 411:35–45. https://doi.org/10.1016/j.foreco.2017.12.049

Pandey PC, Anand A, Srivastava PK (2019) Spatial distribution of mangrove forest species and biomass assessment using field inventory and earth observation hyperspectral data. Biodivers Conserv 28(8–9):2143–2162. https://doi.org/10.1007/S10531-019-01698-8/METRICS

Paul DK (2021) Respondent analysis in context to impact of climate change on the regulating services of mangrove vegetation. J Mech Continua Math Sci

Pham TD, Yokoya N, Bui DT, Yoshino K, Friess DA (2019) Remote sensing approaches for monitoring mangrove species, structure, and biomass: opportunities and challenges. Remote Sens 11(3):1–24. https://doi.org/10.3390/rs11030230

Rahman AF, Dragoni D, El-Masri B (2011) Response of the Sundarbans coastline to sea level rise and decreased sediment flow: a remote sensing assessment. Remote Sens Environ 115(12):3121–3128. https://doi.org/10.1016/j.rse.2011.06.019

Ratnasermpong S (2004) Coastal zone environment management with emphasis on mangrove ecosystem, a case study of Ao-Sawi Thung Khla, Chumphon, Thailand, gisdevelopment.net

Reshma S, Komali K, Chandradeo B, Upaka R, Michael F (2024) Mangrove mapping and monitoring using remote sensing techniques towards climate change resilience. Dent Sci Rep 14. https://doi.org/10.1038/s41598-024-57563-4

Rizal A, Sahidin A, Herawati H (2018) Economic value estimation of mangrove ecosystems in Indonesia. Biodivers Int J 2(1):98–100

Ruwaimana M, Satyanarayana B, Otero V, M. Muslim A, Syafiq A, M. Ibrahim S, Raymaekers D, Koedam N, Dahdouh-Guebas F (2018) The advantages of using drones over space-borne imagery in the mapping of mangrove forests. PLOS ONE 13(7):e0200288. https://doi.org/10.1371/journal.pone.0200288

Salum RB, Souza-Filho PWM, Simard M, Silva CA, Fernandes ME, Cougo MF, Rogers K (2020) Improving mangrove above-ground biomass estimates using LiDAR. Estuar Coast Shelf Sci 236:106585. https://doi.org/10.1016/j.ecss.2020.106585

Sandilyan S, Kathiresan K (2012) Mangrove conservation: a global perspective. Biodivers Conserv 21:3523–3542. https://doi.org/10.1007/s10531-012-0388-x

Santos LCM, Bitencourt MD (2016) Remote sensing in the study of Brazilian mangroves: review, gaps in the knowledge, new perspectives and contributions for management. J Integr Coast Zone Manag 16(3):245–261. https://doi.org/10.5894/rgci662

Vizcaya-Martínez DA, Flores-de-Santiago F, Valderrama-Landeros L, Serrano D, Rodríguez-Sobreyra R, Álvarez-Sánchez LF, Flores-Verdugo F (2022) Monitoring detailed mangrove hurricane damage and early recovery using multisource remote sensing data. J Environ Manag 320:115830. https://doi.org/10.1016/j.jenvman.2022.115830

Wang Y-S, Gu J-D (2021) Ecological responses, adaptation and mechanisms of mangrove wetland ecosystem to global climate change and anthropogenic activities. Int Biodeterior. Biodegradation 162:105248, ISSN 0964-8305, https://doi.org/10.1016/j.ibiod.2021.105248

Xu C, Wang J, Sang Y, Li K, Liu J, Yang G (2023) An effective deep learning model for monitoring mangroves: a case study of the Indus Delta. Remote Sens 15(9):2220. https://doi.org/10.3390/rs15092220

Zimudzi E, Sanders I, Rollings N, Omlin CW (2021) Remote sensing of mangroves using unmanned aerial vehicles: current state and future directions. J Spat Sci 66(2):195–212. https://doi.org/10.1080/14498596.2019.1627252

Modelling Ecosystem Disruption Along the Odisha Coast of the Bay of Bengal Using Biophysical Parameters

Monalisha Mishra, G. K. Panda, and Snigdharani Panda

Abstract Disruption of ecosystems along the Bay of Bengal coast in Odisha is a pressing environmental challenge, impacting various species and their habitats. Human activities, combined with climate change, rising sea temperatures, and increasing tropical cyclones, exacerbate this crisis. The 480 km coastline of Odisha features diverse ecosystems, including tidal flats, marshes, mangroves, sandy beaches, and estuarine river mouths, which support species like Olive Ridley turtles, saltwater crocodiles, blackbucks, and migratory birds. Critical areas like Chilka Lake and the Little Sundarbans mangroves are particularly vulnerable. This study uses the Coastal Hazard Wheel (CHW) model, developed by UNEP with Danish institutions, to assess the vulnerability and risks of ecosystem disruption. The CHW model evaluates coastal hazards such as erosion, inundation, saltwater intrusion, and flooding, using biophysical parameters include geological features, wave exposure, tidal range, flora and fauna, sediment balance, and storm climate. Multi-date satellite images and geospatial tools were used to map areas at risk. Results indicate that 22% of Odisha's coastline is at very high risk, 74% at moderate risk, and 4% at low risk of ecosystem disruption. The study identifies vulnerable areas and recommends management strategies, including coastal zoning, ecosystem-based management, fluvial sediment management, dune construction, rehabilitation, and coastal setbacks. These strategies, supported by the CHW model, aim to mitigate risks and promote sustainable coastal zone management. Additionally, the study highlights the importance of open-source satellite data and geoinformatics in managing ecosystem disruption in coastal regions.

M. Mishra (✉) · G. K. Panda · S. Panda
School of Tribal Resource Management, KISS Deemed University, Bhubaneswar, Odisha, India
e-mail: monalishaiirs@gmail.com

G. K. Panda
e-mail: gopalkrishna_panda@yahoo.co.uk

S. Panda
e-mail: snigdharani.panda@kiss.ac.in

O. Mutanga et al. (eds.), *Revealing Ecosystem Services Through Geospatial Technologies*, Springer Remote Sensing/Photogrammetry,
https://doi.org/10.1007/978-3-031-98048-0_8

Keywords Ecosystem · Risk assessment · Vulnerability analysis · Geographical Information System · Remote sensing · Coastal management

1 Introduction

Coastal regions have long been some of the most desirable areas for human settlement. However, in recent times, these areas are facing increasing environmental threats, particularly coastal erosion. This growing issue is driven by factors such as global warming, rising sea levels, and human activities that disrupt natural coastal systems (Liew et al. 2020). Coastal ecosystems, vital to both human well-being and biodiversity, are increasingly threatened by the combined impacts of climate change and human activities (IPCC 2022). Using geographic information systems and remote sensing, the primary aim of this project is to model ecological disruption along the Bay of Bengal coast in Odisha. While extensive research has been conducted on coastal erosion and climate change impacts (Nichols et al. 2007; IPCC 2021), a significant gap remains in understanding how these processes specifically disrupt coastal ecosystems in terms of biophysical parameters. This study seeks to address this gap by integrating high-resolution satellite imagery and the Coastal Hazard Wheel (CHW) Model to assess ecological disruption at a localized scale. The findings will contribute to coastal management efforts by offering insights into ecosystem vulnerability, potential restoration strategies, and the role of remote sensing in long-term monitoring.

Human activity on the atmosphere, particularly the release of carbon dioxide (CO_2) and other greenhouse gases, has caused climate change, which has resulted in global warming. According to the Intergovernmental Panel on Climate Change (IPCC), one of the most evident signs of climate change has been the rise in land and ocean temperatures during the last several decades, as well as the adaptation of habitats and ecosystems to these changes. Climate change has a profound impact on coastal regions and communities (Nichols et al. 2007). Approximately half of the world's population lives within 200 km of the coastline, and this figure is anticipated to double by 2025 (Population Reference Bureau (PRB) 2023). Coastal zones, rich in biodiversity and pristine environments, are important natural resources. However, these zones are also vulnerable to a wide range of natural hazards, both geological and meteorological, as well as human-induced environmental challenges. Natural forces and human activities, or a combination of both, are key contributors to coastal dynamics. Geological events arise from the natural processes of the Earth, meteorological events stem from global weather variations, and biological events are driven by the actions of living organisms. Human interference in the coastal zone, often through competing uses and resource exploitation, has significantly altered the coastal landscape. Climate change-induced coastal hazards have further exacerbated environmental degradation in these areas. The IPCC's forecasts of climate change add new risk factors to human activities in coastal regions. While coastal dynamics have historically been relatively stable and predictable, rapid changes are now expected due to sea-level rise, increased ocean temperatures, and shifting patterns in tropical storms and precipitation. According to the IPCC (2013), these changes are likely to accelerate under current climate conditions. Previous coastal trends cannot be predicted for the future because coastal ecosystems will respond to these changes in different ways and at different speeds, depending on their bio-geophysical characteristics. Thus, to determine how various coastal habitats would respond to the predicted changes in climate and related dangers, integrated coastal zone management is necessary. According to Appelquist and Balstrom (2014), this approach will improve coastal communication, help identify suitable methods, and clarify pertinent management options. The study models ecological disruption along the Bay of Bengal coast in Odisha using biophysical features. It illustrates how to use the Coastal Hazard Wheel Model, Geographic Information Systems (GIS), and remote sensing data, to generate comprehensive data and maps related to ecosystem disruption. Furthermore, the study explores potential management strategies and ecological restoration options. The need for this research is pressing, as coastal ecosystems are rapidly disappearing due to development, coastal flooding, saltwater intrusion, and erosion. Integrated coastal zone management strategies are critical to understanding how these ecosystems will respond to predicted climate changes and associated hazards, enabling the development of effective mitigation measures (Appelquist and Balstrom 2014). By modelling ecosystem disruption and evaluating the effectiveness of restoration strategies, this study aims to fill a knowledge gap by using remote sensing and GIS to study ecosystem changes along the Odisha coast. By analysing satellite images taken at different times, it will help us understand how the coastal environment is changing. The findings will support better coastal management, help in conservation efforts, and contribute to climate resilience.

2 Review of Literature

Ecosystem disruption has emerged as a critical environmental issue, driven by both natural and human-induced factors. A comprehensive review of the existing literature is essential to understand the underlying causes, patterns, and implications of this growing threat. The Gulf of Mannar has experienced significant changes in its coastal landscape, coral reefs, and seafloor over time, as revealed by the study of Thanikachalam and Ramachandran (2003). By using the remote sensing information from IRS LISS-III (1998), IRS LISS-II (1988), Survey of India Topographic sheets (1969), and bathymetric data (1999), and conducting GIS analysis, they identified the shoreline erosion and accretion, and from it, it was found that 4.34 km^2 mainland coast and 4.14 km^2 island coast have undergone erosion. Meanwhile, 23.49 km^2 mainland coast and 3.31 km^2 island coast have been undergoing accretion. The study also highlighted that in a period of ten years, there was a loss of 26 km^2 of coral reefs and 2 km^2 of coastal vegetation due to increased suspended sediment due to erosion and tectonic uplift. Additionally, a decrease in seafloor depth of 0.51 m on average was observed over a period of 24 years. These findings were supported by ground truth data and are consistent with other worldwide investigations on coastal erosion, damage to reefs, and the effects of tectonic activity on coast ecosystems. The study highlights the critical significance of using remote sensing and GIS for monitoring and understanding changes in marine and coastal ecosystems, which play a pivotal role in effective environmental management and conservation measures. Climate change adaptation ecosystem-based approaches (EbA) and ecosystem-based disaster risk reduction (Eco-DRR) have become popular sustainable strategies with the potential of merging environmental restoration and conservation into development goals. Bhardwaj and Gupta (2024) highlight the multifaceted benefits of these approaches, emphasizing their cost-effectiveness, flexibility, and ability to address the increasing risks of disasters exacerbated by climate change, land-use changes, and ecosystem degradation. These interventions align with developmental planning to achieve economic efficiency and environmental sustainability. The evolution of India's policy landscape reflects significant progress in incorporating ecosystem-based strategies into key sectors, including climate change adaptation, disaster management, and developmental policies. The authors provide a critical review of the strengths and gaps within existing policy frameworks, underscoring the need for customization and enhanced integration to mainstream ecosystem-based approaches. By examining case studies and sector-specific policies, the study underscores the importance of embedding ecosystem management principles in national strategies to build resilience against disasters and promote sustainable development A study by Panda et al. (2022) revealed the environmental degradation risks along the Puri coast of the Mahanadi delta using geospatial technology and the Coastal Hazard Wheel (CHW) model where they examined coastal hazards such as erosion, flooding, saltwater intrusion, and ecosystem disruption by analysing bio geophysical parameters like geomorphological structure and coastal vegetation. Utilizing multi-date remote sensing data and GIS, they found that 78% of the coastline is highly vulnerable, with coastal erosion and saltwater intrusion as significant threats. The study highlights the Puri coast's increasing vulnerability, providing valuable insights for coastal management and showcasing the efficacy of open-source satellite data in hazard assessment. The Coastal Hazard Wheel (CHW) model offers several benefits, such as its comprehensive approach to assessing multiple coastal hazards like erosion, flooding, saltwater intrusion, and ecosystem disruption by considering various bio geophysical parameters. Its ability to integrate multi-date remote sensing data and GIS enables detailed hazard mapping, providing valuable insights for coastal management. However, the CHW model has some limitations, including the potential for data inconsistencies when applying remote sensing data over large and dynamic coastal regions, and the model's reliance on accurate and up-to-date environmental data, which may not always be available in remote areas. Other studies have applied the CHW model to assess coastal vulnerability, such as those focusing on regions like the Sundarbans and the east coast of India. These studies have similarly highlighted significant risks of erosion, saltwater intrusion, and flooding, reinforcing the model's utility in identifying vulnerable coastal zones and supporting effective management strategies.

Micallef et al. (2017) conducted an assessment of erosion risk along the Maltese coast by applying the Coastal Hazard Wheel model. This approach facilitated the analysis of a broader spectrum of inherent coastal hazards, including ecosystem disruption, gradual inundation, salt water intrusion, and flooding Mishra et al. (2021) employed a robust suite of GIS and RS techniques to assess the impact of Cyclone Fani on the Olive ridley nesting grounds. This likely included analyzing pre- and post-cyclone high-resolution satellite imagery (e.g., Landsat, Sentinel) to detect changes in shoreline position, vegetation cover, and coastal morphology. Techniques such as image classification, change detection analysis, and NDVI were likely utilized. Furthermore, the study likely incorporated Digital Elevation Models (DEMs) to accurately map shorelines and analyze topographic features, providing insights into erosion vulnerability. GIS software played a crucial role in overlaying and analyzing spatial datasets, enabling the researchers to map the extent of shoreline erosion, identify changes in habitat suitability, and visualize the potential future impacts of such extreme weather events on this critical nesting ground. The study by Mishra et al. (2022) assesses the vulnerability of the Mahanadi delta's coastal environment to inherent hazards using the Coastal Hazard Wheel (CHW) model. It evaluates risks such as saltwater intrusion, coastal erosion, and ecosystem disruption, employing bio-geophysical parameters and open-source remote sensing data. The study identifies high-risk zones along the delta and highlights the increasing vulnerability compared to previous assessments. Its findings offer vital insights for integrated coastal zone management and propose sustainable interventions for environmental protection. Shoreline changes, particularly along the Ganjam Coast in Odisha, India, are significant due to various coastal processes. Mishra and Panda (2018) utilized geospatial techniques, including Survey of India topographic maps and Landsat imagery, to investigate these changes. Erosion, especially in the Ramayapatam and Podampeta sectors, is influenced by wave action and climate change. The study highlights the importance of continuous monitoring to understand the dynamics of coastal features and the impacts of low-pressure systems and rising sea levels. The study by Su et al. (2021) highlights the application of the Coastal Hazard Wheel (CHW) for assessing coastal multi-hazards and management in the Guangdong-Hong Kong-Macao Greater Bay Area (GBA), which faces threats from rising sea levels, extreme events, and human interventions. Using CHW, the authors classified GBA's coasts, identified hazard hotspots, and analyzed changes in hazards between 2010 and 2020. The results show that the most common coastal types are hard rock slopes and delta/low estuarine islands, and that GBA is extremely highly susceptible to flooding, gradual inundation, saltwater intrusion, and ecosystem loss. While most high-risk hazards decreased by 2020, flooding risks increased slightly, with areas like the Pearl River Estuary and Dapeng Bay exhibiting the highest vulnerabilities. The interdependence of hazards underscores the need for integrated management approaches, including soft measures like coastal zoning, tsunami warning systems, wetland restoration, and hazard simulations. This study demonstrates CHW's utility in hazard assessment and provides valuable insights for sustainable coastal management in the GBA.

3 Study Area and Coastal Hazards

The topography of Odisha is typified by varied landforms, from the rich coastal plains on the Bay of Bengal in the eastern side to the highlands and plateau country that occupy most of the state. Rolling uplands feature the western and northwestern regions, and large floodplains characterize the state's river systems, adding to its distinct geomorphological character (Government of Odisha 2023). The Utkal Plains or Odisha Coastal Plains are the latest formed sedimentary landmasses. They belong to Paleogene and Neogene epochs (about 65–2.6 million years ago) geologically. They have an extremely straight shoreline in the east and a border to the west at roughly 250 ft in the eastern ghats. This region stretches from the north up to the Subarnarekha basin and goes south up to the Rushikulya basin. The Odisha state, which is located on the eastern coast of India, is surrounded by the Bay of Bengal and has a coastline of over 480 km. This coastal area comprises the deltas of Mahanadi, Brahmani, and Baitarani rivers, and floodplains of Subarnarekha, Budhabalanga, and Rushikulya rivers. The coastline of Odisha extends across six districts and 22 administrative blocks. These coastal segments include approximately 88 km in Baleshwar, 52 km in Bhadrak, 83 km in Kendrapara, 59 km in Jagatsinghpur, the longest stretch of around 137 km in Puri, and 61 km along the Ganjam district (Fig. 1). The coast is narrow in the northern part, widest in the middle, narrows near Chilika Lake, and broadens again towards the south. The coastal plain has been shaped by rivers such as the Subarnarekha, Budhabalanga, Baitarani, Brahmani, Mahanadi, and Rushikulya (Fig. 1). Distributaries from the Baitarani to the north and the Mahanadi to the south have encroached, significantly disrupting the Brahmani delta's development. The coastline in this area has a convex shape while the stretch from Chandabali to the Subarnarekha River mouth is concave. (Source: iczmpodisha.org, project report). Significant sediment transport occurs along the coast due to factors such as littoral drift, tidal inflows, and longshore drift, which together create a depositional environment ideal for the formation of backwaters, spits, bars, hooks, lagoons, sandy beaches, and dunes. However, this once accretional environment is now shifting towards an erosional phase. Coastal erosion and submergence have already displaced several coastal communities, with many villages and beaches facing imminent threats of submersion. This pattern is intensifying during the southwest monsoon season, which is made worse by cyclones making landfall as well as storm surges, high tides, and low-pressure systems across the Bay of Bengal. The Odisha Coastal Plains are young, depositional landforms that are part of the Post-Tertiary Period. Their western boundary is defined by the 75-m contour line, which also sets them apart from the Middle Mountainous Region. From the River Subarnarekha in the north to the south of River Rushikulya, this area forms the boundary of West Bengal. This area is the union of a number of deltas of different sizes and shapes created by the great rivers of Odisha, including the Subarnarekha, the Budhabalanga, the Baitarani, the Brahmani, the Mahanadi, and the Rushikulya. Thus, the "Hexadeltaic region" or the "Gift of Six Rivers" are such names given to the coastal plain of Odisha. It stretches along the coast of the Bay of Bengal having the maximum width in the Middle Coastal

Plain (the Mahanadi Delta), narrow in the Northern Coastal Plain (Baleshwar Plain) and narrowest in the Southern Coastal Plain (Ganjam Plain) (Government of Odisha Report 2023). A number of unique natural landmarks may be found along the coast of Odisha, such as Chilika Lake, a well-known bird sanctuary and Asia's largest lagoon ecosystem. It also includes the second-largest mangrove habitat on India's eastern coast, the Bhitarakanika mangrove forest. The area is also home to wildlife reserves like the saltwater crocodile habitat in Bhitarakanika and the spotted deer sanctuary at Balukhanda along the Puri-Konark coast, as well as the internationally notable Olive Ridley turtle nesting grounds at Gahiramatha. However, a number of devastating cyclones, such as the 1999 super cyclone, Strom Phailin in 2013, Hudhud in 2014 and hazardous cyclone Fani in 2019, have had a significant impact on this region. Coastal floods, erosion, saltwater intrusion, vegetation degradation, and infrastructural damage have all resulted from these disasters. The coastline of Odisha is increasingly under risk from climate change and global warming, it has garnered international attention. The World Bank has extended its support through the Integrated Coastal Zone Management program to help mitigate these challenges.

Chilika Lake in Odisha is one of India's key biodiversity hotspots and ranks among the world's largest brackish water lakes. The endangered Irrawaddy dolphin calls it home as well. In the meantime, Asia's second-largest mangrove ecosystem is found in the Bhitarkanika wetlands. Both geographic regions are home to sizable populations whose livelihoods depend on coastal resources. However, ongoing port development poses additional pressure on these ecosystems. Odisha's beaches are also critical nesting grounds for hundreds of thousands of endangered Olive Ridley Sea turtles each year. Unfortunately, these turtles face threats from illegal mechanized fishing in restricted zones, the failure to implement turtle excluder devices, and poorly managed tourism activities. Furthermore, due to ongoing coastal erosion, the primary mass nesting site along the Gahirmatha coastline has shifted northward over the past two decades.

Parts of Odisha's mangrove forests have been cleared for agriculture, fuelwood collection, timber extraction, and large-scale shrimp farming. Bhitarkanika's fragile ecosystem is also under significant threat due to changes in freshwater inflows caused by the construction of upstream hydrological structures. Odisha's coastline experiences extreme tidal variations, with sea levels rising by up to 4 m in some areas, leading to the inundation of coastal land extending as far as 3 km inland. The livelihoods of many impoverished farmers and fishermen are also disrupted by frequent storms, which do significant damage to farmland and fisheries. An additional factor affecting the shoreline is a significant littoral drift that moves around approximately 1.5 million tons of sand shift annually in a north-eastward direction from the southwestern region. Although industrial pollution is minimal in the area, the disposal of untreated or partially treated sewage from coastal settlements, along with effluents from fertilizer plants and industries at Paradip port, remains a concern that needs to be addressed (Banerjee 2012).

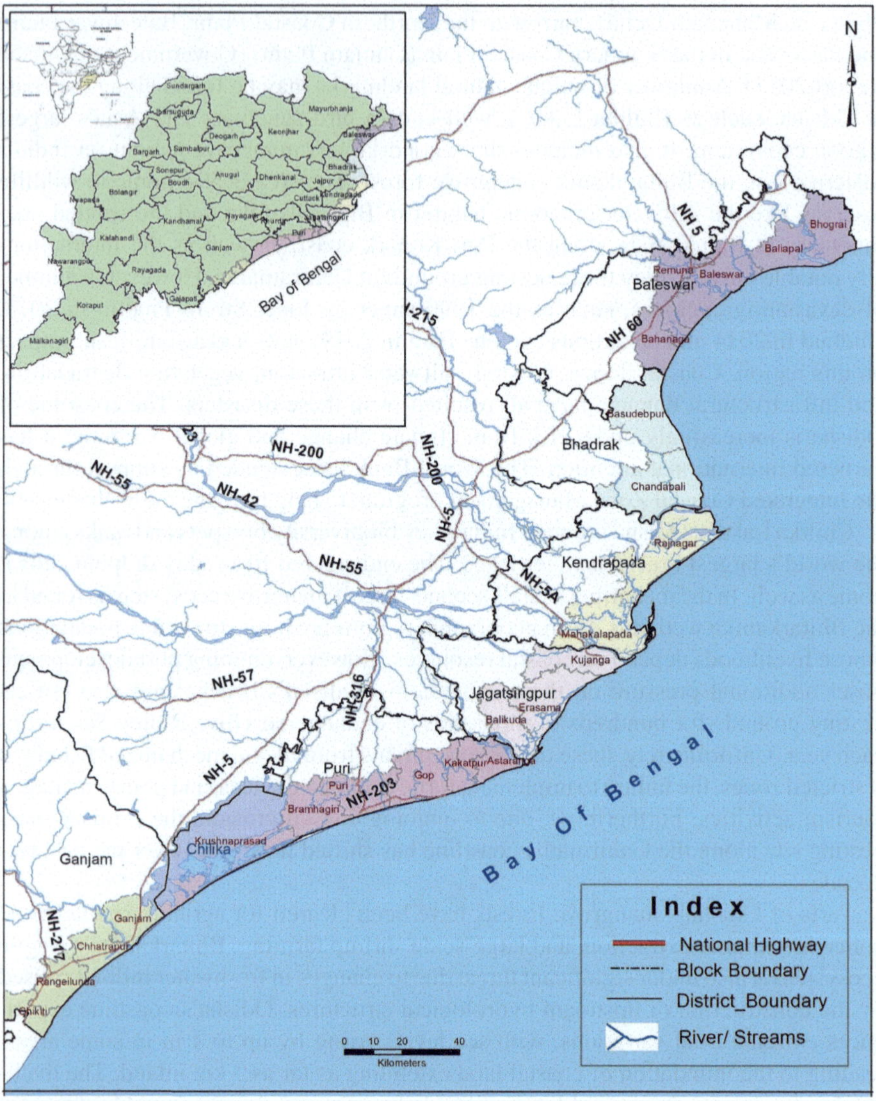

Fig. 1 Location map along with coastline and coastal administrative blocks

This chapter aims to consolidate information on the critical coastal regions of Odisha and analyse the disruption of their ecosystems. Additionally, it seeks to demonstrate the application of ecosystem-based management and ecological risk assessment approaches specifically tailored for the Odisha coastline.

4 Materials and Methods

Coastal mapping science has advanced significantly in the past 50 years, largely due to data from satellite, high-speed geospatial data processing tools and technological advancements. These methods provide an edge over traditional field techniques by offering cost-effectiveness, reducing manual errors, eliminating subjective decision-making, and enabling repeated coverage. The Coastal Hazard Wheel Decision Support System (CHW) developed by the Danish Institute of Hydrology and Danish Technology University (Appelquist and Halsnaes 2015) is being utilized to assess the extent, presence, and intensity of disruption to coastal ecosystems along the Odisha coast.

The current study uses remote sensing data from the public domain, the GIS-based Coastal Hazard Wheel Framework, and geospatial technology to evaluate the risk of ecosystem damage. Spatial data utilized in this assessment include a geological map of the Odisha coast and an administrative map delineating district, block, and village boundaries. Information on wave characteristics, tidal range, and the presence of storm climate is obtained from the model integrated into the original CHW framework (Masselink and Hughes 2003). The study utilizes satellite data from Google Earth, incorporating satellite images with timeline and ground elevation functions (Google Earth 2012), Bing maps, and Rapid Eye satellite images (GRAS 2012). For every coastal classification circle in the CHW framework, a range of geographical and non-spatial variables are used (Table 1).

4.1 Model of the Coastal Hazard Wheel (CHW)

A very flexible and standardized instrument for evaluating the degree of environmental damage along coastal areas is the Coastal Hazard Wheel (CHW, version 3.0). Recommended by the United Nations Environment Programme (UNEP) as a risk assessment tool, it helps coastal managers, planners, and policymakers evaluate how various coastal regions may be impacted by different levels of climate change-induced risks (Appelquist 2012). Initially developed to enhance decision-making, particularly in developing countries (Appelquist and Balstrom 2014), the CHW can be utilized without requiring extensive digital data (UNEP 2012). In its most basic form, the CHW is useful in situations where data availability is limited, enabling a preliminary evaluation of hazards and possible management strategies. However, incorporating field observations and remotely sensed data can yield more accurate results (intermediate level). A more refined, localized assessment of hazards can be achieved with high-resolution, in situ studies, enhancing the results from the first two categories. In this study, the intermediate level of accuracy was achieved through a combination of field observations and satellite imagery.

The CHW is composed of successive layers, each representing various configurations for six key coastal traits (Fig. 2). The wheel begins with the coastline, extending

Table 1 The CHW application uses data sets to map the bio-geophysical characteristics of the coastline of Odisha and their corresponding sources

Sl. No	Bio-geophysical parameters	Data sources and datasets	Spatial resolution
1	Geological features	Elevation profiles derived from Google Earth, 2022 landsat satellite imagery, and a 1:50,000 scale geology map were employed to analyse the study region	Google Earth—0.8 m
2	Wave exposure	Data were obtained from Google Earth (2022) provided the data as well as basic information on the local wind climate. The work of Davies (1980), which was subsequently refined by Masselink and Hughes (2003) and Rosendahl Appelquist (2014), was also cited in relation to global wave environments	
3	Tidal-range	Masselink and Hughes (2003) modified Davies' (1980) mapping of the world's tidal settings. Data on local tide tables was obtained from Chandabali, Gopalpur, and Paradip Port	
4	Coastal vegetation	Using a visual interpretation of the coastline in a Google Earth image from 2022, data from NATMO district planning maps, the coastal vegetation is identified using secondary data from the Forest and Environment Department of the Government of Odisha, 2019 and topographic maps from the Survey of India at a scale of 1:50,000	Google Earth—0.8 m
5	Sedimentary balance	Both Google Earth's timeline function andv satellite imagery were utilized. Depending on whether the geological layout is classified as hard rock or sedimentary/soft rock, there are two methods for evaluating the sediment balance. It is established whether a given coastline has a surplus or balance of sediments for all sedimentary/soft rock coastlines. The timeline feature in Google Earth is used for this, enabling an assessment of the variations in shoreline development over time. Appelquist (2012)	Google Earth—0.8 m

(continued)

Table 1 (continued)

Sl. No	Bio-geophysical parameters	Data sources and datasets	Spatial resolution
6	Storm climate	To analyse the coastal dynamics of Odisha, multiple datasets were utilized. The Cyclone E Atlas of India, published by the India Meteorological Department (IMD), provided essential information regarding the landfall locations of cyclonic disturbances along the Odisha coastline. This dataset was integrated with high-resolution satellite imagery obtained from Google Earth (2022) for spatial referencing. Additionally, global wave environment data based on the models of Davies (1980), as further modified by Masselink and Hughes (2003) and Appelquist (2014), were referenced to understand wave conditions affecting the coast. The approach aligns with the methodology adopted by Panda et al. (2022) in their coastal erosion risk assessment study	Google Earth—0.8 m

up to 200 m inland, and progresses through five rows describing different hazard levels for five coastal hazards sensitive to climate change (Fig. 2). At the centre of the CHW, the starting point features eight potential geological layouts for the coastal area under study: sloping hard rock, sloping soft rock coast, flat hard rock coast, coral island, sedimentary plain, barrier island, delta/low estuary island, and tidal inlet/sand spit/river mouth (Micallef et al. 2017). The Coastal Hazard Wheel (CHW) framework systematically addresses a range of coastal characteristics, progressing outward from the center. These include factors such as wave exposure—categorized as exposed, moderately exposed, or sheltered—along with tidal range, classified as micro (< 2 m), meso (around 2–4 m), or macro (> 4 m). It also considers ecological features like mangroves, tidal flats, intermittent marshes, coral reefs, and areas with or without vegetation.

Additionally, the model evaluates sediment dynamics (whether there's a surplus, deficit, or stable beach condition) and the influence of tropical cyclones on sediment patterns. Each concentric layer of the CHW incorporates elements from geological structure, wave and tidal conditions, to biological cover, sediment status, and prevailing storm regimes. Finally, at the outermost edge, inherent hazard levels (ranging from low to very high) are identified for ecosystem disruption, gradual inundation, saltwater intrusion, erosion, and flooding (Micallef et al. 2017).

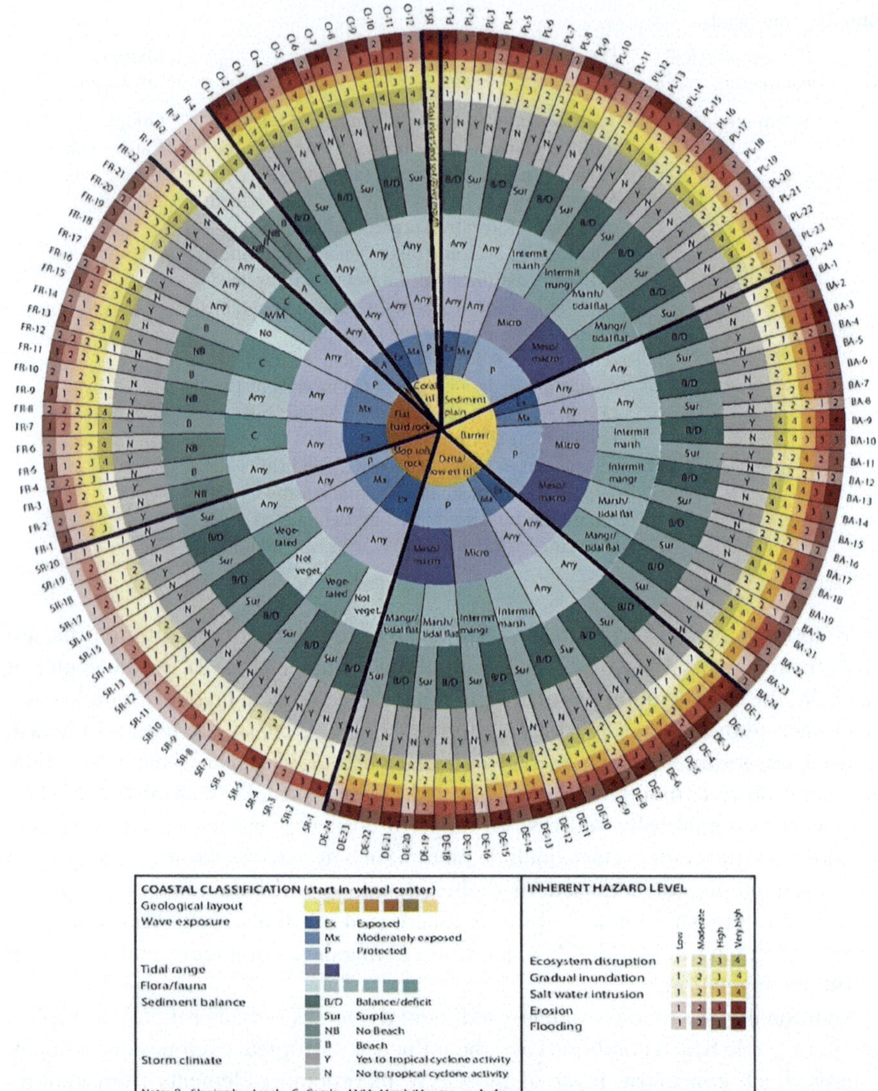

Fig. 2 The Coastal Hazard Wheel (CHW) version 3.0 consists of five hazard zones, six concentric rings for coastal classification, and a set of coastal classification codes. The classification process begins at the centre of the wheel and progresses outward, following the structure proposed by Appelquist (2012) and further refined by Appelquist and Halsnaes (2015)

4.2 Data Layers and Analytical Techniques

ArcGIS Hybrid Bing Maps are used for coastal classification and hazard assessment. The entire categorization process is carried out on the Quick Bird (2022) photos since their spatial resolution (0.65) is higher than that of Bing Maps. First, data on coastal categorization and environmental disturbance levels are stored in a geo-database created in ArcGIS. An updated shoreline map of Odisha is then digitized and stored in this geo-database, using the WGS1984 coordinate system and UTM Zone 45. The coastline of Odisha was manually digitized using the polyline tool, aligning with the approximate mean sea level (MSL) at a zoom scale ranging from 2 to 4 km. It is important to note that the satellite imagery used for this process was captured at various stages of the tidal cycle, generating potential deviations from the real MSL, these changes are considered negligible for the purpose of this assessment, which primarily demands an accurate and current coastal description. With purposeful gaps at tidal inlets and river mouths, the digitizing procedure attains an accuracy of 5–10 m. Islands are represented digitally as separate entities. Subsequent coastal classification and the creation of hazard maps are based on this digitised line.

Using the Quick Bird satellite (spatial resolution 0.65 m), different line features are created to aid in the assessment of sediment balance and coastline slope. In any location where the slope surpasses 3–4%, a sequence of shore-parallel line sections is drawn landward from the shoreline for the purpose of evaluating the slope. When performing coastal classification, this method makes it simple for users to determine whether a coastline area has a slope. By manually moving the pointer over the first 200 m inland from the beach in the picture and noting the elevation levels shown at the bottom of the image window, the slope of each region is determined. Using the Quick Bird satellite imagery, viewed at a zoom scale between 2 and 4 km, this procedure is performed for every 200-m section of the shoreline.

A non-continuous line is drawn along the approximate coastal vegetation line in order to assess the sediment balance. Utilizing the satellite image timeline capability in Google Earth and Arc GIS 10.8 version, sediment stability is assessed when doing coastal classification by looking at satellite photos taken at various intervals to see how the shoreline has changed in comparison to the most current digitalized coastline. Beach width is not considered a reliable indicator of sediment balance, as the images were captured at varying times of the year and during different tidal stages. The vegetation line is seen to be a more accurate indicator for overall evaluation. The initial linear feature is divided into many portions using polylines, each of which represents a distinct coastline type as specified by the CHW framework. Critical bio-geophysical characteristics of the coast, such as vegetation, sediment balance, tidal range, wave exposure, geological structure, storm activity, and others, have been mapped by the CHW-3 during this procedure. These parameters and their subcategories are used to characterize the generic coastal environment (Table 1).

Because the attribute table can only include integers, the coastal environment type is then entered into the ID column in the associated polyline's attribute table. According to the CHW framework, coastal settings are given values between 1 and

131; CHW type PL-1 is given a value of 1 (Fig. 3). Before completing, the user must choose an appropriate type of coast and its extent for each polyline after each coastline stretch has been classified based on CHW and the input data has been listed. Occasionally, a shoreline may cover comparable features over long distances, resulting in polygon lengths that range from a few meters to many kilometers. The original digitized coastline is then divided into segments using these polylines, each of which represents a distinct coastal category (Fig. 3). Following that, hazards levels specified in the CHW document (Appelquist 2012) are entered into an independent attribute table that is connected to the attribute table of the Coastal Classification File. Then, for each of the hazard kinds, five different hazard maps are created, including different hazard levels that are encoded and calculated.

5 Result and Discussion

Applying the CHW framework to Odisha's coastline results in the creation of a summary table outlining the distribution of different coastal ecosystems, along with a series of hazard maps at regional and sub-regional scales. The lengths and percentages of the four most prevalent coastal types in Odisha in relation to the entire coastline are shown in Table 2. The total coastline length of Odisha, as determined by this study, is 480 km, which is far longer than several estimations seen in the literature.

5.1 Circle-1—Geological Layout

Table 2 shows the top four most common coastal types and their twenty-one subtypes based on the geological and geomorphological characteristics of the Odisha coast. The length of these coastal sections is classified using the coastal hazard wheel (Fig. 4), and each zone is designated by a particular set of bio-geophysical criteria that determine its basic characteristics, as well as its sensitivity to ecosystem disruption and the level of risk. In order to comprehend the nature of ecosystem disruption along the Odisha coast, the next section goes into additional detail about the thorough analysis of bio-geophysical parameters and their categorization in the Coastal Hazard Wheel, as well as the geological outline for determining susceptibility and risk. The geological layout types that were relevant to the Odisha coast were selected for analysis from those listed in the standard CHW. These types include slope soft rock (SR), sediment plain (PL), delta/low estuary island (DE), and tidal inlet/sand spit/ river mouth (Fig. 4).

The application of the Coastal Hazard Wheel (CHW), considering the twenty-one coastal types in combination with factors such as wave exposure, tidal range, the presence or absence of coastal vegetation, and storm climate characteristics, has provided the following hazard profile for the Odisha coastline (Table 2).

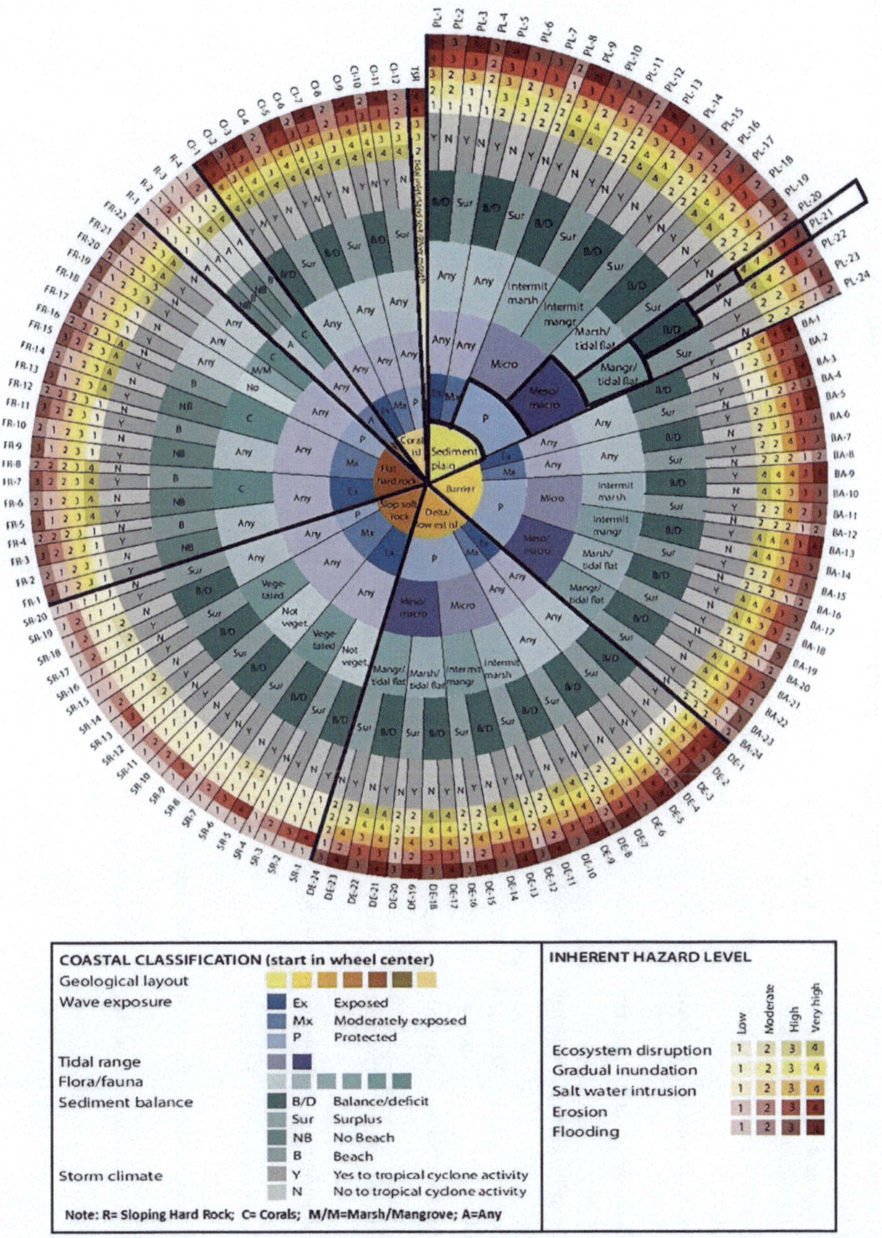

Fig. 3 CHW classification for PL-21, labelled as a sediment plain, indicating a wave-protected area with meso/micro tidal conditions. It is identified as a mangrove tidal flat, experiencing either sediment balance or deficit. Additionally, it is exposed to potential cyclonic storms and is characterized by a very high level of coastal ecosystem disruption

Table 2 Outlines the various regions of the Odisha coast, detailing their biophysical attributes and associated hazard risks, categorized as low (1), moderate (2), high (3), and very high (4)

Sl. No.	CHW	Geological composition	Wave exposure	Tidal range	Flora/fauna	Sediment balance	Storm climate	Ecosystem	Length in km	% of coastline
1	DE-13	Delta/low estuary island	Protected	Micro	Intermitted mangrove	Balance/ deficit	Yes	4	3	1
2	DE-15	Delta/low estuary island	Protected	Micro	Intermitted mangrove	Surplus	Yes	2	9	2
3	DE-16	Delta/low estuary island	Protected	Micro	Intermitted mangrove	Surplus	No	2	41	8
4	DE-21	Delta/low estuary island	Protected	Meso/Micro	Mangrove/ tidal flat	Surplus	Yes	2	6	1
5	DE-5	Delta/low estuary island	Moderately exposed	Any	Any	Balance/ deficit	Yes	2	22	4
6	PL-1	Sediment plain	Exposed	Any	Any	Balance/ deficit	Yes	1	5	1
7	PL-11	Sediment plain	Protected	Micro	Intermitted mangrove	Surplus	Yes	2	6	1
8	PL-13	Sediment plain	Protected	Micro	Intermitted mangrove	Balance/ deficit	Yes	4	17	4
9	PL-16	Sediment plain	Protected	Micro	Intermitted mangrove	Surplus	No	2	5	1
10	PL-17	Sediment plain	Protected	Meso/macro	Marsh/tidal flat	Balance/ deficit	Yes	4	17	4
11	PL-19	Sediment plain	Protected	Meso/macro	Marsh/tidal flat	Surplus	Yes		24	5

(continued)

Table 2 (continued)

Sl. No.	CHW	Geological composition	Wave exposure	Tidal range	Flora/fauna	Sediment balance	Storm climate	Ecosystem	Length in km	% of coastline
12	PL-20	Sediment plain	Protected	Meso/macro	Marsh/tidal flat	Surplus	No	2	26	5
13	PL-21	Sediment plain	Protected	Meso/macro	Mangrove	Balance/deficit	Yes	4	17	3
14	PL-23	Sediment plain	Protected	Meso/macro	Mangrove	Surplus	Yes	2	25	5
15	PL-5	Sediment plain	Moderately exposed	Any	Any	Balance/deficit	Yes	2	103	22
16	PL-8	Sediment plain	Moderately exposed	Any	Any	Surplus	No	2	4	1
17	PL-9	Sediment plain	Protected	Micro	Intermitted marsh	Balance/deficit	Yes	4	48	10
18	SR-10	Slope soft rock	Moderately exposed	Any	Not vegetated	Balance/deficit	No	1	12	3
19	SR-13	Slope soft rock	Moderately exposed	Any	Vegetated	Balance/deficit	Yes	2	14	3
20	SR-9	Slope soft rock	Moderately exposed	Any	Not vegetated	Balance/deficit	Yes	1	2	1
21	TSR	Tidal inlet/Sand spit/River mouth						2	75	16
		Total							480	100

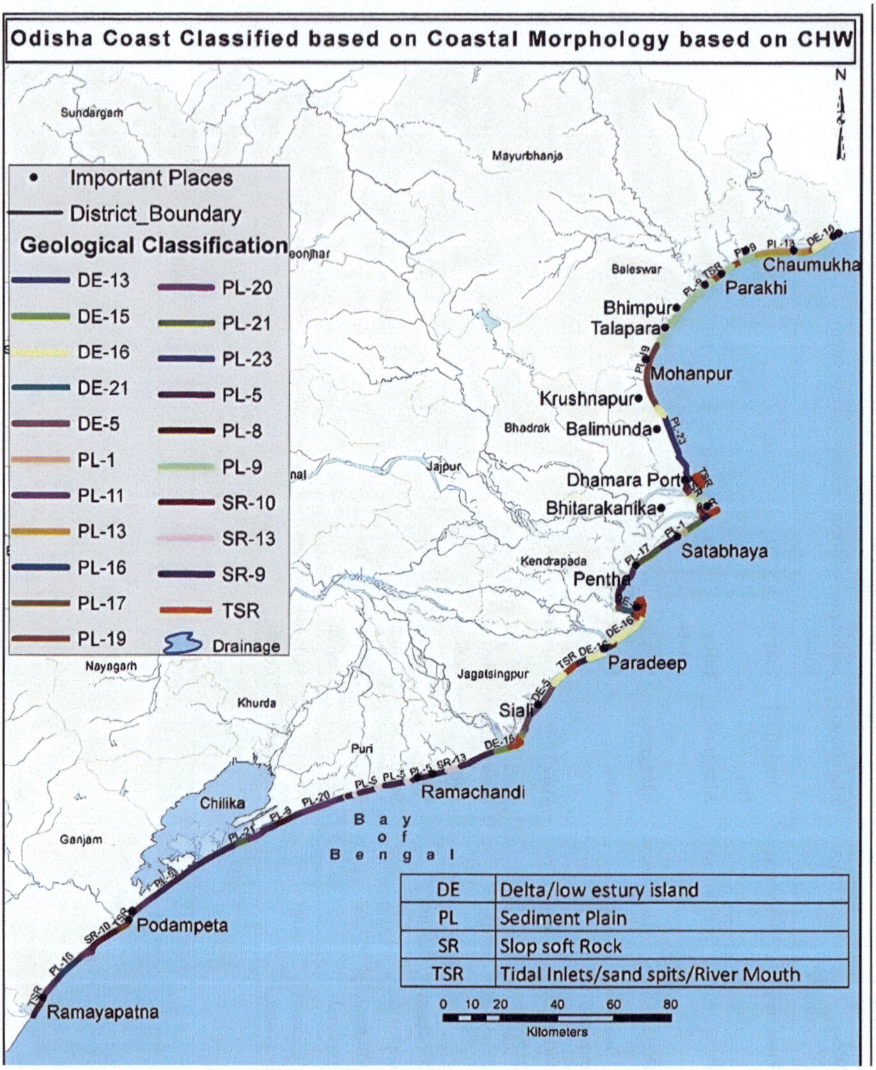

Fig. 4 Coastal Morphology classification based on CHW for Odisha

5.2 *Circle-2—Wave Exposure*

Wave exposure serves as the primary energy source within shore environments and plays a crucial role in shaping coastal morphodynamics. The intensity of wave energy varies, with higher wave heights signifying greater energy levels compared to lower ones. The wave exposure of a coastal region is influenced by the geomorphological configuration of the coastline and the prevailing wind patterns. Coastlines situated in regions affected by swell waves are generally categorized as moderately exposed.The CHW uses a wave environment approach, classifying coastlines as exposed, moderately exposed, or protected. The wave climate is determined using wave maps from the original CHW paper (Appelquist 2012; Masselink and Hughes 2003). Since Odisha lies outside regions with a monsoon climate, the exposure level depends on factors such as free fetch and wind speed. This analysis utilized Google Earth to assess the free fetch for each coastal stretch, with distances less than 10 km, between 10 and 100 km, and greater than 100 km serving as the thresholds for protected, moderately exposed, and exposed coastlines, respectively, in the Coastal Hazard Wheel framework (Fig. 2). In general, the outer sections of Odisha's coastline, directly facing the Bay of Bengal, are classified as exposed or moderately exposed, while the coastlines within inner estuaries are categorized as protected. Following the model, the coastline of Odisha in the southern and central regions faces the full impact of waves from the Bay of Bengal, classifying it as exposed. In contrast, the northern Odisha coast, particularly in the districts of Baleswar and Bhadrak, experiences moderate wave exposure, with tidal influences and significant wave heights generally under one meter. The observed trends in wave exposure along Odisha's coastline are likely influenced by a combination of geomorphological factors, such as the coastline's orientation to the Bay of Bengal, and local climatic conditions, including wind patterns and fetch distances. The southern and central regions, directly facing the Bay of Bengal, experience higher wave energy, resulting in their classification as exposed. In contrast, the northern Odisha coast, shielded by estuaries and affected by lower wave heights, falls under moderate exposure. These findings align with similar studies, such as Appelquist (2012) and Masselink and Hughes (2003), which emphasize the role of wave climate and fetch distance in determining wave exposure levels. Additionally, studies of coastal morphology in other regions have similarly highlighted the influence of offshore conditions, such as swell waves and local wind patterns, on coastal dynamics.

5.3 Circle-3—Tidal Range

The coastal landscape can be significantly affected by tides, serving as an indication of the gravitational pull of the moon and sun on the Earth's hydrosphere. This results in the formation of ocean waves with wavelengths spanning thousands of miles, leading to periodic fluctuations in coastal water levels. Tides follow a daily fluctuation pattern, including diurnal, midday, and mixed tidal cycles (Appelquist 2012). The determination of the tide range is based on tide range maps found in the original CHW documents (Appelquist 2012; Masselink and Hughes 2003). More precise local tidal range data has been collected, despite the fact that Odisha's tides fall between the micro and meso tide types. According to this data, the Mahanadi delta is classified as a micro tide, with the tidal range growing as one moves northward in the state. For coastal classification, this category was selected as a result (Fig. 2). On the delta's present north side, mesotidal conditions might exist. However, it is anticipated that the tidal range at these sites will continue to be close to the line separating micro and meso tide levels. As a result, the entire delta can be classified under the micro and meso tide classifications.

5.4 Circle-4—Flora and Fauna

The coastal environment is influenced by flora and fauna, which in turn affect coastal sediments, erosion, and the coastal ecosystem. The Coastal Hazard Wheel (CHW) classifies natural vegetation into nine distinct types, including intermittent wetlands, intermittent mangroves, swamp/tidal flats, mangrove/mudflats, swamp/mangrove vegetation, barren areas, coral, and a general category encompassing all. Sedimentary plains along the coast often contain intermittent swamps and tidal wetlands. In sheltered, low-energy coastal areas, herbaceous plants that thrive in salty, moist conditions flourish. Wetlands form gradually through the accumulation of sediment, offering protection against coastal flooding and saltwater intrusion. Flora and fauna are evaluated by visually inspecting the coastline on Google Earth, while considering its geographical position and the overall condition of the reef. Odisha's tropical climate is sheltered by coastal lowlands, with mangrove vegetation acting as a barrier, though these barriers are often incomplete due to insufficient suffocation. The delta region lacks coral reefs, so the reef classification does not apply to any part of the delta.

5.5 Circle-5—Sediment Balance

The formation of coastlines is significantly impacted by the balance of sediment. This model classifies sediment balance into two main types: deficit or surplus sediment, particularly for soft rock coastlines. Hard rock coastlines are further categorized into two specific groups: those with beaches and those without. When it comes to coastlines with sedimentary soft rocks, changes in the coastal vegetation line are used to assess the sediment balance. Google Earth data and imagery are used to assess the sediment balance along the coast, making use of its timeline feature to compare the coastline over the past decade with topographic maps. Generally, coastal stretches are assumed to have a surplus of sediment to prevent underestimating certain hazard levels. With Google Earth's images over the past five to ten years, it has been reasonably possible to derive an accurate representation of the sediment balance for Odisha's exposed littoral coasts because changes in the vegetation line are frequently easily observable. However, for protected coastlines, visually assessing smaller temporal changes based on satellite images has proven to be challenging, leading to many of these coasts being categorized as balanced or having a deficit. Apart from the widespread difficulties of predicting sediment balance, Google Earth's timeline function has significant flaws that prevent temporal assessments because some regions are only covered by a single satellite image. This is observed for the coastline from the south of Vanshadhara river to the south of Devi River mouth (PC22), the south of Devi River mouth to the Spit of Bhitarakanika (PC 23), the spit of Bhitarakanika to the mouth of Paga Nadi (PC24), the mouth of Paga Nadi to the south of Subarnarekha River (PC25), the south of Subarnarekha to the north of Rasalpur (PC26), and these coastlines have thus been classified as balanced or having a deficit. On Google Earth, there are occasionally only two images accessible, separated by a few years, which causes uncertainty in the assessments. In these situations, the existing LISS III images are used to augment the necessary data.

5.6 Circle-6—Storm Climate

The arrival of tropical cyclones in coastal regions, along with their strong winds, waves, and rainfall, can significantly impact the shape of the coast and have negative effects on the coastal environment. The model differentiates between areas that experience tropical cyclones and those that do not, regardless of how often they occur. The impact of tropical cyclones on coastal regions (Masselink and Hughes 2003) is depicted in Fig. 2. In regions identified as 'tropical cyclone' areas, the classification indicates 'yes' for tropical cyclone activity and 'no' for areas outside of these regions. This is determined using wave and storm maps from the original CHW publication (Appelquist 2012; Masselink and Hughes 2003). The Odisha coast is categorized as being subject to a storm climate since cyclonic storms occur there frequently. However, the 480 km coastline has been further categorized at the block level based on the presence or absence of storm activity, determined by the passage of cyclones and their landfall points between 1900 and 2019.

5.7 The Inherent Hazard Level for Ecosystem Disruption

In general, coastal regions with inadequately consolidated sedimentary features disintegrate more quickly than those with solid structure. Furthermore, low-lying coasts are more susceptible to rapid erosion, primarily depending on the type of rock (IPCC 2007). Along the Odisha coast, 108 km stretch, which makes up 22% of the total coastline, is labelled as at very high risk for ecosystem disruption. Meanwhile, 74% of the coastline, spanning 354 km, is classified as a moderate-risk zone, and the 18 km stretch, covering 4%, is considered at lower risk for coastal ecosystem disruption (Table 3). The high-risk area for ecosystem disruption is associated with the mangrove regions of Bhitarakanika, Gahirmatha Beach, and the mouth area of Chilika Lagoon. This is particularly associated with the barrier coast and protected estuary coast, which are expected to face additional risk due to rising sea levels from long-term climate change. A portion of the coast is covered by the moderate-risk zones

Table 3 Lists the different types of hazards and their risk levels as a percentage (%) of the entire length of the Odisha coastline

Hazard level	Low (length in km)	%	Moderate (length in Km)	%	High (length in km)	%	Very high (length in Km)	%
Ecosystem disruption	18	4	354	74	–		108	22

for ecosystem disruption. The level of risk indicates potential long-term harm to the coastal ecosystem as a result of climate change and rising sea levels (Fig. 5).

The application model identifies the Chilika mouth region between Satapada and Humma as having a high risk of ecosystem damage. This area, which has a casuarina plantation, is 25 km long and 1 km wide (Fig. 6). With a tidal range of over 1–2 m and a storm climate, this section of the coast is becoming a sediment plain, which is relatively exposed and shows a deficit or sediment balance.

6 Assessment of Coastal Environment Hazards at a Regional Level

At the district and block levels, a comprehensive regional analysis of ecosystem disruption is carried out to pinpoint vulnerable areas and their associated coastal segments. This analysis demonstrates the model's ability to effectively assess hazards on a regional scale, while also identifying the specific locations of these hazards, facilitating targeted intervention. Figure 7 shows the findings of the ecological disruption assessment conducted in the six coastal districts of Ganjam, Puri, Jagatsinghpur, Kendrapara, Baleshwar, and Bhadrak in Odisha. The maps examine the use of the coastal hazard wheel framework for mapping regional hazards, providing a clear view of hazard-prone areas and the varying degrees of risk that necessitate focused attention. These insights serve as a valuable foundation for conservation planning and management strategies, particularly in the context of climate change and its associated impacts in the area. The coastline of the Baleshwar district predominantly falls within a very high-risk zone for ecosystem disruption. The vulnerability of this coast to very high risk is influenced by several factors, including high tides ranging from 2 to 4 m, expansive tidal and mudflats, low-lying topography, tropical cyclones, a gentle coastal slope, and the absence of coastal vegetation. Bhadrak district, on the other hand, faces a moderate risk of ecosystem disruption (Fig. 7). This district shares similar vulnerabilities due to high tides, low-lying coastal areas, and cyclone activity. However, the presence of coastal plantations provides a protective barrier against shore processes that could otherwise impact the coastal zone. Kendrapara's coastline, stretching 83 km, experiences very high hazard levels in four key areas: the Bhitarkanika mangrove and estuarine crocodile sanctuary, the Gahirmatha coast, renowned as the world's largest nesting site for Olive Ridley turtles, the heavily

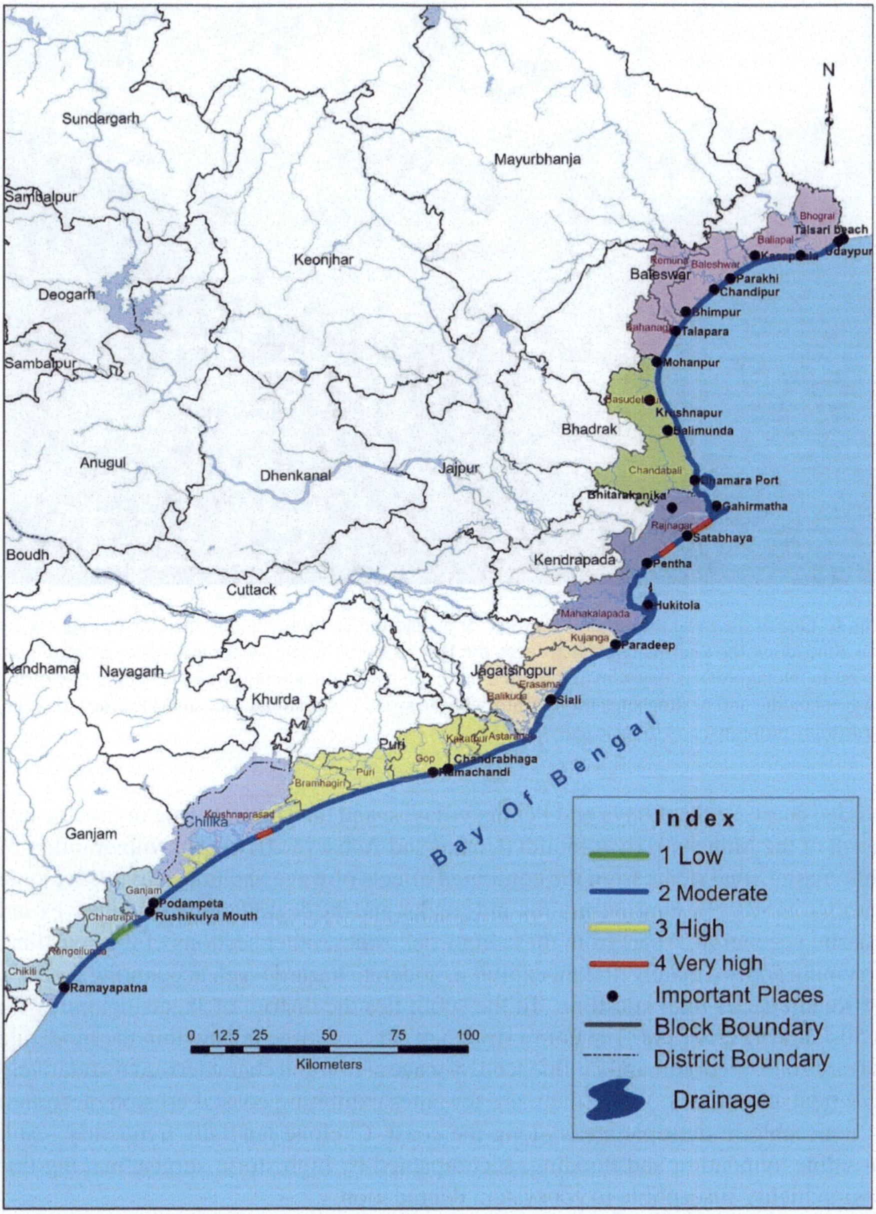

Fig. 5 The map illustrates the coastline of Odisha, categorized by the level of hazard associated with coastal ecosystem disruption

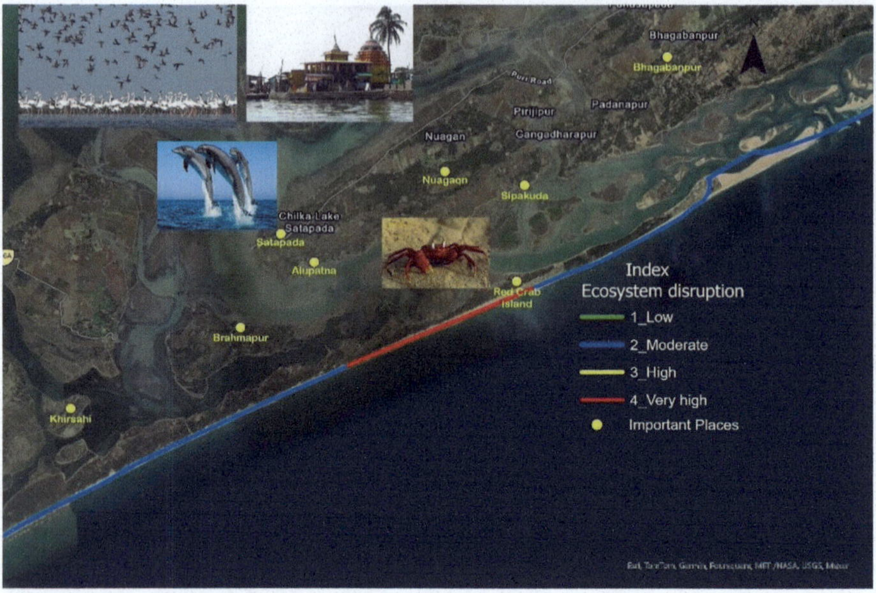

Fig. 6 One of the most important locations for ecosystem disruption along the Odisha coast is the red zone along the Chilika coast that faces the Bay of Bengal. This disruption poses a significant threat to the lagoon's ecosystem and its rich biodiversity. Chilika is renowned for being one of the largest landing and nesting sites for migratory birds, as well as a habitat for estuarine crabs, prawns, Irrawaddy dolphins, and diverse saltwater weeds

eroded coasts of Satabhaya and Pentha villages, and the Hukitola Bay region located north of the Mahanadi river mouth (Chand and Acharya 2010). The vulnerability of this coastal zone stems from the combined effects of wave and tide activity, cyclonic events, numerous estuarine mouths and creeks, and the degradation of mangroves and coastal vegetation. Apart from these high-risk zones, other sections of the coastline, spanning approximately 46 km, exhibit a moderate hazard level, accounting for 55% of Kendrapara's total coastline. To the south lies the district of Jagatsinghpur, with a 59-km-long coastline. The entire stretch of this coastline falls within the moderate hazard zone, predominantly influenced by wave action and characterized by relatively low tidal amplitudes. While there are few sites exhibiting coastal erosion, accretion is noticeable in multiple areas along the coast. Cyclone landfalls frequently result in saline inundation and flooding, accompanied by high storm surges, making this region highly susceptible to ecosystem degradation.

Puri, with a coastline of 137 km, faces moderate vulnerability to coastal hazards. The Chilika coast, however, is extremely prone to ecosystem disruption, in contrast to the rest of the coastline, which faces moderate risk (Fig. 7). High sand dunes and beach plantations at various points along the coast offer some defense against storm surges and flooding. In the southernmost district of Ganjam, which has a 61-km coastline, the coastal zone is characterized by high wave energy and a low tidal

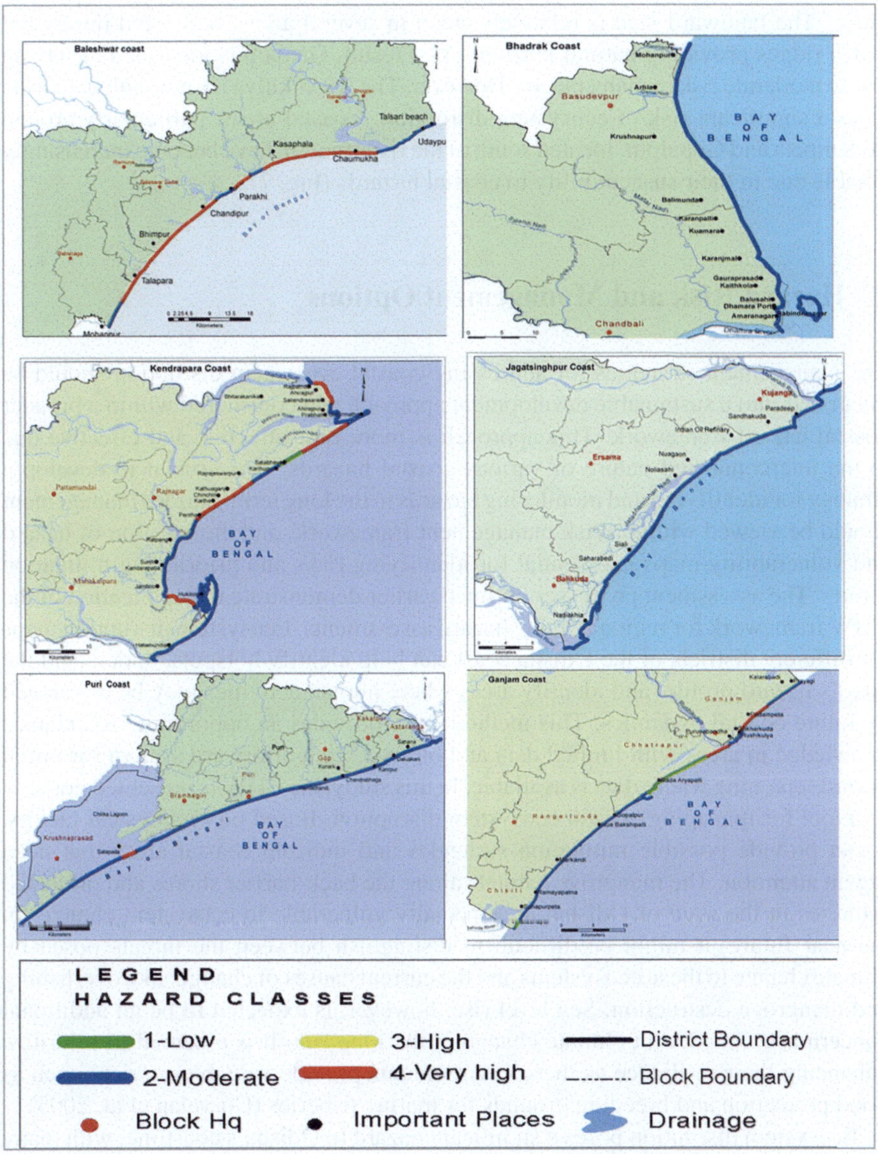

Fig. 7 District-level vulnerability and risk assessment of ecosystem disruption along the Odisha coast utilizing the CHW model

range. The landward side is relatively steep in several areas, with sand dunes and beach ridges providing natural barriers. As a result, Ganjam is identified as having low to moderate risk in many places. However, The Rushikulya river mouth in Odisha faces a significant risk of ecosystem disruption. Coastal areas, particularly around Podampeta and Gopalpur, located south of the river mouth, have become increasingly notable due to their susceptibility to coastal hazards (Fig. 7).

7 Hazard Risk and Management Options

The management strategies for addressing coastal ecosystem disruption should be integrated into a sustainable development approach and considered within a broader coastal hazard framework. This approach is more cost-effective and effective due to the interconnected nature of various coastal hazards. It is crucial to develop a strategy for identifying and monitoring hazards in the long term. Hazard management should be viewed within a risk management framework, and the creation of hazard and vulnerability maps is essential for identifying risks and prioritizing mitigation efforts. The assessment processes outlined earlier demonstrate the application of the CHW framework for regional multi-hazard assessments. Ecosystem disruption maps for different districts of the Odisha coast can help identify hotspots, understand the coast's hazard profile, and identify areas where human activities may be threatened by future coastal dynamics. This methodology provides an opportunity to enhance knowledge in areas with limited data and offers a straightforward system for initial hazard screening where data is available. In this study, the CHW is presented not only as a tool for managing coastal ecosystem disruption. Based on biophysical factors, it can provide possible mitigation strategies and indicate coastal areas that need urgent attention. The mangrove habitats along the back-barrier shores and protected estuaries in the state of Odisha are especially vulnerable to ecosystem change. In the near future, it might be difficult to distinguish between the threats posed by climate change to these ecosystems and the current causes of change, like overfishing and mangrove destruction. Sea level rise, however, is expected to be an additional concern associated with climate change in the long run. It is essential to prioritize enhancing their resilience as these environments provide valuable services such as flood protection and breeding grounds for marine fisheries (Corvalan et al. 2005).

Ecosystem disruption poses a significant hazard to Odisha's coastline, with many areas already affected. While the state has limited experience in coastal zoning, options like dune construction, ecosystem-based management, fluvial sediment management, groundwater management, and wetland restoration could effectively mitigate hazards in densely populated coastal regions. The Odisha coast faces risks related to climate change, but these vary by location within the multi-hazard framework. Management priorities should focus on areas with high and very high-risk levels. Table 4 identifies villages categorized as moderate, high, and very high risk, along with their geological backgrounds and potential management strategies at the district level. In Baleshwar district, five villages in Bhogarai, Baliapal, and Baleshwar

Sadar have been prioritized for coastal zoning, fluvial sediment management, and wetland restoration due to their characteristics of high tide, low estuarine islands, beaches, tidal inlets, and tidal flats.

Bhadrak district faces moderate ecosystem disruption risks, particularly in the Basudevpur and Chandbali blocks, with vulnerable villages such as Amarnagar, Panditanilakanthnagar, and Rabindranagar. The coastline features delta regions, low estuary islands, and coastal plains, necessitating soft management solutions like beach nourishment and wetland restoration.

Kendrapada district also shows vulnerability to ecosystem disruption, with Rajnagar and Mahakalpara blocks experiencing low, moderate, and very high risks. Coastal villages in Rajnagar (18) and approximately 10 additional villages in coastal plains and estuaries are classified under moderate to very high risk, highlighting the need for similar management strategies, including dune rehabilitation and ecosystem-based management. In Jagatsinghpur district, ten villages within the Kujanga, Ersama, and Balikuda blocks are at risk, requiring prioritized management options like beach nourishment and wetland restoration. These areas encompass high tide and low estuarine islands and coastal plains.

Puri district is particularly vulnerable, with high and very high-risk levels affecting blocks like Astaranga and Gop. Vulnerable coastal villages (over 40) across multiple blocks necessitate comprehensive management strategies similar to those in other districts.

Ganjam district, while at low to moderate risk, includes vulnerable coastal villages in Ganjam, Rangeilunda, and Chikiti blocks, emphasizing the need for beach nourishment and ecosystem-based management to mitigate potential hazards effectively.

8 Limitations of the Study

The study has certain limitations, primarily due to its reliance on publicly available geophysical data and remote sensing from open sources, which are constrained by limited data availability. While the assessment process is straightforward, human error may arise due to the subjective nature of the methodology. However, a more thorough grasp of the coastal issues would be possible by adding socioeconomic data, information on infrastructure, human activities, land use/land cover, and field observations to the CHW evaluation. This method would support coastal stakeholders in developing successful plans and choices for mitigation and intervention activities.

Table 4 Management options for the coastal ecosystem disruption of the CHW for Odisha

District	Block	Villages	Geological background	Management options
Baleshwar	Bhograi	Udaypur	DE-16 Delta and Low estuary island and TSR (Tidal inlet, Sandspit and River mouth)	Beach nourishment, ecosystem based management
	Baliapal	Chaumukh		Fluvial sediment management, Dune construction/ rehabilitation
	Baleshwar Sadar	Kasafal, Parakhi and Srikona		
Bhadrak	Chandbali	Amarnagar, Panditanilakanthnagar, Rabindranagar	DE-5 and 16(Delta and Low estuary island), PL-1,5,8 and 21 (Coastal Plain) and TSR (Tidal inlet, Sandspit and River mouth)	Beach nourishment,
				Ecosystem Based Management,
				Fluvial sediment management, Dune construction/ rehabilitation, Wetland restoration
Kendrapara	Rajnagar	All villages of Rajnagar	DE-16- Delta and Low estuary island, PL-5 and 23 (Coastal Plain) and TSR (Tidal inlet, Sand spit and River mouth)	Beach nourishment
				Ecosystem Based Management,

(continued)

Table 4 (continued)

District	Block	Villages	Geological background	Management options
	Mahakalapara	Banapara, Oriyasal, Baro, Paunsiapal, Kandarapatia, Jamboo, Bhitarakharanasi and Batighara of Mahakalpara block		Fluvial sediment management, Dune construction/ rehabilitation, Wetland restoration
Jagatsinghpur	Kujanga	Sandhakuda	DE-5 and 16-Delta and Low estuary island, PL-5 (Coastal Plain) and TSR (Tidal inlet, Sandspit and River mouth)	Beach nourishment
	Ersama	Abhayachandapur, Kansaipatia, Dinkia, Jatadhara Padampur, Saharabedi, Harisipurgar of Ersama		Ecosystem based management
	Balikuda	Nadiakhia of Balikuda		Fluvial sediment management, Dune construction/ rehabilitation, Wetland restoration
Puri	Astaranga	All villages of Astaranga	Delta and Low Estuary Island (DE-5), Coastal Plain (PL-5 and PL-20), Sloping Soft Rock Coast (SR-13), and Tidal Inlet, Sandspit, and River Mouth (TSR)	Beach nourishment
	Gop	Konark, Madhipur, Salakata, Sahu Khanata and Gadabangara		Ecosystem Based Management,

(continued)

Table 4 (continued)

District	Block	Villages	Geological background	Management options
	Kakatpur	Guhalpur, Godhanpara, Jamra		Fluvial sediment management, Dune construction/ rehabilitation, Wetland restoration
	Puri	Chakratirtha, Bankimuhan, Gandhighat, Balukhand, Balisahi, Baliapanda, Beladala and Mohinipur		
	Brahmagiri	Sipasurubili		
	Krushna prasad	All Blocks of Krushnaprasad		
Ganjam	Ganjam	All blocks of Ganjam	PL-1,5,8,16, and PL-20 (Coastal Plain), SR-9 and 10(Sloping soft rock coast) and TSR	Beach nourishment
	Rangeilunda	Kirtipur, Laudiguan, Paphasinuapada	(Tidal inlet, Sand spit and River mouth)	Ecosystem based management
		Markandi, Indrakhi, Dhepanuapada of Rangeilunda		Fluvial sediment management, Dune construction/ rehabilitation, Wetland restoration
	Chikiti	All blocks of Chikiti		

9 Conclusion

The provided data sets are valuable resources that offer essential guidance for coastal managers, civil protection authorities, and decision-makers in implementing suitable measures to protect and mitigate the vulnerability of the Odisha coastline to hazards. Future research could consider additional parameters for characterizing coastal environments, especially those with regional significance. Factors such as high humidity can impact weathering rates, leading to increased susceptibility to ecosystem disruption. Furthermore, the level of human presence and intervention in coastal areas can influence the inherent hazard levels. For instance, the construction of coastal protection structures, like seawalls, will undoubtedly affect the geological layout parameter currently under consideration. Future studies may benefit from incorporating additional parameters relevant to the characterization of coastal environments, particularly those of regional significance, when selecting specific methods and strategies for particular locations. Furthermore, it is crucial to consider the "level of human intervention or presence" in coastal environments, as it can significantly influence inherent hazard levels. This aspect represents a limitation of the current study in the CHW-based decision-making process. Utilizing the CHW model, this assessment of coastal vulnerability and risk analysis concerning ecosystem disruption along the Odisha coast especially in the context of climate change—has been effectively highlighted. Integrating local observational data with high-resolution satellite imagery can enhance the accuracy of the model's outputs. The management strategies derived from this study provide valuable insights for decision-making processes and can be further refined by accounting for the results of human interventions along the coast and the socioeconomic vulnerabilities of coastal communities.

The CHW technique has been recognized as a highly effective tool in the risk analysis process. Considering vulnerability as the product of hazard intensity and susceptibility, the calculation and identification of coastal hazard intensity are crucial for assessing vulnerability and, ultimately, risk.

References

Appelquist LR (2012) Generic framework for meso-scale assessment of climate change hazards in coastal environments. J Coast Conserv 17(1):59–74

Appelquist L, Balstrom T (2014) Application of the coastal hazard wheel methodology for coastal multi-hazard assessment and management in the state of Djibouti. Clim Risk Manag 3:79–95

Appelquist L, Halsnaes K (2015) The coastal hazard wheel system for coastal multi-hazard assessment and management in a changing climate. J Coast Conserv 19(2):157–179

Banerjee A (2012) Protecting India's coastline: Odisha. https://www.worldbank.org/en/news/feature/2012/10/11/protecting-indias-coastline-odisha

Bhardwaj S, Gupta AK (2024) Disaster risk reduction through climate adaptive development: strategies and road ahead. In: Gupta AK, Gupta A, Acharya P (eds) Disaster risk and management under climate change. Springer, pp 631–648. https://doi.org/10.1007/978-981-99-4105-6_31

Chand P, Acharya P (2010) Shoreline change and sea level rise along coast of Bhitarkanika Wildlife Sanctuary, Orissa: an analytical approach of remote sensing and statistical techniques. Int J Geomat Geosci 1(3):436–455

Corvalan C et al (2005) Ecosystems and human well-being: health synthesis, a report of the millennium ecosystem assessment

Davies JL (1980) Geographical variation in coastal development (2nd ed). London, Longman

Government of Odisha Report (2023) https://odisha.gov.in/odisha-profile/topography#:~:text=The%20Odisha%20Coastal%20Plains%20This%20region%20is,the%20Brahmani%2C%20the%20Mahanadi%2C%20and%20the%20Rushikulya

IPCC (2007) In: Parry ML, Canziani OF, Palutikof JP, van der Linden PJ, Hanson CE (eds) Climate change 2007. Impacts, adaptation and vulnerability: contribution of working group II to the fourth assessment report of the intergovernmental panel on climate change. Cambridge University Press, Cambridge, UK, p. 976

IPCC (2013) Summary for policymakers. In: Stocker TF, Qin D, Plattner G-K, Tignor M, Allen SK, Boschung J, Nauels A, Xia Y, Bex V, Midgley PM (eds) Climate change 2013: the physical science basis. Contribution of working group I to the fifth assessment report of the intergovernmental panel on climate change. Cambridge University Press, Cambridge

IPCC (2021) Sixth assessment report (AR6). Intergovernmental Panel on Climate Change

Liew M et al (2020) Prevention and control measures for coastal erosion in northern high-latitude communities: a systematic review based on Alaskan case studies. Environ Res Lett 15:093002

Masselink G, Hughes MG (2003) Introduction to coastal processes and geomorphology. London, Hodder Arnold

Micallef S et al (2017) Application of the coastal hazard wheel to assess erosion on the Maltese coast. Ocean Coast Manag 156:209–222

Mishra M, Acharya T, Kar D (2021) Vulnerability of nesting ground of olive Ridley turtle to severe cyclonic storm Fani: a case study from Rushikulya rookery, East coast of India. J Coast Conserv 25:39

Mishra M, Panda GK (2018) Coastal erosion and shoreline change in Ganjam coast along east coast of India. J Earth Sci Climatic Change

Mishra M, Panda GK, Dandapat K, Chatterjee U (2022) Vulnerability assessment of the inherent hazards of climate change on the coastal environment of the Mahanadi delta, east coast of India. In: Chatterjee U, Akanwa AO, Kumar S, Singh SK, Dutta Roy A (eds) Ecological footprints of climate change. Springer Climate. Springer, Cham

Nichols RJ, Hanson S, Herweijer C, Patmore N, Hallegatte S, Corfee-Morlot J, Chateau J, Muir-Wood R (2007) Ranking port cities with high exposure and vulnerability to climate extremes—exposure estimates. OECD environmental working paper no. 1, Organisation for Economic Co-operation and Development (OECD), Paris

Panda GK, Chatterjee U, Mishra M, Mohanty A (2022) Vulnerability assessment of environmental degradation due to coastal hazards using geospatial technology—a study of Puri coast along the Mahanadi Delta, India. Arab J Geosci 15. Article 1322

Population Reference Bureau (2023) Global population trends and coastal settlement. Population Reference Bureau

Ramachandran S, Sundaramoorthy S, Krishnamoorthy R, Devasenapathy J, Thanikachalam M (1998) Application of remote sensing and GIS to coastal wetland ecology of Tamil Nadu and Andaman and Nicobar group of islands with special reference to mangroves. Curr Sci Assoc 75(3):236–244

Rao KN, Subraelu P, Rao TV et al (2008) Sea-level rise and coastal vulnerability: an assessment of Andhra Pradesh coast, India through remote sensing and GIS. J Coast Conserv 12:195–207

Rosendahl Appelquist L (2014) Coastal vulnerability and the role of environmental narratives in governance: the case of climate change and coastal erosion in Southern Scandinavia

Su Q, Li Z, Li G, Zhu D, Hu P (2021) Application of the coastal hazard wheel for coastal multi-hazard assessment and management in the Guang-Dong-Hongkong-Macao Greater Bay Area. Sustainability 13(22):12623. https://doi.org/10.3390/su132212623

Thanikachalam M, Ramachandran S (2003) Shoreline and coral reef ecosystem changes in Gulf of Mannar, Southeast coast of India: a remote sensing and GIS approach. J Indian Soc Remote Sens 31(3):157–173. https://doi.org/10.1007/BF03030823

Wetland Vegetation Species Diversity and Productivity Assessment: Implications on Ecosystem Services

Timothy Dube and Tatenda Musasa

Abstract Wetlands are engines of the earth as they supply provide ecological, economic and cultural benefits. However, these invaluable landscapes are increasingly threatened by expanding human footprint, droughts as well as the changing climate change. This chapter provides a comprehensive documentation of wetlands status, highlighting the profound impacts of alterations in precipitation and temperature patterns, sea level rise, and carbon dynamics on these ecosystems. The chapter further provides an overview of the diverse and complex challenges faced by wetlands and reflects on possible adaptive and mitigation strategies that contribute to sound environmental performance. It is clear that, wetlands are under threat from human activities, which calls for the need to harness the potential of modern technology for instance remote sensing. By understanding and addressing climate change impacts, efforts to preserve wetlands and their vital functions for future generations will be put in place to prevent further degradation. Lastly, the chapter provides recommendation to proffer solutions that counteract the emerging threats to enhance ecosystem resilience.

Keywords Conservation · Climate change and variability · Ecosystem services · Remote sensing · Vegetation wetlands systems

1 Introduction

Globally, wetlands are regarded as vital ecosystems that offer necessary products and services that are vital to all living things. Globally, wetlands are estimated to occupy 6% of the total land surface (Erwin 2009; Junk et al. 2013). In terms of geographical

T. Dube (✉) · T. Musasa
Institute for Water Studies, University of the Western Cape, Cape Town, South Africa
e-mail: tidube@uwc.ac.za

T. Musasa
Department of Geography Environmental Sustainability and Resilience Building, Midlands State University, Gweru, Zimbabwe

O. Mutanga et al. (eds.), *Revealing Ecosystem Services Through Geospatial Technologies*, Springer Remote Sensing/Photogrammetry,
https://doi.org/10.1007/978-3-031-98048-0_9

distribution, literature shows that about 50% of the global wetlands lie in high North Latitude between 50° and 70 N° characterized by permanent permafrost conditions (Davidson et al., 2018).

Generally, wetlands can manifest in several classes with the common ones being marine/coastal, inland, and those that are man-made wetlands (Erwin 2009; Rebelo et al. 2010; Davidson et al. 2018). However, their classification is further refined based on the prevailing nature of environmental and climatic conditions. The resultant classification includes more common categories such as bogs, dambo, floodplains, marshes, and swamps (Alexander and McInnes 2012; Davidson et al. 2018). There is also the Danube River that passes through Europe forming the Danube Delta, which is an integral ecosystem (Davies et al. 2008).

Wetlands are home to aquatic plants due to their natural occurring conditions (Grasel et al. 2021). This role is important since it can show the health condition of the wetlands (Davies et al. 2008). Numerous animal species find a home in the U.S. state of Florida which is a unique wetland ecosystem, with sawgrass marshes, pine fatwoods, and coastal mangroves (Fig. 1).

Wetlands play an essential role in ecosystem functioning. In the Okavango Delta in Botswana, it is not a coincidence that a substantial proportion (67%) of the living creatures in the ecosystems closely live in wetlands and rely on vegetation. This clearly indicates that wetland vegetation contributes immensely to ecosystem functioning as it acts as a host for various key species (Darko and Mbaiwa 2014). In Zimbabwe and Zambia, grasslands have been primarily used for livestock grazing which has exponentially increased over the past decade which thus contributes to ecosystem functioning (Davies et al. 2008; Kotze and Malan 2010). However, wetlands are

Fig. 1 Everglades wetland in view with the varying types of vegetation. Adapted from Davidson et al. (2018)

subjected to immense pressure from human activities and if not monitored they will be loss and possibility of drying up. Wetlands are threatened by human activities such as urbanization, cultivation which all result in vegetation clearance yet it is an important aspect (Socolar et al. 2016).

2 Wetland Vegetation Diversity

It is important to recognize that wetlands serve as crucial reservoirs of plant diversity, hosting unique, rare, and endangered species as well as distinctive ecological communities (Davies et al. 2008; Oliveira-Filho et al. 2021). These wetlands have variations in terms of classification and also share a particular range of abiotic features for instance geomorphological and hydrological (Keddy 2010; Draper et al. 2018; Socolar et al. 2016; Darwall et al. 2018). It is clear from the analysis that wetlands are crucial for ecosystems functioning as they support a diversity of vegetation (Fig. 2).

Wetlands are an important ecosystem that is a renowned rich reservoir of biodiversity that are vital for humanity and nature (Darwall et al. 2018). Research estimates show that about 40% of the world's plant and animal species depend on wetlands, inclusive of 30% of all known fish species (Keddy 2010; Draper et al. 2018). To add on, wetlands are home to a vast amount of biodiversity, including thousands of plant species, and are essential for the survival of many species. For example, fens reduce the risk of floods and also improve water quality which is key to communities.

Fig. 2 Wetland vegetation diversity (fern, evergreen shrubs and water lilies)

3 Methods for Assessing Wetland Ecosystems Productivity

Wetlands can be regarded as one of the most important ecosystems of the earth. In light of this several methods have been applied to assess the productivity in wetlands across several countries with varying degrees of success. Wang et al. (2022), reiterates that experimental methods can be used to assess productivity of wetlands. Ten randomly chosen plants were used to determine plant height in a study by Wang et al. (2022). Using this method, the number of individuals in each quadrat was used to determine the plant diversity. To ascertain plant coverage, observations were done based on transect walks. In this instance, every plant in every quadrat was taken out at ground level and gathered for analysis (Wang et al. 2022).

The adoption of socio-cultural methods has also brought a paradigm shift to wetland ecosystem service assessment (Kotze and Malan 2010). This however, excludes economic value which is often used to value wetland services. Wang et al. (2022) revealed that socio-cultural methodologies are employed to quantify and delineate fundamental regulating, provisioning and cultural services inherent to the ecosystem. This can be done by use of "traditional" socio-cultural data collection techniques. For example, Adekola et al. (2008) used these to assess local people perceptions on the nature of ecosystem services and observed that like any other ecosystem, wetland provide numerous services and goods which local communities have been depending upon.

Wetlands Ecosystem Services Protocol (WESP) is also another typical method that is used to assess wetland productivity. In actual sense the WESP is a standardized method that adopts field survey methodologies and office questions to assess a wetland's functions and attributes (Kotze and Malan 2010). In this case, WESP can be used to compare wetlands in a region, prioritize conservation, and evaluate wetland restoration projects based on the potential benefits of the wetland. In British Columbia, the tool is being developed by the BCWF's Watershed Team in partnership with First Nations across the province. Currently, land managers and decision makers have no standardized tools to measure the relative functional attributes of wetlands which leads to discrepancies and subjectivity in decisions related to protection and mitigation measures. Functional assessment tools are becoming a standardized practice in many other jurisdictions in North America (Davies et al. 2008). Although the model itself is not a regulatory tool, it can support individuals involved in decision making processes. Therefore, it is clear that although the method is robust it also has its share of weakness which may need to be complemented by other key methods and approaches.

Wetland productivity has also been assessed using rapid assessment methods. The main advantage of these methods is that they necessitate reduced time in the field and diminished taxonomic skill compared to more quantitative methodologies employed in other contexts. The Rapid Assessment of Wetland Ecosystem Services (RAWES) approach is flexible and can be used to assess different habitat units within a wetland complex. Previous studies e.g. Kotze and Malan have adopted the RAWES approach to assess the productivity of wetlands in semi-arid regions and observed that productivity varies with wetland type. In its strict sense, the benefits derived from a specific wetland can be determined by factors such as accessibility and remoteness as well as the prevailing natural conditions since these drive key processes.

4 Linkages Between Vegetation Diversity and Ecosystem Services

It is crucial to remember that in order to balance the demands, an ecosystem must have a connection between biodiversity, ecological services, and human well-being. As such biodiversity underpins ecosystems, which in turn provide ecosystem services and these services contribute to human well-being especially in resource constrained communities (Millenium Ecosystem Assessment 2005). A study by Kotze and Malan (2010) shows that vegetation plays a key role in various ways as it is essential in the provision of ecosystem services. For example, tree species diversity in temperate forests can promote greater ecosystem service outputs, such as water yield and soil retention. To support this, studies by Mandishona and Knight (2019) and Musasa and Marambanyika revealed that local communities often use vegetation as a source of medicine which clearly demonstrates the fundamental role in as far as service provision is concerned.

Literature shows that the has been a gap between vegetation and ecosystem service provision in wetland across the globe (Adekola et al. 2008; Turpie et al. 2010). For example, vegetation found in wetland areas can be used as an important element for water filtration. A study by Junk et al. (2013) reported that vegetation plays a key role in water filtration. Some types of wetlands are so good at this filtration function that environmental managers construct similar artificial wetlands to treat storm water and wastewater (Chuma et al. 2008).

Wetlands are also important in as far as biological productivity is concerned (Adekola et al. 2008; Chuma et al. 2008). In other cases, vegetation also acts as an important source for water retention. For example, tree species diversity in temperate forests can promote greater ecosystem service outputs, such as water yield and soil retention. Transboundary basin wetlands such as the Kafue Flats, Mafefe and Missavene, provide several benefits such as fish, reed cutting, beekeeping for honey, water for brick moulding, sand extraction, thatch grass (Adekola et al. 2008; Chuma et al. 2008). Just like any other wetland in Africa, the majority of rural people in GaMampa wetland collect reeds between June and July for roofing of houses when they are not working on their fields (Adekola et al. 2008).

5 Impacts of Environmental Change on Wetland Vegetation Species

Global environmental change poses significant threats in as far as wetland vegetation dynamics is concerned. It can manifest in several ways including one of the pressing threats in the form of climate change (Kebtieneh 2017). In the developing countries, where the impact of climate change has been largely felt many wetlands will be destroyed. This will lead to wetland loss and possibly drying up especially in naturally occurring wetlands in semi-arid regions. For example, the Okavango Delta in Botswana often experiences severe environmental change (Darkoh and Mbaiwa 2014).

Literature review shows that, environmental changes in wetlands can be influenced by several factors (Adekola et al. 2008; Kotze and Malan 2010). These changes can manifest as climate change which can be observed by increased carbon dioxide levels, increasing precipitation and higher temperature. These can have either positive or negative impacts depending on the emerging conditions. In wetlands in the United States increased precipitation levels in the Everglades resulted in flourishing on vegetation and increased diversity which contributes to healthy ecosystems (Keddy 2010). Some groups of plants are more responsive to higher CO_2 concentrations than others because of fundamental physiological differences (Davies et al. 2008). For example, when CO_2 levels are increased, photosynthesis in the marshes on the Chesapeake Bay, where the majority of pertinent research has been conducted, rises noticeably. According to this research, in the long run, the more responsive, the less receptive plant species may eventually be displaced. Therefore, analysis of information in the subsequent sections shows that environmental change influences vegetation dynamics in wetlands in various areas.

It is also important to note that, water plants which constitute an integral part of vegetation dynamics in wetlands often rely on a combination of geomorphologic and hydrologic factors which influence ground water availability (Kotze 2013). This implies that when ground water is available these various types of water plants are enriched (Musasa and Marambanyika 2023). On the other hand, wetlands dominated by precipitation are already at the "dry end" of the spectrum of wetland types which also has a compromise the diversity of vegetation in wetlands (Kotze and Malan 2010 Davidson et al. 2018). Therefore, when these types of wetlands are not able to resist fires the vegetation dynamics are affected which may affect the health of the particular ecosystem in the area of jurisdiction.

6 Remote Sensing and Geospatial Techniques for Monitoring Wetland Health

Remote sensing has emerged as one of the plausible solutions to most environmental problems (Meftah et al. 2021). The adoption of any particular aspect of technology is depended on activity and extent covered (Meftah et al. 2021). In this manner the launch of Satellites with varying magnitude have potential to cover much more land space than planes, and can monitor areas on a regular basis. In addition, digital sensors permitted the transmission of imagery from the platform itself to the ground station without a problematic film return system as was used on CORONA and other satellites (Toth and Jóźków 2016).

Ground-based and airborne platforms (including both manned and unmanned aircraft), are rather common, even though satellites are regarded as key platforms since they gather a large portion of the data needed by governmental and academic institutions. Depending on the user's requirements, various platforms can support a variety of sensor types, and each platform has pros and cons with regard to precision, frequency, stability, cost, range, and size.

Vegetation cover and other terrain features, which can all be direct or indirect indicators of the stability and health of ecosystems, have been analysed using multispectral and hyperspectral imaging. An et al. (2018) opine that vegetation density could be tracked using vegetation indices derived from multispectral satellite data. According to the literature, numerous researchers have identified regions with a variety of flora by using multispectral satellite data. Hyperspectral remote sensing, on the other hand, provides more detailed information on vegetation than multispectral imagery. Analysis of information in the subsequent section therefore shows that both hype and multispectral scanners play a formidable role in monitoring vegetation health as they produce plausible results.

Drone technology has also been utilized of late to monitor wetland vegetation. For instance, Mohanty et al. (2024b) adopted this technology since it enables for precise mapping of wetland species, providing data on their extent and the area they cover. They also highlight that by creating a digital database, the technology enables long-term monitoring of changes in wetland species' numbers and distribution, helping to assess trends of increase or decline in freshwater ecosystems. Such strategies are vital for supporting both local and global conservation efforts, offering insights for forward-looking, data-driven preservation initiatives.

7 Case Studies—Assessing Vegetation Diversity in African Wetlands

Literature shows that, wetlands are important and unique water dominated ecosystems that providing habitats (Darkoh and Mbaiwa 2014). This clearly demonstrates that wetlands play a fundamental role in as far as ecosystem functioning is concerned.

Africa is largely endowed with large types of wetlands such as the deltas of Congo, Okavango delta, Kafue flats, Lake Victoria and St Lucia in South Africa (Rebelo et al. 2010; Ndhlovu 2012).

There are various types of wetlands in Africa which have various definitions. In central Southern Africa including parts of Malawi, Angola, Zambia and Zimbabwe dambos cover approximately 10% (Ndhlovu 2012). Jogo and Hassan (2010), observed that 3% of the Limpopo basin is covered by wetlands which contribute to human livelihoods in several ways. Elsewhere in East Africa, approximately 12 million hectares is covered by wetlands and are increasingly converted to agriculture as the main livelihood basis.

Literature clearly demonstrates that African wetlands are largely endowed with diverse vegetation (Adekola et al. 2008). Dambos which are the most common type of wetland are usually waterlogged areas with grass and thick vegetation. These types of wetlands are largely interspersed by Miombo woodlands dominated by *Brachystegia* and *Julbernadia* with a well-developed grass layer. The Miombo woodlands therefore dominate on wetland fringes and catchments. To add on, *Terminalia* vegetation species (*Mususu*) are invading other areas once dominated by *Julbernardia Globiflora* and *Brachystegia Spicifomis*. This therefore shows that, wetland ecosystems constitute diverse flora (Table 1), which is important for the habitat of faunal and avifaunal species.

Vleis which is also a common type of wetland in Africa is also largely endowed with various grass and tree species (Rebelo et al. 2010). Some of the major grass species in vleis countries such as Zambia, Zimbabwe, South Africa Yellow include water lily (*Nymphaea Mexicana*), Blue water lily (*Nymphaea nouchalia*), Common reed (*Phragmites australis*), Water reed (*Schoenoplectus brachyceras*) (Chabwela and Haller 2010). Additionally, another typical wetland type in Africa are dams and lakes sedge/rush (*Cyperus sexangularis*), Sedge (*Eleocharis acutangula*), Leafy juncos (*Juncus lomatophylus*). This clearly demonstrates that wetlands of Africa are indeed endowed with diverse grass and trees which contribute to diversity hence wetlands are an integral element.

8 Implications of Vegetation Changes on Wetland Ecosystem Services

Wetland vegetation is an important aspect in as far as ecosystem service provision is concerned. Nonetheless, human disturbance is the primary cause of the frequent occurrence of vegetation changes in wetlands. Therefore, the process of wetland degradation may be directly characterized by changes in the plant community. These changes resulted in accelerated loss of key vegetation species which could be potential sources of herbs and acting as buffers to flooding hence affecting the provision of goods. Information analysis in this section reveals that wetland plant diversity and richness, as well as endangered species' habitats, are under severe threat.

Table 1 African wetland types and their associated common flora

Wetland type	Grass	Tree	References
Dambo	*Setaria verticillata, Kyllinga erecta*, Love grass, *Fimbristylis dichotoma*, Fleabane (*Conyza albida*) Rapoko grass (*Eleusine indica*) Sawgrass (*Cladium maricus*) Yellow nutsedge (*Cyperus esculentis*) Black-footed water grass (*Eriochloa meyeriana*)	Terminalia vegetation species (*Mususu*) are dominant as they are invading other areas once dominated by *Julbernardia globiflora* and *Brachystegia spicifomis, Cyrtanthus contractus* Fire Lily *Eulophia cucullata*, Muonde (*Ficus capensis*)	Jogo and Hassan (2010), Sakane et al.
Floodplains	Black jack (*Bidens pilosa*), Chickweed (*Galinsoga parviflora*), Salt weed (*Oxalis latifolia*) Mexican marigold (*Tagetes minuta*) Pigweed (*Amaranthus hybridus*), Thatching reed (*Phragmites mauritianus*), Couch grass (*Cynodon dactylon*)	Zambezi teak (*Baikiaea plurijuga*) and Mukwa (*Pterocarpus angolensis*), *Colophospermum mopane*, Monkey ball (*Strychnos spinose*) Mobola plum (*Parinari curatellifolia* Planch. ex Benth.), Cutch (*Acacia polyacantha* Willd.), Balsam (*Colophospermum mopane*) Zebrawood, (*Brachystegia Spiciformis*) Weeping fig (*Ficus*) Large-fruited sycamore fig (*Ficus gnaphalocarpa*)	Adekola et al. (2008), Musasa and Marambanyika
Pans	Yellow water lily (*Nymphaea Mexicana*) Blue water lily (*Nymphaea nouchulia*) Common reed (*Phragmites australis*) Water reed (*Schoenoplectus brachyceras*)	*Ficus Capensis, Ficus natalences Syzigium codedum*	Musasa and Marambanyika, Darko and Mbaiwa (2014)
Dams/lakes	Sedge/rush (*Cyperus sexangularis*) Sedge (*Eleocharis acutangula*) Leafy juncos (*Juncus lomatophylus*)		Musasa and Marambanyika

(continued)

Table 1 (continued)

Wetland type	Grass	Tree	References
Vleis	Yellow water lily (*Nymphaea Mexicana*) Blue water lily (*Nymphaea nouchalia*) Common reed (*Phragmites australis*) Water reed (*Schoenoplectus brachyceras*)	Poplar-leaved fig (*Ficus fischeri* Mildbr. & Burret), Common bushweed (*Flueggea virosa*) Snot-apple (*Azanza garckeana* (F. Hoffm. Exell & Hillc) African sicklebush	Mbereko et al. (2015), Mtambanengwe, Chuma et al. (2008)
Swamp	Mexican clover (*Richardia scabra*) Common corkwood (*Commiphora pyracanthoides*) Goat apple (*Solanum aculeastrum*) Giant reed (*Arundo donax*) Giant sedge (*Cyperus dives*)	Large-leaved guarri (*Euclea natalensis* A. DC.,) The Weeping wattle (*Peltophorum africanum Sond*), Camel-foot (*Bauhinia thonningii* Schumach.), Prickly pear (*Opuntia ficus-indica* (L.) Mill), Bbuffalo thorn (*Ziziphus mucronata*), (*Acacia*)	Millenium Ecosystem Assessment (2005), Mbereko et al. (2015), Mtambanengwe, Rebelo et al. (2010)

Changes in wetland vegetation can also result in alteration of hydrological regimes in surrounding areas. Wetlands are diverse components that respond rapidly to alterations in hydrological regimes. For example, vegetation in wetlands for instance *Pinus*, *Artemisia* and Chenopodiaceae respond to large-scale climate changes, while wetlands respond more strongly to short-term changes in hydrological regimes. A study by Chabwela and Haller (2010) in Zambia reported that loss of vegetation in wet areas resulted in ecosystem imbalances as vegetation which acted as surface area for evaporation was disturbed hence accelerating hydrologic regimes. Elsewhere, Sakane et al. also observed that vegetation changes in wetland areas resulted in disappearance of water as well as deterioration of the quality.

9 Conservation Strategies for Enhancing Wetland Vegetation Resilience

Wetlands are essential ecosystems that drive key processes for both human beings and animals (Alexander and McInnes 2012). Despite this concerning reality, they face a number of risks that could compromise the quality of the water and the size of the wetland. This has provoked wider thinking which has resulted in establishment of several initiatives aimed at ensuring that these ecosystems are protected so that future generations have benefits (Fennessy and Lei 2018). In its strict the management of wetlands is a key process that seeks to preserve and protect these ecosystems which provide crucial services, such as water purification and biodiversity support.

Provision of clean purified water will thus contribute to attainment of Sustainable Development Goals especially SDG 6 on access to clean water and sanitation for all (Mohanty et al. 2024a, b).

Wetlands in developed countries have been exposed to several threats which affect wetland vegetation resilience. Junk et al. (2013) observed that in order to enhance wetland vegetation resilience there have been projects established with the aim to assess wetland sediment and soil pollution. It is important to note that healthy wetlands have diverse vegetation and once pollution is increasing the status is compromised (Fennessy and Lei 2018). There can be implementation of several approaches such as the use of indicators to provide a scientific basis for natural restoration which thus provides a pivotal role. It is clear that if there is sound understanding on the indicators of pollution in ecosystem restoration will assist local communities in several ways. Managing issues to do with water pollution will also contribute towards attainment of SDG 14 on life below water. In its strict sense, reduced pollution levels means there is no disturbance to life that primarily depend on water (Mohanty et al. 2024a, b).

Numerous attempts have been made in moving strides to enhance the vegetation resilience within ecosystems (Kotze and Malan 2010). One way of doing this is to raise awareness among the local communities on the importance of ecosystem components and their contributions. For instance, Musasa et al. (2023) observed that local communities may play an integral role in wetlands management hence they need to be aware since awareness is a key capacity measure in the preservation or success of a particular programme. As such there is need to engage universities and training institutions to cultivate policy makers, researchers, and practitioners who promote nature-based approaches (Chabwela and Haller 2010). This may involve research studies aimed at conserving vegetation resilience using several assessment tools with varying functions across (Alexander and McInnes 2012).

In addition, vegetation resilience to several threats in the natural environment also requires detailed planning for restoration at the catchment scale (Oliveira-Filho et al. 2021). This may include setting a clear line of connection for instance between floodplains with their rivers and streams to restore the hydrologic benefits of wetlands by re-establishing the natural vegetation patterns (Keddy 2010). This also contributes towards attainment of SDG 14 on life below water as well if properly managed. Analysis of information in the subsequent section thus shows that planning is essential as it results in reservation of sensitive areas which will limit exposure thus enhanced vegetation resilience.

To add on, in order to conserve wetlands, there is also need to develop wetland policies as an interagency nexus for coordinated protection (Chabwela and Haller 2010; Musasa and Marambanyika 2023). A study by Marambanyika et al. (2016), revealed that vegetation disappearance in local communities was attributed to lack of stakeholder participation in this case issues to do with government departments yet they have power which often results in loss of vegetation yet it is crucial.

10 Policy Recommendations for Sustainable Wetland Management

Analysis of information from the previous sections shows that vegetation plays an important role in ecosystem service provision. However, wetlands are subjected to pressure from human population growth, clearance of vegetation for settlement establishment as well as natural factors. In trying to address this several strategies have been put forward to address these shortcomings. The main issue is that, in many nations, there are policy flaws that appear to be evident when it comes to the specific protection of wetlands. It is very clear that most of the readily available resources ae governed by pieces of legislation and treaties focused on them. However, what is surprising is the level of success towards protection which implies that policy innovation should be more focused on human consequences of wetland loss. This will enable local people to harness services obtained from wetlands so as to mitigate risks from unprecedented disasters. By so doing it is clear that wetland protection can aid in life saving while at the same time reducing costs of restoration thus promoting long-term sustainability. Additionally, it should be noted that national wetland commissions stress the importance of combining wetland science and policy in order to look for ways to connect wetland protection with other social goals. This will undoubtedly increase confidence and also spur the role of policies and legislation and bolster sound environmental performance at large. It is encouraging that the application of integrated methodologies, which blend several modeling techniques, holds the key to the future of vegetation health modeling. There is need to combine several approaches each with its own share of the strengths and weakness.

11 Future Directions in Wetland Vegetation Research

The chapter provided a comprehensive analysis of literature on vegetation diversity in wetlands and their contribution to ecosystem services. Findings from the reviewed literature demonstrate that wetland vegetation plays an important role in wetland ecosystem functioning (Kotze and Malan 2010; Grasel et al. 2021). This implies that, wetland vegetation boasts as important habitats for key species at the same time purifying water which cushions local communities in resource constrained areas as they do afford to buy disinfectants for accessing clean water. As a result, wetlands and the flora that grows there helps achieve the Sustainable Development Goals, particularly the sixth one, which is about access to clean water and sanitation. This also means that the environment in which wetlands are hosted needs to be protected by encouraging positive minds in people so that they do not become a degrading force. The findings further show that the status of wetlands across the globe, has been compromised by unsustainable utilization mainly attributed to human activities (Musasa and Marambanyika 2023). A combination of human and natural factors was observed to have profound impacts such as rising temperatures, altered precipitation patterns, sea level rise, and changing carbon dynamics which has a significant bearing on these ecosystems. This therefore, calls for an urgent need to come up with several approaches and methods to assess productivity in wetlands so that they continue to support efforts already put in place. One approach that has a promising future is the use of remotely sensed data that can be used to monitor vegetation health and condition since they also contribute to wetland ecosystem health. There is also need to emphasize the importance of integrated water management, community involvement, policy reinforcement, and international cooperation so as to protect various components of the environment including wetlands. By understanding and addressing the consequences of environmental change, we can work towards preserving wetlands and their vital functions for future generations.

References

Adekola O, Morardet S, De Groot R, Grelot F (2008) The economic and livelihood value of provisioning services of the Ga-Mampa wetland, South Africa. In: 13th IWRA world water congress, p 24

Alexander S, McInnes R (2012) The benefits of wetland restoration. Ramsar Briefing Note No. 4. Ramsar Convention Secretariat, Gland, Switzerland. Available at https://www.ramsar.org/resources/ramsar-briefingnotes

An Y, Gao Y, Tong SZ, Lu XG, Wang X, Wang G, Liu XH, Zhang DJ (2018) Variations in vegetative characteristics of Deyeuxia angustifolia wetlands following natural restoration in the Sanjiang Plain, China. Ecol Eng 112:34–40

Chabwela HN, Haller T (2010) Governance issues, potentials and failures of participatory collective action in the Kafue Flats, Zambia. Int J Commons 4(2):621–642

Chuma E, Masiyandima M, Finlayson M, McCartney M, Jogo W (2008) Guideline for sustainable wetland management and utilisation: key cornerstones. CPWF Project Report. CGIAR Challenge Programme on Water and Food, Colombo, Sri Lanka. https://hdl.handle.net/10568/21605

Darkoh MBK, Mbaiwa JE (2014) Okavango delta-a Kalahari Oasis under environmental threats. J Biodivers Endangered Species

Darwall W, Bremerich V, De Wever A et al (2018) The alliance for freshwater life: a global call to unite efforts for freshwater biodiversity science and conservation. Aquat Conserv Mar Freshwat Ecosyst 28:1015–1022. https://doi.org/10.1002/aqc.2958

Davidson NC, Fluet-Chouinard E, Finlayson CM (2018) Global extent and distribution of wetlands: trends and issues. Mar Freshw Resour. https://doi.org/10.1071/MF17019

Davies B, Biggs J, Williams P, Whitfield M, Nicolet P, Sear D, Bray S, Maund S (2008) Comparative biodiversity of aquatic habitats in the European agricultural landscape. Agr Ecosyst Environ 125:1–8. https://doi.org/10.1016/j.agee.2007.10.006

Draper FC, Coronado ENH, Roucoux KH et al (2018) Peatland forests are the least diverse tree communities documented in Amazonia, but contribute to high regional beta-diversity. Ecography 41(1256):1269. https://doi.org/10.1111/ecog.03126

Erwin KL (2009) Wetlands and global climate change: the role of wetland restoration in a changing world. Wetlands Ecol Manage 17(1):71–84

Fennessy SM, Lei G (2018) Wetland restoration for climate change resilience. Ramsar Briefing Note No. 10. Ramsar Convention Secretariat, Gland, Switzerland

Grasel D, Giehl ELH, Wittmann F, Jarenkow JA (2021) Patterns of plant diversity and composition in wetlands across a subtropical landscape: comparisons among ponds, streambanks and riverbanks. Wetlands 41(7):90

Jogo W, Hassan R (2010) Balancing the use of wetlands for economic well-being and ecological security: The case of the Limpopo wetland in southern Africa. Ecol Econ 69(7):1569–1579

Junk WJ, An S, Finlayson CM, Gopal B, Květ J, Mitchell SA, Mitsch WJ, Robarts RD (2013) Current state of knowledge regarding the world's wetlands and their future under global climate change: a synthesis. J Aquat Sci 75(1):151–167

Kebtieneh N (2017) The impact of global climate change on wetlands. Int J Fish Aquat Stud 5(2):498–508

Keddy PA (2010) Wetland ecology: principles and conservation. Cambridge University Press, Cambridge

Kotze D, Malan H (2010) WET-sustainable use: a system for assessing the sustainability of wetland use. Water Research Commission, Pretoria

Kotze DC (2013) Assessing the ecological sustainability of wetland cultivation: experiences from Malawi and Zambia. In: Wood A, Dixon A, McCartney (eds) Wetland management and sustainable livelihoods in Africa. Earth scan, London

Mandishona E, Knight J (2019) Users' perceptions and understanding of two urban wetlands in Harare, Zimbabwe. S Afr Geogr J 101(3):326–348

Marambanyika T, Beckedahl H, Ngetar NS, Dube T (2016) Assessing the environmental sustainability of cultivation systems in wetlands using the WET-health framework in Zimbabwe. J Phys Geogr 38(1):62–82

Mbereko A, Mukamuri BB, Chimbari MJ (2015) Exclusion and contests over wetlands used for farming in Zimbabwe: a case study of broad-ridge and broad-furrow tillage system on Zungwi Vlei. J Polit Ecol 21:322–338

Meftah S, Tan BHM, Mun CF, Aung KMM, Veeravalli B, Chandrasekhar V (2021) Doren: toward efficient deep convolutional neural networks with fully homomorphic encryption. IEEE Trans Inf Forensics Secur 16:3740–3752

Millennium Ecosystem Assessment (2005) Ecosystems and human well-being. Island press Washington, DC

Mohanty S, Srivastava PK, Pandey PC, Singh P, Srivastava S (2024b) Wetland species mapping using advanced technological measurement. Aquat Conserv Mar Freshwat Ecosyst 34(12):e70018

Mohanty S, Pandey PC, Pandey M, Srivastava PK, Dwivedi CS (2024a) Wetlands contribution and linkage to support SDGs, its indicators and targets-a critical review. Sustainable Development

Musasa T, Marambanyika T (2023) Assessing the sustainability of land uses in Driefontein and Intunjambili wetlands, Zimbabwe. Sci Afr 16:e01195

Musasa T, Muringaniza KC, Manyati M (2023) The role of stakeholder participation in wetland conservation in urban areas: A case of Monavale Vlei, Harare. Sci Afr 19:e01574

Ndhlovu N (2012) A preliminary assessment of the wetland biological integrity in relation to land use: a case of the Intunjambili Wetland, Matobo District, Zimbabwe

Oliveira-Filho AT, Dexter KG, Pennington RT, Simon MF, Bueno ML, Neves DM (2021) Onthefloristic identity of Amazonian vegetation types. Biotropica. https://doi.org/10.1111/btp.12932

Rebelo LM, McCartney MP, Finlayson CM (2010) Wetlands of Sub-Saharan Africa: distribution and contribution of agriculture to livelihoods. Wetlands Ecol Manage 18(5):557–572

Socolar JB, Gilroy JJ, Kunin WE, Edwards DP (2016) How should beta diversity inform biodiversity conservation? Trends Ecol Evol 31:67–80. https://doi.org/10.1016/j.tree.2015.11.005

Toth C, Jóźków G (2016) Remote sensing platforms and sensors: a survey. ISPRS J Photogramm Remote Sens 115:22–36

Turpie J, Lannas K, Scovronick N, Louw W (2010) Wetland valuation vol I-wetland ecosystem services and their valuation: a review of current understanding and practice. WRC Report. Water Research Commission (440/09)

Wang L, Wang X, An Y, Song T, Tong S, Wang X (2022) Effects of land use changes on the plant community characteristics in the wetlands of the semi-arid regions. Diversity 14:1049. https://doi.org/10.3390/d14121049

Remote Sensing Vegetation Cover Change and Its Impact on Water Flux Parameters at Catchment Scale: A Data Driven Approach for the Kafue Basin

Abel Chemura, Munyaradzi D. Shekede, Benjamin Adjah Torgbor, and Terence D. Mushore

Abstract Human induced and natural changes in vegetation cover, type and density have significant effects on the role of vegetation to provide critical and diverse ecosystem services. Among the most important regulating services provided by healthy forests, shrubland and grassland is soil conservation and stabilisation of stream flows and runoff. Quantifying vegetation cover changes at catchment scale and linking the changes to water flux parameters is important in informing integrated catchment. In this chapter, we provide a three-step approach to assess vegetation cover dynamics and their influences on water fluxes in the Kafue catchment in Zambia, with the Kafue flats being a Ramsar site (No. 530). First, land cover maps of 2009, 2015 and 2021 were obtained from MODIS land cover product. Then, the amount of precipitation, transpiration, evaporation and interception for each land cover type from the SEBAL energy balance model were estimated for the respective time periods. Shrublands are the dominant land cover type covering up to two-thirds

A. Chemura (✉)
Department of Natural Resources, Faculty of Geo-Information Science and Earth Observation, University of Twente, Enschede, The Netherlands
e-mail: a.chemura@utwente.nl

M. D. Shekede
Department of Geography Geospatial Sciences and Earth Observation, University of Zimbabwe, Harare, Zimbabwe

Department of Geospatial Sciences and Earth Observation, Zimbabwe National Geospatial and Space Agency, Harare, Zimbabwe

B. A. Torgbor
Applied Agricultural Remote Sensing Centre, University of New England, Armidale, Australia
e-mail: atorgbo2@une.edu.au

Forest Commission, Accra, Ghana

T. D. Mushore
Department of Space Science and Applied Physics, Faculty of Science, University of Zimbabwe, Harare, Zimbabwe

O. Mutanga et al. (eds.), *Revealing Ecosystem Services Through Geospatial Technologies*, Springer Remote Sensing/Photogrammetry,
https://doi.org/10.1007/978-3-031-98048-0_10

of the catchment, followed by grassland and forests. Significant changes in land-cover types were observed particularly in forested areas which are declining and being converted primarily into shrubland and grassland with some areas transitioning to croplands in some parts of the catchment. Forests decreased from 18.5% of the catchment in 2009 to 10.1% in 2021. Shrubland accounts for most of the transpiration, evaporation, and interception processes in the catchment. However, this proportion is gradually declining over time. Shrublands contribute about three times to transpiration compared to forests and six times compared to grassland. It is therefore concluded that shrubland should be considered in integrated catchment management in addition to forests as they are important for water flux in the catchment and regulate ecosystem services.

Keywords Land-use land cover · Water availability · MODIS · Water balance

1 Introduction

The intricate relationship between land cover and hydrological processes plays a critical role in shaping water resource dynamics within catchments (de Oliveira Serrão et al. 2022; Yang et al. 2017). One of the greatest footprints of humans on the earth system is their alteration of land cover on a large scale, which is unprecedented and continuing (Estoque et al. 2021; Mu et al. 2022; Venter et al. 2016). These changes in land cover, driven by natural or human factors, significantly alter the water balance by influencing key components such as evapotranspiration, surface runoff, and groundwater recharge (Kayitesi et al. 2022; Neupane and Kumar 2015). Understanding these impacts is essential for sustainable water resource management, particularly in regions experiencing rapid land-use changes and water stress. Assessment of land cover changes at catchment level is thus critical to the identification of change hotspots and their effects on the ability of catchments to provide ecosystem services, particularly regulating services related to water balance (Govender et al. 2022; Hasan et al. 2020).

Land cover changes, such as the conversion of forests to agricultural land or grassland and the expansion of urban areas alter the quantity and timing of hydrological processes by modifying infiltration rates, runoff patterns, and evapotranspiration (Chemura et al. 2020; Guzha et al. 2018). Any changes in vegetated land cover types such as deforestation reduces canopy interception and increases surface runoff (Dwarakish and Ganasri 2015). On the other hand, conversion of forest or grassland to agriculture may increase evapotranspiration during the crop growing season and none during fallow or dry seasons (Marhaento et al. 2018). Cumulatively, these changes have significant impacts on water resources especially in periods or areas of scarcity. They result in cascading effects on water availability and quality, affecting both ecological functioning and human livelihoods in the catchment area and beyond (Camara et al. 2019; Hasan et al. 2020; Mu et al. 2022). Quantifying these impacts and identifying where exactly within the catchment they are occurring is an important first step in developing adaptive water management strategies.

Remote sensing has emerged as a powerful tool for monitoring and analysing land cover changes and their implications on hydrological/aquatic systems (Bedair et al. 2024; Leila et al. 2024). By providing spatially and temporally consistent data, remote sensing facilitates a comprehensive assessment of land cover dynamics and their influence on hydrological processes and other ecosystem services (Gashaw et al. 2018; Hasan et al. 2020; Tena et al. 2019a). Advances in satellite technology and image processing techniques enable the detection of even subtle changes in land cover over time. As such, land cover change monitoring has been one of the primary applications of remote sensing for the past five decades (Cheng et al. 2024; Wulder et al. 2022). When this is coupled with hydrological modelling, it allows for the quantification of water balance components and the identification of where and when intervention is required (Koneti et al. 2018). Such integrated analyses are increasingly recognized as essential for addressing complex environmental challenges, especially in tropical water scarce areas.

The Kafue River catchment, located in Zambia, is a vital hydrological and ecological system supporting diverse socio-economic activities, including agriculture, hydropower generation, and biodiversity conservation (Kasaro et al. 2019). The catchment spans a significant area (156,995 km^2) making it one of Zambia's key watersheds as it covers 20% of the country (Lugomela et al. 2021; Mwamba 2021). Over the past few decades, this catchment has experienced substantial land cover changes due to urbanization, agricultural expansion, and deforestation (Dietz et al. 2023; Liang et al. 2024; Tena et al. 2019b). These changes have raised concerns about their potential impact on the region's water balance and the long-term sustainability of its water resources. Stakeholders, including local communities, government departments and private sector players, among others, are increasingly emphasizing the need to address these challenges (Mbewe and Dixon 2022).

Studies have focused on assessing LULC changes in large catchments, but many of these are either localised to administrative regions such as ward or district boundaries, confined to short time ranges and/or generalised or unmatched to water flux dynamics (Chanza and Musakwa 2022; Dwarakish and Ganasri 2015; Mohanty et al. 2024). This is despite the fact that this is an important boundary layer process with significant impacts on economic activities, biodiversity, ecosystem services and ultimately human wellbeing (Cuo 2016; Kayitesi et al. 2022). By integrating remote sensing data with hydrological models, this research seeks to advance our understanding of the complex interactions between land cover and water resources (Bedair et al. 2024; Chemura et al. 2020; Pandey et al. 2024). This study aimed to evaluate the long-term impact of human-induced vegetation changes on water availability in the Kafue catchment, utilizing remote sensing data products as the primary analytical method. Specific objectives were to (i) assess LULC change in the catchments between three time periods, (ii) quantify the influence of the LULC change on water flux in the catchment and (iii) identify change hotspots and their land cover types in the catchment. It is expected that results from this study will provide opportunities for spatial and temporal water accounting in a changing climate at catchment level by providing quantitative information on which management and policy decisions can be based on.

2 Data and Methods

2.1 Study Area

The Kafue Catchment in Zambia is a vital hydrological and ecological region, encompassing a significant portion of the Kafue River, one of the country's most important water systems (Fig. 1). This catchment spans approximately 157,000 km^2, covering a diverse range of ecosystems from highlands in the northern regions to floodplains and wetlands in the south (Kasaro et al. 2019). The Kafue River itself is a tributary of the Zambezi River, contributing significantly to Zambia's water resources. The catchment supports multiple land uses, including agriculture, hydropower generation, fisheries, and tourism, making it central to the country's socio-economic development (Lehner et al. 2021). Notably, the Kafue Flats, a floodplain within this catchment, is a Ramsar Site (No. 530) recognized for its rich biodiversity and critical role as a habitat for wildlife (https://rsis.ramsar.org/ris/530).

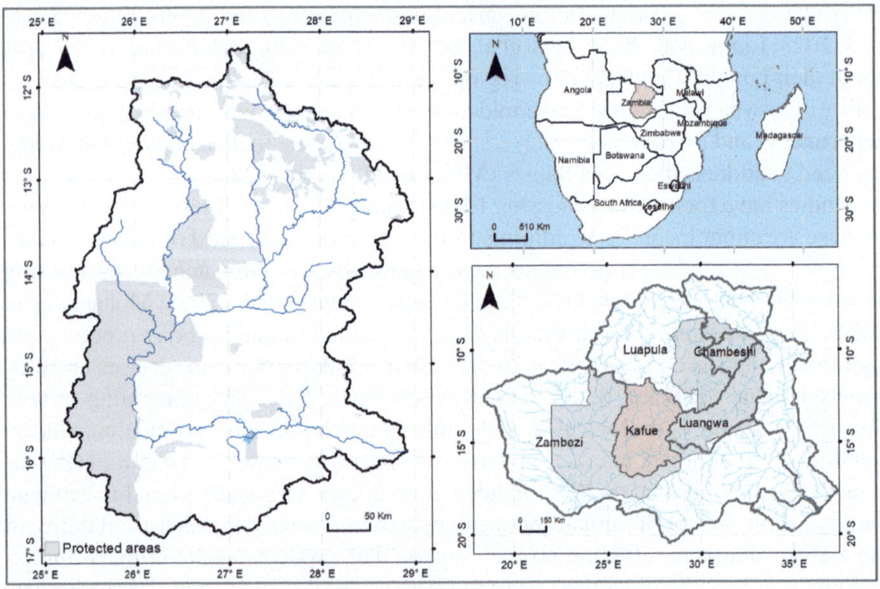

Fig. 1 The location of the study area and the river network. The insets show the location of the Kafue catchment in the Zambezi catchment and in Zambia

2.2 Land Cover Change Assessment

First, we conducted a temporal analysis of land cover dynamics at six-year intervals i.e., 2009, 2015, and 2021 to quantify and characterize transitions within and between land cover classes, offering insights into the spatial and temporal drivers of landscape change in the catchment. To do this, we relied on existing land cover products to enable temporal and spatial comparability of results across the catchment as they use consistent processing algorithms and standardised classification protocols. We used the MODIS land cover collection (MCD12Q1) Version 6 (Sulla-Menashe and Friedl 2018). The MCD12Q1 provides annual global land cover datasets in five classification systems at 500 m spatial resolution. They are developed from the input of observation data from the Terra and Aqua satellites. The five land cover classification systems used by the MCD12Q1 are (i) IGBP global vegetation classification scheme (Land Cover Type 1), (ii) UMD scheme (Land Cover Type 2), (iii) MODIS-derived Leaf Area Index (LAI) and Fraction of Photosynthetically Active Radiation (fPAR) (LAI/fPAR) scheme (Land Cover Type 3), (iv) MODIS-derived Net Primary Production (NPP) scheme (Land Cover Type 4), and (v) Plant Functional Type (PFT) scheme (Land Cover Type 5). The MCD12Q1 product has been shown to be reliable when compared to field data and high resolution remote sensing products owing to its use of standardised land cover classification approaches (Liang et al. 2015; Pandey et al. 2021; Zhang and Roy 2017).

We adopted the Land Cover Type 1: International Geosphere-Biosphere Programme (IGBP) class descriptions. This was selected because it has two cropland categories, which is the focus of the current study and also reported to be reliable across many areas (Liang et al. 2015). A total of 17 land cover classes are included, and of these, 11 are natural vegetation classes, 3 developed and mosaicked land classes, and 3 non-vegetated land classes (Table 1). The MCD12Q1 data was downloaded in hierarchical data format (HDF) with the sinusoidal (SIN) projection and was reprojected to World Geodetic System 1984 (WGS84)/Universal Transverse Mercator (UTM) corresponding to that catchment. Subsequently, the images were subset and reclassified with all processes done in the R environment (R-Core Team 2020).

2.3 Assessing Vegetation-Related Landscape Controls of Water Balance

After obtaining the LULC types, the next step was to determine how much of the water balance components were associated with each of the 6 land cover types that were mapped. To achieve this objective, LULC types were linked to the annual water balance parameters (precipitation, transipiration, interception and evaporation) for each catchment. These were obtained from the FAO Wapor data portal (https://wapor.apps.fao.org/catalog/1). This enabled the quantification of water fluxes associated

Table 1 Description of land cover types that were used for the classification scheme

Land use/cover type	Description of land cover/use classes	IGBP class number
Forest	These closed forests have a canopy cover of over 40% and a density of trees and shrubs growing together	1, 2, 3, 4, 5
Shrubland	Dominated by woody perennials (1–2 m height) > 60% cover or savannas with tree cover of 30–60%	6, 7, 8, 9
Grassland	Areas with open grasses and shrubs or sparse trees of lower density that cannot be considered as forests	10
Cropland	These are areas with at least 60% cultivated cropland or mosaics of small-scale cultivation, 40–60% with natural tree, shrub, or herbaceous vegetation	12, 14
Built-up	Area dominated by artificial surfaces and open land cover but may also include bare areas. At least 30% impervious surface area including building materials, asphalt, and vehicles	13, 16

with each LULC cover with the aim of (i) accounting for pathways of water flux and (ii) identifying opportunities for minimising or optimising loss-related fluxes in the catchment. These FAO WAPOR products are obtained from the use of a combination of remotely sensed data and a process-based SEBAL model applied at a grid scale (Bastiaanssen et al. 1998). The SEBAL model calculates each of the water balance components from satellite images and weather data using the surface energy balance. This energy balance model assumes that incoming net solar radiation drives all energy exchanges on the earth's surface and therefore all water fluxes require/use energy (Ning et al. 2017). The rainfall product used was the Climate Hazards Group Infrared Precipitation Fields (CHIRPS) rainfall at 0.05 degrees (~ 5 km) resolution resampled to ~ 5 km (Funk et al. 2015).

3 Results

3.1 *Land Cover and Land Use Change in Kafue Basin*

The LULC change assessment showed the dominant LULC in the Kafue basin for the 2009, 2015 and 2021 time periods (Fig. 2). Forest areas were concentrated in the northern and north-western regions of the basin, while croplands predominantly occupied the southern and central zones, with additional dispersed agricultural areas throughout the landscape. Grassland were in the southern parts of the basin. Shrubland were the main land cover type, accounting for between 56 and 62% of the catchment area during the period, followed by grassland with 22.2, 23.2 and 25.1% for 2009, 2015 and 2021 respectively (Figs. 2 and 3). The least area was built-up which accounted for less than 1% of the total area of the basin. Visual inspection

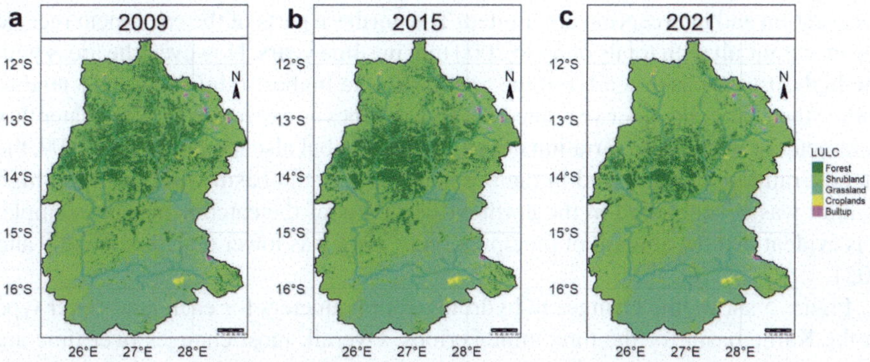

Fig. 2 The land cover classes in Kafue basin for **a** 2009, **b** 2015, **c** 2021

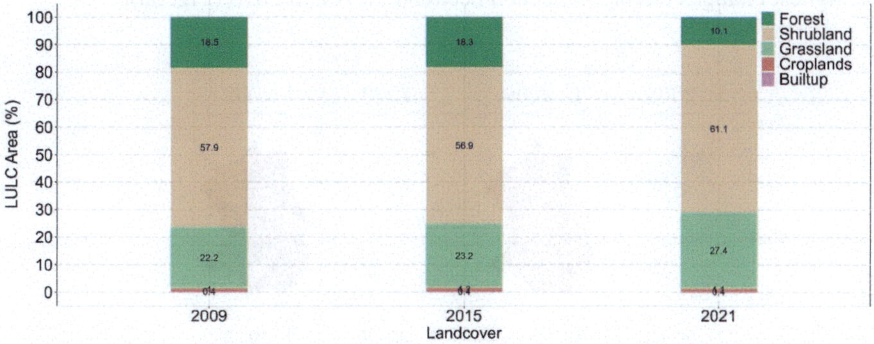

Fig. 3 Percentage area of land cover types in the Kafue basin (2009–2021)

of Fig. 2 shows a decrease in Forest due to replacement mainly by Shrubland from 2009 to 2021 in the northern half of the catchment area.

An assessment of the land cover changes across the three periods showed that the major transitions were from forest to shrubland and grassland in Kafue basin (Figs. 2 and 3). Built-up and croplands also increased by 6 and 12% respectively but this was not as dramatic as the changes in forested area, which reduced from 18.5% in 2009 to 10.1% in 2021, a decrease of almost half (Fig. 3). These changes mainly happened in the northern parts of the catchment as shown by the transitions from forest to shrubland (Fig. 3).

3.2 *Water Flux and Water Flux Changes in Kafue Basin*

The water flux in the Kafue basin is shown in Fig. 4. From these results, the two main water flux parameters in the catchment are precipitation and transpiration while evaporation and interception are limited. The northern parts of the catchment receive the most rainfall with totals of up to 2000 mm in some years. However, the areas with the highest transpiration are not the areas with the highest rainfall as these tend to follow the forest and other vegetated land cover types as expected (Fig. 4). Water flux parameters demonstrate variation not only spatially, but also temporally. In 2009, the highest rainfall was received in the northern parts of the basin while in 2021, high rainfall was concentrated in the north-eastern parts of the catchment. For example, it is evident that the amount of precipitation in 2015 was lower than that in 2009 and 2021.

Figure 5 shows the changes in hydrological parameters for each land cover type in the Kafue basin for the three time periods. Overall, most changes over time are happening in evaporation and transpiration, both of which are decreasing with magnitudes of up to 50% for land cover types. The total contribution of shrubland, forests and grassland to evaporation and transpiration is decreasing in the catchment with the highest changes having happened between 2009 and 2015 (Fig. 5a).

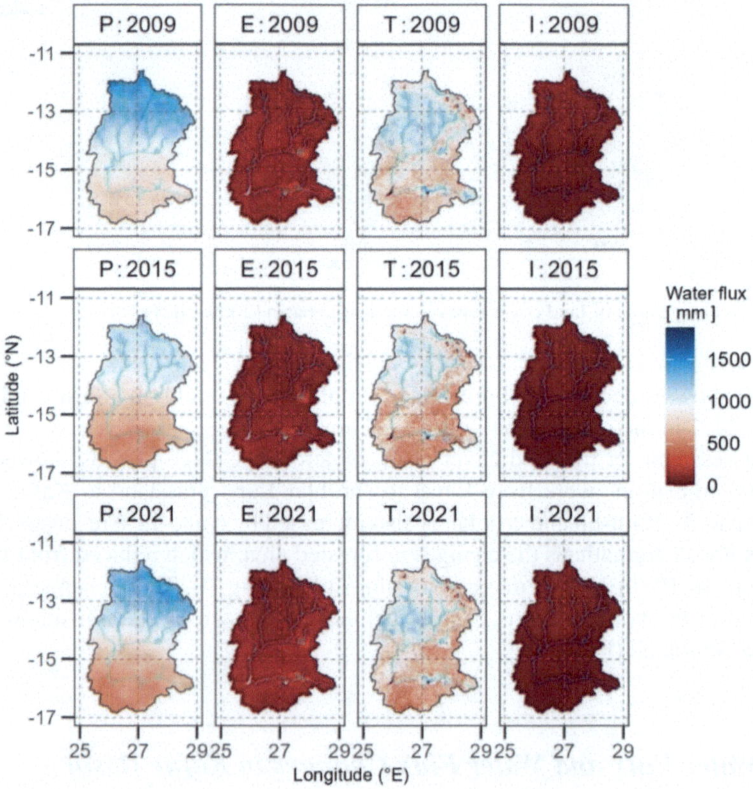

Fig. 4 Water flux shown by each of the four hydrological parameters in the Kafue basin for the period 2009, 2015 and 2021

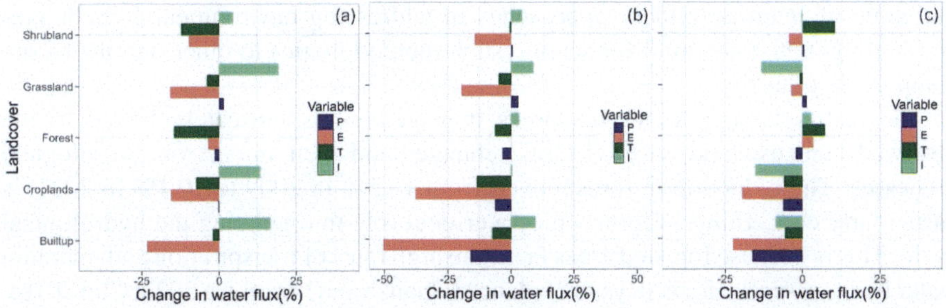

Fig. 5 Changes in water flux parameters in each land cover type **a** between 2009 and 2015, **b** between 2009 and 2021 and **c** between 2015 and 2021

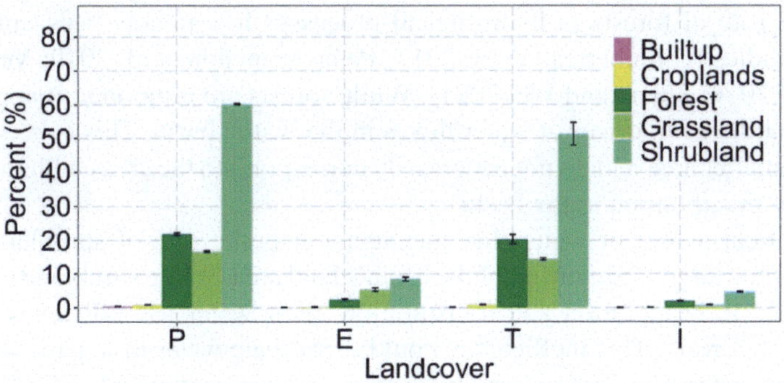

Fig. 6 Average proportion of water flux attributable to each land cover types for the three periods in the Kafue basin

These results indicate that most water flux losses (evaporation, transpiration and interception) are happening in the shrubland, which has values that are almost three times those of forests and six times more than those of grassland (Fig. 6). As expected, built-up and croplands, with lower area, are contributing the least to water flux in the catchment. While differences for interception and evaporation are limited, the differences in transpiration between the land cover type is the most significant. Relatedly, the most rainfall in the catchment is received in the shrubland, followed by the forests, despite grassland having a higher area than that of forest (Fig. 6).

4 Discussion

The goal of this study was to develop an approach that accounts for water fluxes from land cover types and apply it in the Kafue basin, an economically and ecologically sensitive catchment in Zambia. It is anticipated that the presented approach can be used in assessing the impacts of LULC on water availability and underscores the value of interdisciplinary approaches in addressing environmental challenges and highlights the potential for scaling such methodologies to other regions facing comparable issues.

The Kafue Basin's water resources management is intricately linked to the observed land use/land cover (LULC) changes and their impact on hydrological processes. The reduction in forest cover, from 18.5% in 2009 to 10.1% in 2021, is particularly concerning as forests play a critical role in regulating the hydrological cycle. This is because forested areas facilitate high rates of transpiration and maintain water fluxes essential for ecosystem balance (Chanza and Musakwa 2022; Cuo 2016). Their replacement by shrubland and grassland, which have distinct water use characteristics, has likely contributed to a decline in total transpiration and altered precipitation dynamics. These changes underscore the need for forest landscape restoration and conservation as key strategies for maintaining the hydrological integrity of the basin. The role of forests in hydrological processes has already been emphasised in many studies (Guimberteau et al. 2017; Peña-Arancibia et al. 2019; Vergopolan and Fisher 2016; Zhang and Wei 2021). While forests are important, it appears that they are not the main driver of water fluxes in the Kafue basin. This is because they occupy a smaller area and do not necessarily correspond to the areas with the highest amounts of precipitation in the basin.

One critical aspect of water flux management is the role of shrubland, which account for the largest proportion of the basin's land area. While shrubland contribute substantially to evaporation and transpiration, their water use efficiency is lower compared to forests. This inefficiency could exacerbate water loss, particularly in a region already experiencing reductions in rainfall during certain periods (Dietz et al. 2023; Mwamba 2021). Grassland and built-up areas, though contributing less to total water flux, represent areas of potential intervention. Thus, from a management perspective, shrubland should receive the highest priority as they are the dominant factor. This is a shift from most of the current discourse in catchment management that focuses on forests as key drivers. It is therefore argued here that spatial analysis of the cover types in a catchment is required before making blanket generalisation about what is important for management. This is especially so in dynamic tropical catchments with significant human footprint where forests may no longer be the most important cover type.

Recent studies have highlighted that in some biomes and ecotypes, traditional forest-centric paradigms are becoming less applicable. For instance, in alpine grassland areas, forests are no longer the dominant landcover type of concern(Yang et al. 2014; Zhang et al. 2016). While historical land management practices have remained largely persistent, many tropical regions are undergoing significant vegetation transformations in structure and composition. Nevertheless, land managers often continue to prioritize forests in policy and practice, despite evidence of declining forest cover. Further complexities are revealed through the spatial distribution of water fluxes which challenge the assumption that forests are the predominant determinants of hydrological processes . For example, Chemura et al. (2020) observed that in some catchment commercial plantations are the most important land cover type that influence water flux dynamics, and therefore require attention. Combined, these results suggest the need for more adaptive and context-specific landscape management strategies that account for emergent landcover types and ecological realities.

Results further show that the northern and north-eastern parts of the basin, which receive the highest precipitation, also exhibit significant water loss through evaporation and transpiration. This spatial variability is expected for such large catchments (King and Brown 2014; Lugomela et al. 2021) that transcend several rainfall gradients. This pattern suggests that while these areas are crucial for replenishing water resources, they are also vulnerable to shifts in LULC from anthropogenic drivers. The observed temporal variability, such as lower rainfall in 2015 compared to 2009 and 2021, highlights the basin's susceptibility to hydrological and meteorological fluctuations. These dynamics emphasize the importance of spatially explicit water management strategies that account for localized differences in precipitation, land cover, and water flux (Camara et al. 2019; Cuo 2016).

The interplay between land cover transitions and water flux changes has direct implications for water-dependent sectors in the Kafue Basin. Agriculture, hydropower generation, and local livelihoods rely on a stable and predictable hydrological cycle (Kalumba and Nyirenda 2017; King and Brown 2014). The increased conversion of shrubland and forest to cropland and built-up areas, though necessary for economic development, must be managed to minimize adverse effects on water availability and quality. Policies promoting agroforestry, reforestation, and controlled expansion of urban areas could help maintain a balance between development and ecological sustainability (Chomba et al. 2019; Dietz et al. 2023; Mwamba 2021). Moreover, investment in water harvesting and retention infrastructure can help mitigate the impact of reduced rainfall and high evaporation rates.

5 Conclusion

Water resources management in the Kafue Basin requires an integrated approach that considers the intricate relationship between LULC changes, hydrological processes, and socio-economic needs. We present a remote sensing-based approach to assess the impact of LULC on hydrological parameters in the catchment. It is therefore concluded that catchment management should consider the main landcover types and not only forests to preserve water. This is important in addressing the challenges posed by land cover changes and increased water demand, which all requires a combination of conservation efforts, sustainable land-use practices, and adaptive water management strategies. By leveraging data-driven insights and fostering collaboration among stakeholders, the basin can achieve a balance that supports both ecological integrity and socio-economic development. The Kafue Basin thus serves as a model for studying and addressing the complexities of water resources management in dynamic and vulnerable catchments with this approach transferrable to other tropical catchments.

References

Bastiaanssen WG, Menenti M, Feddes R, Holtslag A (1998) A remote sensing surface energy balance algorithm for land (SEBAL). 1 formulation. J Hydrol 212:198–212

Bedair H, El-Makawy AM, Nasser H, Halmy MWA (2024) Application of remote sensing for monitoring aquatic vegetation: a case study of egyptian coastal wetlands using MODIS/terra data. In: Aquatic ecosystems monitoring. CRC Press, pp 133–153

Camara M, Jamil NR, Abdullah AFB (2019) Impact of land uses on water quality in Malaysia: a review. Ecol Process 8(1):1–10

Chanza N, Musakwa W (2022) Ecological and hydrological indicators of climate change observed by dryland communities of Malipati in Chiredzi, Zimbabwe. Diversity 14(7):541

Chemura A, Rwasoka D, Mutanga O, Dube T, Mushore T (2020) The impact of land-use/land cover changes on water balance of the heterogeneous Buzi sub-catchment, Zimbabwe. Remote Sens Appl: Soc Environ 18:100292

Cheng G, Huang Y, Li X, Lyu S, Xu Z, Zhao H, Zhao Q, Xiang S (2024) Change detection methods for remote sensing in the last decade: a comprehensive review. Remote Sens 16(13):2355

Chomba MJ, Hill T, Nkhata B, Nel A (2019) A social exchange analysis of adaptive governance in water allocation processes, the Kafue Flats, Zambia. Int J Commons 13(2)

Cuo L (2016) Land use/cover change impacts on hydrology in large river basins: a review. In: Terrestrial water cycle and climate change: natural and human-induced impacts, pp 103–134

de Oliveira Serrão EA, Silva MT, Ferreira TR, de Ataide LCP, dos Santos CA, de Lima AMM, da Silva VdPR, de Sousa FdAS, Gomes DJC (2022) Impacts of land use and land cover changes on hydrological processes and sediment yield determined using the SWAT model. Int J Sedim Res 37(1):54–69

Dietz J, Treydte AC, Lippe M (2023) Exploring the future of Kafue National Park, Zambia: scenario-based land use and land cover modelling to understand drivers and impacts of deforestation. Land Use Policy 126:106535

Dwarakish G, Ganasri B (2015) Impact of land use change on hydrological systems: a review of current modeling approaches. Cogent Geosci 1(1):1115691

Estoque RC, Ooba M, Togawa T, Hijioka Y, Murayama Y (2021) Monitoring global land-use efficiency in the context of the UN 2030 agenda for sustainable development. Habitat Int 115:102403

Funk C, Peterson P, Landsfeld M, Pedreros D, Verdin J, Shukla S, Husak G, Rowland J, Harrison L, Hoell A (2015) The climate hazards infrared precipitation with stations—a new environmental record for monitoring extremes. Sci Data 2:150066

Gashaw T, Tulu T, Argaw M, Worqlul AW (2018) Modeling the hydrological impacts of land use/ land cover changes in the Andassa watershed, Blue Nile Basin, Ethiopia. Sci Total Environ 619:1394–1408

Govender T, Dube T, Shoko C (2022) Remote sensing of land use-land cover change and climate variability on hydrological processes in Sub-Saharan Africa: key scientific strides and challenges. Geocarto Int 37(25):10925–10949

Guimberteau M, Ciais P, Ducharne A, Boisier JP, Dutra Aguiar AP, Biemans H, De Deurwaerder H, Galbraith D, Kruijt B, Langerwisch F (2017) Impacts of future deforestation and climate change on the hydrology of the Amazon Basin: a multi-model analysis with a new set of land-cover change scenarios. Hydrol Earth Syst Sci 21(3):1455–1475

Guzha A, Rufino MC, Okoth S, Jacobs S, Nóbrega R (2018) Impacts of land use and land cover change on surface runoff, discharge and low flows: evidence from East Africa. J Hydrol: Reg Stud 15:49–67

Hasan SS, Zhen L, Miah MG, Ahamed T, Samie A (2020) Impact of land use change on ecosystem services: a review. Environ Dev 34:100527

Kalumba M, Nyirenda E (2017) River flow availability for environmental flow allocation downstream of hydropower facilities in the Kafue Basin of Zambia. Phys Chem Earth, Parts A/B/C 102:21–30

Kasaro D, Phiri E, Nyambe I (2019) Deforestation impact on ecosystem services in Kamfinsa sub-catchment of Kafue River Basin in Zambia. J Ecol Nat Environ 11(4):33–45

Kayitesi NM, Guzha AC, Mariethoz G (2022) Impacts of land use land cover change and climate change on river hydro-morphology-a review of research studies in tropical regions. J Hydrol 615:128702

King J, Brown C (2014) Determination of "holding" environmental flow requirement for the upper and middle Kafue river. In: Integrating climate change in water resources monitoring, report Nr. GFA/B5/2013

Koneti S, Sunkara SL, Roy PS (2018) Hydrological modeling with respect to impact of land-use and land-cover change on the runoff dynamics in Godavari river basin using the HEC-HMS model. ISPRS Int J Geo Inf 7(6):206

Lehner B, Katiyo L, Chivava F, Sichingabula HM, Nyirenda E, Rivers-Moore NA, Paxton BR, Grill G, Nyoni F, Shamboko-Mbale B (2021) Identifying priority areas for surface water protection in data scarce regions: an integrated spatial analysis for Zambia. Aquat Conserv Mar Freshwat Ecosyst 31(8):1998–2016

Leila S, Zohra A, Islem BM, Omar S, Essalah AM (2024) Aquatic ecosystems monitoring with remote sensing perspective. In: Aquatic ecosystems monitoring, pp 45–64

Liang D, Zuo Y, Huang L, Zhao J, Teng L, Yang F (2015) Evaluation of the consistency of MODIS land cover product (MCD12Q1) based on Chinese 30 m GlobeLand30 datasets: a case study in Anhui Province, China. ISPRS Int J Geo-Inf 4(4):2519–2541

Liang D, Reed J, Fakheran S, Moombe K, Siangulube F, Sunderland T (2024) Monitoring spatiotemporal changes in land use/land cover and its impacts on ecosystem services in southern Zambia. Environ Res Commun 6(4):045004

Lugomela GV, Chavula G, Mwedzi T (2021) Hydrology of the Zambezi river basin. In: Ecological changes in the Zambezi river basin, p 151

Marhaento H, Booij MJ, Hoekstra AY (2018) Hydrological response to future land-use change and climate change in a tropical catchment. Hydrol Sci J 63(9):1368–1385

Mbewe A, Dixon A (2022) Local institutions and the functional landscape approach in the Kafue flats

Mohanty S, Pandey PC, Pandey M, Srivastava PK, Dwivedi CS (2024) Wetlands contribution and linkage to support SDGs, its indicators and targets-a critical review. Sustain Dev

Mu H, Li X, Wen Y, Huang J, Du P, Su W, Miao S, Geng M (2022) A global record of annual terrestrial human footprint dataset from 2000 to 2018. Sci Data 9(1):176

Mwamba B (2021) Water resources management in Zambia: a case of cumulative impacts associated with copper mining in the Upper Kafue Catchment, Copperbelt Province, Zambia

Neupane RP, Kumar S (2015) Estimating the effects of potential climate and land use changes on hydrologic processes of a large agriculture dominated watershed. J Hydrol 529:418–429

Ning J, Gao Z, Xu F (2017) Effects of land cover change on evapotranspiration in the Yellow River Delta analyzed with the SEBAL model. J Appl Remote Sens 11(1):016009

Pandey PC, Koutsias N, Petropoulos GP, Srivastava PK, Ben Dor E (2021) Land use/land cover in view of earth observation: data sources, input dimensions, and classifiers—a review of the state of the art. Geocarto Int 36(9):957–988

Pandey PC, Srivastava PK, Srivastava SK (2024) Aquatic ecosystems monitoring: conventional assessment to advanced remote sensing. CRC Press

Peña-Arancibia JL, Bruijnzeel LA, Mulligan M, Van Dijk AI (2019) Forests as 'sponges' and 'pumps': assessing the impact of deforestation on dry-season flows across the tropics. J Hydrol 574:946–963

R-Core Team (2020) R: a language and environment for statistical programming. In: R foundation for statistical computing, Vienna, Austria. www.R-project.org

Sulla-Menashe D, Friedl MA (2018) User guide to collection 6 MODIS land cover (MCD12Q1 and MCD12C1) product, vol 1. USGS, Reston, VA, USA, p 18

Tena TM, Mwaanga P, Nguvulu A (2019a) Hydrological modelling and water resources assessment of Chongwe river catchment using WEAP model. Water 11(4):839

Tena TM, Mwaanga P, Nguvulu A (2019b) Impact of land use/land cover change on hydrological components in Chongwe river catchment. Sustainability 11(22):6415

Venter O, Sanderson EW, Magrach A, Allan JR, Beher J, Jones KR, Possingham HP, Laurance WF, Wood P, Fekete BM (2016) Sixteen years of change in the global terrestrial human footprint and implications for biodiversity conservation. Nat Commun 7(1):12558

Vergopolan N, Fisher JB (2016) The impact of deforestation on the hydrological cycle in Amazonia as observed from remote sensing. Int J Remote Sens 37(22):5412–5430

Wulder MA, Roy DP, Radeloff VC, Loveland TR, Anderson MC, Johnson DM, Healey S, Zhu Z, Scambos TA, Pahlevan N (2022) Fifty years of Landsat science and impacts. Remote Sens Environ 280:113195

Yang F, Zhang G-L, Yang J-L, Li D-C, Zhao Y-G, Liu F, Yang R-M, Yang F (2014) Organic matter controls of soil water retention in an alpine grassland and its significance for hydrological processes. J Hydrol 519:3086–3093

Yang L, Feng Q, Yin Z, Wen X, Si J, Li C, Deo RC (2017) Identifying separate impacts of climate and land use/cover change on hydrological processes in upper stream of Heihe River, Northwest China. Hydrol Process 31(5):1100–1112

Zhang HK, Roy DP (2017) Using the 500 m MODIS land cover product to derive a consistent continental scale 30 m Landsat land cover classification. Remote Sens Environ 197:15–34

Zhang M, Wei X (2021) Deforestation, forestation, and water supply. Science 371(6533):990–991

Zhang X, Zhao W, Liu Y, Fang X, Feng Q (2016) The relationships between grasslands and soil moisture on the Loess Plateau of China: a review. CATENA 145:56–67

Advancements in Remote Sensing for Monitoring and Assessment of Coastal Mangrove Ecosystems

Mahadev Bera, Malini Roy Choudhury, Sumanta Das, and Suman Dutta

Abstract Coastal mangrove ecosystems play a crucial role in maintaining environmental balance. They offer shelter to a wide range of species, act as natural barriers against coastal erosion, and store large amounts of carbon, helping combat climate change. Despite their importance, mangroves are under serious threat from human activities and the changing climate, which makes effective monitoring essential for their protection. In recent years, remote sensing has emerged as a powerful method for studying mangroves. Its ability to collect detailed information over time and across large areas makes it ideal for tracking changes in these fragile ecosystems. This chapter takes a closer look at how remote sensing technologies have evolved and how they're being used to monitor mangroves more accurately. It starts by outlining why mangroves matter and the many challenges they face, such as deforestation, pollution, and rising sea levels. It then explores the tools researchers use, like passive sensors (such as multispectral and hyperspectral imaging) and active sensors (including LiDAR and SAR), explaining how they work, what they can do, and where they fall short. The chapter also highlights how combining remote sensing with Geographic Information Systems and machine learning is improving how we map and monitor mangrove forests. Tools like satellite imagery and aerial photography help scientists measure changes in mangrove size, species, and overall

M. Bera · M. Roy Choudhury · S. Das (✉)
School of Environment and Disaster Management, Ramakrishna Mission Vivekananda Educational and Research Institute, Kolkata, West Bengal, India
e-mail: sumanta.das@uq.net.au

M. Bera
e-mail: mahadev.bera@gm.rkmvu.ac.in

M. Roy Choudhury
e-mail: m.roychoudhury@uq.net.au

M. Roy Choudhury · S. Das
The University of Queensland, St Lucia, QLD, Australia

S. Dutta
Department of Genetics and Plant Breeding, Ramakrishna Mission Vivekananda Educational and Research Institute, Kolkata, West Bengal, India

O. Mutanga et al. (eds.), *Revealing Ecosystem Services Through Geospatial Technologies*, Springer Remote Sensing/Photogrammetry,
https://doi.org/10.1007/978-3-031-98048-0_11

health. Drones are proving especially useful for gathering high-resolution images of smaller, specific areas. Looking ahead, newer technologies like satellite-based hyperspectral sensors and unmanned surface vehicles (USVs) offer exciting possibilities for advancing mangrove monitoring even further. Ultimately, the chapter emphasizes the need for innovation and collaboration to close existing research gaps and build stronger conservation strategies.

Keywords Mangrove ecosystems · Remote sensing · Ecological balance · Anthropogenic threats · Monitoring techniques · Conservation strategies

1 Introduction

Mangrove ecosystems rank among the most productive and carbon-dense environments on Earth. They play a crucial role in stabilizing the global climate by capturing carbon and limiting greenhouse gas emissions (Alongi 2020). However, despite their ecological significance, widespread deforestation and degradation have caused a dramatic drop in global mangrove carbon reserves. This decline is largely the result of human activity and a general lack of understanding around how mangroves function compared to other types of forests. One major hurdle is accurately measuring mangrove biomass and carbon storage, especially since these ecosystems are often located in hard-to-reach intertidal zones.

Currently, the lack of standardized, scalable, and accurate methods for assessing carbon and biomass makes it difficult to build strong conservation and restoration plans. This is where remote sensing has become increasingly valuable. It allows for consistent tracking of mangrove health, species makeup, physical structure, and carbon trends over time. Technologies like synthetic aperture radar (SAR), multispectral, hyperspectral sensors, and aerial imagery have shown promise in mapping mangrove extent, identifying species, and estimating biomass (Roy et al. 2024; Vizcaya-Martínez et al. 2022; Abbas et al. 2020).

Typically found along tropical and subtropical shorelines, mangroves are natural shields against erosion, storms, and rising seas (Henriques et al. 2021). They also provide essential services—storing carbon in their roots and soil (Alongi 2020; Rahim et al. 2024), offering wood and other resources, and supporting fisheries (Asari et al. 2021; Strain et al. 2022). Unfortunately, they are under increasing pressure from growing populations and coastal migration, which has led to their conversion into farmland, aquaculture ponds, and urban areas (Soares et al. 2021; Tian et al. 2024). Weak policies and disjointed coastal development plans have only made things worse. Between 1980 and 2005, around 3.6 million hectares of mangroves were lost, significantly contributing to carbon emissions, particularly in countries like Indonesia (Friess et al. 2020).

The causes of mangrove loss vary by region, ranging from urban development and overharvesting to salt production and farming (Dayal et al. 2022). One area where research falls short is in understanding the long-term effects of mangrove degradation on their ability to store carbon. There's also a lack of comprehensive data and consistent methods for tracking these changes. Remote sensing can help fill this gap by offering reliable, spatial, and time-based monitoring that's essential for conservation efforts like REDD+ (Köhl et al. 2020; Perosa et al. 2023).

Compared to traditional methods, remote sensing is faster, cheaper, and better suited for large-scale monitoring and mapping (Farzanmanesh et al. 2021; Abu Bakar et al. 2024). New tools like machine learning and data integration are helping researchers process and understand complex data from various sources. These techniques have been used to study structure, estimate biomass, and even distinguish between different mangrove species (Fig. 1). Still, more research is needed to refine these methods and understand their limitations (Yang et al. 2024). In recent years, cutting-edge technologies like drones (UAVs), unmanned boats (USVs), and high-resolution satellite sensors in gathering detailed, site-specific data. UAVs, in particular, are excellent for capturing images in remote or hard-to-reach coastal areas, making them ideal for local assessments and involving communities in management efforts.

This chapter evaluates how remote sensing is being used to study mangrove ecosystems, focusing on carbon storage, biomass, and species identification. It aims to spotlight research gaps and technological challenges while also proposing ways to overcome them. The broader goal is to strengthen cross-disciplinary approaches and international cooperation to improve mangrove conservation. It emphasizes the need for standardized protocols and integrated systems to support better policy decisions, sustainable management, and adaptation strategies in coastal areas.

2 Understanding Coastal Mangrove Ecosystems and Its Degrading Trends

Mangroves represent a distinct and vital component of tropical and subtropical coastal environments, occupying the transitional interface between terrestrial and marine ecosystems. These halophytic, woody plant species are uniquely adapted to saline, anoxic, and waterlogged conditions that most other vegetation cannot tolerate. Distributed predominantly along coastlines and estuarine margins, mangrove forests provide critical ecological functions and serve as biodiversity hotspots that support numerous threatened and endemic species. Despite covering less than 1% of the global tropical forest area, mangroves offer disproportionately high ecological and socio-economic value. They underpin coastal resilience, food security, and the livelihoods of millions, particularly in developing regions (UNEP 2023). Mangrove ecosystems are increasingly vulnerable to a suite of anthropogenic pressures, including unsustainable resource extraction, industrial and agricultural pollution, coastal development, and conversion to aquaculture and agriculture. Furthermore, the escalating impacts of climate change, such as sea-level rise, altered precipitation regimes, and increased storm intensity further threaten their survival. According

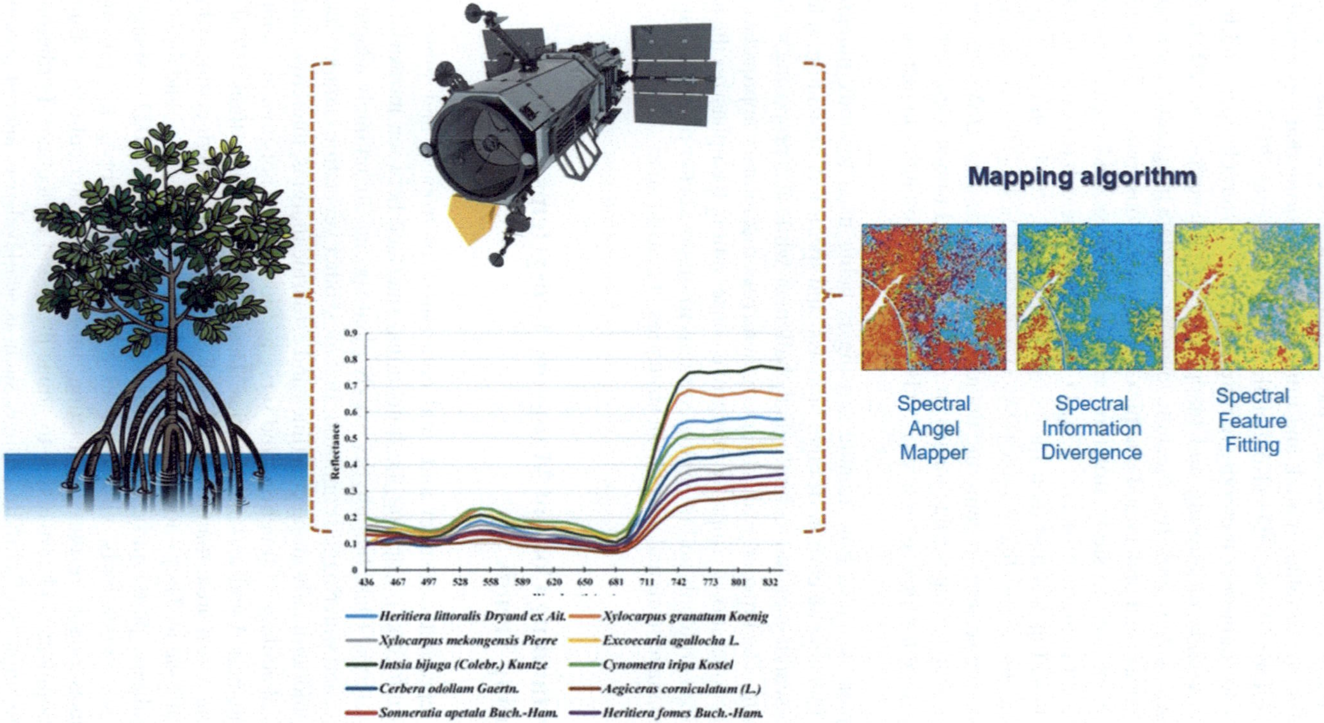

Fig. 1 Mangrove species mapping using remote sensing techniques

to data from Global Mangrove Watch, the global mangrove cover in 2020 stood at approximately 147,359 km^2. Regional distribution was uneven, with 51% located in the Asia–Pacific region, 29% in the Americas, and 20% in Africa. This reflects a net decline of 5245 km^2 (or 3.4%) since 1996. Encouragingly, recent satellite-based assessments indicate a deceleration in the rate of global mangrove loss, particularly near major riverine and deltaic systems, suggesting localized gains through conservation and restoration efforts (UNEP 2023).

Mangroves deliver a wide array of ecosystem services, with an estimated annual economic value ranging between USD 33,000 and USD 57,000 per hectare. These services include provisioning functions such as fisheries, fuelwood, medicinal resources, timber, and building materials. Regulating services include nutrient cycling (e.g., nitrogen), maintenance of water quality, flood mitigation, sediment trapping, and shoreline stabilization. Cultural services are also significant, especially through ecotourism and traditional livelihoods. Notably, mangroves are among the most carbon-dense forest types, sequestering substantial quantities of carbon in both biomass and sediments, making them integral to global climate change mitigation strategies. The degradation of these carbon-rich systems contributes substantially to greenhouse gas emissions and results in economic losses estimated in the billions of dollars annually (UNEP 2023).

Ecologically, mangroves form some of the most diverse and productive coastal habitats. Their aerial and stilt root systems facilitate sediment deposition and water filtration, enhancing shoreline integrity and providing natural defense against tropical cyclones, tsunamis, and storm surges. Mangrove organic matter contributes to nutrient enrichment, while their structural complexity offers refuge and nursery grounds for a range of aquatic and terrestrial fauna (Kanan et al. 2024; Su and Gasparatos 2023; Riska et al. 2023). However, the combination of intensive anthropogenic exploitation and accelerating climate change, including ocean acidification, is severely undermining the structural and functional integrity of these ecosystems.

Globally, mangroves occupy just 0.1% of the Earth's terrestrial surface. Yet, their distribution is highly skewed, with over 75% of mangrove cover concentrated in just 15 countries (Jia et al. 2023). In India, mangrove forests span approximately 4,975 km^2—constituting 0.15% of the national land area, about 3% of the global mangrove area, and 8% of Asia's mangrove cover. The eastern coast adjacent to the Bay of Bengal accounts for over 60% of India's mangroves, followed by the Arabian Sea coastline (27%) and the Andaman and Nicobar Islands (13%) (Roy et al. 2023). Mangrove degradation in India is particularly pronounced in ecologically sensitive and economically significant coastal states such as Tamil Nadu, West Bengal, and the Andaman and Nicobar Islands, where coastal development, aquaculture expansion, and industrial discharges have severely impaired mangrove health (Roy et al. 2023).

The spatial distribution of mangrove forests in India is highly heterogeneous. West Bengal harbors the largest mangrove tract, comprising 42.45% of the national cover, followed by Gujarat with 23.66%. Other states in India such as Goa (0.52%; 26 km^2), Kerala (0.20%; 9 km^2), Karnataka (0.205%; 10 km^2), Tamil Nadu (0.09%; 45 km^2), Daman and Diu (0.03%; 3 km^2), and Puducherry (0.06%; 2 km^2) contribute modestly. In terms of density, West Bengal (996 km^2), the Andaman and Nicobar

Islands (398 km^2), and Odisha (81 km^2) account for the majority of the country's very dense mangrove stands, totaling 1,476 km^2 or approximately 29.66% of the national mangrove cover (Youdon 2020) (Fig. 2).

Alarmingly, global mangrove forests are being lost at an annual rate of 1–2%, a rate that surpasses the decline observed in both tropical rainforests and coral reefs. Since the 1950s, nearly half of the world's mangrove ecosystems have been degraded or destroyed due to insufficient protection and substantial habitat transformation. If

Fig. 2 Distribution of mangrove forests area in India (*Data source* Youdon 2020)

current trends persist, the complete disappearance of mangrove habitats within this century is a conceivable scenario (MOEFCC 2019).

3 Fundamentals of Remote Sensing to Distinguish Mangrove Species

Effective management of coastal mangrove ecosystems requires accurate identification of mangrove species, as this provides essential insights into the associated environmental and geomorphic conditions. However, accurately mapping these species presents several challenges. One significant issue arises from the small size of individual mangrove trees, which may fall below the resolution of available remote sensing data. Additionally, the co-occurrence of multiple mangrove species, often closely related and exhibiting similar spectral characteristics, complicates species identification. Traditional image classification methods have struggled to distinguish between these species due to spectral and geographical limitations. Recent advancements in high spatial resolution data have shown promising results in overcoming these challenges (Table 1). Satellite imagery with resolutions below five meters, such as that from IKONOS, QuickBird, GeoEye, and WorldView, has proven effective for mapping mangrove species (Marshall and Thenkabail 2015; Blasch et al. 2023). Specifically, panchromatic data from GeoEye, QuickBird, IKONOS, and WorldView offer resolutions of 2 m, 0.61 m, 0.83 m, and 0.50 m, respectively.

Recent innovations in object-based image classification, such as image segmentation methods like object-based image analysis (OBIA), have enhanced the ability to map mangrove species (Pasaribu et al. 2021). OBIA techniques have shown superior accuracy in identifying mangrove communities compared to traditional pixel-based methods (Maurya et al. 2021; Lombard and Andrieu 2021). Hyperspectral datasets, including EO-1 Hyperion and CASI-2, have proven effective in distinguishing between different mangrove species (Anand et al. 2020). Combining hyperspectral and high-resolution spatial data with OBIA techniques has consistently outperformed conventional pixel-based methods in mangrove species mapping (Pasaribu et al. 2021). Hybrid approaches, such as those integrating OBIA with lacunarity spatial measurements or machine learning algorithms, have further improved accuracy, achieving over 80% accuracy with medium-resolution imagery from Landsat and over 94% with high-resolution Pléiades-1 imagery (Htitiou et al. 2021). Feature extraction techniques applied to very high-resolution imagery have also significantly improved mangrove species classification, with the grey-level co-occurrence matrix (GLCM) showing promising results, achieving over 81% accuracy with IKONOS data and over 89% with WorldView-2 images (Maurya et al. 2021; Jiang et al. 2021).

In addition to optical imagery, remote sensing technologies beyond optical imaging have proven effective in differentiating mangrove species. For instance, Truong et al. (2018) demonstrated the effectiveness of Advanced Land Observing Satellite Phased Array type L-band Synthetic Aperture Radar (ALOS PALSAR)

Table 1 A summary of the studies across different countries showing remote sensing techniques in distinguishing different mangrove species

Technique used	Sensor	Performance (%)	Location	References
Object-based logistics model tree (LMT) algorithm	ALOS PALSAR and ALOS-2 PALSAR-2	80.2–83.8	Vietnam	Pham et al. (2018a, b)
Pixel-based methods maximum likelihood algorithm	ALOS AVNIR-2	80	East Malaysia	Zhang et al. (2018)
Pixel-based methods Linear spectral unmixing (LSU)	Hyperspectral CASI-2	56	Queensland, Australia	Kamal and Phinn (2011)
Unsupervised ISODATA classifier	IKONOS	78	Guinea, West Africa	Kovacs et al. (2010)
NDVI pixel-based methods	Landsat 8 OLI SPOT-5 Sentinel-2 WorldView-2	64 75 78 93	Mexican Pacific	Valderrama-Landeros et al. (2018)
Visual interpretation methods	Aerial photographs WorldView-2	68 42–58	Darwin, Australia	Heenkenda et al. (2014)
Object-based image classifier	Hyperspectral hyperion	88%	Mai Po Hong Kong	Jia et al. (2014)
Object-based image classifier	Hyperspectral CASI-2	69–76	Southeast Queensland, Australia	Kamal and Phinn (2011)

data in classifying mangrove species in Vietnam, achieving over 80% accuracy using the object-based logistic model tree (LMT) algorithm. Valderrama-Landeros et al. (2018) also showcased the potential of using NDVI-based pixel classifiers for mapping mangrove species in Mexico, employing Landsat-8 OLI, SPOT-5, Sentinel-2A, and WorldView-2 datasets, which demonstrated the importance of using diversified remote sensing sources. Higher spatial resolutions correlated with improved classification accuracy.

Furthermore, advancements in computer vision, pattern recognition, and artificial intelligence (AI) have further enhanced species classification accuracy. Recent machine learning algorithms are capable of utilizing optical, hyperspectral, and synthetic aperture radar (SAR) data for improved mangrove species identification (Table 2). While traditional machine learning techniques, such as decision trees (e.g., CART and C4.5), are commonly used in other ecosystems, they are less frequently applied to mangrove classification. Researchers have instead focused on methods such as random forests (RF), artificial neural networks (ANN), and support vector machines (SVM) due to their ability to create non-parametric classification models that do not assume a specific data distribution (Cao et al. 2018; Shen et al. 2023). For instance, Wang et al. (2018) demonstrated that RF outperformed SVM in object-based image classification using Pléiades-1 data, while SVM yielded the highest accuracy in pixel-based analyses. SVM algorithms, when applied to pan-sharpened WorldView-2 data, achieved classification accuracies exceeding 87%. Additionally, time-series data from Gaofen-1 (GF-1) and SPOT satellites have produced accuracy ranges between 67 and 92%. Machine learning techniques, particularly those combining spectral and textural features, have shown significant potential for differentiating mangrove species. Clustering-based neural networks, for example, have achieved overall accuracies above 90% by integrating textural information from panchromatic bands with high-resolution multispectral images. At the Mai Po Ramsar site in Hong Kong, Wong and Fung (2014) demonstrated that combining radar and optical data improved classification accuracy. Later on, Zhang et al. (2018) utilized a rotation forest (RoF) algorithm and found that merging optical and SAR data enhanced both overall accuracy and the Kappa coefficient for mangrove species recognition, achieving an impressive accuracy of 85.23%.

4 Applications of Remote Sensing in Mangrove Monitoring and Assessment

Remote sensing is a technique that enables the study of distant objects without physical contact, primarily utilizing satellites to measure the radiation reflected and emitted by a region. This allows for the analysis of the physical characteristics of the area. A variety of methods exist to assess the energy reflected and absorbed by the Earth's surface, as well as to capture imagery with suitable spatial and spectral resolutions. Additionally, spectral information can be obtained in a manner tailored to specific data types and their respective applications. Mangrove ecosystems are uniquely identified by the spectral and textural characteristics of their leaves and canopies, which can appear either homogeneous or heterogeneous. These variations depend on factors such as species composition, growth density, height, and spatial distribution. Mapping, monitoring, and extracting key parameters from mangroves are critical for understanding the geomorphology and environmental conditions of these ecosystems (Maurya et al. 2021; Giri 2016). However, the use of remote sensing for mangrove assessment presents several challenges. Mangrove areas may be submerged under water and are often located in coastal regions that are difficult to access. Additionally, identifying individual trees can be problematic when their canopies are smaller than the image resolution, especially in mixed-species communities where closely related species exhibit similar spectral reflectance. Despite these challenges, remote sensing image analysis remains crucial for mangrove monitoring.

Table 2 A summary of the findings where different machine learning models were integrated with remote sensing datasets to classify mangrove species across different regions

Machine learning method	Description	Performance	Data type	Key features	References
Support vector machine (SVM)	Classifies data by finding the optimal hyperplane in high-dimensional space	> 87% accuracy with WorldView-2	Optical, Hyperspectral	Effective for high-dimensional data, robust against overfitting	Huang et al. (2018), Pandey et al. (2019a)
Random forest (RF)	Ensemble method that combines multiple decision trees to improve classification accuracy	85.23% accuracy	Optical, SAR	Handles non-linear relationships well, reduces variance	Rodriguez-Galiano et al. (2015), Pandey et al. (2019a)
Artificial neural network (ANN)	Mimics neural connections in the brain to model complex relationships in the data	Variable (architecture-dependent)	Optical, Hyperspectral	Capable of learning complex patterns, requires larger datasets	Ghorbanian et al. (2022)
Rotation forest (RoF)	An ensemble method that rotates feature subsets to enhance accuracy in classification	85.23% accuracy	Optical, SAR	Improves accuracy through feature diversity	Pham et al. (2018a, b)
Clustering-based neural networks	Utilizes clustering techniques to identify and classify patterns among data points	> 90% accuracy	High-resolution multispectral	Combines unsupervised learning with classification	Kim et al. (2019)

(continued)

Table 2 (continued)

Machine learning method	Description	Performance	Data type	Key features	References
Decision trees (DT)	Tree-based model that splits data based on feature values to make classifications	Variable (often lower accuracy)	Optical, Hyperspectral	Simple interpretation, prone to overfitting	Charbuty and Abdulazeez (2021)
K-nearest neighbors (KNN)	Classifies data points based on the majority class among their nearest neighbors	Variable (depends on k value)	Optical, Hyperspectral	Simple and intuitive, sensitive to irrelevant features	Zhang (2016)

Various techniques are available for extracting and analyzing spectral, spatial, and textural information from these images, which can be categorized through digital image classification methods. A generalized flow model for mangrove mapping and monitoring is depicted in Fig. 3. The primary focus of these methods is classification, which involves processing remote-sensing image data to assign labels to each pixel based on real-world environmental characteristics. By classifying each pixel, these methods facilitate the extraction of valuable information regarding land features and the creation of vegetation and land use/land cover (LULC) maps (Kuenzer et al. 2011).

Digital image classification techniques for mangroves can be divided into several categories: pixel-based and subpixel-based methods, which focus on analyzing pixel properties, object-oriented methods, which focus on delineating objects directly, and knowledge-based approaches that incorporate additional data and specific criteria. These methods encompass both supervised and unsupervised learning, as shown in Fig. 4. The choice of classification technique depends on the spatial and spectral features of the image, and the accuracy of classification may vary based on the available data. A summary of the various remote-sensing techniques applied to mangrove monitoring is provided in Table 3.

sensing data and different digital image classification techniques. Adapted from Maurya et al. (2021).

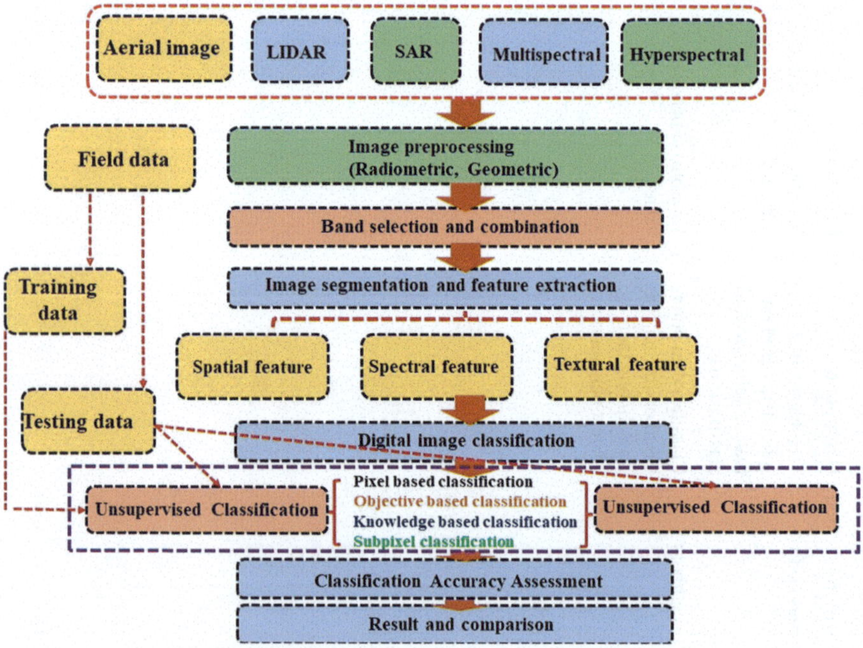

Fig. 3 A generalized flow model for mapping and monitoring mangroves using remote

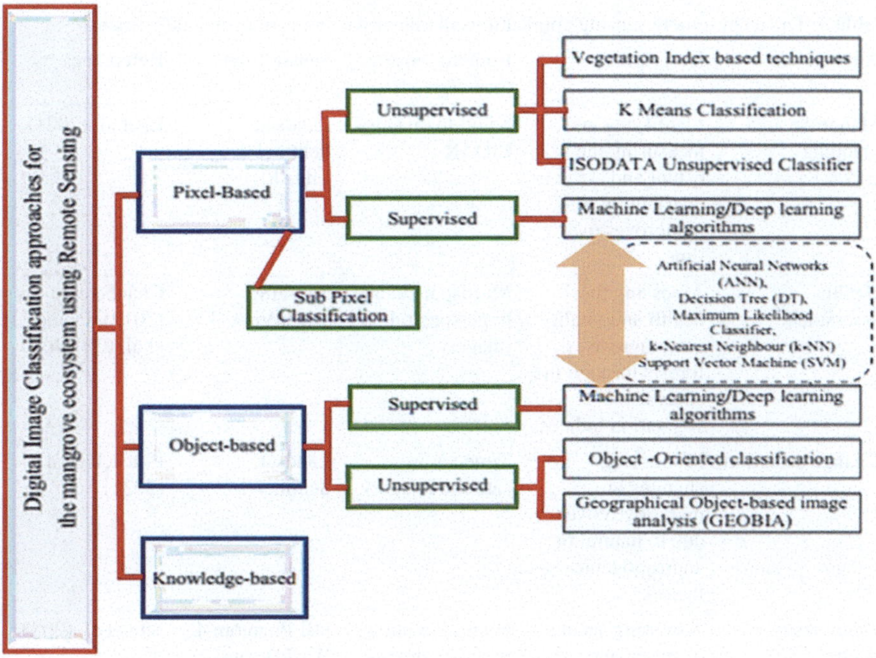

Fig. 4 Approaches for Digital image classification for the mangrove ecosystem using remote sensing. Adapted from Maurya et al. (2021)

Index-based approaches are instrumental in classifying plant cover and distinguishing it from other land cover types through the use of various vegetation indices derived from remotely sensed imagery. A significant body of research has focused on applying these methods to mangrove ecosystems, with some key studies outlined in Table 4. These studies reveal that these techniques are adaptable to different satellite images and sensors, encompassing both hyperspectral and multispectral data. Moreover, combining these indices with other digital image classification techniques often enhances performance. The Normalized Difference Vegetation Index (NDVI) is one of the most prevalent methods used for monitoring vegetation, including mangrove forests (Lemenkova and Debeir 2023). Vegetation indices (VIs), in particular, are widely applied to assess differences in land cover types, making them highly relevant for mangrove studies. High-resolution imagery can also be utilized to delineate vegetation characteristics more precisely. The choice of these indices should be informed by robust validation techniques and ground truth data, as each index is based on distinct mathematical formulas involving various spectral combinations, sensors, platforms, and resolutions.

Table 3 Different remote sensing applications in mangrove monitoring and assessment

Application	Description	Remote sensing technique	Sensor used	References
Mangrove area mapping	Identifying and measuring the extent and distribution of mangroves over time	Satellite imagery, LiDAR	Landsat, Sentinel-2, LiDAR	Li et al. (2021)
Health assessment	Assessing the health and vitality of mangrove forests through the analysis of vegetation indices	Multispectral and hyperspectral imagery	MODIS, WorldView-3	Chaube et al (2019), Pandey et al. (2019b)
Change detection	Monitoring changes in mangrove cover due to natural or anthropogenic factors	Time-series satellite imagery	Landsat, Sentinel-1	Pirasteh et al. (2021)
Biodiversity studies	Assessing species diversity and habitat structure in mangrove ecosystems	High-resolution imagery, drones	DJI Phantom 4, WorldView-2	Sun et al. (2023)
Carbon stock estimation	Estimating carbon storage potential of mangrove forests for climate change studies	Remote sensing combined with ground data	ICESat, Sentinel-2	Yu et al. (2024)
Coastal erosion monitoring	Tracking changes in shoreline and erosion rates affecting mangrove habitats	Aerial imagery, satellite data	UAVs, Google Earth, Landsat	McCarroll et al. (2024)
Hydrological assessment	Analyzing water quality and hydrological conditions influencing mangrove ecosystems	Remote sensing combined with GIS	MODIS, Sentinel-3	Ganguly and Bhan (2023)

(continued)

Table 3 (continued)

Application	Description	Remote sensing technique	Sensor used	References
Climate change impact studies	Evaluating the effects of climate change on mangrove distribution and health	Long-term satellite monitoring	Landsat, Sentinel-2	Chopade et al. (2023)
Pollution assessment	Detecting and monitoring pollution levels affecting mangrove areas, such as oil spills	Hyperspectral imaging	AVIRIS, Hyperion	Lassalle et al. (2023)

5 Ecosystem Services

Mangroves provide a wide array of ecosystem services benefiting both the environment and human communities. These services are categorized into provisioning, regulating, supporting, and cultural services.

- **Provisioning Services**: Mangroves supply resources like timber, fuelwood, honey, and fish, which are essential for the livelihoods of many coastal communities. For example, their rich biodiversity supports fisheries by acting as nurseries for commercially valuable species (Alongi 2012).
- **Regulating Services**: Mangroves are known for their exceptional ability to mitigate climate change. They store significant amounts of carbon in biomass and soil, acting as carbon sinks (Donato et al. 2011). Furthermore, mangroves protect coastlines from erosion and reduce the impact of storm surges and tsunamis, providing vital disaster risk reduction services (Duarte et al. 2013).
- **Supporting Services**: Mangroves support biodiversity by providing habitats for numerous terrestrial and aquatic species. They enhance nutrient cycling and water filtration, contributing to the overall health of coastal ecosystems (Kathiresan and Bingham 2001).
- **Cultural Services**: Mangroves hold cultural, aesthetic, and educational significance. Many coastal communities value mangroves for their spiritual importance and recreational opportunities, which also contribute to ecotourism (Barbier et al. 2011).

Despite their value, mangroves face threats such as deforestation, pollution, and climate change. Protecting these ecosystems is vital to ensure the continuation of their services for future generations.

Table 4 Vegetation index-based techniques for monitoring the mangrove

Vegetation index	Description	Formula	Application in mangrove monitoring	Sensor compatibility	References
Normalized difference vegetation index (NDVI)	Assesses vegetation health by comparing near-infrared and red reflectance, which is useful for evaluating mangrove health	NDVI = (NIR − Red)/(NIR + Red)	Evaluating mangrove vitality and cover assessment	Landsat, Sentinel-2, MODIS	Pamungkas (2023)
Enhanced vegetation index (EVI)	Enhances sensitivity to high biomass regions and minimizes atmospheric effects, which is advantageous for dense mangrove canopies	EVI = G × (NIR − Red)/(NIR + C1 × Red − C2 × Blue + L)	Assessing biomass and productivity in mangrove forests	MODIS, Sentinel-2	Nguyen et al. (2023)
Soil adjusted vegetation index (SAVI)	Adjusts for soil brightness to enhance vegetation detection in mangrove areas, particularly in coastal zones	SAVI = ((NIR − Red)/(NIR + Red + L)) × (1 + L)	Monitoring vegetation in coastal and muddy environments	Landsat, MODIS	Konno and Homma (2023)
Normalized difference water index (NDWI)	The classification and monitoring procedures are complicated due to existing challenges	NDWI = (Green − NIR)/(Green + NIR)	Identifying mangrove wetlands and assessing water quality	Landsat, Sentinel-2	León-Pérez et al. (2023)

(continued)

Table 4 (continued)

Vegetation index	Description	Formula	Application in mangrove monitoring	Sensor compatibility	References
Green normalized difference vegetation index (GNDVI)	Focuses the significance of the green band in enhancing vegetation assessment, especially in densely vegetated areas like mangroves	GNDVI = (NIR − Green)/(NIR + Green)	Monitoring mangrove health and leaf area dynamics	Sentinel-2, Landsat	Anggraini et al. (2023)
Leaf area index (LAI)	Estimates leaf area per unit ground area, crucial for assessing mangrove productivity	LAI = − ln(1 − (NDVI/ NDVImax))	Evaluating leaf density and forest structure in mangroves	MODIS, Landsat	Zhao et al. (2023)

6 Challenges and Future Directions

The monitoring and assessment of coastal mangrove ecosystems through remote sensing face several challenges that must be addressed to improve accuracy and effectiveness. One of the primary limitations lies in the need for high spatial and temporal resolution data to fully capture the complex structure and dynamic nature of mangrove forests. Variability in atmospheric conditions and inconsistencies in sensor quality can significantly affect the precision and reliability of remote sensing outputs. Furthermore, the presence of multiple mangrove species, each exhibiting distinct spectral signatures, complicates classification and monitoring processes. Tidal fluctuations also influence mangrove ecosystems, leading to alterations in vegetation structure that are not easily detected or accurately represented through conventional remote sensing techniques. Additionally, human-induced disturbances such as urbanization, agriculture, and aquaculture alter mangrove habitats and necessitate regular monitoring to track ecological changes over time. The integration of socioeconomic and ground-based data with remote sensing observations remains a critical but often challenging task, essential for producing comprehensive assessments. Moreover, evaluating the long-term impacts of climatic variables—including storm events, temperature shifts, and sea level rise on mangrove health requires sustained, longitudinal monitoring efforts.

To overcome these challenges and enhance future capabilities, several strategic directions are emerging. Recent advancements in satellite and UAV sensor technologies have led to improved spatial resolution and the availability of multispectral datasets, significantly enhancing the ability to monitor mangrove ecosystems with greater detail and frequency. The incorporation of LiDAR and hyperspectral imaging further facilitates high-resolution structural and functional analysis of mangrove stands, contributing to a more nuanced understanding of ecosystem dynamics. Concurrently, the application of machine learning and AI algorithms has shown considerable promise in automating data processing, improving classification accuracy, and enabling predictive modeling of ecological trends. Establishing robust, long-term monitoring networks that integrate in-situ measurements with remote sensing data provides a more holistic perspective on ecosystem health and resilience. Addressing the multifaceted threats to mangroves also necessitates interdisciplinary collaboration, bridging ecological science, geospatial analysis, and socioeconomic research. Moreover, engaging local communities and fostering citizen science initiatives can play a crucial role in enhancing data collection efforts and promoting environmental stewardship. Finally, the strategic use of remote sensing data to inform policy formulation and management frameworks holds significant potential for supporting conservation goals and ensuring the sustainable use of mangrove resources.

7 Conclusion

This chapter highlights the key ecological, climatic, and socio-economic importance of mangrove ecosystems. Mangroves not only serve as natural defenses that shield coastlines from erosion and storm surges, but also contribute importantly to climate change mitigation through their high carbon sequestration potential, frequently being more efficient than many terrestrial forest systems. Additionally, they support the livelihoods of millions living in coastal communities and provide significant contributions to local economies in the form of fisheries, timber, and eco-tourism. Though their critical significance, mangrove forests are becoming increasingly threatened by human pressures like urbanization, aquaculture development, and unsustainable land use. These pressures have contributed to alarming levels of mangrove deforestation, fragmentation, and degradation across the world. Against this backdrop, the chapter highlights the imperative for robust, scalable, and affordable monitoring systems that can deliver accurate measurements of mangrove biomass, carbon stocks, and ecological integrity. New developments in remote sensing technologies, such as the combination of high-resolution multispectral and hyperspectral satellite imagery, SAR, LiDAR data, and the use of AI and machine learning algorithms, have greatly improved our capacity to identify spatial–temporal changes in mangrove extent, species distribution, and biophysical properties. These technologies provide new avenues to break through conventional field-based constraints, allowing large-scale, repeatable, and non-destructive measurements.

Nevertheless, the chapter also acknowledges the continued challenges in synthesizing disparate datasets, establishing standard methodologies, and addressing local heterogeneity in mangrove structure and function. Filling these methodological gaps requires interdisciplinary effort and co-development of integrative frameworks that merge remote sensing outputs with ground-truth data and ecological modeling. Ultimately, this chapter promotes continued research efforts, stakeholder involvement, and policy integration to advance mangrove conservation, restoration, and adaptive management. Protecting mangrove ecosystems is not only vital to the mitigation of climate change but also to ensuring coastal resilience, conservation of biodiversity, and sustainable development. The future that captures the full potential of remote sensing innovations in tandem with inclusive conservation approaches will be key to securing the ecological integrity and socio-economic value of mangroves globally.

References

Abbas S, Wong MS, Wu J, Shahzad N, Muhammad Irteza S (2020) Approaches of satellite remote sensing for the assessment of above-ground biomass across tropical forests: pan-tropical to national scales. Remote Sens 12(20):3351

Abu Bakar NA, Wan Mohd Jaafar WS, Abdul Maulud KN, Muhmad Kamarulzaman AM, Saad SNM, Mohan M (2024) Monitoring mangrove-based blue carbon ecosystems using UAVs: a review. Geocarto Int 39(1):2405123

Alongi DM (2012) Carbon sequestration in mangrove forests. Carbon Management 3(3):313–322

Alongi DM (2020) Global significance of mangrove blue carbon in climate change mitigation. Science 2(3):67

Anand A, Pandey PC, Petropoulos GP, Pavlides A, Srivastava PK, Sharma JK, Malhi RKM (2020) Use of hyperion for mangrove forest carbon stock assessment in Bhitarkanika forest reserve: a contribution towards blue carbon initiative. Remote Sens 12(4):597

Anggraini N, Suardana AAMAP, Nandika MR, Ulfa A, Ginting DNB, Adawiah SW et al (2023) Mangrove damage based on canopy density derived on SPOT 6 at Ngurah Rai forest park, Bali. IOP Conf Ser Earth Environ Sci 1224(No. 1):012017

Asari N, Suratman MN, Mohd Ayob NA, Abdul Hamid NH (2021). Mangrove as a natural barrier to environmental risks and coastal protection. Mangroves Ecol Biodiv Manage 305–322

Barbier EB, Hacker SD, Kennedy C, Koch EW, Stier AC, Silliman BR (2011) The value of estuarine and coastal ecosystem services. Ecol Monogr 81(2):169–193

Blasch G, Anberbir T, Negash T, Tilahun L, Belayineh FY, Alemayehu Y, et al (2023). The potential of UAV and very high-resolution satellite imagery for yellow and stem rust detection and phenotyping in Ethiopia. Sci Rep 13(1):16768

Cao J, Liu K, Liu L, Zhu Y, Li J, He Z (2018) Identifying mangrove species using field close-range snapshot hyperspectral imaging and machine-learning techniques. Remote Sens 10(12):2047

Charbuty B, Abdulazeez A (2021) Classification based on decision tree algorithm for machine learning. J Appl Sci Technol Trends 2(01):20–28

Chaube NR, Lele N, Misra A, Murthy TV, Manna S, Hazra S et al (2019) Mangrove species discrimination and health assessment using AVIRIS-NG hyperspectral data. Curr Sci 116(7):1136–1142

Chopade MR, Mahajan S, Chaube N (2023) Assessment of land use, land cover change in the mangrove forest of Ghogha area, Gulf of Khambhat, Gujarat. Expert Syst Appl 212:118839

Dagar JC, Gangaiah B, Gupta SR (2020) Agroforestry to sustain island and coastal agriculture in the scenario of climate change: Indian perspective. Agroforest Degraded Landscapes Recent Adv Emerg Challenges 1:367–424

Dayal SR, Ravuiwasa K, Bhat JA (2022) Deforestation and degradation in the mangrove ecosystem: implication on environment and livelihoods. In: Land degradation neutrality: achieving SDG 15 by forest management. Springer, Singapore, pp 99–116

Donato DC, Kauffman JB, Murdiyarso D, Kurnianto S, Stidham M, Kanninen M (2011) Mangroves among the most carbon-rich forests in the tropics. Nat Geosci 4(5):293–297

Duarte CM, Losada IJ, Hendriks IE, Mazarrasa I, Marbà N (2013) The role of coastal plant communities for climate change mitigation and adaptation. Nat Clim Chang 3(11):961–968

Farzanmanesh R, Khoshelham K, Thomas S (2021) Technological opportunities for measuring and monitoring blue carbon initiatives in mangrove ecosystems. Remote Sens Appl Soc Environ 24:100612

Friess DA, Krauss KW, Taillardat P, Adame MF, Yando ES, Cameron C (2020) Mangrove blue carbon in the face of deforestation, climate change, and restoration. Annual Plant Rev 3(10.1002):9781119312994

Ganguly S, Bhan U (2023) Remote sensing and GIS based monitoring and management of coastal aquifers and ecosystem. Emerging technologies for water supply, conservation and management. Springer International Publishing, Cham, pp 171–193

Ghorbanian A, Ahmadi SA, Amani M, Mohammadzadeh A, Jamali S (2022) Application of artificial neural networks for mangrove mapping using multi-temporal and multi-source remote sensing imagery. Water 14(2):244

Giri C (2016) Observation and monitoring of mangrove forests using remote sensing: opportunities and challenges. Remote Sens 8(9):783

Heenkenda MK, Joyce KE, Maier SW, Bartolo R (2014) Mangrove species identification: comparing WorldView-2 with aerial photographs. Remote Sens 6(7):6064–6088

Henriques M, Granadeiro JP, Piersma T, Leão S, Pontes S, Catry T (2021) Assessing the contribution of mangrove carbon and of other basal sources to intertidal flats adjacent to one of the largest West African mangrove forests. Mar Environ Res 169:105331

Htitiou A, Boudhar A, Lebrini Y, Lionboui H, Chehbouni A, Benabdelouahab T (2021) Classification and status monitoring of agricultural crops in central Morocco: a synergistic combination of OBIA approach and fused Landsat-Sentinel-2 data. J Appl Remote Sens 15(1):014504–014504

Huang S, Cai N, Pacheco PP, Narrandes S, Wang Y, Xu W (2018) Applications of support vector machine (SVM) learning in cancer genomics. Cancer Genomics Proteomics 15(1):41–51

Jia M, Wang Z, Mao D, Ren C, Song K, Zhao C et al (2023) Mapping global distribution of mangrove forests at 10-m resolution. Sci Bull 68(12):1306–1316

Jia M, Zhang Y, Wang Z, Song K, Ren C (2014) Mapping the distribution of mangrove species in the Core Zone of Mai Po Marshes Nature Reserve, Hong Kong, using hyperspectral data and high-resolution data. Int J Appl Earth Obs Geoinf 33:226–231

Jiang Y, Zhang L, Yan M, Qi J, Fu T, Fan S, Chen B (2021) High-resolution mangrove forests classification with machine learning using worldview and UAV hyperspectral data. Remote Sensing 13(8):1529

Kamal M, Phinn S (2011) Hyperspectral data for mangrove species mapping: a comparison of pixel-based and object-based approach. Remote Sens 3(10):2222–2242

Kanan AH, Masiero M, Pirotti F (2024) Estimating economic and livelihood values of the World's Largest Mangrove Forest (Sundarbans): a meta-analysis. Forests 15(5):837

Kathiresan K, Bingham BL (2001) Biology of mangroves and mangrove ecosystems. Adv Mar Biol 40:81–251

Kim EH, Oh SK, Pedrycz W, Fu Z (2019) Reinforced fuzzy clustering-based ensemble neural networks. IEEE Trans Fuzzy Syst 28(3):569–582

Köhl M, Neupane PR, Mundhenk P (2020) REDD+ measurement, reporting and verification—a cost trap? Implications for financing REDD+ MRV costs by result-based payments. Ecol Econ 168:106513

Konno T, Homma K (2023) Prediction of areal soybean lodging using a main stem elongation model and a soil-adjusted vegetation index that accounts for the ratio of vegetation cover. Remote Sens 15(13):3446

Kovacs JM, de Santiago FF, Bastien J, Lafrance P (2010) An assessment of mangroves in Guinea, West Africa, using a field and remote sensing based approach. Wetlands 30:773–782

Kuenzer C, Bluemel A, Gebhardt S, Quoc TV, Dech S (2011) Remote sensing of mangrove ecosystems: a review. Remote Sensing 3(5):878–928

Lassalle G, Ferreira MP, La Rosa LEC, Scafutto RDPM, de Souza Filho CR (2023) Advances in multi-and hyperspectral remote sensing of mangrove species: a synthesis and study case on airborne and multisource spaceborne imagery. ISPRS J Photogramm Remote Sens 195:298–312

Lemenkova P, Debeir O (2023) Computing vegetation indices from the satellite images using GRASS GIS scripts for monitoring mangrove forests in the coastal landscapes of Niger Delta, Nigeria. J Marine Sci Eng 11(4):871

León-Pérez MC, Reisinger AS, Gibeaut JC (2023) Spatial-temporal dynamics of decaying stages of pelagic *Sargassum* spp. along shorelines in Puerto Rico using Google Earth Engine. Marine Pollution Bull 188:114715

Li Q, Wong FKK, Fung T (2021) Mapping multi-layered mangroves from multispectral, hyperspectral, and LiDAR data. Remote Sens Environ 258:112403

Lombard F, Andrieu J (2021) Mapping mangrove zonation changes in Senegal with Landsat imagery using an OBIA approach combined with linear spectral unmixing. Remote Sens 13(10):1961

Marshall M, Thenkabail P (2015) Advantage of hyperspectral EO-1 Hyperion over multispectral IKONOS, GeoEye-1, WorldView-2, Landsat ETM+, and MODIS vegetation indices in crop biomass estimation. ISPRS J Photogramm Remote Sens 108:205–218

Maurya K, Mahajan S, Chaube N (2021) Remote sensing techniques: mapping and monitoring of mangrove ecosystem—a review. Complex Intell Syst 7(6):2797–2818

McCarroll RJ, Kennedy DM, Liu J, Allan B, Ierodiaconou D (2024) Design and application of coastal erosion indicators using satellite and drone data for a regional monitoring program. Ocean Coast Manag 253:107146

Ministry of Environment, Forest and Climate Change (MOEFCC) (2019) India State of Forest Report 2019. Chapter 3. https://fsi.nic.in/forest-report-2019?pgID=forest-report-2019

Nguyen TPC, Le TH, Nguyen TO, Le CL, Ha TC (2023) Vegetation indices for spatio-temporal analysis of the quality of can gio mangrove for biodiversity and conservation. IOP Conf Ser Earth Environ Sci 1247(1):012003

Pamungkas S (2023) Analysis of vegetation index For Ndvi, Evi-2, and Savi For Mangrove Forest Density Using Google Earth Engine in Lembar Bay, Lombok Island. IOP Conf Ser Earth Environ Sci 1127(1):012034

Pandey PC, Koutsias N, Petropoulos GP, Srivastava PK, Ben Dor E (2019a) Land use/land cover in view of earth observation: data sources, input dimensions, and classifiers—a review of the state of the art. Geocarto Int 36(9):957–988. https://doi.org/10.1080/10106049.2019.1629647

Pandey PC, Anand A, Srivastava PK (2019b) Spatial distribution of mangrove forest species and biomass assessment using field inventory and earth observation hyperspectral data. Biodivers Conserv 28:2143–2162. https://doi.org/10.1007/s10531-019-01698-8

Pasaribu RA, Aditama FA, Setyabudi P (2021) Object-based image analysis (OBIA) for mapping mangrove using unmanned aerial vehicle (UAV) on Tidung Kecil Island, Kepulauan Seribu, DKI Jakarta Province. IOP Conf Ser Earth Environ Sci 944(1):012037

Perosa B, Newton P, da Silva RFB (2023) A monitoring, reporting and verification system for low carbon agriculture: a case study from Brazil. Environ Sci Policy 140:286–296

Pham BT, Shirzadi A, Bui DT, Prakash I, Dholakia MB (2018a) A hybrid machine learning ensemble approach based on a radial basis function neural network and rotation forest for landslide susceptibility modeling: a case study in the Himalayan area, India. Int J Sedim Res 33(2):157–170

Pham TD, Bui DT, Yoshino K, Le NN (2018b) Optimized rule-based logistic model tree algorithm for mapping mangrove species using ALOS PALSAR imagery and GIS in the tropical region. Environ Earth Sci 77(5):159

Pirasteh S, Zenner EK, Mafi-Gholami D, Jaafari A, Kamari AN, Liu G et al (2021) Modeling mangrove responses to multi-decadal climate change and anthropogenic impacts using a long-term time series of satellite imagery. Int J Appl Earth Observation Geoinform 102:102390

Rahim A, Soeprobowati TR, Putranto TT, Al Falah MH, Gell P (2024) Contribution of mangrove forest carbon stocks on climate change mitigation: a case study at Tuntang Estuary, Central Java. J Coast Conserv 28(4):65

Riska JAT, Syukur A, Zulkifli L (2023) Association between mangrove types and some mangrove crab species in West Lombok Sheet Mangrove Ecosystem. Jurnal Penelitian Pendidikan IPA 9(7):5610–5619

Rodriguez-Galiano V, Sanchez-Castillo M, Chica-Olmo M, Chica-Rivas MJOGR (2015) Machine learning predictive models for mineral prospectivity: an evaluation of neural networks, random forest, regression trees and support vector machines. Ore Geol Rev 71:804–818

Roy AD, Arachchige PSP, Watt MS, Kale A, Davies M, Heng JE et al (2024). Remote sensing-based mangrove blue carbon assessment in the Asia-Pacific: a systematic review. Sci Total Environ 938:173270

Roy P, Pal SC, Chakrabortty R, Chowdhuri I, Saha A, Shit M (2023) Effects of climate change and sea-level rise on coastal habitat: vulnerability assessment, adaptation strategies and policy recommendations. J Environ Manage 330:117187

Shen Z, Miao J, Wang J, Zhao D, Tang A, Zhen J (2023) Evaluating feature selection methods and machine learning algorithms for mapping mangrove forests using optical and synthetic aperture radar data. Remote Sens 15(23):5621

Soares MDO, Campos CC, Carneiro PBM, Barroso HS, Marins RV, Teixeira CEP et al (2021) Challenges and perspectives for the Brazilian semi-arid coast under global environmental changes. Perspect Ecol Conserv 19(3):267–278

Strain EMA, Kompas T, Boxshall A, Kelvin J, Swearer S, Morris RL (2022) Assessing the coastal protection services of natural mangrove forests and artificial rock revetments. Ecosyst Serv 55:101429

Su J, Gasparatos A (2023) Perceptions about mangrove restoration and ecosystem services to inform ecosystem-based restoration in Large Xiamen Bay, China. Landscape Urban Plan 235:104763

Sun Y, Ye M, Jian Z, Ai B, Zhao J, Chen Q (2023) Species classification and carbon stock assessment of mangroves in Qi'AO Island with Worldview-3 Imagery. Forests 14(12):2356

Tian P, Liu Y, Li J, Wang H, Zhang H et al (2024). Tracking annual changes of coastal aquaculture ponds in China during 1986–2021. Aquaculture 589:740965

Truong XL, Mitamura M, Kono Y, Raghavan V, Yonezawa G, Truong XQ et al (2018) Enhancing prediction performance of landslide susceptibility model using hybrid machine learning approach of bagging ensemble and logistic model tree. Appl Sci 8(7):1046

UNEP (2023) The importance of mangroves to people: a call to action. https://www.unep.org/resources/report/importance-mangroves-people-call-action

Valderrama-Landeros L, Flores-de-Santiago F, Kovacs JM, Flores-Verdugo F (2018) An assessment of commonly employed satellite-based remote sensors for mapping mangrove species in Mexico using an NDVI-based classification scheme. Environ Monit Assess 190:1–13

Vizcaya-Martínez DA, Flores-de-Santiago F, Valderrama-Landeros L, Serrano D, Rodríguez-Sobreyra R, Álvarez-Sánchez LF, Flores-Verdugo F (2022) Monitoring detailed mangrove hurricane damage and early recovery using multisource remote sensing data. J Environ Manage 320:115830

Wang D, Wan B, Qiu P, Su Y, Guo Q, Wu X (2018) Artificial mangrove species mapping using pléiades-1: An evaluation of pixel-based and object-based classifications with selected machine learning algorithms. Remote Sens 10(2):294

Wong FK, Fung T (2014) Combining eo-1 hyperion and envisat asar data for mangrove species classification in Mai Po Ramsar Site, Hong Kong. Int J Remote Sens 35(23):7828–7856

Yang Y, Meng Z, Zu J, Cai W, Wang J, Su H, Yang J (2024) Fine-scale mangrove species classification based on UAV multispectral and hyperspectral remote sensing using machine learning. Remote Sens 16(16):3093

Youdon C (2020) Climate change impact on mangrove ecosystems in India's coastal regions. Nat Maritime Found 8–60

Yu J, Nie S, Liu W, Zhu X, Sun Z, Li J et al (2024) Mapping global mangrove canopy height by integrating Ice, Cloud, and Land Elevation Satellite-2 photon-counting LiDAR data with multi-source images. Sci Total Environment 173487

Zhang H, Wang T, Liu M, Jia M, Lin H, Chu LM, Devlin AT (2018) Potential of combining optical and dual polarimetric SAR data for improving mangrove species discrimination using rotation forest. Remote Sens 10(3):467
Zhang Z (2016) Introduction to machine learning: k-nearest neighbors. Ann Transl Med 4(11)
Zhao D, Zhen J, Zhang Y, Miao J, Shen Z, Jiang X et al (2023) Mapping mangrove leaf area index (LAI) by combining remote sensing images with PROSAIL-D and XGBoost methods. Remote Sens Ecol Conserv 9(3):370–389

Threats to Ecosystem Services

Land Cover and Land Use Change Impacts on Temperature Extremes in China from 1982 to 2013

Yaqian He, Di Yang, Eungul Lee, and Anni Yang

Abstract Climate regulation is a key ecosystem service. Land cover and land use change (LCLUC), altering ecosystem functioning as well as processes, can influence temperature extremes by altering surface energy fluxes, soil moisture, and surface roughness. This study examines the impacts of LCLUC on ecosystem climate regulation services by exploring its effects on the temperature extremes in China from 1982 to 2013. This study first applied remote sensing-derived LCLU data to examine the LCLUC in China from 1982 to 2013. Then, integrating the LCLU data into the Community Earth System Model 2 (CESM2), we assess the effects of LCLUC on six temperature extreme indices: highest maximum temperature (TXx), lowest minimum temperature (TNn), warm nights (TX90p), cold nights (TN10p), warm spell duration (WSDI), and cold spell duration (CSDI). Our results reveal that LCLUC generally reduces TXx in western China while increasing it in eastern China, and similar regional patterns are observed for TNn and TN10p. Oppositely, LCLUC reduces WSDI and CSDI in eastern China, while increasing them in western China. LCLUC

Y. He
Department of Geosciences, University of Arkansas, Fayetteville, AR, USA

Y. He (✉)
Department of Geography, Indiana University Bloomington, Bloomington, IN, USA
e-mail: yaqhe@iu.edu

D. Yang
Department of Geography, University of Florida, Gainesville, FL, USA
e-mail: yangdi1031@ufl.edu

E. Lee
Department of Geography, Kyung Hee University, Seoul, South Korea

Department of Climate-Social Science Convergence, Kyung Hee University, Seoul, South Korea

E. Lee
e-mail: eungul.lee@khu.ac.kr

A. Yang
Department of Geography and Environmental Sustainability, University of Oklahoma, Norman, OK, USA
e-mail: anni.yang@ou.edu

O. Mutanga et al. (eds.), *Revealing Ecosystem Services Through Geospatial Technologies*, Springer Remote Sensing/Photogrammetry,
https://doi.org/10.1007/978-3-031-98048-0_12

impacts on the TX90p are more uniform, with a general decrease in warm nights across the country. Although these effects are statistically insignificant in most areas, they reflect regional divergent climate responses to land surface dynamics. Analysis of additional climate variables during winter and summer suggests that albedo and evapotranspiration effects dominate in winter and summer, respectively. These findings offer important insights into the impact of LCLUC on ecosystem climate regulation services.

Keywords Ecosystem service · Land cover and land use change · Temperature extreme · Biogeophysical process · Community earth system model · Remote sensing

1 Introduction

Ecosystem services are the benefits of the ecosystem provided to humans, such as providing food and raw materials, supporting recreation and ecotourism, and climate regulation (Sharma et al. 2023). Land cover and land use change (LCLUC) (a terminology adopted from NASA's LCLUC program (NASA 2024)), altering ecosystem structure and functions, can influence all aspects of ecosystem services, including climate regulation (Rimal et al. 2019; Belay et al. 2022). Climate extremes, including rainfall-induced flooding, drought, and heat waves, have been one of the most pressing challenges facing humanity in the contemporary century (IPCC 2023). Among the various factors influencing climate extremes, LCLUC has emerged as an important driver of climate extreme variations across the globe (Pielke et al. 2011; Mahmood et al. 2014). Understanding the complex interactions between LCLUC and climate extremes is crucial for better capturing the roles of LCLUC on ecosystem climate regulation services and developing effective nature-based strategies to alleviate or adapt to climate change, especially in rapidly developing countries like China, where land surface transformations have been occurring at an unprecedented pace (He et al. 2017; Li et al. 2018). China has experienced tremendous LCLUC over the last few decades. Urbanization in eastern China, afforestation in southern China, desertification in western China, and cropland expansion in Northeast China have been observed by satellite data (He et al. 2017). Meanwhile, China experienced extensive climate extremes. With the maximum 2-m air temperature > 40 °C in 2022, over 270 million people experienced heat stress, and hundreds lost their lives to heat stroke (Yin et al. 2023). Thus, there is a pressing necessity to identify how LCLUC affects climate extremes in China.

LCLUC can affect the climate system through biogeochemical (e.g., changing methane (CH_4)) and biogeophysical (e.g., altering surface roughness, albedo, and soil moisture) processes. A growing body of studies has documented the impacts of LCLUC on both mean climate and climate extremes. For example, He et al. (2020) revealed that increased cropland area in Northeast China can reduce both surface and tropospheric mean temperatures, which is mainly because of the increased latent heat flux due to the transition from grassland to cropland in the region. Silva et al. (2018) identified that extensive urban growth can increase the intensity of urban heat islands (UHI) in Paço do Lumiar, Brazil, which is largely attributed to the decrease of green spaces and increasing energy consumption. LCLUC affected climate extremes as well. Liu et al. (2021) showed that cropland expansion in western India negatively influenced the maxima of daily maximum temperature (TXx), which is likely because of evaporative cooling due to cropland-induced intensified evapotranspiration. Boyaj et al. (2020) demonstrated that LCLUC increased heavy rainfall in South India. This is potentially because of the higher surface temperatures due to LCLUC, leading to higher convective available potential energy.

Despite tremendous progress having been made in the understanding of LCLUC and climate interactions, the impacts of LCLUC on temperature extremes in China are rarely explored. Some global modeling studies incorporate the climatic effects of LCLUC in China. Zhang et al. (2024) utilized model outputs from the Land Use Model Intercomparison Project (LUMIP) within the Coupled Model Intercomparison Project Phase 6 (CMIP6) to investigate the global LCLUC influences on climate extremes, including temperature extremes. However, the underlying physical mechanisms are competing in certain geographic regions (Oh and Lee 2023; Im et al. 2024). Generally, two main physical mechanisms are related to the impacts of LCLUC on temperature extremes. The first involves altering surface albedo, where changes in land cover reflect or absorb more solar energy, subsequently affecting temperature extremes (Li et al. 2018). For example, Davin et al. (2014) found that no-tillage could result in an increase in albedo, leading to the cooling of hot extremes over Europe by using a regional climate model (RCM). The second mechanism involves modifying the Bowen ratio, which changes surface energy flux partitioning by altering evapotranspiration rates (Hirsch et al. 2018). Hong et al. (2022) used the LUMIP model outputs to assess the future LCLUC impacts on climate extremes. They found that TXx and maxima of daily minimum temperature (TNx) are projected to increase, mainly because the evapotranspiration decreases with future land use change. The relative importance of these competing biophysical processes remains uncertain, particularly in temperate mid-latitude regions such as China (Findell et al. 2017).

Furthermore, prior modeling studies have often relied on extreme LCLUC scenarios and Land-Use Harmonization (LUH2) LCLU data, a model-derived dataset (Hurtt et al. 2020). For instance, Cherubini et al. (2018) simulated four idealized LCLU transitions, including the abrupt conversion of current forest to herbaceous vegetation or bare land and the transition of contemporary cropland to deciduous broad-leaf forest or evergreen needle-leaf forest, to explore the effects of deforestation and afforestation on regional climate in Europe. Yu and Leng (2022) used the LUH2 data set from 1901 to 2014 to identify the impacts of different types of LCLUC on near-surface air temperature globally. None of them can capture real-world land surface dynamics, over- or underestimating the effects of LCLUC on temperature extremes.

This study seeks to understand LCLUC effects on ecosystem climate regulation services by exploring the effects of LCLUC on the duration, intensity, and frequency of temperature extremes in China, utilizing an annual-scale time series of LCLU data derived from remote sensing. By incorporating this more realistic and dynamic LCLU data into a climate model (i.e., Community Earth System Model 2: CESM2), this study attempts to provide a more accurate evaluation of how human-induced LCLUC has influenced temperature extreme variability in China. Additionally, we aim to offer a thorough understanding of the physical mechanisms driving these impacts. Specifically, the objectives are to: (1) identify the spatiotemporal patterns of LCLUC in China, (2) assess how LCLUC has affected temperature extremes in China, and (3) examine the underlying physical mechanisms associated with these effects.

2 Data and Methods

2.1 Annual Remotely Sensed LCLU Data

Remotely sensed LCLU data in China with a spatial resolution of 0.0833° was developed by He et al. (2017). They used a machine learning algorithm named Random Forest classifier to map the LCLU in China from 1982 to 2013 based on Advanced Very High Resolution Radiometer (AVHRR) Global Inventory Modeling and Mapping Studies (GIMMS) third generation NDVI (NDVI3g) dataset. The classification system is the International Geosphere-Biosphere Programme (IGBP) with 13 LCLU classes, including water, snow and ice, different types of forests (i.e., evergreen needleleaf forest, deciduous broadleaf forest, evergreen broadleaf forest, mixed forest), woody savannas, open shrublands, grasslands, cropland and natural vegetation mosaic, croplands, barren or sparsely vegetated, urban and built-up. The overall accuracy was 73.8% for the 2012 LCLU map, validated against very-high-resolution (VHR) images from Google Earth. This was about 2.5% higher than the accuracy of the LCLU data from Moderate-resolution Imaging Spectroradiometer (MODIS) for 2012 (He et al. 2017). This dataset effectively captured important LCLUC in China, such as afforestation in southern China and cropland expansion in Northeast China (He et al. 2018, 2020), which is also demonstrated by Man et al. (2016) and Tong et al. (2018).

Remote sensing-derived LCLU data cannot be directly applied in a climate model due to two key mismatches. One is the differences in how land surface types are represented, and the other is the disparity in spatial resolution. To implement remotely sensed data into the land model (i.e., Community Land Model (CLM5.0)) in CESM2, we first used a cross-walking method to convert the 13 IGBP types into 15 plant function types (PFTs) plus bare ground following Poulter et al. (2015) and Bonan et al. (2002). This method reclassified IGBP types into physiognomic categories, like forest. These classes were then assigned to each PFT according to climate criteria (Table 1). The second step is to upscale remotely sensed data to a coarse resolution that is used in CLM5.0. In this study, we resampled the converted PFTs and bare ground to the spatial resolution of 1.25° × 0.9°. We chose this resolution because it is commonly used in model output and is a supported resolution in CESM (Zhao et al. 2021; Fan et al. 2022). There are many different resampling methods, such as majority-aggregation and nearest-neighbor methods, which can retain the categorical classes in LCLU maps, and the bilinear interpolation method, which is widely used in resampling interval data (e.g., precipitation) (Santhosh Baboo and Devi 2010). As PFTs and bare ground in CLM5.0 are represented as fractions (i.e., percentages), meaning the fractional abundance of each PFT within a grid cell, the existing resample methods are not suitable to upscale our converted PFTs. We used a fractional method to up-scale the derived PFTs to $1.25° \times 0.9°$ following He et al. (2018). The approach computed the proportion of each LCLU type (e.g., cropland) within the new coarse-scale $1.25° \times 0.9°$ grid cell through computing the ratio of the number of grid cells of each LCLU type in the original resolution (i.e., 0.0833°) to the overall count of grid cells in the original resolution contained in the new coarse-scale grid cell. It is important to note that water, snow and ice, as well as urban and built-up, were not converted using remotely sensed data in this study because they are rarely presented in the data due to the coarse resolution (i.e., 0.0833°).

2.2 CESM Experiments and Extreme Temperature Variables

To examine the impacts of LCLUC on temperature extremes in China, we conducted two CESM experiments: (1) EXP_FULL: a 32-year run using annual remotely-sensed LCLU data from 1982 to 2013 as a time-varying boundary condition; and (2) EXP_LCLU_FIX: same as (1), but only using remotely-sensed LCLU data in 1982 as a fixed boundary condition. Both experiments were run using fully prognostic CAM6.0 and CLM5.0 with time-varying greenhouse gas concentrations from 1982 to 2013 and prescribed ocean and sea ice modes. It is important to mention that both experiments ran from 1979 to 2013, in order to use the identical initial file (i.e., b.e20.BHIST.f09_g17.20thC.297_01_v3.clm2.r.1979–01-01-00000.nc). The initial three years from 1979 to 1981 were excluded for the following analyses.

Given our focus on temperature extremes, we documented the daily maximum and minimum temperature data from EXP_FULL and EXP_LCLU_FIX, respectively. We calculated 6 extreme temperature variables from the daily temperature for each year from 1982 to 2013, to represent the duration, intensity, and frequency of temperature extremes, using the *Climpact R* package (Climpact 2023). These variables align with the definitions outlined by the World Meteorological Organization (Climpact 2023), including TXx, TNn, TX90p, TN10p, WSDI, and CSDI. Table 2 shows the definition for each of the temperature extreme variables.

To uncover the underlying physical mechanisms, we also documented additional variables including leaf area index (LAI), soil moisture (SM), vegetation transpiration (T), ground evaporation (E), sensible heat flux (SH), latent heat flux (LH), and albedo from EXP_FULL and EXP_LCLU_FIX experiments. Since the model output does not directly provide albedo, we calculated it as the ratio of the reflected solar radiation

Table 1 Conversion of IGBP types to PFTs following Bonan et al. (2002) and Poulter et al. (2015)

Plant functional type	IGBP type	Climate criteria
Needleleaf evergreen tree, temperate	Evergreen needleleaf forest, mixed forest	$GDD > 1200$ and $T_{cold} > -19$ °C
Needleleaf evergreen tree, boreal	Evergreen needleleaf forest, mixed forest	$GDD \leq 1200$ or $T_{cold} \leq -19$ °C
Needleleaf deciduous tree, boreal		
Broadleaf evergreen tree, tropical[a]	Evergreen broadleaf forest, mixed forest	$T_{cold} > 15.5$ °C
Broadleaf evergreen tree, temperate	Evergreen broadleaf forest, mixed forest	$T_{cold} \leq 15.5$ °C
Broadleaf deciduous tree, tropical[a]	Deciduous broadleaf forest, mixed forest	$T_{cold} > 15.5$ °C
Broadleaf deciduous tree, temperate	Deciduous broadleaf forest, mixed forest	$GDD > 1200$ and -15 °C $<$ $T_{cold} \leq 15.5$ °C
Broadleaf deciduous tree, boreal	Deciduous broadleaf forest, mixed forest	$GDD \leq 1200$ or $T_{cold} \leq -15$ °C
Broadleaf evergreen shrub	Open shrublands	$GDD > 1200$ and $T_{cold} > 19$ °C and $P_{win} > 2/3\ P_{ann}$ and $P_{ann} > 520$ mm
Broadleaf deciduous shrub, temperate	Open shrublands	$GDD > 1200$ and $T_{cold} > 19$ °C and $(P_{win} \leq 2/3\ P_{ann}$ or $P_{ann} \leq 520$ mm$)$
Broadleaf deciduous shrub, boreal	Open shrublands	$GDD \leq 1200$ or $T_{cold} \leq 19$ °C
C_3 grass, arctic	Grassland, woody savannas,	$GDD < 1000$
C_3 grass[b]	Grassland, woody savannas,	$GDD > 1000$ and ($P_{mon} \leq$ 25 mm and for months with T > 22 °C or $T_{warm} \leq 22$ °C)
C_4 grass[b]	Grassland, woody savannas	$GDD > 1000$ and driest month $P_{mon} > 25$ mm and $T_{cold} > 22$ °C
Crop	Croplands Cropland and Natural Vegetation Mosaic	
Barren	Barren or Sparsely Vegetated	

T_{cold} is the temperature of the coldest month
T_{warm} is the temperature of the warmest month
GDD is the growing-degree days above 5 °C
P_{ann} is the annual precipitation
P_{win} is the winter (November through April) precipitation
P_{mon} is the monthly precipitation
[a] A 1.25° × 0.9° grid cell is assumed to be 50% broadleaf deciduous tree, tropical and broadleaf evergreen tree, tropical when they are converted from mixed forest
[b] A 1.25° × 0.9° grid cell is to be 50% C_3 and C_4 grass if neither the C_3 nor C_4 criteria are met and $GDD > 1000$

Table 2 The 6 temperature extreme variables

Variable	Short name	Definition	Unit
Intensity			
TXx	Highest maximum temperature	Annual maximum value of daily maximum temperature	°C
TNn	Lowest minimum temperature	Annual minimum value of daily minimum temperature	°C
Frequency			
TX90p	Warm nights	Annual percentage of days when daily maximum temperature > 90th percentile	%
TN10p	Cold nights	Annual percentage of days when daily minimum temperature < 10th percentile	%
Duration			
WSDI	Warm spell duration	Annual count of days with at least 6 consecutive days when daily maximum temperature > 90th percentile	Days
CSDI	Cold spell duration	Annual count of days with at least 6 consecutive days when daily minimum temperature < 10th percentile	Days

to the total incoming solar radiation, which is the sum of absorbed and reflected solar radiation.

2.3 Model Evaluation

To assess the model performance, we obtained hourly temperature data from the fifth generation of European Center for Medium-Range Weather Forecasts (ECMWF) Reanalysis (ERA5) from 1982 to 2013 (Hersbach et al. 2020). ERA5 data is produced using a four-dimensional variational (4D-Var) data assimilation technique relying on an Integrated Forecasting System (IFS) that combines vast amounts of observational data from weather stations and satellites (Hersbach et al. 2020). We chose ERA5 data as it performs best compared to other reanalysis data (e.g., the Japanese 55-year Reanalysis (JRA55)) in East Asia (Kim and Lee 2022). Based on the hourly data, we extracted the daily minimum and maximum temperatures, respectively. The spatial resolution of ERA5 data (0.25°) is upscaled to the resolution of 1.25° × 0.9°, to align with the spatial resolution of that from CESM2. We then calculated the same set of 6 extreme temperature variables for comparative evaluation with EXP_FULL. The calculation of these extreme temperature variables was conducted using the *R* package of *Climpact* (Climpact 2023). Finally, we computed the climatological mean difference for each extreme temperature variable from 1982 to 2013 between ERA5 and EXP_FULL. We used a Student's *t*-test to evaluate the significance of the differences.

2.4 LCLUC Detection in China

To explore China's LCLUC changes from 1982 to 2013, we conducted a trend analysis for the three main LCLU types in China: cropland, forest, and grassland. As shown in Table 1, there are multiple tree types (e.g., needleleaf deciduous tree, boreal and needleleaf evergreen tree, temperate) and grass types (e.g., C_3 grass, C_4 grass). We grouped them into forest and grassland, respectively. We performed a regression of cropland, forest, and grass fractional values on time at each $1.25° \times 0.9°$ grid cell using an autocorrelation-corrected linear fit. Autocorrelation-corrected linear fit has been used to detect temporal trends of LCLU (He et al. 2020). Compared to simple linear regression, it considers the temporal autocorrelation shown in the time series data in each grid cell (Santer et al. 2000). We chose $1.25° \times 0.9°$ spatial resolution because it is consistent with the LCLU data that we used in CESM2. When significant autocorrelation (at the 5% level) was detected using *R* function *acf*, we applied a generalized least squares (GLS) regression. The GLS was with an autoregressive moving average process with order $p = 1$ or $p = 2$, meaning that the previous one or two-time point is used in the autoregressive model. We selected $p = 2$ to capture the temporal dependencies (Wilson 1979; Holan et al. 2010). If the model failed, we chose $p = 1$. If autocorrelation was not significant, we utilized ordinary least squares regression instead. We employed a Student's *t*-test to test the statistical significance of the trends.

2.5 LCLUC Impacts on Temperature Extremes

We computed the climatological mean difference for each extreme temperature variable at each grid cell from 1982 to 2013 between EXP_FULL and EXP_LCLU_FIX (Eq. 1), to examine the effects of LCLUC on temperature extremes. Computing the difference between two experiments is a common way to explore LCLUC impacts on climate (Chen and Dirmeyer 2019, 2020; Yu and Leng 2022).

$$TEMPEX_{x,y} = \underline{TEMPEX_{x,y,EXP\_FULL}} - \underline{TEMPEX_{x,y,EXP\_LCLU\_FIX}} \quad (1)$$

where x and y are grid cell coordinates. $\underline{TEMPEX_{x,y,EXP\_FULL}}$ represents the mean extreme temperature variable for the grid cell (x, y) derived from the EXP_FULL experiment. $\underline{TEMPEX_{x,y,EXP\_LCLU\_FIX}}$ represents the mean extreme temperature variable for the grid cell (x, y) derived from the EXP_LCLU_FIX experiment.

To unveil the underlying physical mechanisms, we also calculated the climatological mean difference for leaf area index, surface soil moisture, vegetation transpiration, ground evaporation, sensible heat flux, latent heat flux, and albedo during winter and summer, respectively, as extreme temperatures normally happened during these two seasons. The significance of the differences was also tested by a Student's *t*-test.

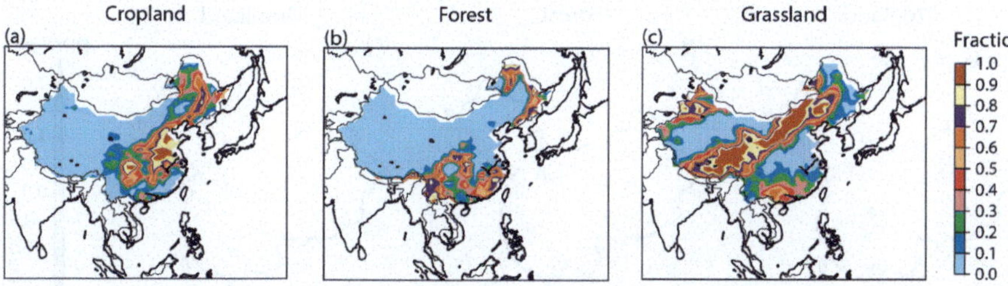

Fig. 1 Climatology mean fraction of three LCLU types from remote sensing data from 1982 to 2013: **a** cropland, **b** forest, and **c** grassland. The color bar represents the three LCLU types' fraction in each 1.25° × 0.9° grid cell

3 Results and Discussion

3.1 Spatiotemporal LCLUC in China

The climatological mean of cropland, forest, and grassland fractional coverage derived from remote sensing data, averaged from 1982 to 2013, is shown in Fig. 1. Cropland is mainly concentrated in Northeast China, the Sichuan Basin, and North China, while forest cover is more prevalent in southern China and along the northeastern borders of China. Grassland, on the other hand, is primarily distributed in a belt stretching from the Tibetan Plateau to the Northeast China Plain. These distribution patterns align with findings from previous studies (Yang and Huang 2021).

Figure 2 demonstrates the trend of each LCLU type in China from 1982 to 2013. Notably, cropland has significantly expanded in Northeast and North China but decreased in southern China. In contrast, forest cover significantly increases in southern China and decreases along the northeastern borders of China. This is likely due to China's land use policy, like the "Grain for Green" afforestation project (Tong et al. 2018). The trend of grassland is more divergent across the country. It significantly decreases in Northeast China and North China and increases in some parts of the Tibetan Plateau and north Xinjiang. These spatial change patterns of LCLU have been revealed in previous studies (Liu et al. 2014; Man et al. 2016; Im et al. 2024).

3.2 Model Evaluation

Overall, the model successfully captures the spatial patterns of six climate extreme variables (Fig. 3). Both the model and ERA5 display a relatively cooler pattern in the Tibetan Plateau and a warmer pattern in southern China for the highest maximum temperature (TXx), though the model slightly overestimates the highest maximum temperature in most parts of China compared to ERA5 (Fig. 3c). Similarly, the

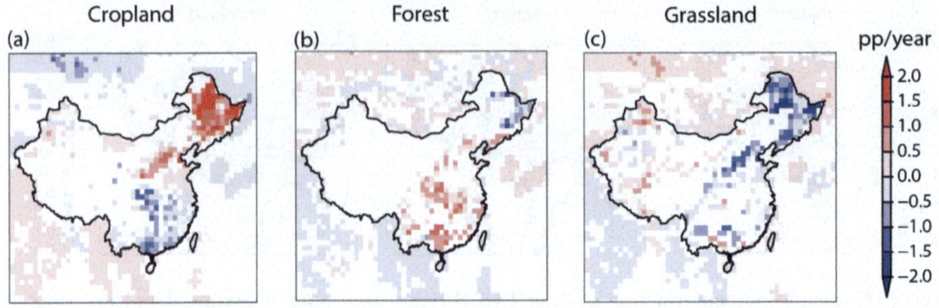

Fig. 2 Temporal trend of three LCLU types from remote sensing data from 1982 to 2013: **a** cropland, **b** forest, and **c** grassland. The color bar means the slope in percentage points (pp) per year. The non-significant areas (p-value > 0.1) are masked out

model accurately represents the lowest minimum temperature (TNn) pattern, showing a relatively warmer pattern in most parts of China and a cooler pattern in southern part of Tibetan Plateau (Fig. 3d, e). However, the spatial consistency between the model and ERA5 is relatively poorer for warm nights (TX90p), cold nights (TN10p), cold spell duration (CSDI), and warm spell duration (WSDI) compared to lowest minimum temperature (TNn) and highest maximum temperature (TXx) (Fig. 3g, h, j, k, m, n, p, q). Nevertheless, all observed differences for these four variables are statistically insignificant (Fig. 3i, l, o, r), indicating only minor discrepancies.

3.3 LCLUC Impacts on Temperature Extremes

LCLUC in China has influenced the duration, frequency, and intensity of temperature extremes divergently (Fig. 4). Generally, LCLUC has decreased the highest maximum temperature (TXx) in western China and the central part of Northeast China, while increasing it in eastern China and the northeastern border (Fig. 4a). For the lowest minimum temperature (TNn), warming effects are also evident in eastern China, while western China shows mixed effects, with both warming and cooling observed (Fig. 4b). LCLUC has generally reduced the warm nights (TX90p) across China (Fig. 4c), but it has increased the cold nights (TN10p) in Northeast China and the southern border (Fig. 4d). In terms of duration, LCLUC has led to a longer warm spell duration (WSDI) in western and central China but a decrease in most other regions (Fig. 4e). Cold spell duration (CSDI) has increased in western China due to LCLUC, while some areas in Northeast China have experienced a decrease (Fig. 4f). The LCLUC-induced decreased warm nights in southern and western China, increased cold nights in southern China and Northeast China, and increased warm spell duration in western China were also identified by Zhang et al. (2024). It is important to mention that most impacts are statistically insignificant (Fig. 4), suggesting that the effects of LCLUC on climate extremes in China are relatively minor. This

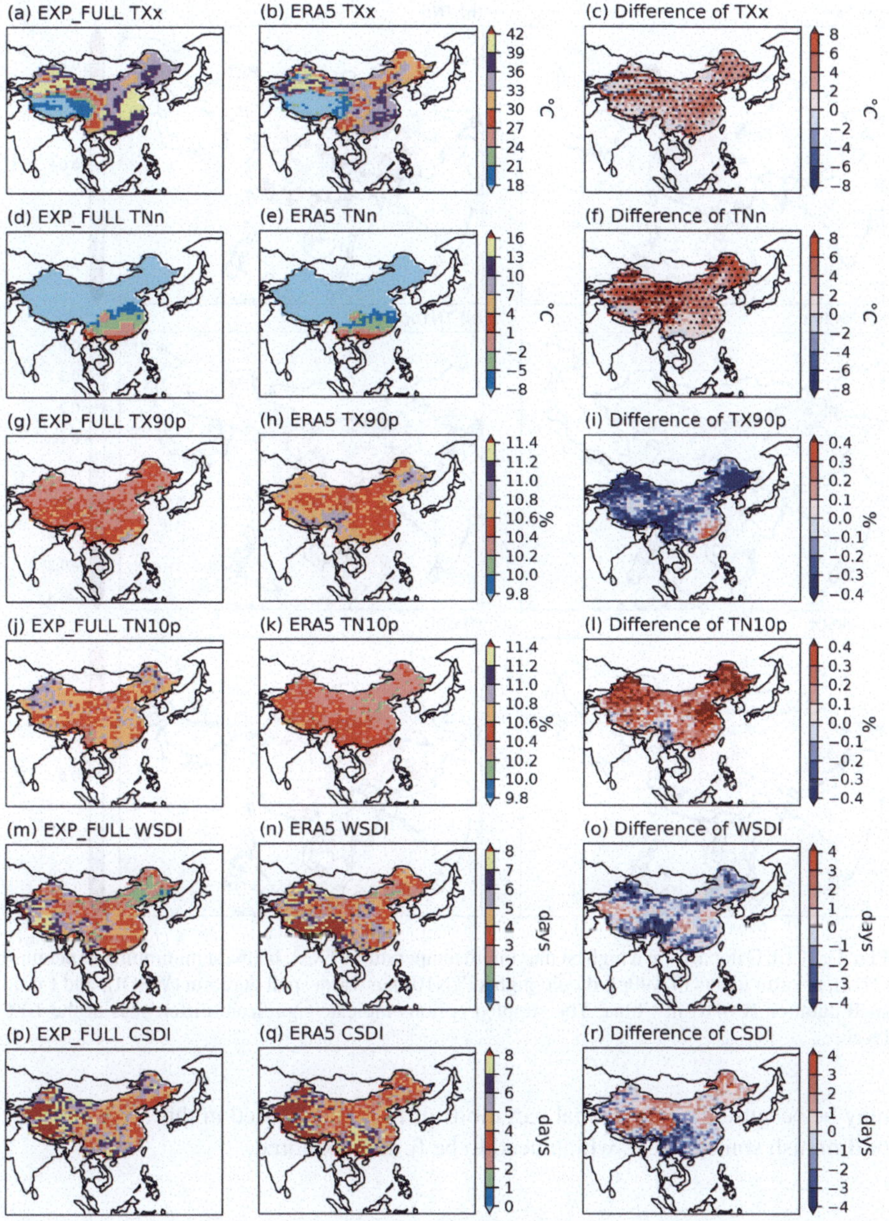

Fig. 3 Climatology mean and difference of **a–c** highest maximum temperature (TXx), **d–f** lowest minimum temperature (TNn), **g–i** warm nights (TX90p), (j-l) cold nights (TN10p), **m–o** warm spell duration (WSDI), and **p–r** cold spell duration (CSDI) derived from EXP_FULL with that from ERA5 from 1982 to 2013. The stippled regions in **(c)**, **(f)**, **(i)**, **(l)**, **(o)**, and **(r)** indicate significant differences at the 10% level

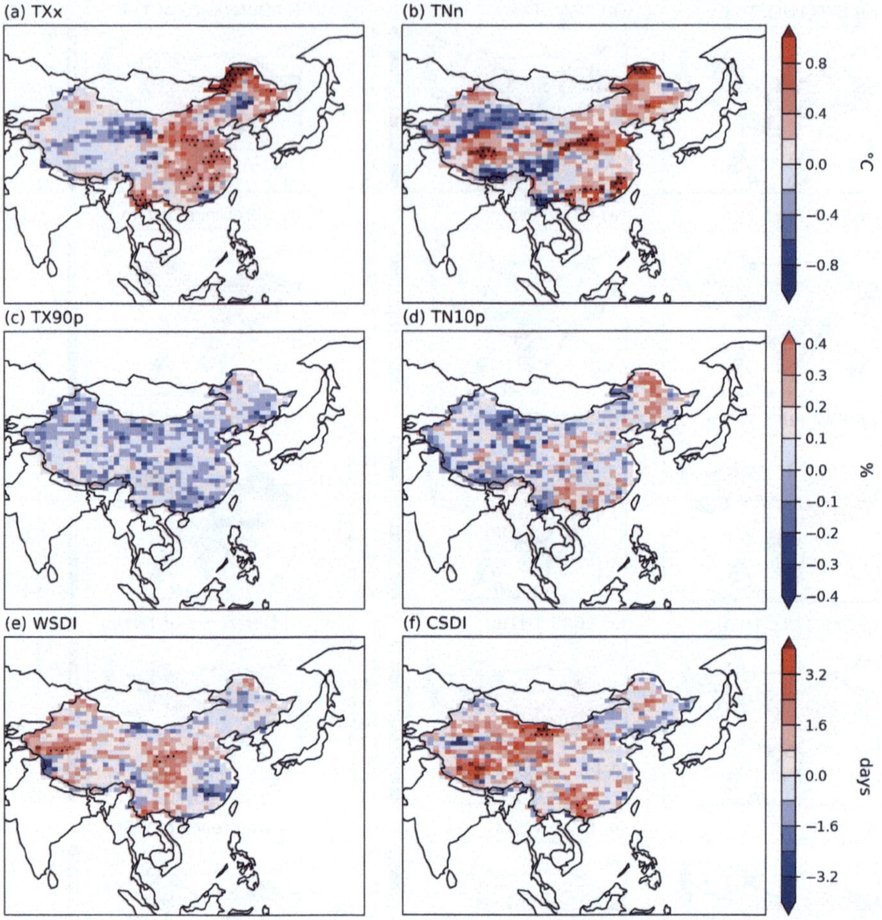

Fig. 4 LCLUC impacts on **a** highest maximum temperature (TXx), **b** lowest minimum temperature (TNn), **c** warm nights (TX90p), **d** cold nights (TN10p), **e** warm spell duration (WSDI), and **f** cold spell duration (CSDI) in China. The stippled regions indicate significant differences at the 10% level

may be because the interannual variations of LCLU data used in this study amplify or diminish some effects, which need to be further explored.

3.4 Physical Mechanisms Underlying the Climatic Impacts of LCLUC

To explore underlying physical mechanisms, we further explore the impacts of LCLUC on various land and climate variables during winter (Fig. 5) and summer

(Fig. 6), when extreme temperatures typically occur. Overall, there are consistent patterns between these two seasons, although the magnitudes of the impacts are stronger during summer than in winter. This is in our expectation, as land and atmosphere interactions are stronger during summertime (Santanello et al. 2018). During winter, LCLUC in China statistically significantly reduces the leaf area index (LAI) in southern China and Northeast China (Fig. 5a). This leads to a statistically significant reduction in transpiration in the corresponding regions (Fig. 5b). Meanwhile, soil moisture significantly increases in southern China and decreases in Northeast China (Fig. 5c). Correspondingly, evaporation significantly increases in southern China and decreases in Northeast China (Fig. 5d). These combined effects of transpiration and evaporation significantly reduce the latent heat flux in Northeast China (Fig. 5f). However, sensible heat flux also significantly decreases in Northeast China and the southwestern border (Fig. 5 (g)), which is likely because of significantly increased albedo in those regions (Fig. 5e). Converting grassland to cropland (Fig. 2) could induce an increase in albedo during winter, as crops are often harvested in October in Northeast China (Xiao et al. 2024), leaving barren lands thereafter. The changes in sensible heat flux due to albedo changes largely lead to variation in the lowest minimum temperature (TNn) (Fig. 4b), excepting for Northeast China. This indicates that the albedo effect dominates in land–atmosphere interactions during winter, which is consistent with previous findings that the albedo effect of vegetation dominates in high-altitude regions in central China (Im et al. 2024).

Like in winter, LCLUC statistically significantly reduces leaf area index (LAI) in southern China, the southern part of the Tibetan Plateau, and Northeast China during summer (Fig. 6a). The reduced LAI results in a statistically significant reduction of transpiration in southern China and Northeast China (Fig. 6b). Soil moisture significantly increases in the southern border and some parts of Northeast China (Fig. 6c). This likely results in statistically significantly increased evaporation in southern and northeastern China during summer (Fig. 6d). These combined effects induced a decrease in latent heat flux in southern China and northern China (Fig. 6f), leading to increases in sensible heat flux in the same regions (Fig. 6g). Albedo

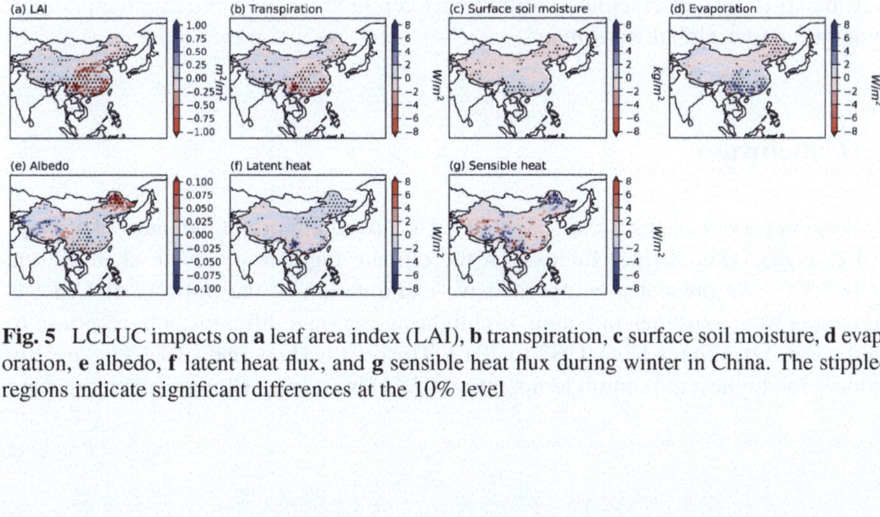

Fig. 5 LCLUC impacts on **a** leaf area index (LAI), **b** transpiration, **c** surface soil moisture, **d** evaporation, **e** albedo, **f** latent heat flux, and **g** sensible heat flux during winter in China. The stippled regions indicate significant differences at the 10% level

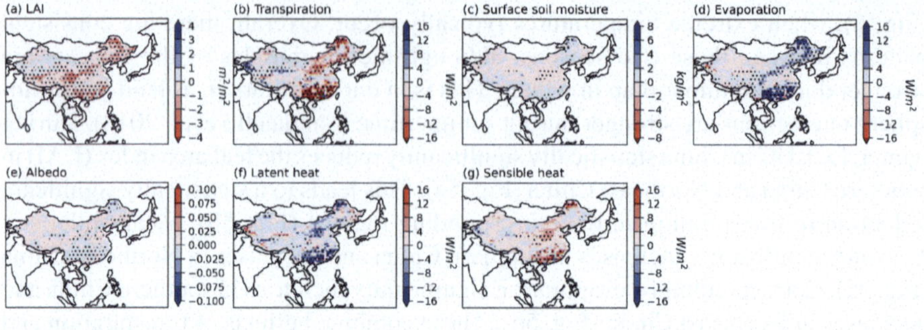

Fig. 6 LCLUC impacts on **a** leaf area index (LAI), **b** transpiration, **c** surface soil moisture, **d** evaporation, **e** albedo, **f** latent heat flux, and **g** sensible heat flux during summer in China. The stippled regions indicate significant differences at the 10% level

statistically decreases in most of southern China and some parts of northeastern China (Fig. 6e). This implies that evapotranspiration effects dominate during summer, which is in line with (Lee et al. 2015). The changes in sensible heat flux due to evapotranspiration result in significant changes in the highest maximum temperature (Fig. 4a), especially in eastern China.

4 Limitations

There are some limitations in the study. First, this study does not include urban land use type. Future research should enhance the representation of urban land areas in the model. Second, land use management (e.g., irrigation) is idealized in the model; future studies should consider more realistic irrigation practices by incorporating farmers' data. Third, idealized model experiments are required to further understand the effects of specific LCLUC types, such as afforestation, on climate extremes. Finally, the interactions between CO_2 emissions and LCLUC are required to be considered to further elucidate the link between land use and extreme temperature dynamics under global warming.

5 Conclusion

Ecosystem service changes caused by LCLUC are an important indicator of ecological changes. This chapter focuses on the climate regulation service changes due to LCLUC. We present a comprehensive examination of the impacts of LCLUC on temperature extremes in China, highlighting regional differences in response to land surface dynamics from 1982 to 2013. The results show that LCLUC generally reduces the highest maximum temperature (TXx) in western China while increasing it in eastern China, and similar regional patterns are observed for the lowest minimum temperature (TNn) and cold nights (TN10p). Oppositely, LCLUC reduces warm and cold spell durations (WSDI, CSDI) in eastern China, while increasing it in western China. LCLUC impacts on the warm nights (TX90p) are more uniform, with a general decrease in warm nights across the country. Despite statistically insignificant effects in most regions, LCLUC impacts are evident in modifying temperature extreme patterns, primarily through changes in surface energy balance by altering albedo in winter and in moisture cycle by altering evapotranspiration in summer.

Acknowledgements We thank the use of computational resources at the NCAR-Wyoming Supercomputing Center provided by the National Science Foundation and the State of Wyoming and supported by NCAR's Computational and Information Systems Laboratory (doi: https://doi.org/10.5065/D6RX99HX). We are also grateful to Justin S. Mankin from the Department of Geography at Dartmouth College for his constructive suggestions to conceive the original idea of the study.

References

Belay T, Melese T, Senamaw A (2022) Impacts of land use and land cover change on ecosystem service values in the Afroalpine area of Guna Mountain, Northwest Ethiopia. Heliyon 8:e12246. https://doi.org/10.1016/j.heliyon.2022.e12246

Bonan GB, Levis S, Kergoat L, Oleson KW (2002) Landscapes as patches of plant functional types: an integrating concept for climate and ecosystem models. Global Biogeochem Cycles 16:5-1–5-23. https://doi.org/10.1029/2000gb001360

Boyaj A, Dasari HP, Hoteit I, Ashok K (2020) Increasing heavy rainfall events in south India due to changing land use and land cover. Q J R Meteorol Soc 146:3064–3085. https://doi.org/10.1002/qj.3826

Chen L, Dirmeyer PA (2020) Distinct impacts of land use and land management on summer temperatures. Front Earth Sci (Lausanne) 8. https://doi.org/10.3389/feart.2020.00245

Chen L, Dirmeyer PA (2019) The relative importance among anthropogenic forcings of land use/land cover change in affecting temperature extremes. Clim Dyn 52:2269–2285. https://doi.org/10.1007/s00382-018-4250-z

Cherubini F, Huang B, Hu X et al (2018) Quantifying the climate response to extreme land cover changes in Europe with a regional model. Environ Res Lett 13:074002. https://doi.org/10.1088/1748-9326/aac794

Climpact (2023) What are extremes indices? https://climpact-sci.org/indices/#index-R95p

Davin EL, Seneviratne SI, Ciais P et al (2014) Preferential cooling of hot extremes from cropland albedo management. Proc Natl Acad Sci 111:9757–9761. https://doi.org/10.1073/pnas.1317323111

Fan X, Duan Q, Shen C et al (2022) Evaluation of historical CMIP6 model simulations and future projections of temperature over the Pan-Third Pole region. Environ Sci Pollut Res 29:26214–26229. https://doi.org/10.1007/s11356-021-17474-7

Findell KL, Berg A, Gentine P et al (2017) The impact of anthropogenic land use and land cover change on regional climate extremes. Nat Commun 8. https://doi.org/10.1038/s41467-017-01038-w

He Y, Lee E, Mankin JS (2020) Seasonal tropospheric cooling in Northeast China associated with cropland expansion. Environ Res Lett 15. https://doi.org/10.1088/1748-9326/ab6616

He Y, Lee E, Warner TA (2017) A time series of annual land use and land cover maps of China from 1982 to 2013 generated using AVHRR GIMMS NDVI3g data. Remote Sens Environ 199:201–217. https://doi.org/10.1016/j.rse.2017.07.010

He Y, Warner TA, McNeil BE, Lee E (2018) Reducing uncertainties in applying remotely sensed land use and land cover maps in land-atmosphere interaction: Identifying change in space and time. Remote Sens (Basel) 10. https://doi.org/10.3390/rs10040506

Hersbach H, Bell B, Berrisford P et al (2020) The ERA5 global reanalysis. Q J R Meteorol Soc 146:1999–2049. https://doi.org/10.1002/qj.3803

Hirsch AL, Guillod BP, Seneviratne SI et al (2018) Biogeophysical impacts of land-use change on climate extremes in low-emission scenarios: results From HAPPI-Land. Earths Future 6:396–409. https://doi.org/10.1002/2017EF000744

Holan SH, Lund R, Davis G (2010) The ARMA alphabet soup: a tour of ARMA model variants. Stat Surv 4. https://doi.org/10.1214/09-SS060

Hong T, Wu J, Kang X et al (2022) Impacts of different land use scenarios on future global and regional climate extremes. Atmosphere (Basel) 13:995. https://doi.org/10.3390/atmos13060995

Hurtt GC, Chini L, Sahajpal R et al (2020) Harmonization of global land use change and management for the period 850–2100 (LUH2) for CMIP6. Geosci Model Dev 13:5425–5464. https://doi.org/10.5194/gmd-13-5425-2020

Im A, Lee E, He Y (2024) Detecting land cover and land use transitions and the associated temperature changes in the East Asian monsoon region. Geogr J. https://doi.org/10.1111/geoj.12590

IPCC (2023) Weather and climate extreme events in a changing climate. In: Climate change 2021—the physical science basis. Cambridge University Press, pp 1513–1766

Kim M, Lee E (2022) Validation and comparison of climate reanalysis data in the East Asian Monsoon Region. Atmosphere (Basel) 13:1589. https://doi.org/10.3390/atmos13101589

Lee E, He Y, Zhou M, Liang J (2015) Potential feedback of recent vegetation changes on summer rainfall in the Sahel. Phys Geogr 36:449–470. https://doi.org/10.1080/02723646.2015.1120139

Li X, Chen H, Wei J et al (2018) Inconsistent responses of hot extremes to historical land use and cover change among the selected CMIP5 models. J Geophys Res Atmos 123:3497–3512. https://doi.org/10.1002/2017JD028161

Liu J, Kuang W, Zhang Z et al (2014) Spatiotemporal characteristics, patterns, and causes of land-use changes in China since the late 1980s. J Geog Sci 24:195–210. https://doi.org/10.1007/s11442-014-1082-6

Liu, Shen W, He Y (2021) Effects of cropland expansion on temperature extremes in Western India from 1982 to 2015. Land (Basel) 10. https://doi.org/10.3390/land10050489

Mahmood R, Pielke RA, Hubbard KG et al (2014) Land cover changes and their biogeophysical effects on climate. Int J Climatol 34:929–953. https://doi.org/10.1002/joc.3736

Man W, Wang Z, Liu M et al (2016) Spatiotemporal dynamics analysis of cropland in Northeast China during 1990–2013 based on remote sening. Trans Chin Soc Agric Eng 32:1–10

NASA (2024) NASA land-cover and land-use change program. https://lcluc.umd.edu/. Accessed 20 Mar 2024

Oh J, Lee E (2023) Competing effects of vegetation on summer temperature in North Korea. Theor Appl Climatol 152:913–931. https://doi.org/10.1007/s00704-023-04413-4

Pielke RA, Pitman A, Niyogi D et al (2011) Land use/land cover changes and climate: modeling analysis and observational evidence. Wires Clim Change 2:828–850. https://doi.org/10.1002/wcc.144

Poulter B, MacBean N, Hartley A et al (2015) Plant functional type classification for earth system models: results from the European Space Agency's land cover climate change initiative. Geosci Model Dev 8:2315–2328. https://doi.org/10.5194/gmd-8-2315-2015

Rimal B, Sharma R, Kunwar R et al (2019) Effects of land use and land cover change on ecosystem services in the Koshi River Basin, Eastern Nepal. Ecosyst Serv 38:100963. https://doi.org/10.1016/j.ecoser.2019.100963

Santanello JA, Dirmeyer PA, Ferguson CR et al (2018) Land-atmosphere interactions: the LoCo perspective. Bull Am Meteorol Soc 99:1253–1272. https://doi.org/10.1175/BAMS-D-17-0001.1
Santer BD, Wigley TML, Boyle JS et al (2000) Statistical significance of trends and trend differences in layer-average atmospheric temperature time series. J Geophys Res Atmos 105:7337–7356. https://doi.org/10.1029/1999JD901105
Santhosh Baboo D, Devi Mr (2010) An analysis of different resampling methods in Coimbatore district
Sharma S, Hussain S, Singh AN (2023) Impact of land use and land cover on urban ecosystem service value in Chandigarh, India: a GIS-based analysis. J Urban Ecol 9. https://doi.org/10.1093/jue/juac030
Silva JS, da Silva RM, Santos CAG (2018) Spatiotemporal impact of land use/land cover changes on urban heat islands: a case study of Paço do Lumiar, Brazil. Build Environ 136:279–292. https://doi.org/10.1016/j.buildenv.2018.03.041
Tong X, Brandt M, Yue Y et al (2018) Increased vegetation growth and carbon stock in China karst via ecological engineering. Nat Sustain 1:44–50
Wilson GT (1979) Some efficient computational procedures for high order ARMA models. J Stat Comput Simul 8:301–309. https://doi.org/10.1080/00949657908810273
Xiao D, Yang X, Bai H et al (2024) Trends and climate response in the yield of staple crops across Northeast China. Front Sustain Food Syst 7. https://doi.org/10.3389/fsufs.2023.1246347
Yang J, Huang X (2021) The 30m annual land cover dataset and its dynamics in China from 1990 to 2019. Earth Syst Sci Data 13:3907–3925. https://doi.org/10.5194/essd-13-3907-2021
Yin Z, Zhou B, Duan M et al (2023) Climate extremes become increasingly fierce in China. The Innovation 4:100406. https://doi.org/10.1016/j.xinn.2023.100406
Yu L, Leng G (2022) Global effects of different types of land use and land cover changes on near-surface air temperature. Agric Meteorol 327:109232. https://doi.org/10.1016/j.agrformet.2022.109232
Zhang M, Gao Y, Wang A et al (2024) Land use change impacts on climate extremes over the historical period. Clim Dyn 62:8993–9011. https://doi.org/10.1007/s00382-024-07375-z
Zhao S, Yu Y, Lin P et al (2021) Datasets for the CMIP6 scenario model intercomparison project (ScenarioMIP) simulations with the coupled model CAS FGOALS-f3-L 38:329–339. https://doi.org/10.1007/s00376

A Systematic Review of the Application of Remote Sensing in Assessing the Impact of Fire on Ecosystem Services

Upenyu Naume Mupfiga and Onisimo Mutanga

Abstract This paper focused on a systematic literature review of the current state of remote sensing utilization in assessing fire's impact on ecosystem services. Specific search strategies were used to search for the literature in Elsevier's Scopus and PubMed databases. The inclusion and exclusion of studies were guided by the Preferred Reporting Items for Systematic Reviews and Meta-analysis (PRISMA) protocol resulting in 23 papers being analyzed. The analysis of the retrieved research papers revealed a gradual increase in the publication of papers on the utilization of remote sensing in assessing the impact of fire on ecosystem services with most studies occurring in Asia. Landsat satellite data was mostly used to analyze the effects of vegetation fires on ecosystem services while MODIS was mostly used to detect the fire occurrence in the landscape. The least utilized satellite data included PlanetScope, AVIRIS, and ASTER. The review revealed that vegetation fires affect forest, grassland, wetland, and soil ecosystems. Additionally, fires disturb the atmosphere through carbon and particulate matter emissions. The forest ecosystems were the most widely studied ecosystem. The study underscores the significance of remote sensing methods in assessing the impact of fires on ecosystem services. The limitations and research gaps within the field were also identified.

Keywords Fire · Vegetation · Ecosystem services · Ecosystems · Remote sensing

U. N. Mupfiga (✉) · O. Mutanga
Discipline of Geography, School of Agricultural, Earth and Environmental Sciences, University of KwaZulu-Natal, Scottsville, Pietermaritzburg, South Africa
e-mail: 218086584@stu.ukzn.ac.za; nyatondou@gmail.com

O. Mutanga
e-mail: MutangaO@ukzn.ac.za

U. N. Mupfiga
Department of Geography, Environmental Sustainability and Resilience Building, Midlands State University, Gweru, Zimbabwe

O. Mutanga et al. (eds.), *Revealing Ecosystem Services Through Geospatial Technologies*, Springer Remote Sensing/Photogrammetry,
https://doi.org/10.1007/978-3-031-98048-0_13

1 Introduction

Vegetation fires are intrinsic to the Mediterranean and Savanna ecosystems, contributing to the shaping of various ecological processes, determining species composition and ultimately influencing the landscape structure (Herawati et al. 2015; Kugbe et al. 2012; Semeraro et al. 2019). The savanna ecosystems are greatly shaped by fire such that the changes in fire regimes greatly modify such ecosystems (Devineau et al. 2010). Uncontrolled fires can, however, disturb ecosystems leading to biodiversity loss, vegetation cover loss, encouraging bush encroachment, and alien species invasion (Molaudzi and Adelabu 2019; Mupfiga et al. 2024; Semeraro et al. 2019). Additionally, vegetation fires are a significant natural disturbance threatening human life, economics, and public health (Chen et al. 2023; Hao et al. 2022).

The variability in climatic conditions has been increasingly contributing to the occurrence of fires by altering fuel availability and flammability (Herawati et al. 2015). Recently, various ecosystems have been affected by the increased frequency and intensity of vegetation fires with likely environmental, economic, and health negative impacts (Souto-Oliveira et al. 2023). Ecosystem services refer to ecosystem-derived social and economic benefits enabled by productive and healthy ecosystems (Marambanyika et al. 2021). They can be classified as regulating, provisioning, supporting, and cultural ecosystem services (Millennium Ecosystem Assessment 2005; Paula et al. 2023). Global projections of climate change impacts of vegetation fires suggest an increase in fire events by up to 50% in fire-prone regions (Bowman et al. 2017). The increased risk of climate change across biomes has greatly exacerbated wildfires' frequency, extent, and severity, threatening various ecosystem services (Qiu et al. 2023). Although wildfires can be beneficial, for example in pest and disease control, they generally have detrimental effects on ecosystem services (Pereira et al. 2021).

Remote sensing methods allow large-scale observations of forest fires at low cost (Graham et al. 2023). Specifically, the brightness temperature of thermal infrared and mid-infrared radiation bands are highly effective in detecting vegetation fires (Hao et al. 2022). Additionally, earth observation methods provide a way of repetitively measuring the impacts of fires on ecosystems at global, regional, and local scales in a synoptic view (He et al. 2021; Semeraro et al. 2019). Satellite-derived vegetation indices are widely used to monitor vegetation status before and after fire occurrence. They are generally derived from the algebraic combination of different spectral band values. Although the use of individual spectral bands gives a simple method to analyze vegetation status, it is difficult to detect the heterogeneous nature of vegetation. The vegetation indices utilize the vegetation spectral properties such as the chlorophyll's strong red-light absorption capability and the ability to highly reflect the near-infrared spectra by the mesophyll layer of the leaf.

The Normalized Difference Vegetation Index (NDVI), for example, is widely utilized in the assessment of vegetation conditions before, during, and after the fire. On the other hand, the Normalized Burn Ratio (NBR), which utilizes the near-infrared (NIR) and the short-wave infrared (SWIR) bands, strongly responds to fire but in the opposite direction of the NDVI. The NIR and SW spectral bands have a strong response to vegetation burning hence the NBR index is more useful for assessing the effect of fire on vegetation because it is highly sensitive to fire damage and vegetation regeneration and hence greatly used for assessing the severity of burning (Hartung et al. 2021).

Several studies have reviewed the utilization of remote sensing methods in mapping wildfire risk in various landscapes (Molaudzi and Adelabu 2019; Szpakowski and Jensen 2019; Vigna et al. 2021). Additionally, some reviews focused on utilizing remote sensing methods in detecting and monitoring vegetation fires (Graham et al. 2023; Kurbanov et al. 2022). Although a few reviews have looked at the utilization of remote sensing in assessing post-fire effects and forest recovery patterns in forest and grassland landscapes (Chu and Guo 2014; Chuvieco 2019; Pérez-Cabello et al. 2021; Stavi 2019), limited studies have looked at the use of earth observation utility in assessing the effect of fire on ecosystem services. The studies done generally focused on a single ecosystem. This study specifically focuses on a comprehensive evaluation of published literature on the use of remote sensing in assessing the effect of vegetation fires on ecosystem services. The current review is based on a broad approach where all ecosystems that are possibly affected by burning were considered.

2 Methods

2.1 Literature Search, Inclusion, and Exclusion Strategy

The systematic review study was based on a thorough examination of research publications. An extensive search for peer-reviewed journal articles was done using the Scopus, Web of Science, and PubMed databases. The search terms which were used included: "fire" AND "remote sensing" AND "ecosystem services", fire AND "remote sensing" AND "ecosystems" AND "threat". To capture all the relevant studies, the time when the research was done was not restricted. The retrieved research articles were aggregated into a final count of research articles for each database. 232, 19 and 100 references were collected from the Scopus, PubMed and Web of Science databases respectively.

The inclusion and exclusion analysis of the research studies was guided by the Preferred Reporting Items for Systematic Reviews and Meta-analysis (PRISMA) statement (Page et al. 2021) due to its comprehensiveness and its ability to follow the scientific method. Figure 1 clearly shows the process of the research article selection based on the PRISMA approach. The articles used in this systematic literature review included all types of peer-reviewed published research papers, including research articles, conference papers, book chapters, and review papers. The subject areas covered by the search criteria included environmental science, agriculture and biological science, earth and planetary science, computer science, and multi-disciplinary. The review excluded papers that were duplicated, unavailable as full texts, not written in English, not GIS or remote sensing-based, and also focusing on non-vegetation fires. Firstly, all 107 duplicated research papers were removed. The screening of titles and abstracts of the remaining papers was done to exclude 133 irrelevant research articles. From the remaining research articles, 47 articles could not be retrieved as full portable document files (pdf). A total number of 25 research articles were finally considered for data extraction.

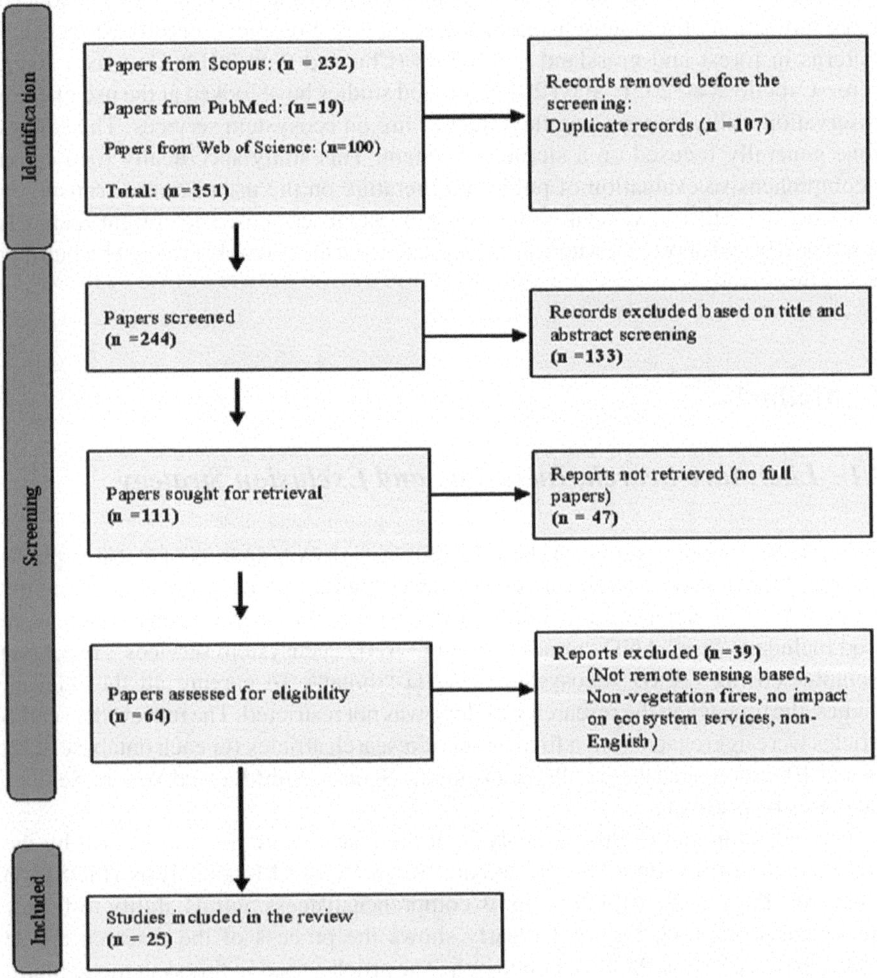

Fig. 1 The schematic diagram of the papers selected based on the Preferred Reporting Items for Systematic Reviews and Meta-Analysis (PRISMA) framework. Adapted from Page et al. (2021)

2.2 Data Extraction and Analysis

The papers identified using the search criteria in the databases were exported as Microsoft Excel CSV files for quantitative and qualitative analysis. The references were exported to Zotero (version 6.0.36) reference management system. The VOSviewer software was utilized in the bibliometric analysis of the research articles to visualize the occurrence and co-occurrence networks of the keywords identified from the retrieved articles (Perianes-Rodriguez et al. 2016; Waltman et al. 2010). This allowed for the identification of interconnections of the keywords within a topic from the published research articles (Han et al. 2020). The VOSviewer platform was used to create a map through a process of choosing the counting method, calculating similarity index, calculating and displaying the most relevant items, and displaying a map created based on the extracted research articles. The input data for the bibliometric analysis in VOSviewer included keywords, abstracts, and titles of the extracted research articles. The output visualization was based on the co-occurrence of the key terms from the retrieved research articles.

The Microsoft Excel CSV file exported from the databases contained bibliographic data including the name of the author/s, year of publication, the title of the research paper, the name of the journal, the digital object identifier (DOI), the keywords, and the uniform resource locator (URL). Additionally, information on the location of the research, in terms of the country and continent was also collected. The quantitative analysis involved the determination of the frequency of research articles published over the study period and the continent where the study was done. The analysis also involved the classification of the studies extracted. The earth observation sensors and remote sensing methods utilized in the assessment of the impacts of vegetation fires on ecosystem services were also documented.

The structure of the review included two sections. The first one explores progress in the application of remote sensing in assessing the impacts of vegetation fires on ecosystem services. The characteristics of the literature search, the temporal trend of the research studies, the satellite data, and the methods applied in the field were detailed in the first section of the review. The second section of the review utilized the findings from the first section to identify research gaps in the application of remote sensing in assessing the impact of vegetation fires on ecosystem services.

3 Results

3.1 Literature Search Characteristics

The analysis of literature in the VOSviewer software resulted in a network map shown in Fig. 2 which grouped the retrieved literature into three clusters of key terms. The keywords within the red cluster included 'forest fires, 'burn severity', 'remote sensing', 'vegetation recovery' and 'ecosystem services'. This cluster shows the link between remote sensing data and forest fires and the utilization of remote sensing methods in assessing burn severity and vegetation recovery as post-fire impacts on ecosystem services. The repetition of the term 'forest fire' in the red cluster is evidence that the impact of fires on forest ecosystem services was highly studied in the reviewed studies. The green cluster is characterized by key terms such as 'wildfire', 'biomass burning', 'air quality', 'particulate matter', 'air pollution', 'atmospheric pollution', 'ecosystem services', and 'forestry'. The interconnections between these keywords show that biomass burning is greatly associated with particulate matter emission resulting in air pollution which eventually affects ecosystem services such as air quality and biodiversity. The inclusion of the term 'forestry' in the green cluster clearly shows that most studies under review focused on the forest ecosystems. The blue cluster generally includes key terms such as 'forest', 'fire', 'ecosystem' 'biodiversity', 'satellite imagery, and 'conservation'. The occurrence of these terms in the blue cluster implies that, from the retrieved studies, the effect of fires on ecosystem biodiversity can be analyzed using satellite imagery and has conservation implications.

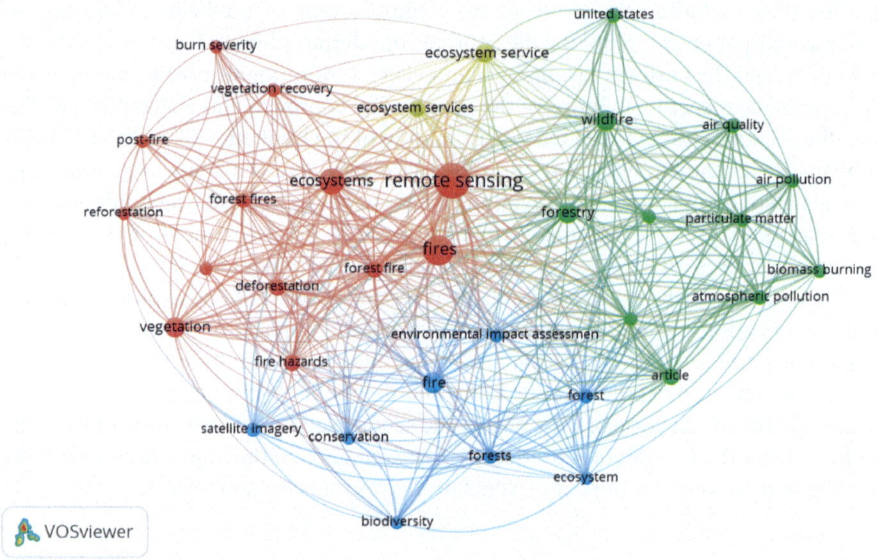

Fig. 2 The major concepts extracted from article titles, abstracts, and keywords

3.2 Progress in the Utilization of Remote Sensing to Assess the Impact of Vegetation Fires on Ecosystem Services

3.2.1 Spatial and Temporal Distribution of Publications

There has been a gradual increase in the publication of studies assessing the impact of vegetation fires on ecosystem services (Fig. 3) with recent years such as 2023 recording the highest number (12) of publications in the field. Notably, the first research study was published in 2010 and there was no noticeable increase in studies before 2015. Between 2015 and 2020 there was no consistent pattern in the trend of publications focusing on the impacts of fires on ecosystem services.

Results from the review show that studies were done across the world with most of the studies generally coming from the United States of America (Figs. 4 and 5). Notably, while just one study was recorded in Africa (Ghana) and only three studies were done in South America (Chile, Brazil and Bolivia), the review recorded no study from Australia. In Asia, the studies included in the analysis were done in China, India, Israel and Indonesia. In Europe, case studies focusing on the impacts of fire on ecosystem services were done in Italy and Greece. Generally, the included case studies came from 12 countries. The nature of ecosystem services affected by fires in the case studies is highlighted in Fig. 6 and Table 1.

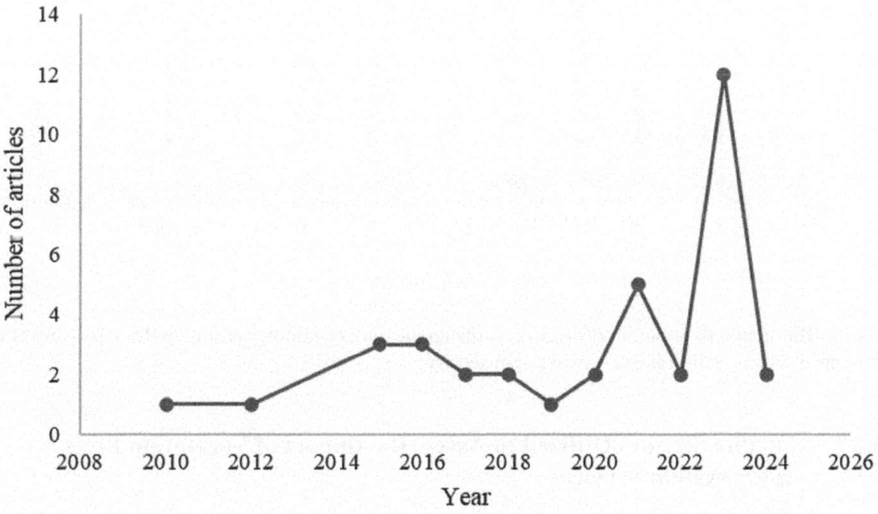

Fig. 3 The temporal distribution of publications on the use of remote sensing technologies to assess the impact of vegetation fires

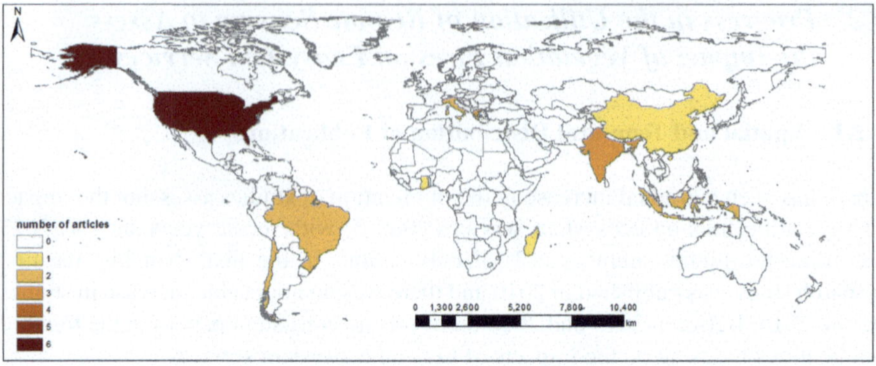

Fig. 4 A map showing the spatial distribution of studies on the application of remote sensing in the assessment of the impact of vegetation fires on ecosystem services

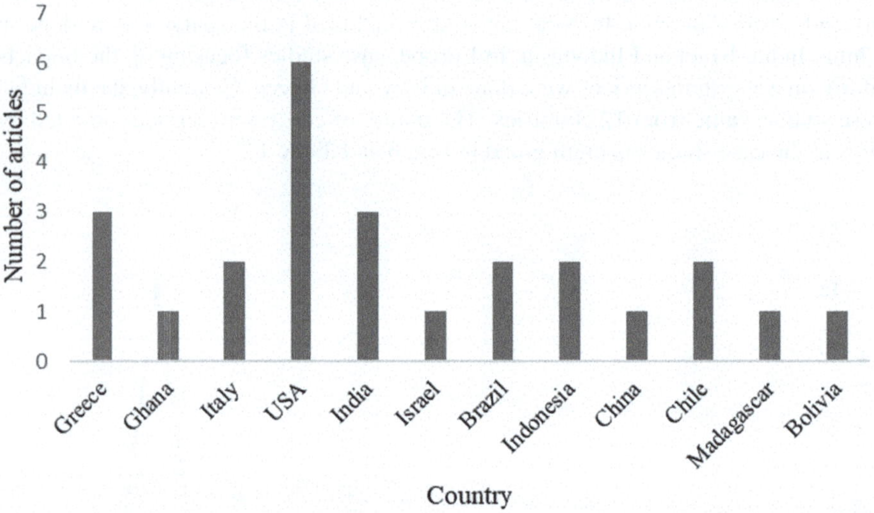

Fig. 5 The spatial distribution of studies on the application of remote sensing in the assessment of the impact of vegetation fires on ecosystem service

3.2.2 Satellite Sensors Utilized to Assess the Impact of Vegetation Fires on Ecosystem Services

Some (38%) of the studies utilized Landsat satellite data to analyze vegetation fires' impact on ecosystem services as shown in Fig. 7. The Moderate Resolution Imaging Spectroradiometer (MODIS) and Visible Infrared Imaging Radiometer Suite (VIIRS) were used in the case studies, mainly as sources of data on the spatiotemporal occurrence of vegetation fires. In some studies, Landsat was used in combination with other datasets such as Sentinel (He et al. 2022; Srivastava et al. 2023; Tselka et al.

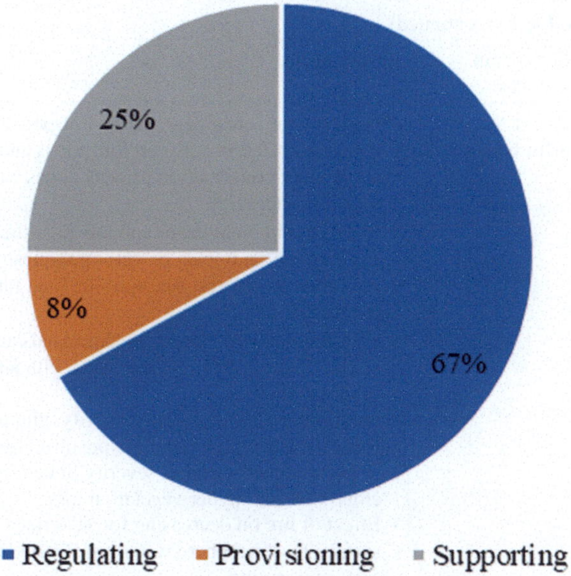

Fig. 6 The nature of the ecosystem services affected by vegetation fires as highlighted in the case studies

Table 1 The summary of studies assessing the impact of vegetation fires on ecosystem services

Ecosystem services	Impact of fire	References
Soil erosion regulation	• Increase in soil loss after a fire event	Stefanidis et al. (2022a) (Sankey et al. (2017), Tselka et al. (2021)
Sediment retention	• Increased water body sedimentation due to post-fire soil erosion	Sankey et al. (2017)
Air quality regulation	• Fire contributed about 1–2.7% of global C emission • Contribution of fires to GHG emission • Fire identified as the major contributor to $PM_{2.5}$ • Increased PM_{10} and $PM_{2.5}$ associated with burning	Kugbe et al. (2012) Souto-Oliveira et al. (2023) Barrera et al. (2018), Chen et al. (2023)
Primary productivity	• Burning induced the regeneration of plants producing better ones • Reduced biomass and gross primary productivity	Semeraro et al. (2019) Pandit et al. (2021)
Biodiversity	• Burning transformed tall grassland ecosystems into shrubs • Assessment of the impact of fires on seasonally dry tropical forests. Woody plants respond to fire by resprouting more than by seedling regeneration • Transformation of forests to shrublands and grasslands	Ratajczak et al. (2016)

(continued)

Table 1 (continued)

Ecosystem services	Impact of fire	References
Forest disturbance	• Detection of forest degradation hotspots due to fires • Strong correlation between forest loss and burned area • Creation of conservation priority zones based on forest loss characteristics • Reduction in plant cover and low post-fire recovery • Burning has an impact on carbon sequestration • Assessment of forest susceptibility and rate of recovery to fires • Transformation of forests to shrublands and grasslands • Fire resilience of forests increases with an increase in fire frequency • Relationship between burn severity with carbon emission and vegetation loss, postfire vegetation and carbon recovery. Higher severity linked to higher emissions and higher vegetation loss • Effect of fire on decreasing forest surface biomass and surface-to-atmosphere water transfer and increased surface warming • Effect of fire on reduction of water use efficiency and growth of pine trees • Assessment of vegetation regeneration after a major fire • Impact of fire severity and climatic factors on post-fire vegetation regeneration. Areas which were most severely burnt had the fastest recovery within 7–10 years	Reddy et al. (2016) Srivastava et al. (2023) Sutomo and van Etten (2023) He et al. (2021) Percival et al. (2024), Reddy et al. (2016) Mancilla-Ruiz et al. (2021) Hartung et al. (2021) Ivo et al. (2020), Michael et al. (2018), Niccoli et al. (2023), Qiu et al. (2023) Hao et al. (2022), Stephens et al. (2020)
Pest and disease control	• Fires increase pests and disease control • Burning reduced tree disease spread and mortality	Pereira et al. (2021) Semeraro et al. (2019)
Natural hazards regulation	• Increased post-fire flooding	Alamanos et al. (2024)

2021) and MODIS (Barrera et al. 2018; Hartung et al. 2021). PlanetScope, AVIRIS (Airborne Visible InfraRed Imaging Spectrometer), and ASTER satellite data were the least utilized by the published studies. Interestingly, only two studies utilized LIDAR data in their analysis (Souto-Oliveira et al. 2023). The fairly recent and high resolution satellite data such as Sentinel and Planet Scope were underutilized in the reviewed studies.

3.3 Impacts of Fire Occurrence on Ecosystem Services

The review of the literature revealed that vegetation fires affect various ecosystem services ranging from forests, grasslands, soil, and atmosphere. The analysis of the

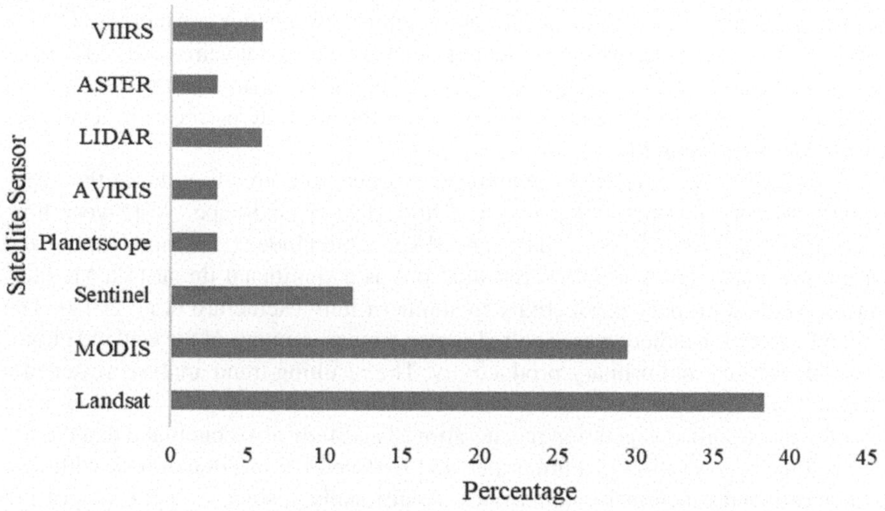

Fig. 7 The earth observation sensors utilized in the studies on the assessment of the impact of vegetation fires on ecosystem services

reviewed studies illustrated that vegetation fires greatly affect various ecosystem services as shown in Table 1. Most of the studies assessed the impacts of burning on forest ecosystem services. Vegetation indices such as Normalized Difference Vegetation Index (NDVI), Enhanced Vegetation Index (EVI), Normalized Difference Water Index (NDWI) and Normalized Burn Ratio (NBR) were widely utilized in the case studies. The Landsat data was mostly utilized to analyze vegetation status.

The selected research papers' analysis revealed that the effects of vegetation fires greatly vary among ecosystems. Forest ecosystems respond differently to burning. For example, the impact of fire frequency on forest resilience was assessed by Hartung et al. (2021) where MCD64A1 MODIS satellite-derived burned area data was utilized to measure the area that was affected by the fire for 20 years. Landsat-derived pre-fire and post-fire Normalized Burn Ratio (NBR) were used as a proxy for vegetation conditions within the burned area. Research findings show that resilience to burning increases with increasing frequency of burning. Burning was also associated with the transformation of forests into shrublands and grasslands thereby affecting the provisioning ecosystem services offered by forests (Ratajczak et al. 2016).

Research also revealed that higher burn severity is greatly associated with higher carbon emission and vegetation loss as observed in a study done in California (Qiu et al. 2023). In the study, MODIS-derived pre-fire and post-fire Enhanced Vegetation Index (EVI) was used to assess the vegetation conditions. Similarly, findings from the review highlight that vegetation burning increases the atmospheric concentration of particulate matter especially PM_{10} and $PM_{2.5}$, and can have far-reaching air pollution effects on areas as far as more than 100 km away from the burned location (Barrera et al. 2018). Based on the utilization of Landsat-derived NBR, shrubs and forest ecosystems were mostly affected by the fires. Additionally, studies done in Alaska and Brazilian forests have indicated the association between fires and increased $PM_{2.5}$ concentration in the atmosphere (Chen et al. 2023; Souto-Oliveira et al. 2023) and also contribute to GHG emission. MODIS satellite fire data has been widely used in these studies. Therefore, clean air as a provisioning ecosystem service is negatively affected by vegetation fires.

According to the reviewed literature, fire occurrence greatly reduces the vegetation's primary productivity across the fire-affected landscape. A 15-year time series analysis of MODIS derived Enhanced Vegetation Index (EVI) and Normalized Difference Water Index (NDWI) revealed fire as a significant threat to a Mediterranean wetland primary productivity in southern Italy (Semeraro et al. 2019). The utilized vegetation indices were applied as proxy measurements of the wetland photosynthetic activity and primary productivity. The resulting trend analysis revealed a drop in both the vegetation indices during the fire event while an increasing trend after the fire signified vegetation regeneration. This study also concluded that the fire induced the regeneration of Phragmites and replaced the old individuals with new structurally and functionally better ones. Additionally, a study on the effect of fire on black pine using an integration of Landsat derived NBR and NDVI and forest surveys in Italy revealed a decline in post-fire pine growth and alteration in the water use efficiency of the burned trees (Niccoli et al. 2023).

Burning has been greatly associated with decreasing forest surface biomass (Ivo et al. 2020). The utilization of Landsat-derived NDVI in the mentioned study has shown that the burning of forests decreases the surface-to-atmosphere water transfer and also tends to increase surface warming. Some of the effects of burning also include increased risk of landslides and flooding in areas outside the burned area as observed in the study by Barrera et al. (2018) in Chile.

Literature has revealed that the severity of burning greatly determines the rate at which post-fire vegetation recovery occurs. Findings from a research study show that areas that were most severely burnt had the fastest recovery (Hao et al. 2022). Vegetation spectral indices have been utilized to measure the severity of burning in various ecosystems. Research done by Hao et al. (2022), for example, utilized MODIS-derived Normalized Difference Vegetation Index (NDVI), Normalized Burn Ratio (NBR), Enhanced Vegetation Index (EVI), and Normalized Difference Moisture Index (NDMI) to detect the severe, moderate, and low severity of burning. High spatial resolution satellite data such as PlanetScope and Sentinel have been utilized in the assessment of urban forest burn severity (using NDVI, gNDVI, GCC) and the spatial distribution of economic loss associated with the burn severity (Michael et al. 2018). Elsewhere, a study done in Greece utilized the Normalized Burn Ratio (NBR) index derived from Sentinel 2 satellite data to detect the vegetation conditions before and after the fire. Based on the study results, burning has been associated with an increased likelihood of post-fire flooding, affecting the regulating ecosystem services (Alamanos et al. 2024).

The assessment of vegetation regeneration after a major fire under different land management systems was also done (Stephens et al. 2020). The Landsat 5 TM and Landsat 7 ETM+ satellite data were utilized to assess forest regeneration, a regulating ecosystem service. The results from the study revealed a higher transformation of forest to shrublands within publicly owned land than in privately owned land. In contrast, greater regeneration of the forest ecosystem was higher in the privately owned forests than in publicly owned ones. The review has also shown that burning has the potential to transform grassland ecosystems (Ratajczak et al. 2016) into shrubs. Utilizing the MODIS burned area maps research findings have shown the impact of burning on grassland ecosystem services.

The review has revealed remotely sensed vegetation indices as key in monitoring changes in wetland primary productivity pre and post-fire. Notably, a study by Semeraro et al. (2019) showed that fire can improve the primary productivity of a wetland ecosystem and regeneration produces better plant individuals. The findings from this study have also shown the association between fire occurrence and vegetation disease control.

While vegetation fires affect the available vegetation, soil regulating ecosystem services are also greatly affected, where burning can be associated with soil loss. For example, a study that coupled Landsat satellite data analysis in a Google Earth Engine platform with the RUSLE model to monitor soil erosion revealed an increase in soil loss after a major fire event in Greece (Stefanidis et al. 2022b). Similarly, the RUSLE model was also utilized in the assessment of soil loss post-fire in a study by Tselka et al. (2021). Soil loss increased by 30% during the first year after a major fire and doubled in 5 years (Tselka et al. 2021). Landsat satellite data was utilized for deriving NBR and dNBR to assess the impact of the fire. Sentinel and Landsat satellite data were used to measure the vegetation cover management Factor (C) in the RUSLE model.

4 Discussion, Limitations and Research Gaps

The gradual increase in the research publications on remote sensing impacts of fire on ecosystem services is associated with the availability of open-source satellite data such as MODIS, VIIRS and Landsat. For instance, the high utilization of Landsat satellite data is mainly due to its free availability and high applicability in vegetation assessment which is the ecosystem generally affected by fires (Graham et al. 2023). Additionally, the increase in expert knowledge enhances the utilization of earth observation data in assessing the impact of vegetation fires on various ecosystem services.

Most of the reviewed studies mainly focused on the impacts of fire on forest ecosystems (He et al. 2021; Michael et al. 2018; Niccoli et al. 2023; Percival et al. 2024; Srivastava et al. 2023; Sudhakar Reddy et al. 2015; Sutomo and van Etten 2023). Forests offer critical regulating, provisioning, supporting, and cultural ecosystem services to many livelihoods including indirect and direct ecosystem contributions to human and animal well-being. (Buthelezi et al. 2024). They are key carbon sinks that absorb carbon dioxide, mitigating climate change's negative impacts. Additionally, forests are global and local climate regulators by influencing temperature and precipitation distribution. The uncontrolled occurrence of fires therefore negatively affects these regulating, supporting, provisioning, supporting, and cultural services. The understanding of the impacts of fires on the various ecosystem services is critical for the sustainable forest management. Prioritization of fire management strategies can also be made.

The review has clearly shown that the ecosystem's primary productivity can be affected by the occurrence of fires. The primary productivity of wetland, forest and grassland ecosystems, for example, tends to drive the regulating and provisioning ecosystem services. While several studies reveal fire occurrence as a threat to forest and grassland ecosystems, a study by Semeraro et al. (2019) indicated that fire induces the regeneration of the wetland ecosystem and replaces old plants with better ones. Contrary to other studies, the wetland primary productivity was higher after the fire than before the disturbance by fire occurrence. The occurrence of fire has a major effect on the primary productivity of such ecosystems and therefore disturbs the flow of the ecosystem services (Semeraro et al. 2019). A study by Semeraro et al. (2019) revealed the utilization of remote sensing methods in monitoring the pre and post-fire primary productivity of a wetland ecosystem.

The literature review has also shown that vegetation indices are widely used in the assessment of the impacts of fires on ecosystem services. Vegetation indices have been widely utilized as indicators of pre- and post-fire vegetation conditions (Graham et al. 2023). While NDVI is highly utilized in fire monitoring, its value increases when regeneration occurs soon after the fire but quickly saturates due to its high sensitivity to leaf-related changes. The Normalized Burn Ratio (NBR), on the other hand, gradually increases with post-fire vegetation regeneration and thus becomes more suitable for this application (Hartung et al. 2021). The assessment of the impacts of burning on vegetation using the NBR index would therefore result in more accurate results (Hartung et al. 2021).

Information about forests' resilience to burning is critical for sustainable forest management. For some forest types, resilience increases with a higher frequency of fire occurrence. For instance, a study carried out in the Chiquitanía dry forest showed a decreasing resilience after the first fire occurrence then increased after the third fire (Hartung et al. 2021). This is associated with the ecological transition from the usual vegetation type to increased abundance in fire-resilient species with the recurrence of burning. Recurrence of fires in such ecosystems tends to increase fire-tolerant plant species (Hartung et al. 2021). Such information on the impacts of fire on specific ecosystems is essential for conservation managers. Prescribed burning can therefore be carefully applied in the relevant ecosystem to harness the benefits of fires on ecosystem services.

Research findings also show that the occurrence of fires induces increased soil loss (Stefanidis et al. 2022a; Sankey et al. 2017; Tselka et al. 2021). This is in line with observations by Pereira et al. (2021) who highlighted that when the available vegetation cover is burned, soil protection is reduced and may lead to the transportation of the produced ash, sediments and soil nutrients into nearby water bodies. Furthermore, reduced soil protection after burning increases the risk of flash floods and landslides (Brogan et al. 2017; Wall et al. 2020). The combination of soil organic carbon consumption and vegetation removal during burning greatly reduces the soil and water regulatory services by the ecosystem. The utilization of remote sensing methods in understanding these critical impacts of fires on ecosystem services is valuable for the sustainable soil conservation efforts.

The systematic literature review has also shown that vegetation fires greatly impact the ecosystem regulatory services. The global occurrence of vegetation fires greatly affects the carbon cycle through carbon and particulate matter emissions (Qiu et al. 2023). Specifically, burning emits carbon in the form of carbon dioxide and greatly influences atmospheric greenhouse gas concentration. Additionally, the removal of vegetation during burning decreases the vegetation's carbon sequestration capacity. On the other hand, the occurrence of fires negatively affects local and global climate regulation as the soil becomes a source of greenhouse gases (Ribeiro-Kumara et al. 2020). The occurrence of fires has been associated with increased atmospheric pollutant emissions as emphasized by Pereira et al. (2021). Furthermore, the contribution of fires to atmospheric pollution and its damaging impacts on ecosystems, highlighted by the research findings is in line with literature (Matz et al. 2020; Nowell et al. 2018). The atmospheric pollutants associated with fire extinguishing during fire suppression may also contain hazardous pollutants which can be persistent in the environment. The post-fire transportation of such materials into water bodies may decrease water quality (Emmerton et al. 2020; Pereira et al. 2021).Based on the study findings, it is clear that fires are a threat to clean air, an important ecosystem service. The valuable information on the contribution of fires to atmospheric pollution is critical for air quality management.

The findings from this review have also shown the association between fire occurrence and vegetation disease control (Semeraro et al. 2019). This is not surprising because several studies have also observed that the occurrence and transmission of diseases are greatly reduced after a fire event (Gleim et al. 2019; Simler-Williamson et al. 2021). Fires are widely known to control the occurrence of pests and reduce forest diseases.

The findings from this review are key to understanding how vegetation fires threaten various ecosystem services. This information is valuable to ecosystem and fire managers to make decisions regarding the protection of ecosystems. The research findings also highlight the importance of remote sensing methods in the long-term monitoring of the effects of burning on ecosystem services. The utilization of remote sensing methods in assessing the impacts of fire on ecosystem services is greatly determined by the availability of satellite data. For example, Chu and Guo (2014) assessed the application of remote sensing techniques in modelling post-fire effects and forest recovery patterns in forest regions. The study findings revealed that remote sensing post-fire effects and forest recovery are mainly limited by gaps in data and the complications of fire regimes, and climatic conditions.

While this review study generally looked at all ecosystems, future review studies should focus on specific ecosystems to deepen the understanding of how fire affects the particular ecosystem. On the other hand, although ecosystem transitions have been associated with fire occurrence, the influence of changing climatic conditions has not been analyzed. Further studies should therefore look at the impact of a combination of fire occurrence and varying climatic conditions on ecosystem services.

The research findings pointed to the gaps in knowledge in terms of the spatial coverage of research done and the satellite data utilized. Specifically, Africa was understudied, with most studies covering Asian and American states. On the other hand, Sentinel satellite data was underutilized in the studies reviewed. There is a need, therefore, for a larger geographic coverage and utilization of modern satellite data in the future studies. Even in geographical locations that were well studied, most studies concentrated on the forest ecosystem. The study findings are key in allowing ecosystem managers to make informed decisions in ecosystem management. The observations enabled by the utilization of remote sensing methods are important for guiding ecosystem conservation initiatives and the adequate allocation of resources. Finally, the findings from this literature review are critical to ecosystem managers in understanding the resilience and vulnerability of ecosystems and their ecosystem services.

The review clearly shows the controversy regarding the impact of fires on ecosystem services. For example, some terrestrial ecosystems are greatly shaped by fire and sustainable fire management ensures the support of ecosystem services. Although the occurrence of fire is intrinsic to ecosystems such as the Mediterranean and Savanna biomes, the high frequency and severity of human-caused fires can greatly harm ecosystems and disturb the proper functioning of ecosystem services.

5 Conclusion

In conclusion, the current study aimed to review literature focusing on the utilization of remote sensing in assessing the effects of vegetation fires on various ecosystems and the flow of ecosystem services. The findings from the review underscore the utility of remote sensing methods in assessing the impacts of fire on ecosystem services. There has been a gradual increase in research papers on the topic and the most utilized satellite data included Landsat and MODIS. The vegetation indices, such as the NDVI and NBR have been highly used to assess the vegetation conditions before and after fire occurrence. The study has shown that the African continent is understudied hence future research work should also focus on the utilization of remote sensing methods in assessing the effects of fires on various ecosystems within the African continent. Notably, the spatial distribution of the studies was biased towards the Asian and American continents. Furthermore, the review has shown that fire affects various ecosystems including forests, wetlands, grasslands, soil, and the atmosphere and most studies focused on forests.

References

Alamanos A, Papaioannou G, Varlas G, Markogianni V, Papadopoulos A, Dimitriou E (2024) Representation of a post-fire flash-flood event combining meteorological simulations, remote sensing, and hydraulic modeling. Land 13. https://doi.org/10.3390/land13010047

Bowman DMJS, Williamson GJ, Abatzoglou JT, Kolden CA, Cochrane MA, Smith AMS (2017) Human exposure and sensitivity to globally extreme wildfire events. Nat Ecol Evol 1(3):58. https://doi.org/10.1038/s41559-016-0058

Brogan DJ, Nelson PA, MacDonald LH (2017) Reconstructing extreme post-wildfire floods: a comparison of convective and mesoscale events. Earth Surf Process Landf 42:2505–2522. https://doi.org/10.1002/esp.4194

Buthelezi MNM, Lottering R, Peerbhay K, Mutanga O (2024) Optimising forest rehabilitation and restoration through remote sensing and machine learning: mapping natural forests in the eThekwini Municipality. Remote Sens Appl: Soc Environ 36:101335. https://doi.org/10.1016/j.rsase.2024.101335

Chen D, Billmire M, Loughner CP, Bredder A, French NHF, Kim HC, Loboda TV (2023) Simulating spatio-temporal dynamics of surface PM2.5 emitted from Alaskan wildfires. Sci Total Environ 898:165594. https://doi.org/10.1016/j.scitotenv.2023.165594

Chu T, Guo X (2014) Remote sensing techniques in monitoring post-fire effects and patterns of forest recovery in Boreal Forest Regions: a review. Remote Sens 6:470–520. https://doi.org/10.3390/rs6010470

Chuvieco E (2019) Global analysis of burned areas for climate assessment: Experiences from the Fire-CCI project. In: International geoscience and remote sensing symposium (IGARSS), Institute of Electrical and Electronics Engineers Inc., pp 9950–9952. https://doi.org/10.1109/IGARSS.2019.8899046

de la Barrera F, Barraza F, Favier P, Ruiz V, Quense J (2018) Megafires in Chile 2017: Monitoring multiscale environmental impacts of burned ecosystems. Sci Total Environ 637–638, 1526–1536. https://doi.org/10.1016/j.scitotenv.2018.05.119

Devineau JL, Fournier A, Nignan S (2010) Savanna fire regimes assessment with MODIS fire data: their relationship to land cover and plant species distribution in western burkina faso (West Africa). J Arid Environ 74:1092–1101. https://doi.org/10.1016/j.jaridenv.2010.03.009

Emmerton CA, Cooke CA, Hustins S, Silins U, Emelko MB, Lewis T, Kruk MK, Taube N, Zhu D, Jackson B, Stone M, Kerr JG, Orwin JF (2020) Severe western Canadian wildfire affects water quality even at large basin scales. Water Res 183:116071. https://doi.org/10.1016/j.watres.2020.116071

Gleim ER, Zemtsova GE, Berghaus RD, Levin ML, Conner M, Yabsley MJ (2019) Frequent prescribed fires can reduce risk of tick-borne diseases. Sci Rep 9. https://doi.org/10.1038/s41598-019-46377-4

Graham E, Dube T, Mpakairi KS (2023) Progress in the remote sensing of veld fire occurrence and detection: a review. Afr J Ecol. https://doi.org/10.1111/aje.13147

Han J, Kang H-J, Kim M, Kwon GH (2020) Mapping the intellectual structure of research on surgery with mixed reality: Bibliometric network analysis (2000–2019). J Biomed Inform 109:103516. https://doi.org/10.1016/j.jbi.2020.103516

Hao B, Xu X, Wu F, Tan L (2022) Long-term effects of fire severity and climatic factors on post-forest-fire vegetation recovery. Forests 13. https://doi.org/10.3390/f13060883

Hartung M, Carreño-Rocabado G, Peña-Claros M, van der Sande MT (2021) Tropical dry forest resilience to fire depends on fire frequency and climate. Front For Glob Change 4. https://doi.org/10.3389/ffgc.2021.755104

He W, Shirowzhan S, Pettit CJ (2022) GIS and machine learning for analysing influencing factors of bushfires using 40-year spatio-temporal bushfire data. ISPRS Int J Geo-Inf 11. https://doi.org/10.3390/ijgi11060336

He Y, Chen G, Cobb RC, Zhao K, Meentemeyer RK (2021) Forest landscape patterns shaped by interactions between wildfire and sudden oak death disease. For Ecol Manag 486:118987. https://doi.org/10.1016/j.foreco.2021.118987

Herawati H, González-Olabarria JR, Wijaya A, Martius C, Purnomo H, Andriani R (2015) Tools for assessing the impacts of climate variability and change on wildfire regimes in forests. Forests 6:1476–1499. https://doi.org/10.3390/f6051476

Ivo IO, Biudes MS, Vourlitis GL, Machado NG, Martim CC (2020) Effect of fires on biophysical parameters, energy balance and evapotranspiration in a protected area in the Brazilian Cerrado. Remote Sens Appl Soc Environ 19:100342. https://doi.org/10.1016/j.rsase.2020.100342

Kugbe JX, Mathias F, Desta TL, Denich M, Vlek PLG (2012) Annual vegetation burns across the northern savanna region of Ghana: period of occurrence, area burns, nutrient losses and emissions. Nutr Cycl Agroecosystems 93:265–284. https://doi.org/10.1007/s10705-012-9514-0

Kurbanov E, Vorobev O, Lezhnin S, Sha J, Wang J, Li X, Cole J, Dergunov D, Wang Y (2022) Remote Sensing of forest burnt area, burn severity, and post-fire recovery: a review. Remote Sens 14. https://doi.org/10.3390/rs14194714

Mancilla-Ruiz D, Barrera FD, González S, Huaico A (2021) the effects of a megafire on ecosystem services and the pace of landscape recovery. Land 10. https://doi.org/10.3390/land10121388

Marambanyika T, Mupfiga UN, Musasa T, Ngwenya K (2021) Local perceptions on the impact of drought on wetland ecosystem services and associated household livelihood benefits: the case of the Driefontein Ramsar Site in Zimbabwe. Land 10. https://doi.org/10.3390/land10060587

Matz, C.J., Egyed, M., Xi, G., Racine, J., Pavlovic, R., Rittmaster, R., Henderson, S.B., Stieb, D.M., 2020. Health impact analysis of PM2.5 from wildfire smoke in Canada (2013–2015, 2017–2018). Sci Total Environ 725:138506. https://doi.org/10.1016/j.scitotenv.2020.138506

Michael Y, Lensky IM, Brenner S, Tchetchik A, Tessler N, Helman D (2018) Economic assessment of fire damage to urban forest in the wildland–urban interface using planet satellites constellation images. Remote Sens 10. https://doi.org/10.3390/rs10091479

Millennium Ecosystem Assessment (2005) Ecosystems and human well-being: wetlands and water synthesis. World Resources Institute, Washington, DC, USA

Molaudzi O, Adelabu S (2019) Review of the use of remote sensing for monitoring wildfire risk conditions to support fire risk assessment in protected areas. South Afr J Geomat 7:222. https://doi.org/10.4314/sajg.v7i3.2

Mupfiga UN, Mutanga O, Dube T (2024) National-scale spatiotemporal patterns of vegetation fire occurrences using MODIS satellite data. PLoS ONE 19:e0297309. https://doi.org/10.1371/journal.pone.0297309

Niccoli F, Altieri S, Kabala JP, Battipaglia G (2023) Fire affects tree growth, water use efficiency and carbon sequestration ecosystem service of pinus nigra Arnold: a combined satellite and ground-based study in central italy. Forests 14. https://doi.org/10.3390/f14102033

Nowell HK, Holmes CD, Robertson K, Teske C, Hiers JK (2018) A new picture of fire extent, variability, and drought interaction in prescribed fire landscapes: insights from Florida Government Records. Geophys Res Lett 45:7874–7884. https://doi.org/10.1029/2018GL078679

Page MJ, McKenzie JE, Bossuyt PM, Boutron I, Hoffmann TC, Mulrow CD, Shamseer L, Tetzlaff JM, Akl EA, Brennan SE, Chou R, Glanville J, Grimshaw JM, Hróbjartsson A, Lalu MM, Li T, Loder EW, Mayo-Wilson E, McDonald S, McGuinness LA, Stewart LA, Thomas J, Tricco AC, Welch VA, Whiting P, Moher D (2021) The PRISMA 2020 statement: an updated guideline for reporting systematic reviews. BMJ 372:n71. https://doi.org/10.1136/bmj.n71

Pandit K, Dashti H, Hudak AT, Glenn NF, Flores AN, Shinneman DJ (2021) Understanding the effect of fire on vegetation composition and gross primary production in a semi-arid shrubland ecosystem using the Ecosystem Demography (EDv2.2) model. Biogeosciences 18:2027–2045. https://doi.org/10.5194/bg-18-2027-2021

Paula A, Figueiredo A, Pulido F, Castro P (2023) Could spaceborne multispectral data be used as SMART proxies' indicators for agroforestry ecosystem services? A systematic review. Ecol Indic 156. https://doi.org/10.1016/j.ecolind.2023.111162

Percival JEH, Sato H, Razanaparany TP, Rakotomamonjy AH, Razafiarison ZL, Kitajima K (2024) Non fire-adapted dry forest of Northwestern Madagascar: escalating and devastating trends revealed by landsat timeseries and GEDI lidar data. PLoS ONE 19:e0290203. https://doi.org/10.1371/journal.pone.0290203

Pereira P, Bogunovic I, Zhao W, Barcelo D (2021) Short-term effect of wildfires and prescribed fires on ecosystem services. Curr Opin Environ Sci Health 22:100266. https://doi.org/10.1016/j.coesh.2021.100266

Pérez-Cabello F, Montorio R, Alves DB (2021) Remote sensing techniques to assess post-fire vegetation recovery. Curr Opin Environ Sci Health 21:100251. https://doi.org/10.1016/j.coesh.2021.100251

Perianes-Rodriguez A, Waltman L, van Eck NJ (2016) Constructing bibliometric networks: a comparison between full and fractional counting. J Informetr 10:1178–1195. https://doi.org/10.1016/j.joi.2016.10.006

Qiu L, Fan L, Sun L, Zeng Z, Feng L, Yue C, Zheng C (2023) Higher burn severity stimulates postfire vegetation and carbon recovery in California. Agric for Meteorol 342:109750. https://doi.org/10.1016/j.agrformet.2023.109750

Ratajczak Z, Briggs JM, Goodin DG, Luo L, Mohler RL, Nippert JB, Obermeyer B (2016) Assessing the potential for transitions from Tallgrass Prairie to Woodlands: are we operating beyond critical fire thresholds? Rangel Ecol Manag 69:280–287. https://doi.org/10.1016/j.rama.2016.03.004

Reddy CS, Manaswini G, Satish KV, Singh S, Jha CS, Dadhwal VK (2016) Conservation priorities of forest ecosystems: evaluation of deforestation and degradation hotspots using geospatial techniques. Ecol Eng 91:333–342. https://doi.org/10.1016/j.ecoleng.2016.03.007

Ribeiro-Kumara C, Köster E, Aaltonen H, Köster K (2020) How do forest fires affect soil greenhouse gas emissions in upland boreal forests? A Review. Environ. Res. 184:109328. https://doi.org/10.1016/j.envres.2020.109328

Sankey JB, Kreitler J, Hawbaker TJ, McVay JL, Miller ME, Mueller ER, Vaillant NM, Lowe SE, Sankey TT (2017) Climate, wildfire, and erosion ensemble foretells more sediment in western USA watersheds. Geophys Res Lett 44:8884–8892. https://doi.org/10.1002/2017GL073979

Semeraro T, Vacchiano G, Aretano R, Ascoli D (2019) Application of vegetation index time series to value fire effect on primary production in a Southern European rare wetland. Ecol Eng 134:9–17. https://doi.org/10.1016/j.ecoleng.2019.04.004

Simler-Williamson AB, Metz MR, Frangioso KM, Rizzo DM (2021) Wildfire alters the disturbance impacts of an emerging forest disease via changes to host occurrence and demographic structure. J Ecol 109:676–691. https://doi.org/10.1111/1365-2745.13495

Souto-Oliveira CE, Marques MTA, Nogueira T, Lopes FJS, Medeiros JAG, Medeiros IMMA, Moreira GA, da Silva Dias PL, Landulfo E, de Andrade MF (2023) Impact of extreme wildfires from the Brazilian Forests and sugarcane burning on the air quality of the biggest megacity on South America. Sci Total Environ 888:163439. https://doi.org/10.1016/j.scitotenv.2023.163439

Srivastava A, Umrao S, Biswas S, Dubey R, Zafar M (2023) FCCC: forest cover change calculator user interface for identifying fire incidents in forest region using satellite data. Int J Adv Comput Sci Appl 14. https://doi.org/10.14569/IJACSA.2023.01407103

Stavi I (2019) Wildfires in grasslands and shrublands: a review of impacts on vegetation, soil, hydrology, and geomorphology. Water 11. https://doi.org/10.3390/w11051042

Stefanidis S, Alexandridis V, Mallinis G (2022a) A cloud-based mapping approach for assessing spatiotemporal changes in erosion dynamics due to biotic and abiotic disturbances in a Mediterranean Peri-Urban forest. Catena 218. https://doi.org/10.1016/j.catena.2022.106564

Stefanidis S, Alexandridis V, Spalevic V, Mincato RL (2022b) Wildfire effects on soil erosion dynamics: the case of 2021 megafires in Greece. Agric for 68:49–63

Stephens CW, Collins BM, Rogan J (2020) Land ownership impacts post-wildfire forest regeneration in Sierra Nevada mixed-conifer forests. For Ecol Manag 468:118161. https://doi.org/10.1016/j.foreco.2020.118161

Sudhakar Reddy C, Vazeed Pasha S, Jha CS, Dadhwal VK (2015) Geospatial characterization of deforestation, fragmentation and forest fires in Telangana state, India: conservation perspective. Environ Monit Assess 187:455. https://doi.org/10.1007/s10661-015-4673-5

Sutomo S, van Etten EJB (2023) Fire impacts and dynamics of seasonally dry tropical forest of East Java, Indonesia. Forests 14. https://doi.org/10.3390/f14010106

Szpakowski DM, Jensen JLR (2019) A review of the applications of remote sensing in fire ecology. Remote Sens 11. https://doi.org/10.3390/rs11222638

Tselka I, Krassakis P, Rentzelos A, Koukouzas N, Parcharidis I (2021) Assessing post-fire effects on soil loss combining burn severity and advanced erosion modeling in Malesina, Central Greece. Remote Sens 13. https://doi.org/10.3390/rs13245160

Vigna I, Besana A, Comino E, Pezzoli A (2021) Application of the socio-ecological system framework to forest fire risk management: a systematic literature review. Sustainability 13. https://doi.org/10.3390/su13042121

Wall SA, Roering JJ, Rengers FK (2020) Runoff-initiated post-fire debris flow Western Cascades, Oregon. Landslides 17:1649–1661. https://doi.org/10.1007/s10346-020-01376-9

Waltman L, van Eck NJ, Noyons ECM (2010) A unified approach to mapping and clustering of bibliometric networks. J Informetr 4:629–635. https://doi.org/10.1016/j.joi.2010.07.002

Land Use Land Cover Changes and Their Implications for Ecosystem Services in a Typical Communal Rangeland Area in Southern Africa

Xolile Zuma, Onisimo Mutanga, Mbulisi Sibanda, Timothy Dube, Tsitsi Bangira, and Tafadzwanashe Mabhaudhi

Abstract Rangelands are globally being transformed drastically by climate change and anthropogenic activities such as the settlement expansion, crop farming and grazing activities. These changes impact ecosystem services by altering land cover, resource availability, and ecological balance. However, the lack of spatially explicit criterion and frameworks for monitoring these changes and their impacts on ecosystem services necessitates timely accurate and effective methods evaluating transformation and fragmentation. This study utilised the random forest classifier

X. Zuma · O. Mutanga
Discipline of Geography and Environmental Science, School of Agricultural Earth and Environmental Sciences, University of KwaZulu-Natal, Scottsville, Pietermaritzburg, South Africa
e-mail: 216004191@stu.ukzn.ac.za

O. Mutanga
e-mail: mutangaO@ukzn.ac.za

M. Sibanda (✉)
Department of Geography, Environmental Studies & Tourism, Faculty of Arts, University of the West-Ern Cape, Bellville, Cape Town, South Africa
e-mail: msibanda@uwc.ac.za

T. Dube
Discipline of Earth Sciences, University of the Western Cape, Bellville, South Africa
e-mail: tidube@uwc.ac.za

T. Bangira · T. Mabhaudhi
Centre for Transformative Agricultural and Food Systems, School of Agricultural, Earth and Environmental Sciences, University of KwaZulu-Natal (UKZN), Scottsville, Pietermaritzburg, South Africa
e-mail: bangi-rat@ukzn.ac.za

T. Mabhaudhi
e-mail: mabhaudhi@ukzn.ac.za

T. Bangira
Centre on Climate Change and Planetary Health, London School of Hygiene and Tropical Medicine, London, UK

O. Mutanga et al. (eds.), *Revealing Ecosystem Services Through Geospatial Technologies*, Springer Remote Sensing/Photogrammetry,
https://doi.org/10.1007/978-3-031-98048-0_14

and Landsat data in Google Earth Engine (GEE) to assess the spatio-temporal variability of rangelands KwaZulu Natal communal areas from 2000 to 2020. Fragmentation levels were also assess using geospatial statistically techniques. In Inhlazuka, grasslands classification achieved overall accuracies of 75%, 79% and 83%, while Vulindlela achieved 89%, 85%, and 89% for the years 2000, 2010 and 2020, respectively. Rangelands in Vulindlela decreased at a rate of 37.52 ha per year between 2000 and 2010 and at 76.46 ha from 2010 to 2020. In contrast, Inhlazuka exhibited a decreased of 40 ha per year between the years 2000 and 2010 followed by an increase of 45.28 ha annually from 2010 to 2020 due to a decline in the forest class. Key spectral features for discriminating the grasslands were Bands 5 (NIR) and 7 (SWIR). Rangeland fragmentation increased, with mean patch sizes in Vulindlela shrinking from 32 ha in 2000 to 9 in 2020, because of urban expansion. In Inhlazuka, mean patch sizes decreased from 2 to 1 ha in 2020. These findings underscore the importance of developing spatially explicit monitoring frameworks to guide sustainable communal rangeland management policies, preserving ecosystem services and mitigating land degradation.

Keywords Rangeland ecosystem services · Random forest classifier · Google Earth Engine · Landsat data · Communal areas

1 Introduction

Rangelands around the world have become one of the most threatened ecosystems due to human activities and climate variability (Abdulahi et al. 2016). These landscapes provide important ecosystem services such as the provision of food, carbon sequestration, biodiversity maintenance, fibre, clean water, recreational space, and wildlife habitat (Sala et al. 2017; Boone et al. 2018; Zhao et al. 2020). A huge portion of the human population, particularly in developing nations, highly depends on rangelands for livelihoods, livestock production and food security. For instance, according to Bedunah and Angerer (2012) about 1.2 billion people who are surviving on less than $1 per day depend on rangelands.

Despite their importance, the limited effective management and monitoring frameworks in developing countries makes it difficult to determine the level of grass degradation and the loss associated ecosystem services (Bedunah and Angerer 2012). As the human population increases, food production is also anticipated to increase by about 75% in the next 30 years (Ceballos et al. 2010; Bedunah and Angerer 2012). This includes the conversion of rangelands into croplands and development of infrastructure and land use activities such as overgrazing and unmanaged farming practices will cause rangeland fragmentation and loss of biodiversity and ecosystem services (Palmer and Bennett 2013). Meanwhile, a large population depends on rangelands for livestock production, food and income generation (Abdulahi et al. 2016). However, the limited monitoring and management strategies of rangelands in the communal areas of developing countries make it difficult to determine the level of grass degradation (Bedunah and Angerer 2012). In South Africa, for example, 60% of the grassland biome has been transformed, with 25% degraded, 15% remaining unchanged and only 2% under formal protection (Little et al. 2015). The degradation of rangelands threatens their ability to sustain ecosystem services, as unmanaged grasslands in unprotected areas lack comprehensive monitoring frameworks for assessing their condition and fragmentation (McGranahan and Kirkman 2013). Conventional methods for rangeland assessment, such as filed observation, are prone to bias and limitations, particularly in communal areas of developing countries as noted by (Al-Bukhari, Hallett and Brewer 2018). This highlights the need for spatially explicit non-invasive techniques to monitor the spatial extent and the magnitude of grassland fragmentation for sustainable utilisation of this natural capital.

Advances in earth-observation (EO) technologies have provided fast, efficient and reliable spatially explicit techniques for mapping landuse and landcover changes (LULC) changes including rangeland fragmentation, with high spatial and temporal accuracy. monitoring the extent of fragmentation and the spatial distribution of grasslands (Jin et al. 2014). EO platforms offer access to long-term high resolution datasets suitable for monitoring rangelands attributes such as biomass (Sibanda et al. 2017), water content (Sibanda et al. 2019) and their phenology (Matongera et al. 2021). Other than monitoring grass attributes, EO datasets have also been widely applied in characterising LULC changes based on a variety of classification algorithms. Among the EO datasets, the Landsat missions have been the most widely in mapping and monitoring LULC changes (Abd El-Kawy et al. 2011). This is attributed to longest uninterrupted record of serving remotely sensed data since 1972 with a spatial resolution of 30 m at a 16 day interval (Loveland and Dwyer 2012; Li et al. 2020). Subsequently, Landsat datasets offer better prospects of objectively detecting, mapping and monitoring the spatial and temporal variations of grassland fragmentation in communal rangelands (Xie et al. 2019).

Literature underscores effectiveness of these datasets when combined with robust algorithms in accurately mapping LULC changes (Ghayour et al. 2021; Kafy et al. 2021). Specifically, machine leaning algorithms such as Random forest (RF) are renowned for their effectiveness in assessing LULC changes due to their robustness (Kafy et al. 2021; Ghayour et al. 2021). RF classifier is renowned for its robustness in managing outliers and noisy datasets, its great performance in dealing with high dimensional and multi-source datasets, achieving superior accuracy when compared to other classifiers such as the Support Vector Machine (SVM) and Maximum Likelihood Classifier (MLC) (Cengiz et al. 2023; Sheykhmousa et al. 2020). RF uses decision trees during classification, such that the trees with the most selections define the class. To date, the algorithm is the most widely used classifier in conjunction with Landsat data in assessing LULC changes (Rodriguez-Galiano et al. 2012; Phan et al. 2020; Thanh Noi and Kappas 2017).

Despite the optimal performance of RF and Landsat data in LULC change assessments for the sustainable management of natural resources, the provision of large-scale land cover change maps of grasslands is often inhibited by elements such as spectral complexity due to the heterogeneity of the environment as well as the limitation of software/hardware resources to store and process large remotely sensed datasets (Rodriguez-Galiano et al. 2012; Wahap and Shafri 2020). Cloud computing platforms such as Google Earth Engine (GEE) have emerged as transformative tools for addressing these data management challenges (Wahap and Shafri 2020). GEE is a free cloud-based platform which can be used for large-scale environmental analysis or mapping (Tamiminia et al. 2020). GEE, has vastly improved the access and processing of satellite imagery making it possible to conduct assessments on the spatial and temporal variations of community-shared natural resources such as the communal rangelands (Tamiminia et al. 2020). The platform provides a large amount of satellite data including Landsat, dating back to the 1970s, while it is capable of administering renowned algorithms such as the RF on the available datasets (Wang et al. 2017; Kumar and Mutanga 2018; Gorelick et al. 2017). Single date Landsat remotely sensed data sets have been widely used in remote sensing LULC (Abd El-Kawy et al. 2011; Langley, Cheshire and Humes 2001). However, grasslands are highly variable ecosystems which are greatly impacted by situational factors such as seasonality of precipitation, and variations in management practices (Cleland et al. 2013). This makes it challenging to characterise the spatial and temporal changes of grasslands based on single-date images (Price et al. 1997). Considering the relative accessibility of high spatial and temporal Landsat data coverage and the robustness of the RF algorithm, all embedded in GEE, presents opportunities for establishing cheap and reliable grassland monitoring techniques (Phan et al. 2020; Thanh Noi and Kappas 2017). Therefore, the objective of this study was to assess the spatio-temporal variability of rangelands within a typical southern African communal area from the year 2000 to 2020 using multi-temporal Landsat datasets in conjunction with random forest. The study also sought to assess the magnitude and extent of grassland fragmentation in these communal rangelands using fragmentation statistics. By providing spatially explicit insights, the findings aim to inform sustainable rangeland management policies and enhance the conservation of ecosystem services provided by these important landscapes.

2 Methods and Materials

2.1 Study Site

This study was conducted in Inhlazuka (centre coordinates 29° 55′ 40″ E and 30° 11′ 34″ S) and Vulindlela (centre coordinates 29° 0.40′ 37.3584″ S and 30° 8′ 13.6572″ E) communal rangelands located in the uMgungundlovu District Municipality under the uMngeni Catchment in the province of KwaZulu-Natal, South Africa (Fig. 1). The catchment hosts the country's second largest economic hub and its largest trade port (Hughes et al. 2018). Activities such as agriculture and urbanisation are causing immense pressure on the catchment's natural resources (Hughes et al. 2018). The catchment is mostly covered by grasslands, with much of its grasses being turned into cultivated land (Hughes et al. 2018). Vulindlela is in the west of Pietermaritzburg and northwest of the Greater Edendale area, while Inhlazuka is located under the Richmond municipality and is a mountainous area with an elevation of 1313 m. The areas of Inhlazuka and Vulindlela are rural settings with high levels of poverty. Vulindlela experiences dry winters and hot wet summers and receives on average an annual rainfall of 979 mm (Sibanda et al. 2021). The growing season in Vulindlela starts in October and ends in April, from late April to August frost conditions begin, where the growing season is restricted (Royimani et al. 2022; Sibanda et al. 2021). The average rainfall received by Inhlazuka is 852 mm which starts in October and ends in April (Mncube 2022). Land use activities in Vulindlela include scattered settlements, grazing land, cultivated lands, pockets of indigenous forest and some major timber plantations while Inhlazuka is predominantly characterised by small and large-scale crop farming lands. Vegetation in Inhlazuka is also characterised by bushveld vegetation, which is heavily invaded by alien vegetation, such as bugweed (*Solanum mauritianum*), Lantana camara and bramble (*Rubus* spp.) which are often cleared but persisting. Overall, the communal rangelands of Inhlazuka and Vulindlela offer vital grassland ecosystem services, including carbon sequestration, soil erosion control, water regulation while supporting local livelihoods through grazing and agriculture. These grasslands also play a critical role in maintaining biodiversity, although they face increasing pressure from urbanisation, agricultural expansion and invasive species like Lantana Kamara, which threaten their ecological integrity (Fig. 2).

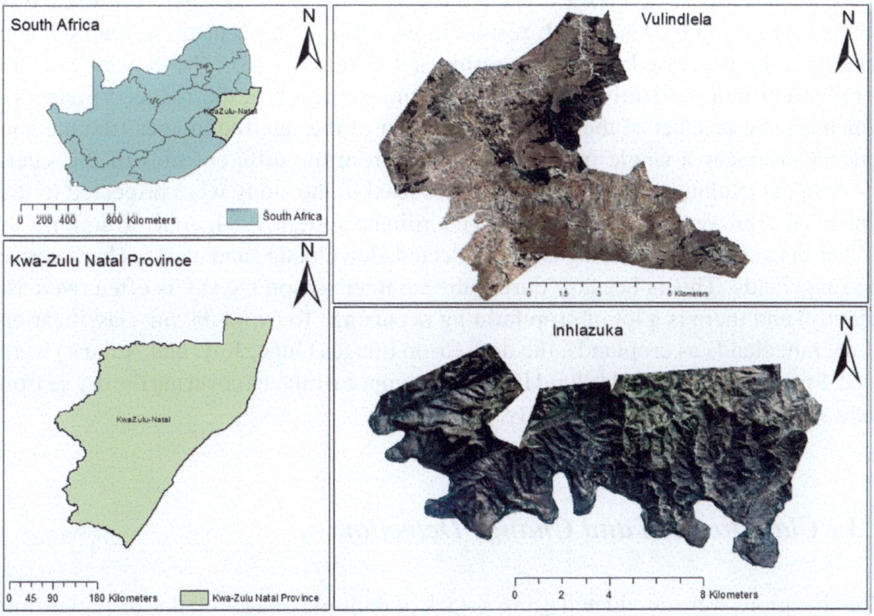

Fig. 1 Inhalzuka and Vulindlela communal areas in Pietermaritzburg, KwaZulu Natal South Africa

Fig. 2 Typical communal rangeland areas in **a** Vulindlela and **b** Inhlazuka

2.2 Satellite Data

The study utilised cloud-free and atmospherically corrected multi-temporal Landsat 7 and Landsat 8 Top of atmosphere (TOA) reflectance remotely sensed data to characterise land covers in the years 2000, 2010, and 2020. The GEE platform provides pre-processed images that will have gone through atmospheric correction. These images were selected and used to conduct the classification in the GEE platform (Gorelick et al. 2017). The study also used the median reducer for image processing. The median reducer applies a title by tile processing method where each scene is divided into several tiles, which results in each tile being sent to various Google servers, to be processed (Sidhu, Pebesma and Câmara 2018). The servers work in parallel and independently to one another and the result is the reduced image(s), which are the product of the tile reconstruction of the satellite images (the median reducer produces a single image constructed from the different tiles of the satellite images) (Sidhu et al. 2018). All images used in the study were projected to the Universal Transverse Mercator (UTM) coordinate system 1984, zone 36 south.

The dry season/winter images were selected, downloaded and used to characterise the rangelands. This is because during the summer season the sky is often overcast (cloudy) and there is a lot of crop farming occurring. To avoid the misclassification of the rangelands as croplands, the dry season images (June, July, and August) were used. Specifically, available cloud free multi-temporal images covering the dry season were selected and used in this study.

2.3 Classification and Change Detection

Classifications were conducted using a stack of multiple images on the GEE platform covering the dry season of each of the years 2000, 2010 and 2020. These images were acquired during dry season months (June, July and August). A median reducer in the

GEE platform was used to compile and average all the images obtained during the dry season of each year. The GEE median reducer function reduces image collections by calculating the median of all values at each pixel across the stack of all matching bands to produce a single image. In addition, this process does not only reduce image data volume but equally produces high accuracies same as the time series data and improves classification accuracies (Phan et al. 2020; Carrasco et al. 2019). The advantage of using the median reducer is that it significantly reduces the volume of data thus making it easier and faster to process and analyse the bulky remotely sensed data (Phan et al. 2020). Above all, the utility of multi-temporal images avoids the possible impact of atmospheric influence on the spectral signatures while addressing the spectral variability of grasses induced by the variability of grass species and their management practices (Carrasco et al. 2019). The Random Forest Classifier was then used to discriminate and map rangelands from other LULC types. In utilising the RF, the three hyperparameters, *ntree*, *mtry* and nodesize were tuned accordingly (Odebiri et al. 2020). Specifically, the *ntree*, which is the number of regression trees used during observation performance was set to be at 500 (Mutanga et al. 2012). After iterations, the optimal ntree identified and used in this study was 100. The *mtry* which is the number of predictors tested at each node was set to a default value which was the square root of the number of variables considered in this study. Lastly, the node size which is the minimal size of the terminal nodes of the trees was set to 1 (Mutanga et al. 2012). Sampling points covering various LULC classes were collected during a field survey. However, sampled points for the previous years (2000 and 2010) were verified using Google Earth Pro (Ghorbani and Pakravan 2013; Pu et al. 2020). Feature classes that were used in the classifications were bare-land (areas that were not covered by any vegetation and were mostly bare soil), grasslands (areas that were covered by grasses only), forests (areas covered by indigenous forests or timber plantations), water (all water bodies such as rivers, and streams) and built-up area (areas where there were buildings and concrete surfaces). Furthermore, vegetation indices were computed and used to characterise the spatial variation of the rangelands. Specifically, the Normalized Difference Vegetation Index (NDVI), the Soil Adjusted Vegetation Index (SAVI), the Green Normalized Difference Vegetation Index (GNDVI), the Normalized Difference Water Index (NDWI), the Infrared Percentage Vegetation Index (IPVI), the Visible Atmospherically Resistant Index (VARI), the Ratio Vegetation Index (RVI), the Difference Vegetation Index (DVI) and the Enhanced Vegetation Index (EVI) were computed and used in this study. These indices were selected and used in this study based on their optimal performance in literature (Xue and Su 2017; Ranjan et al. 2019; da Silva et al. 2020). Vegetation indices were also considered in this study because they have been proven in literature to improve accuracies while resisting the noise from the sun, topography and clouds, atmospheric changes, and soil background signals (Xue and Su 2017).

2.4 Rate of Change

Using the classified image, the spatial extent of each land cover type was computed in a GIS environment. The Rate of change (ROC) in the spatial extent of grassland patches were calculated to understand the magnitude of change in their spatial extent in relation to other land cover types and across different years. Therefore, the rate of change in this study refers to the speed at which grassland areas increased or decreased between the years 2000, 2010, and 2020 in relation to the changes of the other classes, as the changes in each land cover class affects the changes in grassland areas.

The study also sought to assess the extent of grassland cover fragmentation as a measure of rangeland degradation. To conduct this, the classified maps were exported from GEE into ArcMap where the spatial extent (area) of each land cover class was calculated for the three periods. The grassland cover classes were then extracted for fragmentation analysis. Fragstats 4.2 was utilised to compute and characterise the extent of grasslands fragmentation. Fragmentation was calculated at the class level metrics. Specifically, landscape (PLAND), the number of patches (NP), patch density (PD), largest patch index (LPI), patch area (AREA_MN), Euclidean nearest neighbour distance (ENN_MN) and effective mesh size (MESH) were computed and used to assess the extent of fragmentation and as a proxy for rangeland degradation. To measure the degree of fragmentation within the rangelands, the NP, PD, LPI were specifically used and the AREA_MN, ENN_MN, PLAND and MESH were utilised to measure extent of grass patches spatial dispersion (Parker and Mac Nally 2002; Midha and Mathur 2010). These Fragmentation metrics were derived from each of the time periods considered in the study and were compared using the analysis of variance test after the data did not significantly deviate from the normal distribution.

The study went on to utilize rainfall data to assess the influence of rainfall on the rangeland spatial distribution changes. Rainfall data was requested from the South African Weather Services (SAWS). The requested data from SAWS was of the years used for classifications (2000, 2010 and 2020) and nine weather stations surrounding the study areas were used by the study to calculate the rainfall patterns in Inhlazuka and Vulindlela by performing Interpolation with the use of ArcMap 10.6. The type of interpolation that was applied by the study was that of the Inverse Distance Weight (IDW). After the application of interpolation, grassland points collected on the field were then overlaid over the study areas to extract the rainfall data. After the extraction of the rainfall data in ArcMap, rainfall averages of the years 2000, 2010 and 2020 in both study areas were calculated with the use of Excel.

3 Results

3.1 Classifications

The land use land cover changes were classified into five classes which are grasslands, water, forest, built-up area and bareland. In assessing, the performance of RF classification models based on spectral bands, vegetation indices independently and the combination of both vegetation indices, the ANOVA test showed that there was no significant difference in the overall accuracies. Subsequently, the classification based on combined datasets were used in this study and all results presented were based on these models. In classifying grasslands from other land cover classes, overall accuracies of 75%, 79% and 83% were obtained in Inhlazuka while accuracies of 89%, 85%, and 89% were obtained in Vulindlela for the years 2000, 2010 and 2020, respectively (Fig. 3). Meanwhile, kappa accuracy scores of 65%, 68% and 73% were exhibited by the Inhlazuka images classification while Vulindlela exhibited scores of 81%, 75% and 83% for the years 2000, 2010 and 2020, respectively (Fig. 3). Following a variable importance analysis, the most frequently optimal discrimination spectral features selected by RF across all classifications were Bands 5 and 7, the near infrared (NIR) and the short-wave infrared bands (SWIR), respectively (Fig. 4).

A general decrease in spatial extent of grasslands was observed in Vulindlela from the year 2000 to 2020 while in Inhlazuka, a decrease in the areal extent of rangelands is observed between the years 2000 and 2010 and an increase between the years 2010 to 2020 (Figs. 5 and 6). In Vulindlela, between 2000 and 2020, when grasslands were reduced, built up areas increased rapidly. Specifically, grasslands decreased at a rate of 37.52 ha per year between the 2000 and 2010. Then from 2010 to 2020, grasslands further decreased at a rate of 76 ha per year while settlement development continued to increase. Meanwhile, the increase in built-up area from the year 2000 to 2010 was at a rate of 43.15 ha per year and at a rate of 114.55 ha per year from the year 2010

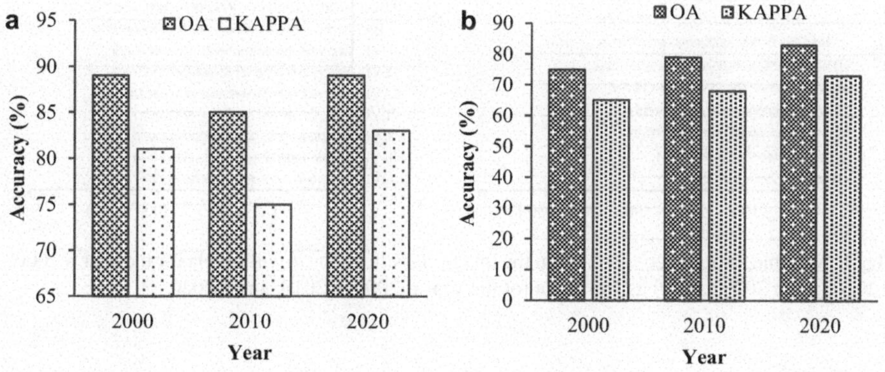

Fig. 3 Classification overall accuracies and kappa scores in the areas of **a** Vulindlela and **b** Inhlazuka

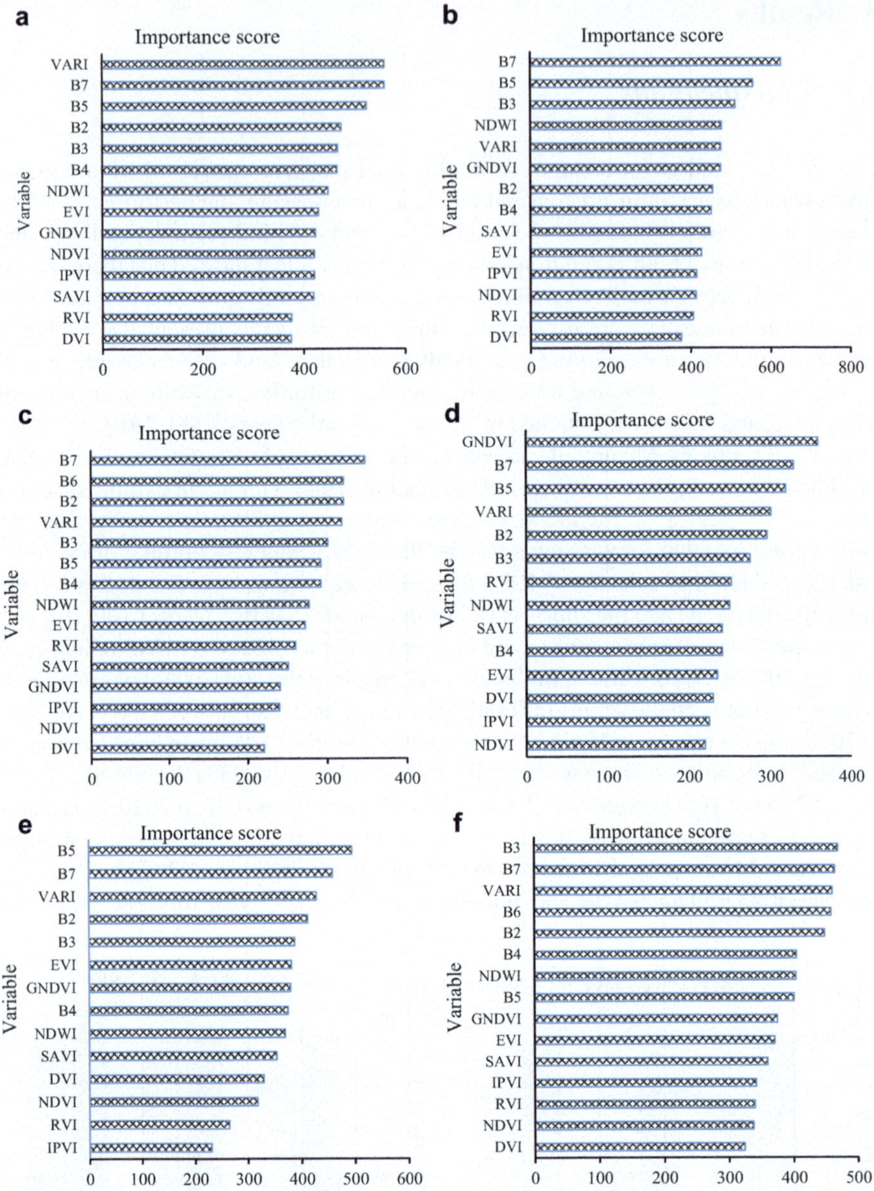

Fig. 4 Variable importance scores from the image classifications in Vulindlela for the year **a** 2000, **b** 2010, and **c** 2020 as well as Nhlazuka for the year **d** 2000, **e** 2010, and **f** 2020

to 2020. In addition, there was also a relative increase in bareland between these periods, although not at the magnitude of increase in spatial extent of built-up areas.

Inhlazuka is mostly characterised by forests, shrubby vegetation and relatively less grasslands compared to Vulindlela. The forest class increased between 2000 and 2010 by 86.54 ha per year thus causing a decline in rangelands. Between 2010 and 2020, there was an increase in rangelands due to a decrease in forests at a rate of 113.84 ha per year. The decrease in rangelands in Inhlazuka between the periods of 2000 and 2010 was at 40.19 ha per year and the increase in rangelands experienced between 2010 and 2020 was at a rate of 45.27 ha per year. In addition, in Inhlazuka there was also an increase in builtup areas at a rate of 41.7 ha/per year from 2010 to 2020. However, the increase of built-up area in Inhlazuka did not cause a significant decrease in rangelands when compared to Vulindlela.

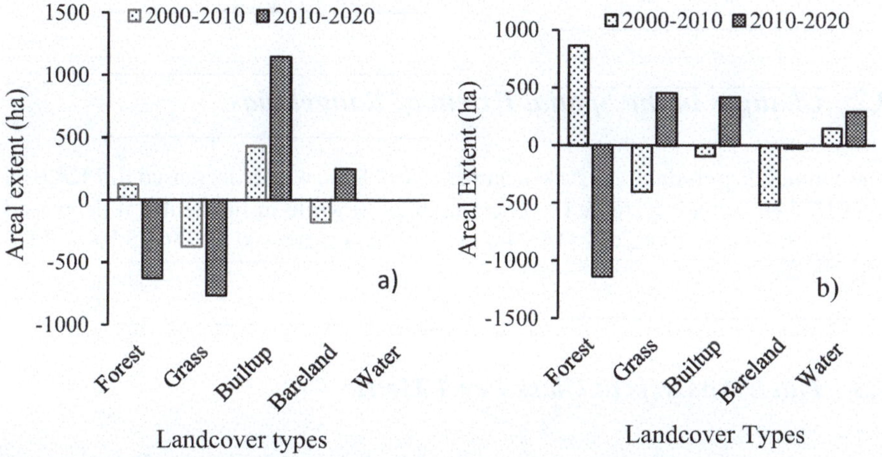

Fig. 5 Changes in the mapped land cover types in **a** Vulindlela and **b** Inhlazuka

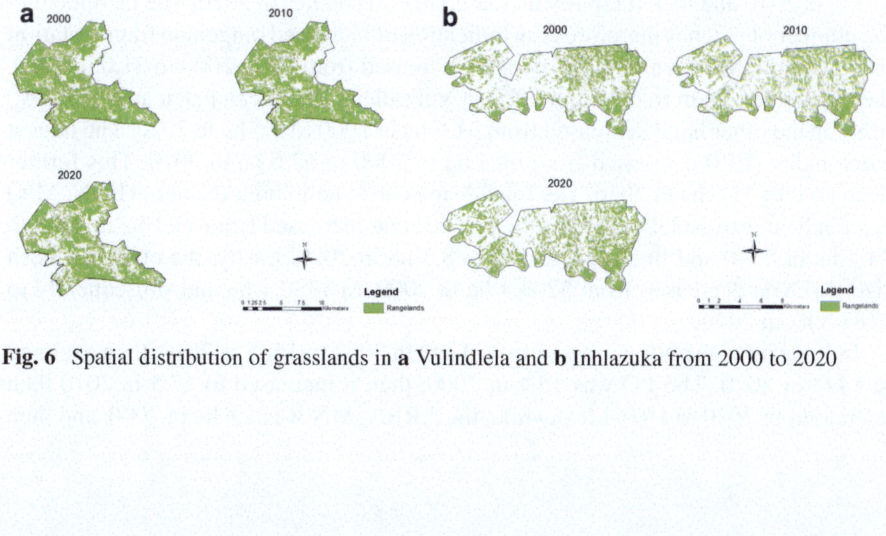

Fig. 6 Spatial distribution of grasslands in **a** Vulindlela and **b** Inhlazuka from 2000 to 2020

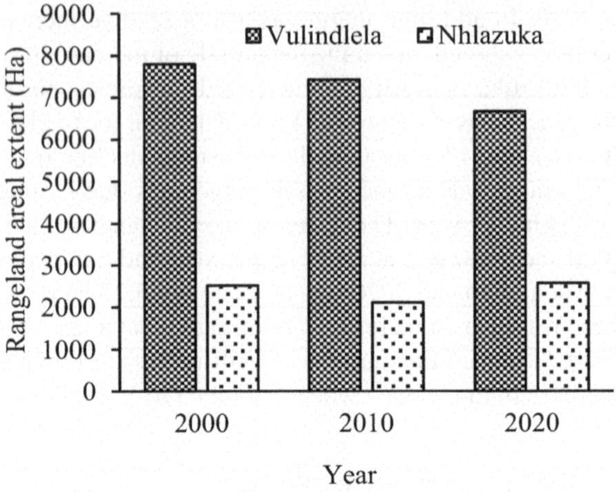

Fig. 7 Changes in the rangeland spatial extent from 2000 to 2020 in Vulindlela and Inhlazuka

3.2 Changes in the Spatial Extent of Rangelands

The Vulindlela grasslands in 2000 occupied 7799.8 ha, which decreased to 7424.6 ha in 2010 then further decreased to 6660 ha in 2020 while in Inhlazuka, it decreased from 2516.7 ha in 2000, to 2114.7 ha in 2010 then increased to 2567.5 ha in 2020 (Figs. 6 and 7).

3.3 Patch Analysis at Class Level Metric

Table 1 shows fragmentation statistics in Vulindlela and Inhlazuka. Percentage of Landscape (PLAND) decreased from 69.9, 59.6 between 2000 and 2020 indicating an increase in grassland fragmentation. The number of patches (NP) increased from 244 to 342 in 2010 and then it rapidly increased to 780 patches in 2020. The increment in the number of patches therefore is an indication of increased rangeland fragmentation in Vulindlela. Meanwhile, patch density increased from 21 in 2000 to 31.0 in 2010, then increased again to 6.9 ha in 2020 in Vulindlela. The mean patch area (AREA_MN) on the other hand decreased from 31.9 ha in 2000 to 8.5 ha in 2020. The largest patch index (LPI) decreased from 68.3 ha in 2000 to 62.6 ha in 2010. This further decreased to 43.0 ha in 2020. The Euclidean nearest neighbour distance (ENN_MN) is an indicator of isolation within the patches that increased from 71.1 ha in 2000 to 74.7 ha in 2010 and finally increased to 8.5 ha in 2020. Lastly, the effective mesh size (MESH) decreased from 5204.4 ha in 2000, to 4380.4 ha and subsequently to 2188.8 ha in 2020.

In Inhlazuka, the NP increased from 1392 in 2002 to 1581 in 2010, then declined to 1443 in 2020. The PD was 15.4 in 2000, then it increased to 17.5 in 2010 then decreased in 2020 to 16.0. Meanwhile, the AREA_MN was 1.8 ha in 2000, and then

Table 1 Class level fragmentation metrics and their descriptions

Name	Abbreviation	Description
Percentage of landscape	PLAND	It is classified as an 'Area and Edge metric,' quantifies the percentage of the landscape occupied by each class. It equals 0 when the proportional area of a class diminishes and reaches 100 when the landscape consists of a single, large patch of that class (McGarigal and Marks 1995; McGarigal 2015)
Number of patches	NP	Determines the number of subpopulations in a spatially dispersed population (number of patches can be used to determine fragmentation; the higher the number of patches, the more fragmented the class) (McGarigal and Marks 1995; McGarigal 2015)
Patch density	PD	Refers to the number of patches of corresponding patch type divided by total area (the higher the density of a particular class, the more fragmented it is) (Midha and Mathur 2010; McGarigal and Marks 1995; McGarigal 2015)
Patch area	Area MN	Is the average size of patches in a particular of a specific land cover type. It can be computed at the class level, or at the landscape level (McGarigal and Marks 1995; McGarigal 2015)
Largest patch index	LPI	Measures the percentage of the total landscape area compromised by the biggest patch (McGarigal and Marks 1995; McGarigal 2015)
Effective mesh size	MESH	Measures the extent of landscape fragmentation by calculating the likelihood that two randomly selected points within a region fall within the same contiguous, unbroken area. A larger mesh size indicates a less fragmented habitat (McGarigal and Marks 1995; McGarigal 2015)
Euclidean nearest neighbor distance	ENN_MN	The total distance (in meters) from each patch to its closest neighboring patch of the same type, measured from edge to edge, is summed up and then divided by the total number of patches of that type (Midha and Mathur 2010; (McGarigal and Marks 1995)

Table 2 Vulindlela and Inhlazuka grassland patch metrics variations between 2000 and 2020

	GRASS	PLAND	NP	PD	LPI	AREA_MN	ENN_MN	MESH
Vulindlela	2000	69.9	244	2.1	68.3	31.9	71.1	5204.4
	2010	66.6	342	3.0	62.6	21.7	74.7	4380.4
	2020	59.6	780	6.9	43.0	8.5	73.0	2188.8
Inhlazuka	2000	28.0	1392	15.4	7.1	1.8	79.5	55.5
	2010	23.5	1581	17.5	3.5	1.3	79.3	20.8
	2020	28.6	1443	16.0	5.4	1.7	75.0	35.4

it decreased to 1.3 ha in 2010, and then increased in 2020 to 1.7 ha. On the other hand, the ENN_MN was higher in 2000 and 2010 when compared to 2020, indicating that the grasses were more isolated in 2000–2010. The LPI decreased from 7.1 ha in 2000 to 3.5 ha in 2010 then increased to 5.4 ha in 2020, while the PLAND and MESH of grassland were relatively lower in 2010. The MESH size results also show that the rangelands were more fragmented in 2010 and in 2020 as compared to 2000 (Table 2).

4 Discussion

This study sought to assess the spatial variation in the areal extent of the rangelands as well as their magnitude of fragmentation using a Landsat time series in conjunction with random forest in a typical communal rangeland of southern Africa between the years 2000, 2010 and 2020. Grasslands, which provide critical ecosystem services such as forage for livestock, carbon sequestration, biodiversity support and soil stabilisation were the primary focus. Results of this study showed that the spatial variability of rangelands and other land cover types could be optimally mapped in Vulindlela and Inhlazuka, respectively based on the NIR (Bands 5) and short-wave infrared (SWIR) (Band 7) spectral variables in conjunction with RF in GEE platform. Specifically, overall accuracies (OA) of 75%, 79% and 83% in Inhlazuka as well as OAs 89%, 85%, and 89% in Vulindlela for the years, 2000, 2010, and 2020, respectively, were obtained in this study. The optimal performance of the NIR and short wave infrared (SWIR) bands could be attributed to their sensitivity to grass biochemical and physical properties such as foliar nitrogen, lignin and starch (Ramoelo et al. 2015; Zhao et al. 2022). These properties directly influence grassland productivity, a crucial component of forage quality and provision services. The NIR section of the electromagnetic spectrum is highly sensitive to the biochemical and biophysical properties of vegetation (Matongera et al. 2017; Mngadi et al. 2021). Specifically, the foliar spongy mesophyll facilitates the high reflectance of the NIR energy by plants with a high vigour (Wang et al. 2008; Kumar, Arya and Jain 2022).

The study results also showed that in Vulindlela, rangelands were rapidly being fragmented and converted into built-up and bare areas and a significant change in the spatial configuration of the grasslands was observed. This directly comprises ecosystem services such as carbon storage, soil retention, and habitat provision. This could be explained by population growth and the demand for land to develop residential areas as well as increasing agricultural production as the major drivers of rangeland fragmentation in rural areas (Abdulahi et al. 2016). According to the Municipality (2016) the population of Vulindlela in 2013 was 161 562 with an estimate of 85 033 structures. The population was estimated to grow at 2% each year to about 183 583, in 2020. The population increase could explain the rapid decrease in grasslands, which are being converted into built-up areas (Fig. 6) to accommodate the burgeoning population. There is growing body of literature which shows that activities such as housing development and overgrazing are the major anthropogenic activities that have extensively degraded rangelands globally (Lu et al. 2017; Dowdy et al. 2019). Also, the lack of effective livestock management practices, generally results in increased grazing activities with limited breaks given to the grassland ecosystem to self-regulate and recover from over grazing.

Meanwhile, in Inhlazuka, the spatial extent of rangelands decreased while the areal extent of built-up areas increased between 2000 and 2010. The decrease in the spatial extent of forest between 2010 and 2020 could be attributed to the increase in population and demand for residential areas. The increase in rangelands in Inhlazuka in 2020 could be explained by clearance of shrubby and tree like alien invasive species which is a typical exercise in Inhlazuka. Specifically, *Lantana camara* is the most dominant alien invasive plant in the rangelands of Inhlazuka. Lantana is said to be the most widespread alien invasive plant which invades agricultural and natural ecosystems (Parveen, Ravinder and Daizy 2011). For instance, Dube et al. (2022) makes an example of the Bushbuckridge area where the grasslands have been replaced by the alien plant lantana thus decreasing available forage for livestock and wildlife (Terefe 2015). To circumvent this, uMgungundlovu district municipality implements continuous alien invasive plant removal programmes regulating the variation in the spatial extent of grasslands especially in Inhlazuka. Such management practices restore the provisioning and support services of grasslands by improving forage availability and enhancing biodiversity. This then explains in the decreases and increases associated with the rangelands in Inhlazuka (Figs. 6 and 7) between 2000, 2010 and 2020. If the alien plant species are not controlled, the communal rangelands of Inhlazuka may be at a risk of being replaced by the Lantana alien plant.

Results also showed that barelands in Vulindlela were increasing. This could be explained by the limited rangeland management interventions which are conducive for unregulated administration of fires. Fire as a management tool in rangelands, is used to improve forage quality and production, to manipulate plant populations (plant composition and structure) and maintain habitats for grazing animals (Little et al. 2015). However, if fire is frequently and improperly administered, it becomes destructive. Specifically, it reduces the seedling recruitment of grasses, while burning and depleting the seedbanks on the topsoil (Ghasempour et al. 2022). This exposes the ground as bareland, reducing its capacity to provide regulating services such as water retention and erosion control. Subsequently, the ground is exposed as bareland. Community members in Vulindlela customarily burn grass during the winter season as a tool to promote its growth as feed for livestock (Cho et al. 2021). However, this results in grass failing to recruit seedlings and replenish its populations as the seed banks will be depleted and the topsoil crusted by regular burns (Fidelis et al. 2010; Ghasempour et al. 2022). Literature also illustrates that in the forest and grasslands biomes in Southern Africa, the dominancy of grasses or trees (forests) is a function of fire and precipitation (Baudena et al. 2015; Accatino et al. 2010).

Rainfall plays a significant role in the growth of vegetation (Beier et al. 2012; Lipiec et al. 2013). The lack of rainfall in rangelands results in the decrease of forage and rangeland resources therefore, negatively impacting biomass productivity (Derner and Augustine 2016; Sibanda et al. 2021). This comprises the provisioning and supporting services of grasslands, including their role in maintaining biodiversity and regulating local climates. For example, in Vulindlela, the spatial extent of the rangelands in the year 2000 was 7799.8 ha and the average rainfall received was 61.79 mm. The rainfall received in the year 2000 may have played a role around rangelands in the year 2000. From the year 2000 to the year 2010, the rangeland area decreased to 7424.6 ha and the average rainfall received in 2010 was 61.74 mm, between the year 2000 and 2010 there is not much difference in the average rainfall received, however, there was a decrease of 375.2 ha in the rangelands. Then in 2020, the average rainfall received in Vulindlela was 67.09 mm which is an increase in the rainfall received in 2010. The increase in rainfall in 2020 did not result in the increase of rangelands, instead further decrease was experienced from 7424.6 ha in 2010 to 6660 ha in 2020. Based on the rainfall patterns in Vulindlela, rainfall has limited influence on the declining rate of the communal rangelands. Therefore, the increase of bareland and builtup areas in Vulindlela are the main drivers of the depletion of the rangelands in combination with other land use activities taking place in Vulindlela. In the area of Inhlazuka, the average rainfall of the year 2000 was 65.6 mm, and the area of rangelands was 2516.7 ha. In 2010, rainfall received decreased to 52.34 mm and the spatial extent of rangelands in 2010 also decreased from 2516.7 ha in 2000 to 2114.7 ha in 2010. Rainfall received in the year 2000 in Inhlazuka decreased in 2010. The decrease in rainfall in 2010 may have played a role in the decrease of the spatial extent of rangelands in 2010 as the decrease in rainfall resulted in a decrease in the area of rangelands. In 2020, the average rainfall of 2020 was 69.63 mm, this is an increase from the rainfall received in 2010. The increase in rainfall resulted in an increase in the areal extent of rangelands in 2020 from 2114.7 ha in 2010 to 2567.5 ha in 2020. The interaction between rainfall patterns and anthropogenic interventions, such as alien plant removal, illustrates the complex drivers of grassland ecosystem services in communal rangelands.

5 The Implications on Grassland Ecosystem Services

The concept of ecosystem services provides a framework for understanding the relationship between biodiversity, ecosystem functioning, and human well-being (Zhao et al. 2020). Grassland ecosystem services encompass the various benefits that grassland biodiversity offers to support human survival (Zhao et al. 2020; Masenyama et al. 2022). Recent research has shown that human activities such as reclamation, deforestation, and overgrazing have led to the global degradation and desertification of grasslands, severely threatening the regional ecosystem services and functions they provide (Wang et al. 2017b). The growing global population has exacerbated the decline of grassland ecosystem services, with domestic overgrazing being a primary driver, particularly in developing countries (Gang et al. 2014). As a result, the monitoring of land-use and land-cover (LULC) changes is crucial for understanding vegetation dynamics and promoting the sustainable use of natural resources (Corner et al. 2014; Olorunfemi et al. 2020). This is essential for informing policy development aimed at improving sustainable rangeland management and rehabilitation practices (Onyango et al. 2022). Such efforts are key to enhancing the protection of natural resources, building rangeland resilience to climate change, and supporting the livelihoods of pastoral communities (Onyango et al. 2022; Reed et al. 2015).

The value of grassland ecosystem services remains under-recognized, as these services are often classified as public goods with no market value and receive limited attention in private land management decisions (Liu et al. 2022). To ensure the sustainability of grasslands and human welfare, it is crucial to consider the full potential of grassland ecosystem services in real-world management practices (Liu et al. 2022). Furthermore, for the sustainable management of grasslands, both global and regional policies must acknowledge the role of grasslands as providers of diverse ecosystem services and hotspots for biodiversity (Bardgett et al. 2021; Prangel et al. 2023). These policies should also establish clear targets for grassland protection and sustainable management (Bardgett et al. 2021). Grassland management approaches may include the explicit inclusion of natural and semi-natural grasslands in conservation laws and policies, while ensuring that such policies do not inadvertently harm grasslands, such as through the promotion of fire suppression or afforestation (Tamburini et al. 2022; Li et al. 2024). Additionally, new policies are needed to support and fund the appropriate management and restoration of grasslands, alongside penalties for activities that degrade these ecosystems, such as overgrazing (Li et al. 2024; Tamburini et al. 2022).

Furthermore, to promote the sustainable management of grasslands, it is crucial to understand the impacts of both climate change and human activities on grassland degradation and ecosystem services, as current studies predominantly focus on the effects of climate change (Schirpke et al. 2017; Wang et al. 2022). It is also important to recognize that grasslands provide a wide range of ecosystem services, which means that the restoration efforts should not only focus on a single aspect of restoration (Lyu et al. 2022; Wen et al. 2013; Wang et al. 2022). Instead, the complexity of grassland ecosystems must be acknowledged, with a focus on comprehensive restoration and the management of trade-offs between multiple ecosystem services (Lyons et al. 2023). Additionally, the limited availability of policies, funding, and human resources poses a challenge for large-scale ecological restoration within a short time frame (Lyu et al. 2022). As there is a need to consider governance issues, identify priority areas for grassland restoration, and develop region-specific strategies for implementation (Lyu et al. 2022).

6 Conclusion

The aim of this study was to characterise rangelands and quantify their variability in spatial extent and fragmentation over the period of 20 years in communal areas. Grasslands which provide vital ecosystem services such as carbon sequestration, soil erosion control, water regulation, and forage for livestock were the primary focus of this study. Based on the findings of this study, it can be concluded that LULCCs can be optimally characterised using Landsat's bands in combination with the vegetation indices. In addition, the NIR and the SWIR bands were the variables of importance thus, mapping the rangeland changes at high accuracies. Settlement expansion and an increase in crop fields are the main drivers of rangeland decline. This decline in grasslands threatens essential ecosystem services such as food provision, biodiversity conservation, and climate regulation. Fragmentation of rangelands is increasing with time as more grassland patches are getting more isolated and relatively smaller in spatial extent with the increase in built-up area. As fragmentation increases, it undermines the ability of grasslands to provide critical ecosystem services like carbon storage, water filtration, and forage for livestock. The findings of this study are a step towards building robust spatially explicit quantitative techniques for monitoring the spatio-temporal variations of grasslands. These techniques are crucial for understanding how land-use changes affect grassland ecosystem services and for informing sustainable land management strategies. This is a very important step required if a geospatial Framework for Monitoring grassland ecosystems that will provide actionable information services for grassland assessment and monitoring across different key land management areas is to be realised.

Acknowledgements The authors would like to acknowledge Dr. Omosalewa Odebiri, Dr. Mthembeni Mgadi, Anita Masenyama, Mohamed Vawda, Adeola Arougande and Duduzile Diza for their assistence in this this project.

Funding This research was funded by The Water Research Commission of South Africa under the WRC Project No. CON2020/2021–00490, "Geospatial modelling of rangelands productivity in wa-ter-limited environments of South Africa" and funded in part by the National Research Foundation of South Africa (NRF) SARChI Chair in Land Use Planning and Management—UKZNunder grant number 84157.

References

Abd El-Kawy O, Rød J, Ismail H, Suliman A (2011) Land use and land cover change detection in the western Nile delta of Egypt using remote sensing data. Appl Geogr 31:483–494

Abdulahi MM, Hashim H, Teha M (2016) Rangeland degradation: extent, impacts, and alternative restoration techniques in the rangelands of Ethiopia. Trop Subtrop Agroecosystems 19

Accatino F, De Michele C, Vezzoli R, Donzelli D, Scholes RJ (2010) Tree–grass co-existence in savanna: interactions of rain and fire. J Theor Biol 267:235–242

Al-Bukhari A, Hallett S, Brewer T (2018) A review of potential methods for monitoring rangeland degradation in Libya. Pastoralism 8:1–14

Bardgett RD, Bullock JM, Lavorel S, Manning P, Schaffner U, Ostle N, Chomel M, Durigan G, Fry LE, Johnson D, Lavallee JM (2021) Combatting global grassland degradation. Nat Rev Earth Environ 2(10):720–735

Baudena M, Dekker SC, van Bodegom PM, Cuesta B, Higgins SI, Lehsten V, Reick CH, Rietkerk M, Scheiter S, Yin Z (2015) Forests, savannas, and grasslands: bridging the knowledge gap between ecology and dynamic global vegetation models. Biogeosciences 12:1833–1848

Bedunah DJ, Angerer JP (2012) Rangeland degradation, poverty, and conflict: how can rangeland scientists contribute to effective responses and solutions. Rangeland Ecol Manage 65:606–612

Beier C, Beierkuhnlein C, Wohlgemuth T, Penuelas J, Emmett B, Körner C, de Boeck H, Christensen JH, Leuzinger S, Janssens IA (2012) Precipitation manipulation experiments—challenges and recommendations for the future. Ecol Lett 15:899–911

Boone RB, Conant RT, Sircely J, Thornton PK, Herrero M (2018) Climate change impacts on selected global rangeland ecosystem services. Glob Change Biol 24:1382–1393

Carrasco L, O'Neil AW, Morton RD, Rowland CS (2019) Evaluating combinations of temporally aggregated Sentinel-1, Sentinel-2 and Landsat 8 for land cover mapping with Google Earth Engine. Remote Sens 11:288

Ceballos G, Davidson A, List R, Pacheco J, Manzano-Fischer P, Santos-Barrera G, Cruzado J (2010) Rapid decline of a grassland system and its ecological and conservation implications. PLoS ONE 5:e8562

Cengiz A, Budak M, Yağmur N, Balçik F (2023) Comparison between random forest and support vector machine algorithms for LULC classification. Int J Eng Geosci 8:1–10

Cho MA, Onisimo M, Mabhaudhi T (2021) Using participatory GIS and collaborative management approaches to enhance local actors' participation in rangeland management: the case of Vulindlela, South Africa. J Environ Plan Manage 1–20

Cleland EE, Collins SL, Dickson TL, Farrer EC, Gross KL, Gherardi LA, Hallett LM, Hobbs RJ, Hsu JS, Turnbull L (2013) Sensitivity of grassland plant community composition to spatial vs. temporal variation in precipitation. Ecology 94:1687–1696

Corner RJ, Dewan AM, Chakma S (2014) Monitoring and prediction of land-use and land-cover (LULC) change. In: Dhaka megacity: geospatial perspectives on urbanisation, environment and health, pp 75–97

da Silva VS, Salami G, da Silva MIO, Silva EA, Monteiro Junior JJ, Alba E (2020) Methodological evaluation of vegetation indexes in land use and land cover (LULC) classification. Geol Ecol Landscapes 4:159–169

Derner JD, Augustine DJ (2016) Adaptive management for drought on rangelands. Rangelands 38:211–215

Dowdy AJ, Ye H, Pepler A, Thatcher M, Osbrough SL, Evans JP, Di Virgilio G, McCarthy N (2019) Future changes in extreme weather and pyroconvection risk factors for Australian wildfires. Sci Rep 9:10073

Dube T, Maluleke XG, Mutanga O (2022) Mapping rangeland ecosystems vulnerability to Lantana camara invasion in semi-arid savannahs in South Africa. Afr J Ecol

Fidelis AT, Delgado Cartay MD, Blanco CC, Muller SC, Pillar VP, Pfadenhauer JS (2010) Fire intensity and severity in Brazilian Campos grasslands. Interciencia: revista de ciencia y tecnologia de america. Caracas. 35(10):739–745

Gang C, Zhou W, Chen Y, Wang Z, Sun Z, Li J, Qi J, Odeh (2014) Quantitative assessment of the contributions of climate change and human activities on global grassland degradation. Environ Earth Sci 72:4273–4282

Ghasempour M, Erfanzadeh R, Török P (2022) Fire effects on soil seed banks under different woody plant species in Mazandaran province, Iran. Ecol Eng 183:106762

Ghayour L, Neshat A, Paryani S, Shahabi H, Shirzadi A, Chen W, Al-Ansari N, Geertsema M, Pourmehdi Amiri M, Gholamnia M (2021) Performance evaluation of sentinel-2 and landsat 8 OLI data for land cover/use classification using a comparison between machine learning algorithms. Remote Sens 13:1349

Ghorbani A, Pakravan M (2013) Land use mapping using visual vs. digital image interpretation of TM and Google earth derived imagery in Shrivan-Darasi watershed (Northwest of Iran). Europ J Experim Biol 3:576–582

Gorelick N, Hancher M, Dixon M, Ilyushchenko S, Thau D, Moore R (2017) Google Earth Engine: planetary-scale geospatial analysis for everyone. Remote Sens Environ 202:18–27

Hughes C, De Winnaar G, Schulze R, Mander M, Jewitt G (2018) Mapping of water-related ecosystem services in the uMngeni catchment using a daily time-step hydrological model for prioritisation of ecological infrastructure investment—part 2: outputs. Water SA 44:590–600

Jin Y, Yang X, Qiu J, Li J, Gao T, Wu Q, Zhao F, Ma H, Yu H, Xu B (2014) Remote sensing-based biomass estimation and its spatio-temporal variations in temperate grassland, Northern China. Remote Sens 6:1496–1513

Kafy A-A, Shuvo RM, Naim MNH, Sikdar MS, Chowdhury RR, Islam MA, Sarker MHS, Khan MHH, Kona MA (2021) Remote sensing approach to simulate the land use/land cover and seasonal land surface temperature change using machine learning algorithms in a fastest-growing megacity of Bangladesh. Remote Sens Appl Soc Environ 21:100463

Kumar L, Mutanga O (2018) Google Earth Engine applications since inception: usage, trends, and potential. Remote Sens 10:1509

Kumar S, Arya S, Jain K (2022) A SWIR-based vegetation index for change detection in land cover using multi-temporal Landsat satellite dataset. Int J Inf Technol 14:2035–2048

Langley SK, Cheshire HM, Humes KS (2001) A comparison of single date and multitemporal satellite image classifications in a semi-arid grassland. J Arid Environ 49:401–411

Li W, Dong R, Fu H, Wang J, Yu L, Gong P (2020) Integrating Google Earth imagery with Landsat data to improve 30-m resolution land cover mapping. Remote Sens Environ 237:111563

Li Y, Xiong K, Song S, Zhang W (2024) Realizing the value of grassland ecosystem services: global practice and its inspiration for the karst desertification control area. Front Sustain Food Syst 7:1240431

Lipiec J, Doussan C, Nosalewicz A, Kondracka K (2013) Effect of drought and heat stresses on plant growth and yield: a review. Int Agrophys 27:463–477

Little IT, Hockey PA, Jansen R (2015) Impacts of fire and grazing management on South Africa's moist highland grasslands: a case study of the Steenkampsberg Plateau, Mpumalanga, South Africa. Bothalia-Afri Biodivers Conserv 45:1–15

Liu H, Hou L, Kang N, Nan Z, Huang J (2022) (2022) The economic value of grassland ecosystem services: a global meta-analysis. Grassland Res 1(1):63–74

Loveland TR, Dwyer JL (2012) Landsat: building a strong future. Remote Sens Environ 122:22–29

Lu X, Kelsey KC, Yan Y, Sun J, Wang X, Cheng G, Neff JC (2017) Effects of grazing on ecosystem structure and function of alpine grasslands in Qinghai-Tibetan Plateau: a synthesis. Ecosphere 8:e01656

Lyons KG, Török P, Hermann JM, Kiehl K, Kirmer A, Kollmann J, Overbeck GE, Tischew S, Allen EB, Bakker JD, Brigham C (2023) Challenges and opportunities for grassland restoration: a global perspective of best practices in the era of climate change. Global Ecol Conserv 46:e02612

Lyu X, Li X, Dang D, Dou H, Wang K, Gong J, Wang H, Liu S (2022) A perspective on the impact of grassland degradation on ecosystem services for the purpose of sustainable management. Remote Sens 14(20):5120

Masenyama A, Mutanga O, Dube T, Bangira T, Sibanda M, Mabhaudhi T (2022) A systematic review on the use of remote sensing technologies in quantifying grasslands ecosystem services. Gisci Remote Sens 59:1000–1025

Matongera TN, Mutanga O, Dube T, Sibanda M (2017) Detection and mapping the spatial distribution of bracken fern weeds using the Landsat 8 OLI new generation sensor. Int J Appl Earth Obs Geoinf 57:93–103

Matongera TN, Mutanga O, Sibanda M, Odindi J (2021) Estimating and monitoring land surface phenology in rangelands: a review of progress and challenges. Remote Sens 13:2060

McGarigal K (2015) FRAGSTATS help. University of Massachusetts, Amherst, MA, USA, p 182

McGarigal K, Marks BJ (1995) Spatial pattern analysis program for quantifying landscape structure. Department of Agriculture, Forest Service, Pacific Northwest Research Station, 10, pp.1–122

McGranahan DA, Kirkman KP (2013) Multifunctional rangeland in Southern Africa: managing for production, conservation, and resilience with fire and grazing. Land 2:176–193

Midha N, Mathur P (2010) Assessment of forest fragmentation in the conservation priority Dudhwa landscape, India using FRAGSTATS computed class level metrics. J Indian Soc Remote Sens 38:487–500

Mncube NL (2022) Investigating the contribution of social cash transfers to the food security situation of agricultural-based rural households of Nhlazuka, Richmond Municipality, South Africa.

Mngadi M, Odindi J, Peerbhay K, Mutanga O (2021) Examining the effectiveness of Sentinel-1 and 2 imagery for commercial forest species mapping. Geocarto Int 36:1–12

Municipality, M. Vulindlela Local Area Plan: Spatial Framework. 2016. Pietermaritzburg, KwaZulu-Natal

Mutanga O, Adam E, Cho MA (2012) High density biomass estimation for wetland vegetation using WorldView-2 imagery and random forest regression algorithm. Int J Appl Earth Obs Geoinf 18:399–406

Odebiri O, Mutanga O, Odindi J, Peerbhay K, Dovey S (2020) Predicting soil organic carbon stocks under commercial forest plantations in KwaZulu-Natal province, South Africa using remotely sensed data. Gisci Remote Sens 57:450–463

Olorunfemi IE, Fasinmirin JT, Olufayo AA, Komolafe AA (2020) GIS and remote sensing-based analysis of the impacts of land use/land cover change (LULCC) on the environmental sustainability of Ekiti State, southwestern Nigeria. Environ Dev Sustain 22:661–692

Onyango V, Masumbuko B, Somda J, Nianogo A, Davies J (2022) Sustainable land management in rangeland and grasslands. Food & Agriculture Organization

Palmer AR, Bennett JE (2013) Degradation of communal rangelands in South Africa: towards an improved understanding to inform policy. Afr J Range Forage Sci 30:57–63

Parker M, Mac Nally R (2002) Habitat loss and the habitat fragmentation threshold: an experimental evaluation of impacts on richness and total abundances using grassland invertebrates. Biol Conserv 105(2):217–229

Parveen KD, Ravinder KK, Daizy RB (2011) Impact of Lantana camara L. invasion on riparian vegetation of Nayar region in Garhwal Himalayas (Uttarakhand, India). J Ecol Nat Environ 3:11–22

Phan TN, Kuch V, Lehnert LW (2020) Land cover classification using google earth engine and random forest classifier—the role of image composition. Remote Sens 12:2411

Prangel E, Reitalu T, Neuenkamp L, Kasari-Toussaint L, Karise R, Tiitsaar A, Soon V, Kupper T, Meriste M, Ingerpuu N, Helm A (2023) Restoration of semi-natural grasslands boosts biodiversity and re-creates hotspots for ecosystem services. Agr Ecosyst Environ 374:109139

Price KP, Egbert SL, Nellis MD, Lee R-Y, Boyce R (1997) Mapping land cover in a high plains agro-ecosystem using a multidate Landsat thematic mapper modeling approach. Trans Kans Acad Sci 1903:21–33

Pu D, Sun J, Ding Q, Zheng Q, Li T, Niu X (2020) Mapping urban areas using dense time series of landsat images and google earth engine. Int Arch Photogram Remote Sens Spatial Inform Sci 42:403–409

Ramoelo A, Cho M, Mathieu R, Skidmore AK (2015) Potential of Sentinel-2 spectral configuration to assess rangeland quality. J Appl Remote Sens 9:094096–094096

Ranjan R, Chandel AK, Khot LR, Bahlol HY, Zhou J, Boydston RA, Miklas PN (2019) Irrigated pinto bean crop stress and yield assessment using ground based low altitude remote sensing technology. Inform Process Agric 6:502–514

Reed MS, Stringer LC, Dougill AJ, Perkins JS, Atlhopheng JR, Mulale K, Favretto N (2015) Reorienting land degradation towards sustainable land management: linking sustainable livelihoods with ecosystem services in rangeland systems. J Environ Manage 151:472–485

Rodriguez-Galiano VF, Ghimire B, Rogan J, Chica-Olmo M, Rigol-Sanchez JP (2012) An assessment of the effectiveness of a random forest classifier for land-cover classification. ISPRS J Photogramm Remote Sens 67:93–104

Royimani L, Mutanga O, Odindi J, Sibanda M, Chamane S (2022) Determining the onset of autumn grass senescence in subtropical sour-veld grasslands using remote sensing proxies and the breakpoint approach. Eco Inform 69:101651

Sala OE, Yahdjian L, Havstad K, Aguiar MR (2017) Rangeland ecosystem services: nature's supply and humans' demand. Rangeland Syst Processes Manage Challenges 467–489

Schirpke UKM, Leitinger G, Fontana V, Tasser E, Tappeiner U (2017) Future impacts of changing land-use and climate on ecosystem services of mountain grassland and their resilience. Ecosyst Serv 26:79–94

Sheykhmousa M, Mahdianpari M, Ghanbari H, Mohammadimanesh F, Ghamisi P, Homayouni S (2020) Support vector machine versus random forest for remote sensing image classification: A meta-analysis and systematic review. IEEE J Selected Top Appl Earth Observ Remote Sens 13:6308–6325

Sibanda M, Mutanga O, Dube T, Vundla TS, Mafongoya PL (2019) Estimating LAI and mapping canopy storage capacity for hydrological applications in wattle infested ecosystems using Sentinel-2 MSI derived red edge bands. Gisci Remote Sens 56:68–86

Sibanda M, Mutanga O, Rouget M, Kumar L (2017) Estimating biomass of native grass grown under complex management treatments using worldview-3 spectral derivatives. Remote Sens 9:55

Sibanda M, Onisimo M, Dube T, Mabhaudhi T (2021) Quantitative assessment of grassland foliar moisture parameters as an inference on rangeland condition in the mesic rangelands of southern Africa. Int J Remote Sens 42:1474–1491

Sidhu N, Pebesma E, Câmara G (2018) Using Google Earth Engine to detect land cover change: Singapore as a use case. Europ J Remote Sens 51:486–500

Tamburini G, Aguilera G, Öckinger E (2022) Grasslands enhance ecosystem service multifunctionality above and below-ground in agricultural landscapes. J Appl Ecol 59(12):3061–3071

Tamiminia H, Salehi B, Mahdianpari M, Quackenbush L, Adeli S, Brisco B (2020) Google Earth Engine for geo-big data applications: a meta-analysis and systematic review. ISPRS J Photogramm Remote Sens 164:152–170

Terefe D (2015) Impacts of invasive plants (*Lantana camara* and *Parthenium hysterophorus*) on livestock and rangeland production in pastoral and agro pastoral areas of Somali Region, Ethiopia. In: 32nd annual research and extension workshop, Haramaya University, Alemaya, pp 24–32

Thanh Noi P, Kappas M (2017) Comparison of random forest, k-nearest neighbor, and support vector machine classifiers for land cover classification using Sentinel-2 imagery. Sensors 18:18

Wahap N, Shafri HZ (2020) Utilization of Google Earth Engine (GEE) for land cover monitoring over Klang Valley, Malaysia. IOP Conf Ser Earth Environ Sci 012003

Wang J, Xiao X, Qin Y, Dong J, Geissler G, Zhang G, Cejda N, Alikhani B, Doughty RB (2017) Mapping the dynamics of eastern redcedar encroachment into grasslands during 1984–2010 through PALSAR and time series Landsat images. Remote Sens Environ 190:233–246

Wang L, Qu JJ, Hao X, Zhu Q (2008) Sensitivity studies of the moisture effects on MODIS SWIR reflectance and vegetation water indices. Int J Remote Sens 29:7065–7075

Wang Z, Ma Y, Zhang Y, Shang J (2022) Review of remote sensing applications in grassland monitoring. Remote Sens 14(12):2903

Wang Z, Deng X, Song W, Li Z, Chen J (2017) What is the main cause of grassland degradation? A case study of grassland ecosystem service in the middle-south Inner Mongolia

Wen L, Dong S, Li Y, Li X, Shi J, Wang Y, Liu D, Ma Y (2013) Effect of degradation intensity on grassland ecosystem services in the alpine region of Qinghai-Tibetan Plateau, China. PLoS ONE 8(3):e58432

Xie Z, Phinn SR, Game ET, Pannell DJ, Hobbs RJ, Briggs PR, McDonald-Madden E (2019) Using Landsat observations (1988–2017) and Google Earth Engine to detect vegetation cover changes in rangelands—a first step towards identifying degraded lands for conservation. Remote Sens Environ 232:111317

Xue J, Su B (2017) Significant remote sensing vegetation indices: a review of developments and applications. J Sens

Zhao Y, Liu Z, Wu J (2020) Grassland ecosystem services: a systematic review of research advances and future directions. Landscape Ecol 35:793–814

Zhao Y, Zhu W, Wei P, Fang P, Zhang X, Yan N, Liu W, Zhao H, Wu Q (2022) Classification of Zambian grasslands using random forest feature importance selection during the optimal phenological period. Ecol Ind 135:108529

Spatiotemporal Urban Growth Assessment in a Coastal City from 1992 to 2042 Using MLC and MLP-MCA Algorithms

Shobhit Chaturvedi, Prince Patel, Riya Makwana, Jayket Sachapara, and Naimish Bhatt

Abstract Urbanization presents profound challenges to coastal region infrastructure, environmental health, and ecological balance. Analyzing Land Cover (LC) changes is crucial for urban planning efforts and forecasting future growth trends. This study presents an integrated geospatial framework to analyze the spatiotemporal dynamics of urban growth in Panaji, Goa from 1992 to 2042, utilizing the Maximum Likelihood Classification (MLC) algorithm to develop LC Maps for 1992, 2003, 2013, and 2022 across four main classes—Built-Up, Vegetation, Water body, and Barelands. The Kappa values for MLC mapping showed consistently high accuracy: 0.889 (1991), 0.886 (2001), 0.890 (2011), and 0.900 (2021). During 1992–2022, Panaji experienced substantial land cover changes: built-up areas expanded from 12.97 to 46.60 km^2 (a 259.24% increase), vegetation decreased to 79.11 km^2 (a 22.7% decrease), barelands decreased to 25.38 km^2 (a 15.63% decrease), and water bodies decreased to 76.90 km^2 (a 6.89% decrease). Additionally, leveraging past LC maps from 2002 and 2012 and urban growth drivers (such as slope, elevation, and proximity), a Multi-Layer Perceptron–Markov Chain (MLP-MC) model was employed to predict future urban growth. The model incorporated drivers such as proximity to built-up areas, bare land, water bodies, vegetation, ground elevation, and slope to predict the 2022 LC map, achieving over 96.42% validation accuracy. Further, MLP-MCA projections for 2022–2042 indicate substantial growth, with built-up areas expected to extend to 72.97 km^2, marking a 56.53% increase from 2022. Conversely, vegetation is forecasted to decrease by 20.71%, water bodies by 8.07%, and barelands to diminish by 12.32% during the same period. These findings underscore the critical need for computational techniques in sustainable urban planning strategies and that balance urban development with environmental conservation in coastal areas.

S. Chaturvedi (✉) · P. Patel · R. Makwana · J. Sachapara · N. Bhatt
Department of Civil Engineering, School of Technology, Pandit Deendayal Energy University, Gandhinagar, India
e-mail: shobhitchaturvedi101@gmail.com; shobhit.chaturvedi@sot.pdpu.ac.in

O. Mutanga et al. (eds.), *Revealing Ecosystem Services Through Geospatial Technologies*, Springer Remote Sensing/Photogrammetry,
https://doi.org/10.1007/978-3-031-98048-0_15

Keywords Coastal-Cities · RS-GIS · Maximum likelihood classification · Multi-layer perceptron—Markov chain model · Panaji

1 Introduction

Land Use and Land Cover (LULC) dynamics are critical for understanding the evolving surface characteristics of coastal cities (Kaliraj et al. 2017; Zhu et al. 2022). Land Cover (LC) includes forests, wetlands, beaches, and urban structures, while Land Use (LU) encompasses human activities like tourism, urban development, and conservation. LULC data plays a pivotal role in urban planning, coastal conservation, disaster management and climate change mitigation in coastal cities (AlDousari et al. 2022; Rasool et al. 2021). LULC changes in coastal cities arise from a combination of natural, anthropogenic, and socio-economic factors. Natural factors, such as climate variability, coastal erosion, and sea-level rise, along with extreme events like floods and cyclones, can cause abrupt landscape changes. Anthropogenic influences such as urbanization, tourism infrastructure expansion, and intensive agriculture contribute to habitat loss and ecosystem transformations. Socio-economic factors like population growth, economic activities, and policy decisions also play significant roles in shaping land use patterns and impacting coastal environments (Mohamed and Worku 2020; Vimala 2020).

Urbanization in coastal cities presents both opportunities and challenges. Economic growth and tourism development attract businesses, creating employment opportunities, but they also strain infrastructure and contribute to issues such as pollution, water scarcity, housing affordability challenges, and inadequate social services (Chettry 2022; Nath et al. 2020). This rapid development often adversely affects crucial ecosystem services provided by coastal forests and wetlands, including storm protection, carbon sequestration, and biodiversity conservation (Adnan et al. 2020). Urbanization frequently leads to habitat loss and degradation, exacerbating environmental concerns such as the Urban Heat Island (UHI) effect and rising Land Surface Temperatures (LST). Additionally, coastal erosion and saltwater intrusion further threaten ecosystems and human health, compounding the multifaceted challenges faced by urbanized coastal cities (Das and Das 2020; Shukla and Jain 2021). A comprehensive understanding of LULC dynamics is crucial for mitigating these issues, including habitat preservation, coastal protection, and sustainable urban expansion strategies.

Advanced Remote Sensing (RS) technologies and Geographic Information Systems (GIS) can play crucial roles in effectively monitoring LULC and environmental changes in coastal areas. Satellite imagery from platforms like LANDSAT, Sentinel-2, and ASTER can be processed to create baseline maps and track temporal land cover variations (Paradis 2022; Wang et al. 2022). GIS offers robust analytical tools for processing spatial data and understanding relationships among LULC, LST and environmental parameters (Chettry 2022; Sannigrahi et al. 2021). Integration of Machine Learning and Artificial Intelligence enhances the accuracy of

LULC mapping and enables predictive modeling of future urban growth, supporting informed decision-making for sustainable urban development (Islam et al. 2023; Rajakumari et al. 2020).

Panaji, Goa, situated along the Arabian Sea on the west coast of India, has experienced significant population growth and urbanization, leading to socio-economic and environmental challenges. This study adopts an integrated RS-GIS framework to enhance understanding of past and current LULC dynamics and to model future urban growth. The study aims to achieve three main objectives:

RO 1. Generate LULC maps for Panaji spanning from 1992 to 2022 utilizing the Maximum Likelihood Classification (MLC).

RO 2. Examine the temporal evolution of LULC patterns over this timeframe.

RO 3. Forecast urban expansion patterns from 2022 to 2042 employing Multi-Layer Perceptron-Markov Chain Analysis (MLP-MCA).

The findings will offer vital insights for urban planning and policy formulation in Panaji, enabling proactive measures to handle evolving land use patterns, address urban heat island effects, and promote sustainable development along the coastal city. The paper is structured as follows: Sect. 2 presents a concise literature review on methods for analyzing historical and future land use dynamics. Section 3 outlines the research methodology. Section 4 provides the results, while Sect. 5 offers a discussion of the findings. Section 6 presents the conclusion, discussing the study's limitations and suggesting directions for future research.

2 Literature Review

Coastal conservation has become an active area of research, with numerous studies analyzing the negative impacts of rapid urbanization on coastal city infrastructure, environment, and ecology (Islam et al. 2023). For instance, Misra and Balaji (2015) studied LULC changes in Surat, Navsari, and Valsad districts, Gujarat, India, from 1990 to 2014 using LANDSAT TM, ETM, and OLI imagery. They employed band ratios and hybrid classification (supervised and unsupervised) with 90–95% accuracy. Significant increases were observed in aquaculture and urban built-up areas, while mudflats decreased due to urbanization. Shoreline erosion was prominent. Kaliraj et al. (2017) examined LULC changes on the South West coast of Kanyakumari from 2000 to 2011 using LANDSAT ETM+ and TM images with the MLC algorithm. They identified significant shifts from beachface land cover, plantations, and barren lands to built-ups, reflecting urbanization. Abdullah et al. (2019) utilized eXtreme Gradient Boosting (XGBoost) for feature selection and Random Forest (RF) classification to analyze 28 years of coastal LULC changes in Bangladesh with LANDSAT imagery, addressing landscape heterogeneity and spectral complexities. Ngondo et al. (2021) used LANDSAT 5 TM (1990) and LANDSAT 8 OLI (2018) images for the Wami–Ruvu Basin, Tanzania. An Artificial Neural Network (ANN) classifier with training data generated LULC maps for 1990 and 2018 through supervised classification. Sathya Bama et al. (2020) evaluated LULC changes and shoreline dynamics along Tamil Nadu's Vedaranyam swamp from 1978 to 2017. Shoreline analysis revealed varying accretion and erosion rates, impacting vulnerability assessments. Islam et al. (2020) studied coastal erosion and LULC changes in Barguna and Patuakhali districts, Bangladesh, from 2000 to 2017, using LANDSAT imagery, field data, and forest carbon inventory, applying K-means clustering and MLC algorithms in ERDAS Imagine 2014. Ziaul Hoque et al. (2022) analyzed land use and LULC changes and ecosystem service values (ESVs) in coastal Bangladesh (1999–2019) using LANDSAT data and economic valuation. They identified increases in built-up areas, mangroves, and aquaculture, with reductions in agriculture and bare land.

Researchers also employed sophisticated machine learning algorithms to predict prospective urban growth and environmental challenges. Qiang and Lam (2015) studied the Lower Mississippi River Basin using ANN and Cellular Automaton models to analyze LULC changes from 1996 to 2006. Incorporating 15 variables, they simulated future scenarios, projecting a 263 km^2 wetland loss by 2016. Adnan et al. (2020) investigated poverty dynamics in Bangladesh's southwestern embanked coastal area concerning LULC changes and flooding. Utilizing logistic regression, cellular automata, and Markov Chain models, they predicted future LULC and flood risks. Zhu et al. (2022) analyzed LULC changes and their impact on carbon storage in China's coastal areas using CA-Markov and InVEST models. Projections for 2050 suggest that ecological conservation measures can mitigate further carbon loss. Rajakumari et al. (2020) used RS-GIS to study 4.5 decades of LULC changes in Deshapran, West Bengal. They employed CA-MARKOV modeling to predict future trends, highlighting expanding aquaculture posing threats to crop lands and coastal dynamics. These studies emphasized the necessity for managing coastal LULC changes effectively, supported by robust computational techniques to address current and future sustainability challenges.

3 Research Methodology

Study Area

Panaji, Goa, situated along the Arabian Sea coast (15.4909° N, 73.8278° E), boasts a attractive landscape characterized by coastal plains and hills. The city spans approximately 36 square kilometers, nestled at an average elevation of 7 m above sea level. It experiences a tropical monsoon climate, with hot and humid conditions from March to May, peaking at around 35 °C, and mild winters from December to February, with temperatures ranging from 20 to 32 °C. The southwest monsoon, from June to September, brings heavy rainfall, averaging about 2500 to 3000 mm annually. Panaji's coastal location influences its climate, contributing to its seasonal weather patterns and environmental dynamics. Panaji has experienced rapid urban growth driven by economic opportunities, improved transportation networks, and the development of commercial and residential areas like Miramar, Dona Paula, Caranzalem, Panaji's historic downtown, Taleigao, and Ribandar. Its status as a tourism hotspot, both domestically and internationally, has amplified this growth, drawing visitors to its beaches, cultural sites, and local festivities.

However, this urbanization has led to significant challenges in infrastructure, including traffic congestion and inadequate waste management, as well as environmental concerns such as water resource strain and habitat degradation. Balancing development with environmental sustainability is a critical priority for Panaji as it seeks to preserve its coastal ecosystem while managing urban expansion and tourism pressures. Figure 1 illustrates the administrative map of India, highlighting Panaji, Goa, along with its surrounding study area covering 229 km^2, chosen for detailed analysis of ongoing and future developments. Figure 2 depicts the structured framework established for the spatiotemporal assessment and future prediction of urban growth in Panaji, Goa.

Data Acquisition and Analysis

Several mid-resolution (30 × 30 m) LANDSAT datasets from the United States Geological Survey (https://earthexplorer.usgs.gov/) were initially downloaded for the study period, detailed in Table 1. Specifically, LANDSAT 4–5 (TM) for 1992, LANDSAT 7 (ETM +) for 2002, LANDSAT 8 (TIRS and OLI) for 2012, and LANDSAT 9 (TIRS and OLI) for 2022 were acquired, all corresponding to the (10–20) May period. Additionally, Goa's digital elevation model data for 2022 was obtained. All images were geo-referenced to the Universal Transverse Mercator (UTM) projection system and clipped to a common area of interest using the same shapefile in Quantum GIS software. Subsequently, image sharpening, smoothing,

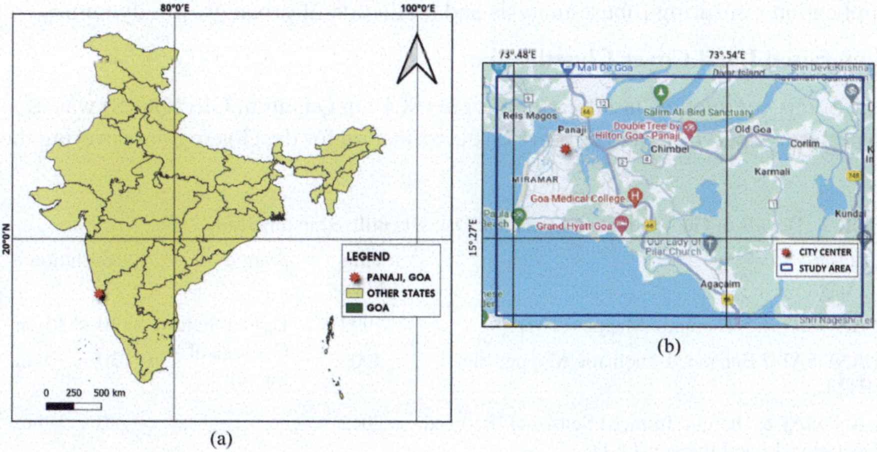

Fig. 1 **a** Administrative map of India showing Panaji, Goa; **b** Study area around Panaji city for this research

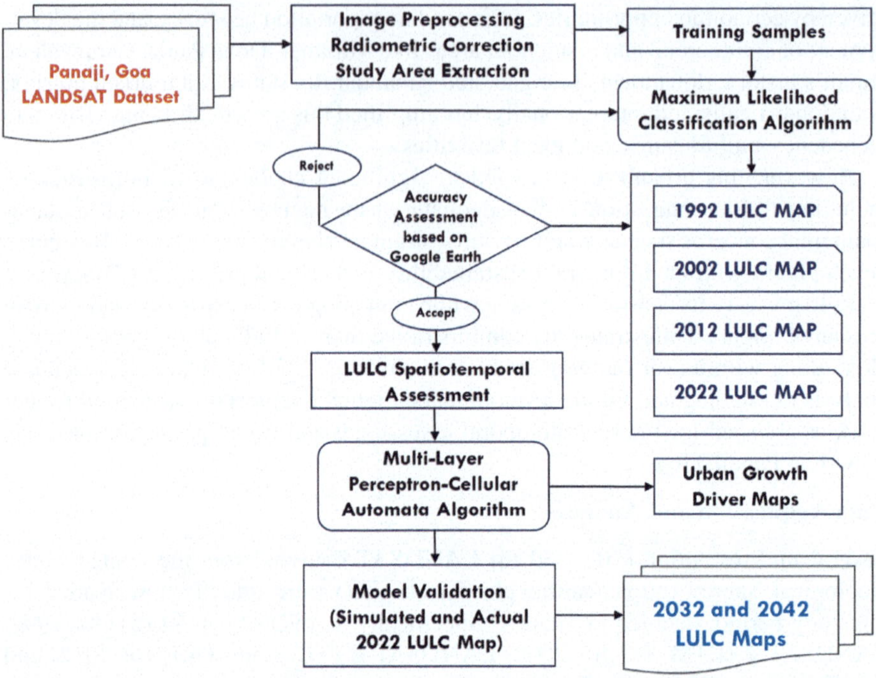

Fig. 2 Research methodology for spatiotemporal assessment and future prediction of urban growth in Panaji, Goa

and atmospheric corrections were applied. These datasets were selected based on their proven accuracy in land cover classification and environmental monitoring applications, ensuring robust analysis and prediction of urban growth dynamics.

Supervised Land Cover Classification

The Semi-Automatic Classification Plugin (SCP) in Quantum GIS (QGIS) was used to perform supervised land cover (LC) classification for the Goa region, covering the

Table 1 Details of Geospatial and Statistical Datasets utilized in this study

Dataset	Time Stamp	Source	Resolution
LANDSAT 4-5 Thematic Mapper (TM)	1992	United States Geological Survey	(30 × 30) m
LANDSAT 7 Enhanced Thematic Mapper Plus (ETM+)	2002		(30 × 30) m
LANDSAT 8 Thermal Infrared Sensor (TIRS) and Operational Land Imager (OLI)	2012		(30 × 30) m
LANDSAT 9 (TIRS and OLI)	2022		(30 × 30) m
SRTM digital elevation model	2022	NASA	(30 × 30) m

years 1992, 2002, 2012, and 2022. The Maximum Likelihood Classification (MLC) algorithm was employed, which assigns pixels to specific LC classes based on user-defined training data and spectral attributes. The classification process included three key steps: data preparation, parameter estimation, and classification.

Data Preparation Training samples were selected to capture representative spectral signatures for each LC class. These signatures were then used to train the classification model.

Parameter Estimation Key statistical parameters, including the mean vector and covariance matrix for each LC class, were calculated as follows:

$$\text{Mean vector}(\mu\_k) := \frac{1}{n_k}\sum_{i=1}^{n_k} X_i \tag{1}$$

$$\text{Covariance matrix}(\Sigma\_\text{k}) := \frac{1}{n_k}\sum_{i=1}^{n_k}(X_i - \mu_k).(X_i - \mu_k)^T \tag{2}$$

Here, n_k is the number of training samples, and X_i represents the spectral feature vector for the ith sample.

Classification Each pixel's likelihood of belonging to a specific LC class was determined using the following equation:

$$\mathrm{L}_{\mathrm{k(x)}} = \frac{1}{\left((2\pi)^{\frac{d}{2}}.(|\Sigma_\mathrm{k}|)^{\frac{1}{2}}\right).e^{\left(-0.5(X_i-\mu_k)^T.\Sigma_\mathrm{k}^{-1}.(X_i-\mu_k)\right)}} \tag{3}$$

In this equation, d is the spectral feature vector's dimensionality, $|\Sigma_\mathrm{k}|$ is the determinant of the covariance matrix, and Σ_k^{-1} is its inverse.

The classification resulted in four primary LC classes:

1. **Built-up**: Includes infrastructure such as residential, institutional and commercial buildings.
2. **Water Bodies**: Covers rivers, lakes, and wetlands.
3. **Vegetation**: Encompasses green areas such as forests, gardens, and agricultural land.
4. **Barelands:** Represents open areas, barren plots, uncultivated land, and landfills.

Accuracy Assessment The LC maps for 1992, 2002, 2012, and 2022 were evaluated for accuracy using kappa statistics in Eq. (4):

$$KS = (Po - Pe)/(1 - Pe) \tag{4}$$

where *Po* is the observed agreement, and *Pe* is the expected agreement by chance. High kappa values indicated a strong classification agreement. Ground control polygons (80 per class, totaling 300) were generated using Google Earth to validate the classification. A kappa value of at least 0.80 was required for the classification results to be deemed acceptable.

Change Detection Analysis LC change detection maps were created for 1992–2002, 2002–2012, and 2012–2022. These maps revealed transitions from vegetation, bare lands, and water bodies to built-up areas, quantifying the extent of urban expansion and its encroachment on natural and open spaces during each time period.

Land Cover Prediction

The Multi-Layer Perceptron with Markov Chain Analysis (MLP-MCA) model was applied to analyze long-term spatiotemporal land use changes and forecast future urban growth in Panaji, Goa. This integrated framework combines the predictive capabilities of the MLP neural network with the stochastic transition probability estimation of Markov Chain Analysis, enabling detailed modeling of complex land cover dynamics. The MLP component establishes relationships between driving factors and land cover changes, while the Markov Chain component estimates transition probabilities to simulate future states. Together, these elements provide a robust methodology for forecasting land cover changes in response to environmental and anthropogenic factors. The implementation of MLP-MCA involved the following steps:

1. **Data Collection and Preparation**: Historical land cover maps L_t and relevant explanatory variables X_t, such as slope, elevation, road proximity, built-up areas, and socioeconomic indicators, were collected and prepared for analysis.
2. **Training the MLP Model**: The MLP neural network was trained using the collected data to establish relationships between X_t and L_t. The association strengths between driving factors and land cover changes were assessed using the Cramer V statistic. Factors with Cramer V values exceeding 0.15 were retained for further analysis.
3. **Scenario Generation and Markov Chain Analysis**: Future scenarios were generated as per Eq. (5) by inputting projected values of explanatory variables X_{t+1} into the model. Markov Chain Analysis estimated transition probabilities P_{ij} of land cover changes based on current L_t and predicted L_{t+1}.

$$P_{ij} = P\left(L_{t+1} = j | L_t = i\right) \tag{5}$$

4. **Predicting Future Land Cover**: The model integrated MLP predictions with Markov Chain transition probabilities as per Eq. (6) to generate future land cover maps Lt + 1L_{t + 1}Lt + 1.

$$L_{t+1} = MLP(X_{t+1}).\ P_{ij} \tag{6}$$

5. **Final Model Validation and Forecasting**: The final step involved validating the model's overall performance by comparing predicted 2022 land cover maps against actual data, achieving a high predictive accuracy (>90%). Once validated,

the model confidently forecasted land cover for 2032 and 2042, incorporating insights from regression analysis on population growth and urban expansion correlations.

4 Results

Land Cover Mapping

Figure 3 depicts the land cover maps for the Goa region in 1991, 2001, 2011, and 2021, while Table 3 details the absolute coverage amounts (in sq km.) for each land cover class. The assessment includes Built-up, Vegetation, Water body, and Bareland classes, evaluated using the Classification Accuracy tool in the QGIS SCP toolbox to compute Kappa Statistics. The Kappa coefficients indicate the accuracy and reliability of classifying water bodies, built-up areas, vegetation, and barelands over time Table 2 presents classwise and overall Kappa values for Panaji, Goa's land cover maps spanning 1992 to 2022, demonstrating high accuracy in classifying water bodies, built-up areas, vegetation, and barelands over time. Water bodies consistently show high accuracy, with Kappa values ranging from 0.922 in 1991 to 0.941 in 2022. Built-up areas maintain stable accuracy, varying between 0.923 and 0.921. Vegetation exhibits slight variability with values from 0.861 to 0.848, while barelands improve from 0.851 to 0.889. These results consistently surpassed the 0.85 threshold for satisfactory classification, affirming the reliability of the land cover mapping process for Panaji, Goa, over the thirty-year period.

LC Change Assessment

In the Panaji, Goa region, significant land cover transitions occurred from 1992 to 2022. The most notable changes from 1992 to 2002 were the conversion of vegetated areas to built-up zones covering 8.71 km^2, bare lands transitioning to built-up areas with 2.24 km^2, and water bodies transforming into built-up areas totaling 0.83 km^2. Moving to 2002 to 2012, the trend continued with vegetated areas increasing their conversion to built-up zones to 10.66 km^2, followed by bare lands transitioning to built-up areas with 4.60 km^2, and water bodies converting to built-up areas covering 0.62 km^2. From 2014 to 2022, vegetated areas saw a significant increase in conversion to built-up zones with 15.77 km^2, followed by bare lands continuing their transition to built-up areas totaling 5.86 km^2, and water bodies further transforming into built-up areas with 0.59 km^2. These transitions highlight the evolving urbanization patterns observed over the thirty-year period in Panaji, Goa.

Predicting Future Urban Expansion

The Multi-Layer Perceptron and Markov Chain Analysis (MLP-MCA) model in the Terrset software package was employed to predict Panaji, Goa region's urban growth. Alongside past LC Maps (2002, 2012), the MLP-MCA model integrated Urban Growth Drivers (UGDs) depicted in Fig. 6, including ground elevation, slope, road proximity, water body proximity, built-up proximity, and vegetation proximity,

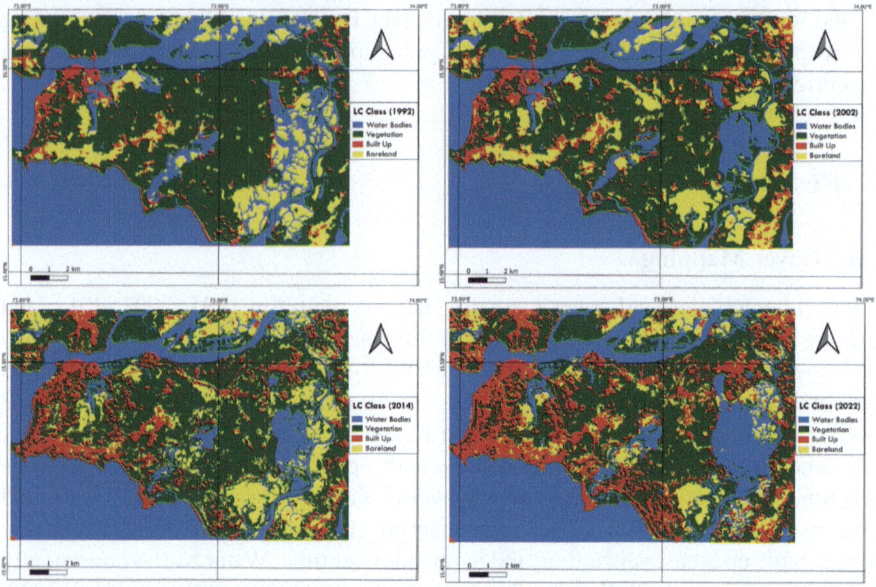

Fig. 3 Land cover maps of the Panaji, Goa region for 1992, 2002, 2012, and 2022

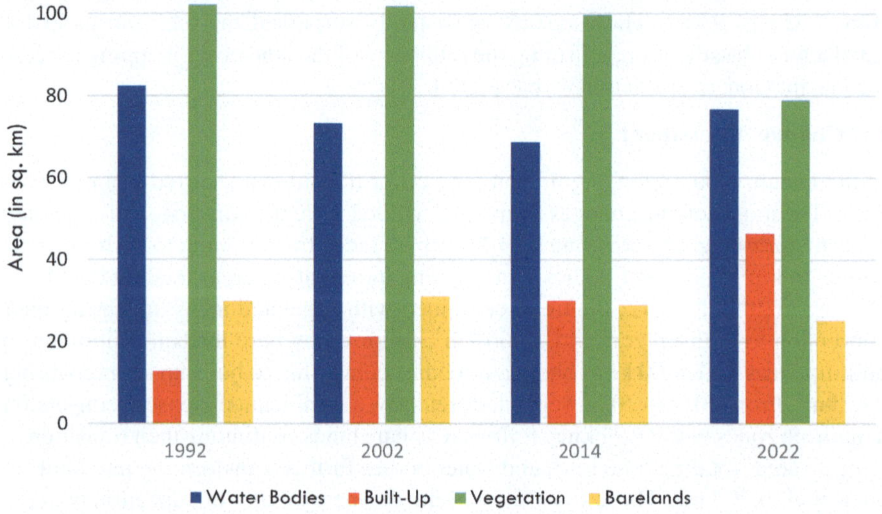

Fig. 4 Built-up, vegetation, waterbody and bareland coverages in Panaji, Goa from 1991–2021

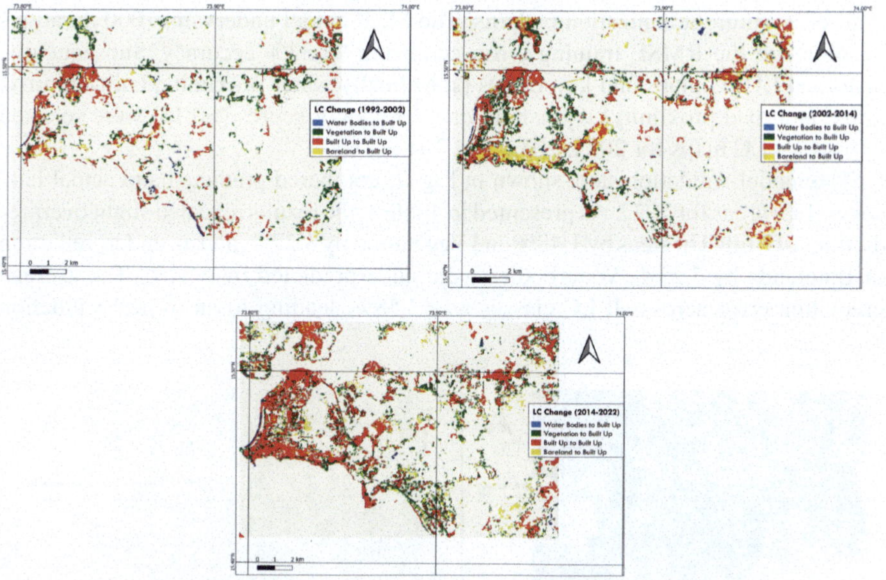

Fig. 5 Transition map showing barelands, water bodies, and vegetation converted to built-up areas in Panaji, Goa (1992–2022)

Table 2 Claswise and overall kappa values for LC maps from 1992–2022

LC class	Kappa Coefficient			
	1992	2002	2012	2022
Vegetation	0.861	0.855	0.857	0.848
Water body	0.922	0.916	0.933	0.941
Built-up	0.923	0.905	0.916	0.921
Bareland	0.851	0.867	0.852	0.889
Overall	0.889	0.886	0.890	0.900

Table 3 Land Cover coverages (in km^2) in Panaji, Goa during 1992–2022

Land cover class	Area (in km^2)			
	Year			
	1992	2002	2014	2022
Water bodies	82.61	73.38	68.81	76.9
Built-up	12.97	21.43	30.17	46.60
Vegetation	102.33	101.89	99.85	79.11
Barelands	30.08	31.290	29.16	25.38
Total	228	228	228	228

to forecast future urbanization patterns. The MLP model underwent 1000 iterations to minimize the RMSE training error, achieving 94.33% accuracy. Subsequently, historical LULC maps and key UGDs (specifically water proximity, built-up proximity, and road proximity) with significant Cramer V scores (>0.15) were selected to forecast LC maps for 2022, 2032, and 2042.

The model validation step, shown in Fig. 7, compared predicted and actual land cover (LC) areas for 2022. As presented in Table 4, the results revealed slight overpredictions of Built-Up areas by 1.48% and Vegetation by 3.27%, and an underprediction of Barelands by 3.23%. Waterbodies were underpredicted by 5.32%. The average prediction error across all LC classes was 3.58%, leading to an overall validation

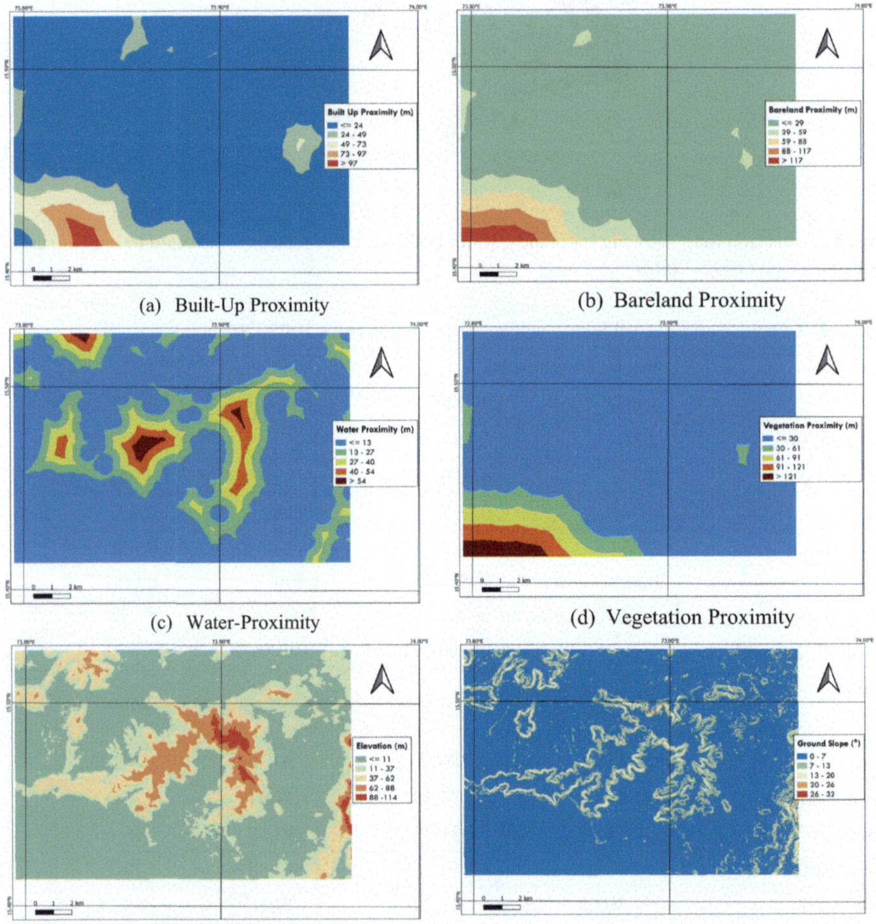

(a) Built-Up Proximity (b) Bareland Proximity (c) Water-Proximity (d) Vegetation Proximity (e) Ground Elevation (f) Ground Slope

Fig. 6 Urban growth drivers considered in predicting urban growth in Goa

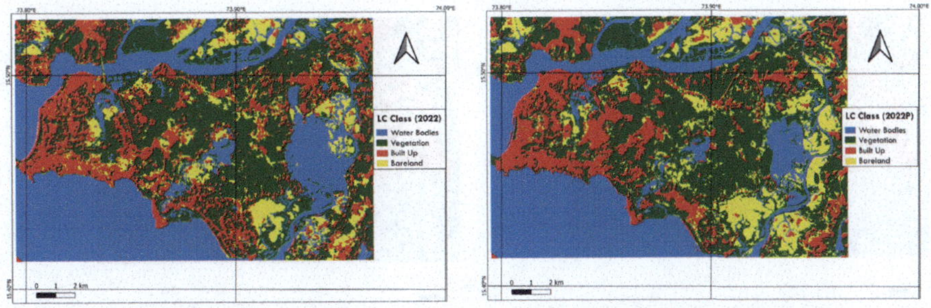

(a) Actual Land Cover Map (b) Predicted Land Cover Map

Fig. 7 The actual and predicted land cover map for Panaji, Goa in 2022

Table 4 MLP-MCA model validation—Comparison of predicted and actual LC areas (in km^2) for 2022

LC class	Actual	Predicted	Error (%)
Waterbodies	76.89	72.8	− 5.32
Vegetation	79.11	81.7	3.27
Built-up area	46.6	47.29	1.48
Bare land	25.38	26.2	3.23

accuracy of 96.42%. This demonstrates the model's robustness and its capability to forecast LC maps for 2032 and 2042 effectively.

After achieving satisfactory validation accuracy, the MLP-MCA model was used to predict LC maps for 2032 and 2042, as shown in Fig. 8. Table 5 presents the predicted LC coverages for these periods. The predictions suggest significant land cover changes in Panaji, Goa, by 2032 and 2042. Built-Up areas are projected to expand substantially, increasing by approximately 29.5% from 2022 to 2032 and an additional 20.92% from 2032 to 2042. In contrast, Vegetation is anticipated to decline by about 10.6% from 2022 to 2032 and a further 11.3%from 2032 to 2042. Water Bodies are predicted to decrease by roughly 5.3% from 2022 to 2032 and an additional 2.9% from 2032 to 2042, while Barelands are expected to see marginal reductions of 4.8% from 2022 to 2032 and a further 7.9% from 2032 to 2042. These projections highlight the dynamic urban growth and environmental dynamics shaping Panaji's landscape over the coming decades.

5 Discussion

The rapid urban growth projected for Panaji, Goa, driven by tourism and construction activities, presents significant challenges to urban infrastructure, environmental sustainability, and ecological integrity. The LULC maps generated using the MLC algorithm from 1992 to 2022 show a clear trend of urban sprawl, with built-up

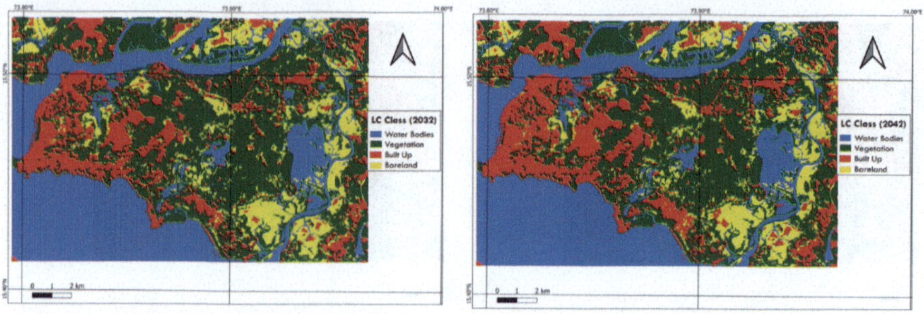

Fig. 8 The predicted land cover maps for Panaji, Goa region during 2032 and 2042

Table 5 Predicted Land Cover coverages (in km^2) in Panaji, Goa during 2032–2022

LC class	2032	2042
Water bodies	72.80	70.70
Built-up	60.33	72.97
Vegetation	70.71	62.71
Barelands	24.15	22.25

areas expanding significantly at the expense of green spaces, water bodies, and vegetation. This expansion raises concerns over UHI effects, reduced air quality, and increased energy consumption as more natural landscapes are converted into impervious surfaces. Additionally, the expansion has reduced groundwater recharge capacity and altered local hydrological patterns, which are likely to increase flood risks. Ecologically, Panaji faces significant biodiversity loss and habitat fragmentation, as urban sprawl encroaches on natural habitats and disrupts crucial ecological corridors necessary for species migration. The reduction in green spaces and natural ecosystems not only diminishes carbon sequestration capabilities but also heightens vulnerabilities to climate change, impacting local flora and fauna and undermining overall ecological resilience. Projections based on MLP-MCA forecast continued urban expansion in Panaji from 2022 to 2042, further straining transportation networks, water supply systems, and waste management capacities. This growth is expected to increase the demand for energy and services, which could compromise the city's livability and resilience, while continuing to intensify environmental degradation and reduce ecological sustainability.

Recent studies across India's coastal regions, summarized in Table 6, highlight critical issues such as green space reduction, rising surface temperatures, and ecological degradation, offering valuable insights for sustainable urbanization in Panaji. Sundar and Deka demonstrated that RF models outperformed other machine learning methods in LULC classification for the Vembanad Lake System in Kerala, stressing the need for accurate tools to mitigate ecosystem impacts. Dhanaraj and Angadi identified significant urban expansion along transport corridors in Mangaluru, advocating for integrating green spaces and conserving agricultural land in urban planning.

Similarly, Abijith and Saravanan reported substantial land use transitions in northern Tamil Nadu, where built-up areas replaced water bodies, emphasizing the urgency of climate-resilient urban policies. S. et al. analyzed changes in a coastal town in Western India, noting a correlation between reduced vegetation and increased land surface temperatures (LST), underscoring the need for eco-friendly development. In the Sundarbans Biosphere Reserve, Thakur et al. documented mangrove loss and a rise in non-vegetated areas, recommending mangrove restoration and stringent land-use regulations to preserve biodiversity and control temperature rise.

To address the challenges of urbanization and its environmental impacts in Panaji, Goa, an integrated approach to coastal zone management is crucial. Sustainable urban planning should prioritize eco-friendly tourism, restrict construction in ecologically sensitive areas, and preserve natural habitats like wetlands and mangroves, which are crucial for flood mitigation and carbon sequestration. Reforestation and mangrove restoration are vital for enhancing coastal resilience, while protecting green spaces within urban areas will improve quality of life and mitigate heat island effects. Additionally, adopting flood-resistant infrastructure and resilient building codes will reduce the impact of extreme weather events. The use of advanced geospatial tools,

Table 6 Key insights from recent studies focussed on on LULC changes and urban growth in coastal regions across India

Authors	Location	Methodology	Key findings
Sundar and Deka	Vembanad Lake System, Kerala	Machine learning, Random Forest (RF), CART, SVM on Google earth engine	RF models exhibited superior performance in classifying barren and built-up areas, emphasizing the need for precise LULC classification to mitigate ecosystem impacts
Dhanaraj and Angadi	Mangaluru, India	Remote sensing, GIS, Landsat imagery, change detection, Shannon entropy, spatial trend analysis	Urban expansion concentrated along transport corridors, leading to a substantial increase in built-up areas. Suggests integrating green spaces into urban planning
Abijith and Saravanan	Northern Tamil Nadu Coast, India	Google earth engine, TerrSet, GIS, RF, CA–Markov model	Conversion of water bodies and vegetation to built-up areas is significant. Recommends sustainable coastal zone planning and climate-resilient urban policies
S. et al.	Coastal town in Western India	Landsat imagery analysis, NDBI, NDVI, land surface temperature	Reduction in vegetation and barren land correlates with increasing built-up areas and LST. Advocates eco-friendly developments to manage environmental stress
Thakur et al.	Sundarbans Biosphere Reserve, India	Remote sensing, GIS, Landsat TM, ETM+, OLI images, NDVI, LST analysis	Mangrove loss and expansion of non-vegetated areas elevated LST. Proposes mangrove restoration and stringent land use regulations to mitigate biodiversity loss

such as remote sensing and GIS, will facilitate effective land-use monitoring and adaptive management strategies by providing real-time data on urban growth and environmental changes. Engaging local communities in conservation efforts is also key, as their involvement ensures long-term success. By combining these strategies—eco-friendly tourism, habitat restoration, flood-resistant infrastructure, and community engagement—Panaji can achieve sustainable growth, balancing urban development with ecological preservation.

6 Conclusion

This study presented an integrated geospatial framework to analyze the spatiotemporal evolution of urban growth in Panaji, Goa from 1992 to 2042. The framework applied the Maximum Likelihood Classification (MLC) algorithm to generate Land Cover (LC) maps for 1992, 2002, 2012, and 2022, categorizing the landscape into Built-Up, Vegetation, Water Bodies, and Bareland classes. Future urban growth projections for 2032 and 2042 were conducted using a Multi-Layer Perceptron-Markov Chain Analysis (MLP-MCA) model. The overall Kappa values across these periods, measuring the accuracy and reliability of the MLC mapping process, consistently demonstrated high performance: 0.889 (1992), 0.886 (2002), 0.890 (2012), and 0.900 (2022).

Future urban growth projections were developed using a Multi-Layer Perceptron-Markov Chain Analysis (MLP-MCA) model. By incorporating past LC maps (2002 and 2012) and urban growth drivers such as proximity to built-up areas, bare land, water bodies, vegetation, elevation, and slope, the MLP-MC model achieved over 95% validation accuracy, reflecting its effectiveness in predicting spatial dynamics. Projections for 2032 and 2042 indicate significant urban expansion, with built-up areas growing to 60.33 km^2 by 2032 (29.5% increase from 2022) and 72.97 km^2 by 2042 (17.63% increase from 2032). Vegetation is forecasted to decline by 10.6% from 2022 to 2032 and by an additional 11.3% by 2042, while water bodies and barelands are expected to decrease by 5.3% and 2.9%, and 4.8% and 7.9%, respectively, during these periods. These results highlight the pressing need for sustainable urban planning to mitigate environmental and ecological impacts, ensuring balanced development in Panaji.

The rapid urban growth projected for Panaji, Goa, presented substantial challenges related to urban infrastructure, environmental integrity, and ecological sustainability, particularly due to the city's vulnerability as a coastal region. The urban expansion placed immense pressure on existing infrastructure, with rising concerns over coastal erosion, flooding, and the degradation of natural habitats. To address these challenges, effective urban planning strategies were required, focusing on balancing development with conservation efforts. Key recommendations included promoting eco-friendly tourism, restricting construction in environmentally sensitive coastal areas, and enhancing green infrastructure to mitigate flood risks and increase climate resilience. Additionally, it was essential to adopt pollution control measures, restore critical habitats, and develop climate-resilient infrastructure to better withstand the impacts of climate change. Given the unique vulnerabilities of the coastal environment, strengthening coastal zone management and preventing habitat loss became crucial. Advanced geospatial tools such as remote sensing and GIS can support these strategies by providing real-time data on urban growth and environmental changes. Finally, engaging local communities and reinforcing environmental policies are crucial to ensuring that Panaji's urbanization remains sustainable, resilient, and preserves the region's ecological balance amid future stresses.

This study faced certain limitations due to the use of mid-resolution (30 m × 30 m) LANDSAT data for land use and land cover mapping. Future research could improve accuracy by incorporating higher-resolution satellite data from platforms like Sentinel-2 (10 m × 10 m) and WorldView, enabling more detailed land cover mapping. The MLP-MCA model utilized historical urban growth trends and drivers to forecast future urban expansion. Incorporating socio-economic variables, such as population growth and economic factors, would enhance the model's predictive accuracy. Additionally, integrating dynamic land-use simulation models, climate change scenarios, and advanced machine learning algorithms could provide a more comprehensive understanding of long-term urban growth patterns and their environmental impacts.

References

Abdullah AYM, Masrur A, Gani Adnan MS, Al Baky M A, Hassan QK, Dewan A (2019) Spatio-temporal patterns of land use/land cover change in the heterogeneous coastal region of Bangladesh between 1990 and 2017. Remote Sens 11(7). https://doi.org/10.3390/rs11070790

Adnan MSG, Abdullah AYM, Dewan A, Hall JW (2020) The effects of changing land use and flood hazard on poverty in coastal Bangladesh. Land Use Policy 99(March):104868. https://doi.org/10.1016/j.landusepol.2020.104868

AlDousari AE, Kafy AA, Saha M, Fattah MA, Almulhim AI, Faisal AA, Al Rakib A, Jahir DMA, Rahaman ZA, Bakshi A, Shahrier M, Rahman MM (2022) Modelling the impacts of land use/land cover changing pattern on urban thermal characteristics in Kuwait. Sustain Cities Soc 86(July):104107. https://doi.org/10.1016/j.scs.2022.104107

Chettry V (2022) Peri-urban area delineation and urban sprawl quantification in Thiruvananthapuram Urban Agglomeration, India, from 2001 to 2021 using geoinformatics. Appl Geomatics 14(4):639–652. https://doi.org/10.1007/s12518-022-00460-0

Das M, Das A (2020) Assessing the relationship between local climatic zones (LCZs) and land surface temperature (LST)—A case study of Sriniketan-Santiniketan Planning Area (SSPA), West Bengal, India. Urban Clim 32(April 2019):100591. https://doi.org/10.1016/j.uclim.2020.100591

Islam MM, Rahman MS, Kabir MA, Islam MN, Chowdhury RM (2020) Predictive assessment on landscape and coastal erosion of Bangladesh using geospatial techniques. Remote Sens Appl: Soc Environ 17:100277. https://doi.org/10.1016/j.rsase.2019.100277

Islam MS, Crawford TW, Shao Y (2023) Evaluation of predicted loss of different land use and land cover (LULC) due to coastal erosion in Bangladesh. Front Environ Sci 11(April):1–12. https://doi.org/10.3389/fenvs.2023.1144686

Kaliraj S, Chandrasekar N, Ramachandran KK, Srinivas Y, Saravanan S (2017) Coastal landuse and land cover change and transformations of Kanyakumari coast, India using remote sensing and GIS. Egypt J Remote Sens Space Sci 20(2):169–185. https://doi.org/10.1016/j.ejrs.2017.04.003

Misra A, Balaji R (2015) Decadal changes in the land use/land cover and shoreline along the coastal districts of southern Gujarat, India. Environ Monit Assessment 187(7). https://doi.org/10.1007/s10661-015-4684-2

Mohamed A, Worku H (2020) Simulating urban land use and cover dynamics using cellular automata and Markov chain approach in Addis Ababa and the surrounding. Urban Clim 31(October 2019):100545. https://doi.org/10.1016/j.uclim.2019.100545

Nath B, Wang Z, Ge Y, Islam K, Singh RP, Niu Z (2020) Land use and land cover change modeling and future potential landscape risk assessment using Markov-CA model and analytical hierarchy process. ISPRS Int J Geo-Inf 9(2). https://doi.org/10.3390/ijgi9020134

Ngondo J, Mango J, Liu R, Nobert J, Dubi A, Cheng H (2021) Land-use and land-cover (Lulc) change detection and the implications for coastal water resource management in the wami–ruvu basin, tanzania. Sustainability (Switzerland), 13(8). https://doi.org/10.3390/su13084092

Paradis E (2022) Probabilistic unsupervised classification for large-scale analysis of spectral imaging data. Int J Appl Earth Obs Geoinf 107:102675. https://doi.org/10.1016/j.jag.2022.102675

Qiang Y, Lam NSN (2015) Modeling land use and land cover changes in a vulnerable coastal region using artificial neural networks and cellular automata. Environ Monitor Assessment 187(3). https://doi.org/10.1007/s10661-015-4298-8

Rajakumari S, Sundari S, Meenambikai M, Divya V (2020) Impact analysis of land use dynamics on coastal features of Deshapran block, Purba East Medinipur, West Bengal. J Coastal Conserv 24(2). https://doi.org/10.1007/s11852-020-00737-9

Rasool R, Fayaz A, ul Shafiq M, Singh H, Ahmed P (2021) Land use land cover change in Kashmir Himalaya: linking remote sensing with an indicator based DPSIR approach. Ecol Indicators 125(August 2020):107447. https://doi.org/10.1016/j.ecolind.2021.107447

Sannigrahi S, Pilla F, Zhang Q, Chakraborti S, Wang Y, Basu B, Basu AS, Joshi PK, Keesstra S, Roy PS, Sutton PC, Bhatt S, Rahmat S, Jha S, Singh LK (2021) Examining the effects of green revolution led agricultural expansion on net ecosystem service values in India using multiple valuation approaches. J Environ Manage 277(January 2020):111381. https://doi.org/10.1016/j.jenvman.2020.111381

Sathiya Bama VP, Rajakumari S, Ramesh R (2020) Coastal vulnerability assessment of Vedaranyam swamp coast based on land use and shoreline dynamics. Nat Hazards 100(2):829–842. https://doi.org/10.1007/s11069-019-03844-5

Shukla A, Jain K (2021) Analyzing the impact of changing landscape pattern and dynamics on land surface temperature in Lucknow city, India. Urban for Urban Green 58:126877. https://doi.org/10.1016/j.ufug.2020.126877

Vimala (2020) Unsupervised ISODATA algorithm classification used in the LANDSAT image for predicting the expansion of Salem urban, Tamil Nadu. Ind J Sci Technol 13(16):1619–1629. https://doi.org/10.17485/ijst/v13i16.271

Wang Q, Wang H, Chang R, Zeng H, Bai X (2022) Dynamic simulation patterns and spatiotemporal analysis of land-use/land-cover changes in the Wuhan metropolitan area, China. Ecol Model 464(December 2021):109850. https://doi.org/10.1016/j.ecolmodel.2021.109850

Zhu L, Song R, Sun S, Li Y, Hu K (2022) Land use/land cover change and its impact on ecosystem carbon storage in coastal areas of China from 1980 to 2050. Ecol Indicators 142(July). https://doi.org/10.1016/j.ecolind.2022.109178

Ziaul Hoque M, Islam I, Ahmed M, Shamim Hasan S, Ahmed Prodhan F (2022) Spatio-temporal changes of land use land cover and ecosystem service values in coastal Bangladesh. Egypt J Remote Sens Space Sci 25(1):173–180. https://doi.org/10.1016/j.ejrs.2022.01.008

Phenotyping Complex Maize Diseases Using Sensor-Based Approaches: Challenges and Prospects

Walter Chivasa, Onisimo Mutanga, and Boddupalli M. Prasanna

Abstract Advances in remote sensing is critical in assessing and monitoring ecosystem services in the context of agricultural productivity as affected by the impact of climate change that favours outbreak of new plant diseases. Remote sensing enables the quantification of threats to ecosystem support services in plant breeding. These include field-based phenotyping of maize diseases, constituting the identification, classification, quantification, and prediction. However, such support through sensor-based approaches are faced with multiple challenges when phenotyping diseases that appear simultaneously as co-infections on the same plant. Empirical evidence of sensor-based phenotyping and unmixing co-infections is non-existent. Furthermore, symptoms that cannot be categorized as co-infections also affect sensor-based phenotyping, e.g., symptoms similarities, different symptoms by same pathogen, and asymptomatic diseases. We review and discuss some of the challenges associated with sensor-based phenotyping of maize diseases under co-infections. We further suggest some possible practical application of unmixing co-infections using machine learning algorithms and suggest future efforts that may be possible in maize disease phenotyping using sensor-based approaches. The increasing capabilities of hyperspectral algorithms offer prospects for unmixing co-infections using sensor-based approaches. The thermal portion of the electromagnetic spectrum remains an active research area in phenotyping co-infections.

Keywords Maize complex diseases phenotyping · Co-infection; machine learning · Hyperspectral unmixing algorithms

W. Chivasa (✉) · B. M. Prasanna
International Maize and Wheat Improvement Center (CIMMYT), Nairobi, Kenya
e-mail: w.chivasa@cgiar.org

B. M. Prasanna
e-mail: b.m.prasanna@cgiar.org

O. Mutanga
School of Agricultural, Earth and Environmental Sciences, University of KwaZulu-Natal, Pietermaritzburg, South Africa
e-mail: MutangaO@ukzn.ac.za

O. Mutanga et al. (eds.), *Revealing Ecosystem Services Through Geospatial Technologies*, Springer Remote Sensing/Photogrammetry,
https://doi.org/10.1007/978-3-031-98048-0_16

1 Introduction

Maize (*Zea mays* L.) disease phenotyping using non-invasive sensor-based techniques has received considerable attention in the last 20 years and has become a critical tool in assessing and monitoring ecosystem services in the context of agricultural productivity and food security. Sensor-based high-throughput phenotyping has made the process faster, cost-effective, more accurate, sensitive, consistent, and reliable (Golhani et al. 2018; Mahlein et al. 2018). Sensor-based disease phenotyping approaches also offer real-time trait measurements at various spatial scales (Zhang et al. 2020), including disease identification (Su et al. 2018; Selvaraj et al. 2019), classification (Dang et al. 2018; Chivasa et al. 2020), quantification (Lin et al. 2019), and prediction (Chivasa et al. 2021). These advances have provided an opportunity to accelerate contemporary crop improvement in plant breeding programs. Barbedo (2013) undertook a detailed literature survey on image-based techniques to detect, quantify and classify biotic stresses. However, ecosystem support services through sensor-based plant disease phenotyping are faced with challenges when phenotyping complex diseases that appear simultaneously as co-infections on the same plant in the region of interest (ROI), especially due to spectral mixing. Previous and current sensor-based disease phenotyping studies focused on spectral response to specific diseases when they appear as a single infection and rarely considered spectral unmixing under co-occurrence of different diseases.

Most of the sensor-based maize disease phenotyping studies have been conducted under controlled conditions. Under field conditions, pathogen-by-genotype-by-environment interactions influence quantitative traits in a complex and dynamic manner. Controlling artificially for all other diseases except for one may defeat the purpose of understanding dynamic interactions between pathogens and their effects. Singh et al. (2021) confirmed the need for future field-based phenotyping studies to focus beyond single disease infection. In a nutshell, there exists a major phenotyping bottleneck of complex plant diseases that hinders further advancement of knowledge-based and application-oriented research using remote sensing approaches. We use "complex plant diseases" to denote co-occurrence of or co-infections by multiple pathogens appearing simultaneously on plant canopy in the ROI. The ability to non-destructively unmix complex diseases beyond binary classification (disease or no disease), identification, quantification (severity), and prediction would represent a key breakthrough in maize disease sensing systems that may use visible (VIS, 0.4–0.7 μm), near-infrared (NIR, 0.7–1.0 μm), and short-wave infrared wavelengths (SWIR 1.0–2.5 μm) reflectance indices.

Infectious pathogens could be viruses, fungi, bacteria, and nematodes (Zhang et al. 2020), which can cause symptoms that are mild to severe or even plant death (Shurtleff et al. 2020). The suite of symptoms differs according to the host-by-pathogen-by-environment interactions; for example, bacteria (necrosis, tumors, soft rot, and vascular wilt), virus (discoloration, necrosis, malformations, stunting or dwarfing), fungi (hyperplasia, wilting, and necrosis), and nematodes (discoloration, growth decline, decay, and wilting) (Shurtleff et al. 2020). The simultaneous presence of more than one disease in the ROI impacts the quality of sensor-based phenotypic data in the field. Empirical studies on unmixing spectral dynamics under co-infection by different diseases are non-existent (to the best of our knowledge), yet co-infections are common under field conditions.

Unmixing spectral dynamics under co-infections using remote sensing as an ecosystem service is of practical importance for both precision agriculture and breeding programs. For precision agriculture, specific procedures (e.g., insecticide and fungicide applications) are required for treating different diseases (Yuan et al. 2014). In breeding programs, disease discrimination is crucial in assessing the resistance of genotypes under different environmental conditions. Common maize diseases include, but not limited to, Grey Leaf Spot (*Cercospora zea-maydis*; GLS), Common rust (*Puccinia sorghi*), Turcicum Leaf Blight (TLB; *Exserohilum turcicum*), Maize streak virus (MSV), Phaeosphaeria Leaf Spot (*Phaeosphaeria maydis*—PLS), and Maize lethal necrosis (MLN). Co-infections in maize hybrids can be found between GLS and TLB, Rust and GLS, TLB and Common rust, TLB and Maize Streak Virus (MSV), Rust and PLS, or any combination of the above with possibilities of more than three diseases appearing in the ROI. Breeding for disease resistance requires efficient evaluation of the host-by-pathogen-by-environment interactions to test the breeding materials' response to specific diseases (Mahlein 2016). The next-generation sensor-based plant disease phenotyping platforms should be robust, flexible and capable of spectral unmixing under co-infections on a single leaf or plant canopy. The objective of this review is to explore some challenges associated with phenotyping maize diseases under co-infections and spectral unmixing using remotely sensed data, current development trends, and future outlook in contemporary crop breeding programs. Maize is one of the most widely cultivated and consumed crops globally, playing a vital role in supporting human well-being and ecosystem services. As a key crop, maize provides numerous benefits to both humans and the environment, making it an essential component of sustainable agriculture and ecosystem management. Despite its importance, maize production faces numerous challenges due to climate change that favors proliferation of new pests and diseases. To address these challenges, contemporary plant breeding programs should develop new improved maize varieties that are tolerant to diseases while giving high yields, therefore, improve ecosystem services and ensure food security. The importance of phenotyping complex maize diseases cannot be overemphasized. Our hypothesis is that the fast-evolving hyperspectral imaging (HSI), deep learning approaches, and hyperspectral unmixing algorithms offer solutions to phenotyping diseases that appear simultaneously as co-infections on the same plant.

2 Challenges in Phenotyping Complex Maize Diseases

Much of the current sensor-based plant phenotyping work is ground-based or using aerial multispectral, hyperspectral, fluorescence, and thermal sensors or imaging devices. Surveys on the sensors and platforms used in plant disease phenotyping are given (Mahlein 2016; Deery et al. 2016; Li et al. 2021). Plant phenotyping is measuring plant traits and physiology in a rigorous and highly efficient manner under distinct environmental conditions (Mahlein 2016; Hickey et al. 2019). The relative effort associated with the measurements is referred to as high throughput. Disease phenotyping should be accurate and precise (Bock et al. 2016). Advances in sensors, high throughput plant phenotyping (HTPP), data analytics, and machine learning have ushered in a new pathway to solve plant disease phenotyping challenges (Singh et al. 2016).

Maize disease symptoms are easier to phenotype when they appear as a single infection in the ROI (Fig. 1). Several studies exploring individual diseases have shown the potential of different optical and non-invasive sensors in HTPP at different disease and plant phenological stages (e.g., Chivasa et al. 2020). However, these studies explored one disease at a time rather than under co-infections, and mostly conducted under controlled conditions where each plot is inoculated with one disease and adequate distance is maintained between plots to avoid cross infection (Yuan et al. 2014; Chivasa et al. 2020).

Fig. 1 A panel of six photos showing different maize leaf diseases. (**a**) Turcicum Leaf Blight (TLB): Leaf with large, elongated elliptical, grayish to tan lesions. (**b**) Grey Leaf Spot (GLS): Leaf with elongated, rectangular, grey to tan spots. (**c**) Maize streak virus (MSV): Leaf with fine, yellow to white parallel streaks of discoloration. (**d**) Phaeosphaeria Leaf Spot (PLS): Leaf with small, oval white spots scattered across the surface. (**e**) Maize lethal necrosis (MLN): Leaf with yellowing or irregular chlorotic streaks. (**f**) Rust: Leaf with numerous small, reddish-brown pustules

In studies where co-infections were included in the investigations in other crops, the symptoms were only compared with single infection treatments rather than unmixing of symptoms within the co-infection complex. For example, Hillnhütter et al. (2012) differentiated between symptoms of *Heterodera schachtii*, *Rhizoctonia solani*, and a combination of *H. schachtii* and *R. solani* in sugar beet under artificial inoculation using imaging spectroscopy. Gold et al. (2020) managed to differentiate late blight (*Phytophthora infestans*), early blight (*Alternaria solani*), and a combination of early and late blight in potato using *in-situ* spectroscopy under controlled environment. Mahlein et al. (2013) managed to discriminate *Cercospora* leaf spot, powdery mildew, and sugar beet rust at different crop phenological stages using a non-imaging spectroradiometer with optimized disease indices constructed from hyperspectral wavelengths extracted using RELIEF-F algorithm. Mahlein et al. (2010) used the pigment-specific simple ratio (PSSRa and b) and other spectral vegetation indices to discriminate *Erysiphe beticola, Uromyces betae,* and *Cercospora beticola* on sugar beet. Berdugo et al. (2014) applied fusion of sensor data and a discriminant analysis approach and demonstrated the possibility of discriminating three diseases in cucumber: cucumber green mottle mosaic virus (CGMMV), cucumber mosaic virus (CMV), and powdery mildew caused by *Sphaerotheca fuliginea*. Zhang et al. (2012) used hyperspectral-derived data to differentiate nutrient stresses from yellow rust disease. Rumpf et al. (2010) detected and differentiated three diseases of sugar beet: (*Cercospora* leaf spot, leaf rust, and powdery mildew) using spectral vegetation indices (SVI) and Support Vector Machines (SVM). Yuan et al. (2014) used hyperspectral sensor systems at leaf level to discriminate powdery mildew (*Blumeria graminis*), and yellow rust (*Puccinia striiformis* f. sp. *tritici*) infestation in wheat. Susic et al. (2018) applied imaging spectroscopy to differentiate water deficiency stress from root-knot nematode symptoms in tomato plants.

The aforementioned studies focused on the binary classification between healthy and stressed plants or differentiation of different diseases due to singly present stress or comparing symptoms of a single infection to that of co-infections. Therefore, they provide no insights into how spectral unmixing can be achieved under co-infection. The primary challenge in HTPP using remotely sensed data remains the ability to unmix spectrally between different diseases under co-infections (Singh et al. 2021; Mahlein et al. 2010). Current spectral vegetation indices (SVIs) are not disease specific (Mahlein et al. 2010). The plant-disease interaction process and changes in pathogens themselves can be shown as color changes in the plant tissue, canopy morphology, rate of transpiration, plant density, the shape of the leaf, and this action of biochemical transformation should be reflected in certain reflectance wavebands (Zhang et al. 2020). Individual host-disease-environment interactions have distinct spatial and temporal dynamics and should potentially have different spectral signatures as influenced by these processes. There is a need to explore the possibility of spectral unmixing under co-infections in disease phenotyping (Yuan et al. 2014; Barbedo 2016; Singh et al. 2021). Sensing diseased plant tissues under co-infections results in spectral confusion that requires spectral unmixing techniques. Under co-infections, interpreting the data obtained is a challenge because spectral reflectance patterns represent the integrated effects of the diseases present in the ROI influenced by multiple species and functional groups.

2.1 *Challenges Caused by Co-infections by Different Diseases*

Co-occurrence of different diseases is known by farmers and breeders to often cause greater crop losses than a particular single stress condition. Before estimating the intensity of the disease symptoms, it is important to compare the spectral reflectance and detect spectral features of individual diseases that distinguish these conditions to avoid potential spectral mixing. Currently, phenotyping of co-occurrence (Fig. 2) of different diseases is performed manually, with low throughput, inefficient, and low accurate due to recorder fatigue (Großkinsky et al. 2015; Qiu et al. 2018). When symptoms of different diseases appear simultaneously in the ROI, but with clear physical separation, analysis can be done on each affected area separately by assigning a unique label (Barbedo 2016). This approach is difficult when the disease's symptoms are minute lesions or powder-like or when there are occlusions, which are difficult to deal with individually. It becomes even difficult when co-infections produce hybrid symptoms that differ dramatically from those expected from the original individual disease symptoms.

Fig. 2 A grid of eight close-up photos of plant leaves, each showing different disease symptoms. The images are labeled: (**a**) TLB + MSV, (**b**) GLS + PLS, (**c**) GLS + Rust, (**d**) GLS + TLB, (**e**) TLB + PLS, (**f**) TLB + Rust, (**g**) GLS + MSV, and (**h**) MSV + Rust + TLB. Each leaf displays various patterns of discoloration, spots, and lesions, indicating different combinations of plant diseases. TLB = Turcicum Leaf Blight; MSV = Maize Streak Virus; GLS = Grey Leaf Spot; PLS = Phaeosphaeria leaf spot

The evolution of sensor-based techniques from multispectral to hyperspectral and the forthcoming ultraspectral sensors that provide data covering a wide spectral range (0.35–2.5 μm) at high spectral resolution (< 1 nm) may resolve the challenge of spectral mixing caused by co-infections and may allow spectral unmixing, which is hardly possible with non-imaging sensors. Research using high spectral and spatial resolution hyperspectral imaging systems and spectral unmixing under complex diseases is still non-existent or at nascent stage. The entire spectrum (0.35–2.5 μm) may not be relevant because data from hyperspectral bands are highly correlated, therefore, few bands may be useful (Mewes et al. 2011). Hyperspectral sensors record large amounts of data and its analysis is a complex domain that need powerful signal processing algorithms for spectral unmixing. Besides co-occurrence, other challenges affect sensor-based phenotyping of plant diseases under field conditions, for example, symptoms similarities, different symptoms by same disease, and asymptomatic diseases.

2.2 *Similar Symptoms Caused by Different Pathogens*

Convergence among different disease symptoms poses major challenges in their phenotyping (Ollinger 2011). Many diseases produce symptoms and signs that are visually similar but are a result of different infections (Baranowski et al. 2015) and are difficult to distinguish spectrally (Stafford 2000). Sensing diseased plant tissues in this case results in similar spectral signatures because the diseases produce similar symptoms and plant physiology (Stafford et al. 2000; Nutter et al. 2010). This limits the utility of sensor-based methods compared to traditional molecular or serological approaches (Gold et al. 2020). For example, Ghosal et al. (2018) found misclassifications among bacterial pustule, *Septoria* brown spot, and bacterial blight in soybean due to similarities in symptoms expression. These confounding effects become even more difficult under co-infections. In maize, symptom similarity is common between GLS and bacterial leaf streak, TLB and Bacterial wilt, Curvularia leaf spot and Southern rust diseases.

2.3 *Challenges of Detecting Different Symptoms Caused by the Same Pathogen*

The dynamic variation of the phenotypes and phenotyping individual disease symptoms that co-vary with a suite of plant properties, environment, or varieties is a challenge using sensor-based approaches (Ollinger 2011). The same disease may exhibit different symptoms (color, shape, and size) due to host-by-pathogen-by-environment interactions (Fig. 3). Symptoms may change as the infection progresses (Bock et al. 2008). Some diseases simply produce highly heterogeneous symptoms (Camargo

Fig. 3 Three-panel image showing close-up photographs of plant leaves with varying patterns of discoloration caused by same disease (GLS). Panel (**a**) displays a leaf with yellow streaks and brown spots. Panel (**b**) shows a leaf with alternating pale brown grey stripes. Panel (**c**) features a leaf with brown-grey patches

and Smith 2009), while others produce a hybrid of symptoms under co-infections (Barbedo 2016). Differences in varietal reaction, leaf age (Zhou et al. 2015), and environment (Zhou et al. 2014) may alter symptoms (color, shape, and size) caused by the same disease. A possible solution is to acquire imagery, which covers the complete variety of variations so that the algorithms used may be adequately trained (Barbedo 2016). However, the task is difficult and laborious because it not only requires the appropriate timing, but it is practically impossible to tell when the entire range of symptoms has been imaged. Phenotypic plasticity, which can occur when a genotype is grown under varied environmental conditions, can be a stumbling block in combined analyses of many datasets, as phenotypic plasticity is a complex and quantitative attribute.

2.4 Challenges Caused by Asymptomatic Diseases

While spectral imaging is expected to be useful for examining plant-by-pathogen-by-environment interactions, sensor-based methodologies may not be sufficient for phenotyping diseases that are asymptomatic (Mutka and Bart 2015), or symptoms that appear only when it is too late for sensor-based phenotyping. *Fusarium* spp., for example, cause maize ear rot disease, and some environmental conditions encourage asymptomatic *Fusarium* infections with hazardous mycotoxins nevertheless accumulating in the host tissue (Mesterházy et al. 2012). Imaging techniques are unlikely to detect such diseases in their early stages. Becker et al. (2014) discovered differences in a collection of volatile sesquiterpenes generated by maize plants infected with *Fusarium* vs uninfected plants. Remote sensing of these many volatile chemical profiles might theoretically be used to detect *Fusarium* infection early. Future options include use of thermal portion of the electromagnetic spectrum with sensors to detect unique temperature characteristics due to reaction of the host to infection.

3 Phenotyping Complex Diseases and Machine Learning

Machine learning (ML) approaches are gaining popularity in the plant disease analytics domain (Singh et al. 2021). Recent reviews of ML usage for HTPP are given (Demidchik et al. 2020; Singh et al. 2021). Singh et al. (2018) indicated that deep learning (DL—a subset of ML approaches) can be deployed to identify, classify, quantify, and predict (ICQP) plant diseases through feature extraction from given imagery. However, studies on the use of ML to unmix disease types under co-infections are non-existent. Here we propose a fifth category, unmixing to Singh et al. (2018)'s ICQP paradigm (Fig. 4). Previous studies have adequately addressed the ICQP; however, no studies to the best of our knowledge have utilized sensor-based approaches and machine learning for spectral unmixing under co-infections. Solving this bottleneck has significant implications in the application of sensor-based HTPP.

Recently, DL has shown some promise in plant diseases symptoms recognition (Zhang et al. 2020). DL network requires a significant amount of imagery for training

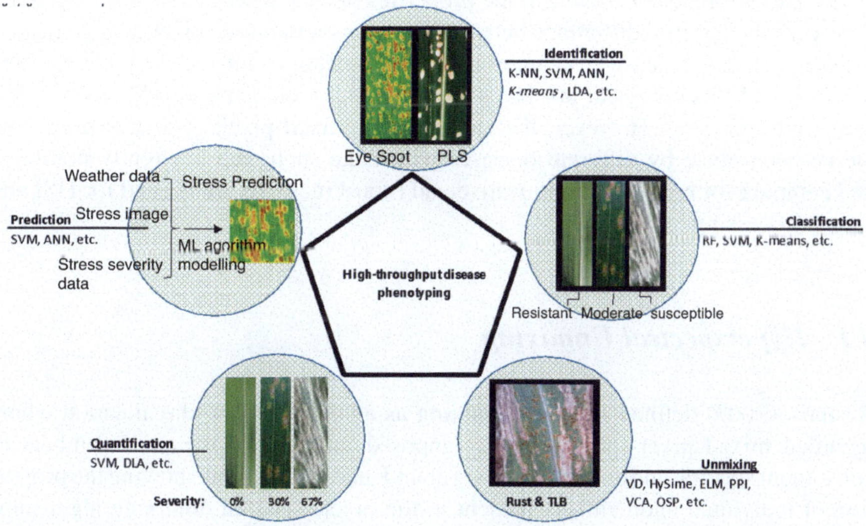

Fig. 4 Figure illustrating high-throughput disease phenotyping. Central pentagon labeled "High-throughput disease phenotyping" connects to five circular nodes. Top left node: "Prediction" with methods like SVM and ANN. Top right node: "Identification" with methods such as K-NN, SVM, and K-means. Middle right node: "Classification" with methods like RF and SVM. Bottom right node: "Unmixing" with methods including VD and HySime. Bottom left node: "Quantification" with severity levels 0%, 30%, and 67%. Each node contains images of plant leaves with varying disease symptoms

to be able to recognize a series of symptoms empirically. However, previous studies using DL explored plant diseases under single infections (Brahimi et al. 2019; Toda and Okura 2019) than co-infections. Machine vision algorithms cluster symptoms according to lesion color, shape, size, texture, etc. The fusion of different features of each disease's symptoms can be used for symptoms discrimination. Zhang et al. (2020) indicated that the utility of image classification models depends on their ability to capture each symptom's distinctive abstract features, when the algorithm learns such features using relatively large set of labelled training data. Therefore, can high resolution hyperspectral imaging (HIS) in combination with spectral unmixing algorithms resolve the challenge of phenotyping complex diseases in a reliable and accurate manner under co-infections?

4 Hyperspectral Imaging and Signal Processing

Hyperspectral sensors with high resolution (~1 nm) can detect minute subtle differences in disease symptoms at macro- and microscopic levels and detect stress signatures before visual symptoms appear (Nagasubramanian et al. 2019; Gold et al. 2020). Hyperspectral sensors cover a large spectral range at narrow intervals and are capable of sensing plant morphology, physiology, and chemistry (Oerke et al. 2019) and might be able to match the properties of each disease's spectral signature to discern the spectral difference between diseases on the same plant. Hyperspectral data can detect the pathogen changes in the host–pathogen interaction processes and the temporal bioecological attributes of individual pathogens to achieve effective symptom unmixing. However, the challenge of mixed pixels, which results from the co-occurrence by different diseases will make such data inherently nonlinear and complex for processing. Innovative and robust methods are needed for HSI and signal processing.

4.1 Hyperspectral Unmixing

Keshava (2003) defined spectral unmixing as an extension of classification where recorded mixed pixel spectra are decomposed into sets of their endmembers or constituent spectra, and a set of matching abundances or fractions showing the proportion of individual endmembers present in the pixel. Spectral unmixing algorithms have been developed to solve the problem of spectral mixtures in hyperspectral images (Boardman et al. 1995; Li et al. 2015). Detailed surveys of spectral unmixing algorithms are given (Keshava 2003; Ghamisi et al. 2017; Wei and Wang 2020). Spectral unmixing algorithms utilize various mathematical techniques in estimating endmembers and their abundances in a sequence of consecutive stages (Keshava 2003): dimension reduction; endmembers determination; and inversion.

Spectral unmixing can be achieved using linear (Li et al. 2015; Bioucas-Dias and Nascimento 2008; Luo et al. 2013; Thouvenin et al. 2016) or nonlinear spectral unmixing approaches (Marinoni et al. 2015). Determination of pure spectral endmembers in HSIs is not an easy task (Ghamisi et al. 2017). Various algorithms have been applied in other studies. For example, Chang and Du (2004) introduced the virtual dimensionality (VD) method to determine the signal sources using distinct spectral properties in selected spectral bands. Bioucas-Dias and Nascimento (2008) used minimum error (HySime) to infer the signal subspace in HSI without the need for any parameter tuning. The eigenvalue likelihood maximization (ELM) approach by Luo et al. (2013) uses a modification of Chang and Du (2004)'s VD method on the observation that the eigenvalues that correspond to noise data have identical covariance and correlation matrices, and the ones consistent with the signal (the endmembers) have a higher correlation than covariance matrix.

Pixel purity index (PPI) has been one of the frequently used endmember extraction algorithms but involves many parameters (Boardan et al. 1995). The N-FINDR has been proposed as an alternative, which assumes that pure pixels exist in the original HSI (Winter 2003). Other algorithms for endmember extraction include orthogonal subspace (OSP), vertex component analysis (VCA) (Nascimento and Bioucas-Dias 2005), and multiple endmember spectral mixture analysis (MESMA) (Somers et al. 2009). In the case where the data lack pure signatures, Ghamisi et al. (2017) suggested fitting a simplex of minimum volume if enough spectral vectors on the facets are available, a concept different from the N-FINDR approach but similar to iterative constrained endmember (ICE) (Berman et al. 2004; Zare and Gader 2007) and minimum volume constrained non-negative matrix methods (Li et al. 2015; Miao and Qi 2007; Chan et al. 2009; Hendrix et al. 2012). In highly mixed data, methods based on a geometrical framework give poor results as compared to statistical methods and are associated with high computational complexity (Chen and Zhang 1999). Inclusion of spatial data in the endmember identification, algorithms can improve endmember extraction (Plaza et al. 2002), although these algorithms use a different approach (Rogge et al. 2007; Zortea and Plaza 2009). After the endmember signatures have been determined, various methodologies for estimating their fractional abundances can be applied (Chang 2003).

Sub-pixel and object-based classification has been used for spectral unmixing in other studies (Blaschke 2010). Sub-pixel approach unmixes pixels into their components classes and extracts the proportion of the component from the mixture (Wang et al. 2014). Object-based approach requires assemblages of homogeneous and contiguous pixels and is better than the pixel-based approach (Hossain and Chen 2019; Ma et al. 2019). However, no spectral unmixing studies (known to the authors) have been reported in complex plant disease phenotyping. There is a need to test these algorithms in complex disease phenotyping, particularly in plant breeding where fine pixel sizes are required. The review of spectral unmixing algorithms available is beyond the scope of this chapter. For a detailed survey on spectral unmixing techniques in other studies, readers are referred to Ghamisi et al. (2017) Keshava (2003), and Wei and Wang (2020). However, not all will be applicable to spectral unmixing under maize disease co-infections. The examples cited here and,

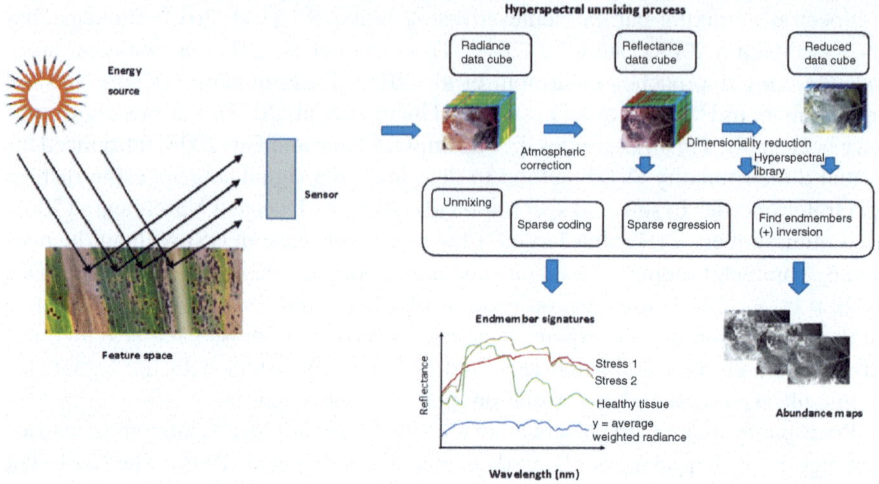

Fig. 5 Flow chart illustrating the hyperspectral unmixing process. Key elements include energy source, sensor, hyperspectral data cubes, endmember signatures, and abundance maps

in the literature, give the readers interested in using spectral unmixing algorithms some reasonable prospects of commonly used algorithms and approaches that can be applied in complex maize disease phenotyping.

4.2 *Hyperspectral Unmixing Process*

Bioucas-Dias et al. (2012) provided a three-stage process for spectral unmixing (Fig. 5) that might be applicable in phenotyping complex diseases in plant breeding programs. The stages entail atmospheric correction (where applicable), dimensionality reduction, and unmixing. The process requires endmember determination and inversion (often implemented simultaneously), sparse regression, or coding techniques. Detailed characteristics of each process are given in Bioucas-Dias et al. (2012), however, the application of spectral unmixing algorithms has its own set of challenges.

4.3 *Hyperspectral Unmixing Challenges*

While the spectral unmixing algorithms discussed above were developed for different applications, the principle might apply in complex disease phenotyping. The main challenges consist of extending these methodologies to spectral unmixing under co-infections in plant phenotyping. The accuracies of extracted phenotypic traits may be affected by pixel sizes in the imagery (Hu et al. 2019). Bioucas-Dias et al. (2012) indicated that spectral unmixing is a challenging, ill-posed inverse problem due to data noise, environmental factors, model inaccuracies, data set size, endmember variability, and pixel size.

The size of the pixel is the main determinant of accuracy in spectral unmixing problems. Smaller-sized pixels give more detailed information and higher accuracy (Hsieh et al. 2001; Hengl 2006; Jin et al. 2017). Therefore, high-resolution imaging is needed, especially under co-infections. The impact of pixel size in image analysis has been given (Peña et al. 2015; Deery et al. 2016). With unmanned aerial vehicles (UAV), smaller pixels require low flight heights and high-resolution sensors, but this affects aerial coverage per unit of flight time, thereby increasing labor and costs of phenotyping (Hu et al. 2021). Conversely, large pixel sizes affect the quality of phenotypic data due to existence of other data classes (e.g., background noise) other than the traits of interest making it even more difficult to unmix. It becomes a challenge to balance costs, flight attributes, and pixel size of images when using UAV-based HTPP under co-infections. Sufficient temporal and spatial resolution are required for spectral unmixing under co-infections. The higher the flight altitudes the better for large area disease detection in precision agriculture (Hunt et al. 2018), contrary to the high accuracy required in small breeding plots (Hu et al. 2019), where fine pixel sizes are required. Under field conditions, image acquisition is further affected by environmental factors, sensor type, flight configurations, and plant phenological stages.

5 Ecological Services and Implications

Phenotyping complex maize diseases using remote sensing and machine learning approaches has significant implications for both ecological services, agricultural management and food security. To date, remote sensing techniques have enabled the identification, classification, quantification and prediction of disease symptoms on crops, providing valuable data on plant health and disease progression. In this Chapter, we argued that apart from identification, classification, quantification and prediction, there is an urgent need for phenotyping crop diseases using remote sensing to be able to unmix diseases under co-infections.

By identifying diseases early and being able to quantify the severity, farmers can make informed decisions about the type of pesticides to use to control the disease or other crop management practices that reduce the need for chemical pesticides and minimizing their environmental footprint. In addition, being able to unmix or identify the types of different diseases in the area of interest will enable farmers and researchers alike to apply the right types of pesticide mixtures for maximum control of the diseases with minimum impact to the environment. These approaches help optimize resource use, improving overall ecological sustainability by protecting ecosystems and reducing their environmental footprint. The integration of remote sensing and machine learning in maize disease phenotyping not only boosts agricultural productivity but also enhances the efficiency and sustainability of farming practices and research, supporting ecological services.

6 Concluding Remarks and Future Prospects

In reviewing current knowledge about sensor-based phenotyping of complex maize diseases, several gaps exist. Co-infections, symptoms similarity, different symptoms from the same disease, hybrid symptoms, and asymptomatic diseases affect spectral signatures in ways that are difficult to unmix using current approaches. The spectral confusion due to co-infections is a challenge in understanding individual spectral signatures. The ability to unmix co-infections using sensor-based phenotyping techniques is still lacking, however, hyperspectral unmixing algorithms offer prospects for the next-generation HTPP.

The complexity of phenotyping co-infections reflects the growing sophistication of an area that is quickly evolving at the intersection of numerous disciplines: signal and image processing, physical modelling, linear algebra, and computing advances. Collaborative research efforts are needed to deploy field-based HSI techniques and hyperspectral unmixing algorithms to solve these challenges. Such trans-disciplinary approaches will usher in new impetus in plant phenotyping science by bringing complementary skills: pathologists, physiologists, computer scientists, sensor development specialists, agricultural scientists (breeders and agronomists), remote sensing specialists, data mining, and informatics to study previously unexplored domains of complex plant diseases phenotyping.

The fast-evolving HSI, deep learning approaches, hyperspectral unmixing algorithms, and potentially other options yet to be discovered may offer solutions to the challenge and find their application in areas with high societal impacts: crop genetic improvement and precision agriculture. There is a need for future research primarily focusing on hyperspectral imaging data, advanced signal processing, and hyperspectral unmixing algorithms to solve the challenges of phenotyping co-infections. We hope that this review will stimulate interest in exploring complex disease phenotyping using multi-disciplinary approaches in the next-generation HTPP. We conclude by giving a list of recommendations for future research:

1. Studies are urgently needed to explore hyperspectral remote sensing and unmixing algorithms that decompose mixed pixel spectra into a set of their constituent spectra, or endmembers, and a set of matching fractions or abundances, which show the percentage of individual endmembers existing in the pixel for complete and comprehensive complex disease phenotyping.
2. There is a need to build complete spectral libraries of imagery of co-infections under different environmental settings, including similar symptoms caused by different diseases, different symptoms caused by the same diseases, and possibly asymptomatic diseases.

3. The thermal portion of the electromagnetic spectrum remains an active research area in phenotyping co-infections. The hypothesis is that different disease infections will result in unique temperature characteristics on the host plants; therefore, by developing algorithms that can detect subtle and matching temperature variabilities will assist in phenotyping co-infections.
4. The phenotyping of co-infectious diseases can be approached as a new type of hybrid stress rather than simply as the sum of different diseases. To this end, we foresee the need for studies focusing on different disease combinations (disease matrix) to explore their potential interactions and implications on sensor-based phenotyping.

Acknowledgements This study was made possible through funding from the Bill & Melinda Gates Foundation and the USAID Feed the Future Program through the Accelerating Genetic Gains in Maize and Wheat (AGG) Project [INV-003439]; CGIAR Research Program Maize Agri-Food Systems (MAIZE). MAIZE receives Windows 1&2 support from the Governments of Australia, Belgium, Canada, China, France, India, Japan, Korea, Mexico, the Netherlands, New Zealand, Norway, Sweden, Switzerland, the UK, the USA, the World Bank; and the NRF SARChI Chair on Land Use Planning and Management (Grant number 84157). This work was funded in part by the United States Agency for International Development (USAID) Bureau for Resilience and Food Security, as a part of Feed the Future activity, for the (AGG Project. Any opinions, findings, conclusions, or recommendations expressed here are those of the authors alone. The authors wish to acknowledge Wilson Mwaura who shared with us MLN photograph.

References

Baranowski P, Jedryczka M, Mazurek W, Babula-Skowronska D, Siedliska A, Kaczmarek J (2015) Hyperspectral and thermal imaging of oilseed rape (Brassica napus) response to fungal species of the genus Alternaria. PLoS ONE 10:e0012313-e0122913

Barbedo JGA (2016) A review on the main challenges in automatic plant disease identification based on visible range images. Biosys Eng 144:52–60

Barbedo JGA (2013) Digital image processing techniques for detecting, quantifying and classifying plant diseases. Springerplus 2(660):1–12

Becker EM, Herrfurth C, Irmisch S, Kollner TG, Feussner I, Karlovsky P, Splivallo R (2014) Infection of corn ears by *Fusarium* spp. Induces the emission of volatile sesquiterpenes. J Agric Food Chem 62:5226–5236

Berdugo CA, Zito R, Paulus S, Mahlein AK (2014) Fusion of sensor data for the detection and differentiation of plant diseases in cucumber. Plant Pathol 63:1344–1356

Berman M, Kiiveri H, Lagerstrom R, Ernst A, Dunne R, Huntington JF (2004) A statistical approach to identifying endmembers in hyperspectral images. IEEE Trans Geosci Remote Sens 42(10):2085–2095

Bioucas-Dias J, Plaza A, Dobigeon N, Parente M, Du Q, Gader P, Chanussot J (2012) Hyperspectral unmixing overview: Geometrical, statistical, and sparse regression-based approaches. IEEE J Sel Topics Appl Earth Observ Remote Sens 5(2):354–379

Bioucas-Dias JM, Nascimento JMP (2008) Hyperspectral sub-space identification. IEEE Trans Geosci Remote Sens 46(8):2435–2445

Blaschke T (2010) Object based image analysis for remote sensing. ISPRS J Photogramm Remote Sens 65:2–16

Boardman JW, Kruse FA, Green RO (1995) Mapping target signatures via partial unmixing of AVIRIS data. In: Proceedings summary, 5th annual jet propulsion laboratory airborne earth science. Workshop, 1995, pp 23–26

Bock CH, Hotchkiss MW, Wood BW (2016) Assessing disease severity: accuracy and reliability of rater estimates in relation to number of diagrams in a standard area diagram set. Plant Pathol 65:261

Bock CH, Parker PE, Cook AZ, Gottwald TR (2008) Visual rating and the use of image analysis for assessing different symptoms of citrus canker on grapefruit leaves. Plant Dis 92(4):530–541

Brahimi M, Mahmoudi S, Boukhalfa K, Moussaoui A (2019) Deep interpretable architecture for plant diseases classification. arXiv:1905.13523

Camargo A, Smith JS (2009) An image-processing based algorithm to automatically identify plant disease visual symptoms. Biosys Eng 102:9–21

Chan TH, Chi CY, Huang YM, Ma WK (2009) A convex analysis-based minimum-volume enclosing simplex algorithm for hyperspectral unmixing. IEEE Trans Signal Process 57(11):4418–4432

Chang C-I, Du Q (2004) Estimation of number of spectrally distinct signal sources in hyperspectral imagery. IEEE Trans Geosci Remote Sens 42(3):608–619

Chang C-I (2003) Hyperspectral imaging: techniques for spectral detection and classification. Kluwer, Norwell, MA

Chen C, Zhang X (1999) Independent component analysis for remote sensing study. In: Proceedings of SPIE image and signal processing remote sensing, vol 3871, pp 150–158

Chivasa W, Mutanga O, Burgueno J (2021) UAV-based high-throughput phenotyping to increase prediction and selection accuracy in maize varieties under artificial MSV inoculation. Comput Electron Agric 2021(184):106128

Chivasa W, Mutanga O, Biradar C (2020) UAV-based multispectral phenotyping for disease resistance to accelerate crop improvement under changing climate conditions. Remote Sens 12:2445. https://doi.org/10.3390/rs12152445

Dang LM, Hassan SI, Suhyeon I, Sangaiah AK, Mehmood I, Rho S, Seo S, Moon H (2018) UAV based wilt detection system via convolutional neural networks. Sustain Compu. Inform Syst 100250

Deery DM, Rebetzke GJ, Jimenez-Berni JA, James RA, Condon AG, Bovill WD, Hutchinson P, Scarrow J, Davy D, Furbank RT (2016) Methodology for high-throughput field phenotyping of canopy temperature using airborne thermography. Front Plant Sci 7:1808

Demidchik VV, Shashko AY, Bandarenka UY, Smolikova GN, Przhevalskaya DA, Charnysh MA, Pozhvanov GA, Barkosvkyi AV, Smolich II, Sokolik AI, Yu M, Medvedev SS (2020) Plant phenomics: fundamental bases, software and hardware platforms, and machine learning. Russ J Plant Physiol 67(3):397–412

Gold KM, Townsend PA, Chlus A, Herrmann I, Couture JJ, Larson ER, Gevens AJ (2020) Hyperspectral measurements enable pre-symptomatic detection and differentiation of contrasting physiological effects of late blight and early blight in potato. Remote Sens 12:286

Golhani K, Balasundram SK, Vadamalai G, Pradhan B (2018) A review of neural networks in plant disease detection using hyperspectral data. Inf Process Agric 5:354–371. https://doi.org/10.1016/j.inpa.2018.05.002

Großkinsky DK, Svensgaard J, Christensen S, Roitsch T (2015) Plant phenomics and the need for physiological phenotyping across scales to narrow the genotype-to-phenotype knowledge gap. J Exp Bot 66:5429–5440

Ghamisi P, Yokoya N, Li J, Liao W, Liu S, Plaza J, Rasti B, Plaz A (2017) Advances in hyperspectral image and signal processing: a comprehensive overview of the state of the art. IEEE Geosci Remote Sens Mag 0274–6638/17

Ghosal S, Blystone D, Singh AK, Ganapathysubramanian B, Singh A, Sarkar S (2018) An explainable deep machine vision framework for plant stress phenotyping. PNAS 115(18):4613–4618

Hendrix EMT, Garcia I, Plaza J, Martin G, Plaza A (2012) A new minimum-volume enclosing algorithm for endmember identification and abundance estimation in hyperspectral data. IEEE Trans Geosci Remote Sens 50(7):2744–2757

Hengl T (2006) Finding the right pixel size. Comput Geosci 32:1283–1298. https://doi.org/10.1016/j.cageo.2005.11.008
Hickey L, Hafeez AN, Robinson HM, Jackson SA, Leal-Bertioli SCM, Tester M, Gao C, Godwin ID, Hayes BJ, Wulff BBH (2019) Breeding crops to feed 10 billion. Nat Biotechnol 37:744–754
Hillnhütter C, Mahlein A-K, Sikora RA, Oerke E-C (2012) Use of imaging spectroscopy to discriminate symptoms caused by Heterodera schachtii and Rhizoctonia solani on sugar beet. Precision Agric 13:17–32
Hu P, Guo W, Chapman SC, Guo Y, Zheng B (2019) Pixel size of aerial imagery constrains the applications of unmanned aerial vehicle in crop breeding. ISPRS J Photogramm Remote Sens 154:1–9. https://doi.org/10.1016/j.isprsjprs.2019.05.008
Hu P, Chapman SC, Zheng B (2021) Coupling of machine learning methods to improve estimation of ground coverage from unmanned aerial vehicle (UAV) imagery for high-throughput phenotyping of crops. Funct Plant Biol 48:766–779. https://doi.org/10.1071/FP20309
Hunt ER, Horneck DA, Spinelli CB, Bruce AE, Gadler DJ, Brungardt JJ, Hamm PB (2018) Monitoring nitrogen status of potatoes using small unmanned aerial vehicles. Precision Agric 19:314–333. https://doi.org/10.1007/s11119-017-9518-5
Hossain MD, Chen D (2019) Segmentation for Object-Based Image Analysis (OBIA): a review of algorithms and challenges from remote sensing perspective. ISPRS J Photogramm Remote Sens 150:115–134
Hsieh P-F, Lee LC, Chen N-Y (2001) Effect of spatial resolution on classification errors of pure and mixed pixels in remote sensing. IEEE Trans Geosci Remote Sens 39(2):2657–2663. https://doi.org/10.1109/36.975000
Jin Z, Azzari G, Burke M, Aston S, Lobell DB (2017) Mapping smallholder yield heterogeneity at multiple scales in Eastern Africa. Remote Sens 9:931. https://doi.org/10.3390/rs9090931
Keshava N (2003) A survey of spectral unmixing algorithms. Lincoln Laboratory J 14(1):55–78
Li D, Quan C, Song Z, Li X, Yu G, Li C. Muhammad, A (2021) High-throughput plant phenotyping platform (HT3P) as a novel tool for estimating agronomic traits from the lab to the field. Front Bioeng Biotechnol 8:623705
Li J, Agathos A, Zaharie D, Bioucas-Dias JM, Plaza A, Li X (2015) Minimum volume simplex analysis: a fast algorithm for linear hyperspectral unmixing. IEEE Trans Geosci Remote Sens 53(9):5067–5082
Lin K, Gong L, Huang Y, Liu C, Pan J (2019) Deep learning-based segmentation and quantification of cucumber powdery mildew using convolutional neural network. Front Plant Sci 10:155. https://doi.org/10.3389/fpls.2019.00155
Luo B, Chanussot J, Doute S, Zhang L (2013) Empirical automatic estimation of the number of endmembers in hyperspectral images. IEEE Geosci Remote Sens Lett 10(1):24–28
Ma L, Liu Y, Zhang X, Ye Y, Yin G, Johnson BA (2019) Deep learning in remote sensing applications: a meta-analysis and review. ISPRS J Photogramm Remote Sens 152:166–177. https://doi.org/10.1016/j.isprsjprs.2019.04.015
Mahlein A-K, Kuska MT, Behmann J, Polder G, Walter A (2018) Hyperspectral sensors and imaging technologies in phytopathology: state of the art. Annu Rev Phytopathol 56:535–558. https://doi.org/10.1146/annurev-phyto-080417-050100
Mahlein A-K (2016) Plant disease detection by imaging sensors—Parallels and specific demands for precision agriculture and plant phenotyping. Plant Dis 100:241–251
Mahlein A-K, Rumpf T, Welke P, Dehne H-W, Plümer L, Steiner U, Oerke EC (2013) Development of spectral indices for detecting and identifying plant diseases. Remote Sens Environ 128:21–30
Mahlein AK, Steiner U, Dehne HW, Oerke EC (2010) Spectral signatures of sugar beet leaves for the detection and differentiation of diseases. Precis Agric 11:413–431
Marinoni A, Plaza J, Plaza A, Gamba P (2015) Nonlinear hyperspectral unmixing using nonlinearity order estimation and polytope decomposition. IEEE J Sel Topics Appl Earth Observ Remote Sens 8(6):2644–2654
Mesterházy Á, Lemmens M, Reid LM (2012) Breeding for resistance to ear rots caused by *Fusarium* spp. In maize—A review. PlantBreed 131:1–19

Mewes T, Fanke J, Menz G (2011) Spectral requirements on airborne hyperspectral remote sensing data for wheat disease detection. Precis Agric

Miao L, Qi H (2007) Endmember extraction from highly mixed data using minimum volume constrained nonnegative matrix factorization". IEEE Trans Geosci Remote Sens 45(3):765–777

Mutka AM, Bart RS (2015) Image-based phenotyping of plant disease symptoms. Front Plant Sci 5:734. https://doi.org/10.3389/fpls.2014.00734

Nagasubramanian K, Jones S, Singh AK, Sarkar S, Singh A, Ganapathysubramanian B (2019) Plant disease identification using explainable 3D deep learning on hyperspectral images. Plant Methods 15:98

Nascimento J, Bioucas-Dias J (2005) Vertex component analysis: a fast algorithm to unmix hyperspectral data. IEEE Trans Geosci Remote Sens 43(4):898–910

Nutter F, van Rij N, Eggenberger SK, Holah N (2010) Spatial and temporal dynamics of plant pathogens. In Oerke EC, Gerhards R, Menz G, Sikora RA (eds) Precision crop protection—The challenge and use of heterogeneity. Springer, Dordrecht, The Netherlands, pp 27–50

Oerke E-C, Leucker M, Steiner U (2019) Sensory assessment of Cercospora beticola sporulation for phenotyping the partial disease resistance of sugar beet genotypes. Plant Methods 15:133

Qiu R, Wei S, Zhang M, Li H, Sun H, Liu G, Li M (2018) Sensors for measuring plant phenotyping: a review. Int J Agric Biol Eng 11:1–17

Ollinger SV (2011) Sources of variability in canopy reflectance and the convergent properties of plants. New Phytol 189:375–394

Peña JM, Torres-Sánchez J, Serrano-Pérez A, de Castro AI, López-Granados F (2015) Quantifying efficacy and limits of unmanned aerial vehicle (UAV) technology for weed seedling detection as affected by sensor resolution. Sensors 15:5609–5626

Plaza A, Martinez P, Perez R, Plaza J (2002) Spatial/spectral endmember extraction by multidimensional morphological operations. IEEE Trans Geosci Remote Sens 40(9):2025–2041

Rogge DM, Rivard B, Zhang J, Sánchez A, Harris J, Feng J (2007) Integration of spatial–spectral information for the improved extraction of endmembers. Remote Sens Environ 110(3):287–303

Rumpf T, Mahlein AK, Steiner U, Oerke EC, Dehne HW, Plümer L (2010) Early detection and classification of plant diseases with support vector machines based on hyperspectral reflectance. Comput Electron Agric 74:91–99

Selvaraj MG, Vergara A, Ruiz H, Safari N, Elayabalan S, Ocimati W, Blomme W (2019) AI-powered banana diseases and pest detection. Plant Methods 15:92. https://doi.org/10.1186/s13007-019-0475-z

Shurtleff MC, Pelczar MJ, Kelman A, Pelczar RM (2020) Plant disease. In: Plant pathology; Encyclopedia Britannica: Chicago, IL, USA. Available online https://www.britannica.com/science/plant-disease. Accessed on 2 May 2021

Singh A, Jones S, Ganapathysubramanian B, Sarkar S, Mueller D, Sandhu K, Nagasubramanian K (2021) Challenges and opportunities in machine-augmented plant stress phenotyping. Trends Plant Sci 26(1):110–124

Singh KA, Ganapathysubramanian B, Sarkar S, Singh A (2018) Deep learning for plant stress phenotyping: trends and future perspectives. Trends Plant Sci 23(10):883–898

Singh A, Ganapathysubramanian B, Singh AK, Sarkar S (2016) Machine learning for high-throughput stress phenotyping in plants. Trends Plant Sci 21:110–124

Stafford JV (2000) Implementing precision agriculture in the 21st century. J Agric Eng Res 76:267–275

Su J, Liu C, Coombes M, Hu X, Wang C, Xu X, Li Q, Guo L, Chen WH (2018) Wheat yellow rust monitoring by learning from multispectral UAV aerial imagery. Comput Electron Agric 155:157–166. https://doi.org/10.1016/j.compag.2018.10.017

Susic N, Žibrat U, Širca S, Strajnar P, Razinger J, Knapic M, Voncina A, Urek G, Geric SB (2018) Discrimination between abiotic and biotic drought stress in tomatoes using hyperspectral imaging. Sens Actuat B Chem 273:842–852

Somers B, Delalieux S, Stuckens J, Verstraeten WW, Coppin P (2009) A weighted linear spectral mixture analysis approach to address endmember variability in agricultural production systems. Int J Remote Sens 30(1):139–147

Thouvenin P-A, Dobigeon N, Tourneret J-Y (2016) Hyperspectral unmixing with spectral variability using a perturbed linear mixing model. IEEE Trans Signal Process 64(2):525–538

Toda Y, Okura F (2019) How convolutional neural networks diagnose plant disease. Plant Phenomics 9237136

Wang H, Shao Y, Kennedy LM (2014) Temporal generalization of sub-pixel vegetation mapping with multiple machine learning and atmospheric correction algorithms. Int J Remote Sens 35:7118–7135. https://doi.org/10.1080/01431161.2014.965288

Wei J, Wang X (2020) An overview on linear unmixing of hyperspectral data. Math Prob Eng 3735403

Winter ME (2003) N-FINDR: an algorithm for fast autonomous spectral end-member determination in hyperspectral data. In: Proceedings of society of photographic instrumentation engineers, imaging spectrometry V, vol 3753, pp 266–277

Yuan L, Huang Y, Loraamm RW, Nie C, Wang J, Zhang J (2014) Spectral analysis of winter wheat leaves for detection and differentiation of diseases and insects. Field Crop Res 156:199–207

Zare A, Gader P (2007) Sparsity promoting iterated constrained endmember detection in hyperspectral imagery. IEEE Geosci Remote Sens Lett 4(3):446–450

Zhang J, Pu R, Huang W, Yuan L, Luo J, Wand J (2012) Using in-situ hyperspectral data for detecting and discriminating yellow rust disease from nutrient stresses. Field Crops Res 134:165–174. https://doi.org/10.1016/j.fcr.2012.05.011

Zhang N, Yang G, Pan Y, Yang X, Chen L, Zhao C (2020) A review of advanced technologies and development for hyperspectral-based plant disease detection in the past three decades. Remote Sens 12:3188. https://doi.org/10.3390/rs12193188

Zhou R, Kaneko S, Tanaka F, Kayamori M, Shimizu M (2015) Image-based field monitoring of Cercospora leaf spot in sugar beet by robust template matching and pattern recognition. Comput Electron Agric 116:65–79

Zhou R, Kaneko S, Tanaka F, Kayamori M, Shimizu M (2014) Disease detection of Cercospora Leaf Spot in sugar beet by robust template matching. Comput Electron Agric 108:58–70

Zortea M, Plaza A (2009) Spatial preprocessing for endmember extraction. IEEE Trans Geosci Remote Sens 47:2679–2693

Mapping Potential Risk Areas for Ecosystem Service Loss: An Assessment of Human Induced Vegetation Degradation in Greater uMngeni Catchment

Dadirai Matarira, Onisimo Mutanga, and Marco Vizzari

Abstract Comprehensive and up-to-date vegetation degradation assessments addressing the dual challenges of climate variability and human actions in ecologically sensitive catchments are limited. Vegetation degradation assessment is crucial in establishing locally valid causal factors of vegetation demise and identifying areas at risk of ecosystem service loss. This study investigates human-induced vegetation loss within the greater uMngeni Catchment, South Africa, using a spatio-temporal approach that integrates land use transitions and the residual trend (RESTREND) technique over the period 2000–2023. The main objective is to identify human induced degradation hotspots to establish potential areas for ecosystem service loss, thereby highlighting the need for targeted restoration efforts. Land use/land cover (LULC) data were obtained from multispectral Landsat images, with land use classes generated using the maximum likelihood classification algorithm in ArcGIS 10.5. The contribution rates of precipitation and soil moisture variations were assessed using 250-m MODIS NDVI time-series data, CHIRPS rainfall data, and NASA datasets. Additionally, the intensity analysis approach quantified and analyzed transitions in LULC over the period. Intensity analysis results revealed a 25% decline in natural vegetation, particularly forests and grasslands, attributed to urban and cropland expansion. RESTREND analysis identified approximately 17.1 and 16.2% of the catchment area as significantly degraded, with severe degradation observed in rainfall- and moisture-adjusted models' outputs, respectively. Key hotspots were

D. Matarira (✉) · O. Mutanga
School of Agriculture, Earth and Environmental Science, University of KwaZulu-Natal, Scottsville, Pietermaritzburg, South Africa
e-mail: matarirad@ukzn.ac.za

O. Mutanga
e-mail: mutangao@ukzn.ac.za

M. Vizzari
Department of Agricultural, Food and Environmental Sciences, University of Perugia, Perugia, Italy
e-mail: marco.vizzari@unipg.it

O. Mutanga et al. (eds.), *Revealing Ecosystem Services Through Geospatial Technologies*, Springer Remote Sensing/Photogrammetry,
https://doi.org/10.1007/978-3-031-98048-0_17

identified in Durban Metro and Pietermaritzburg, where 32.7 and 29% of the municipality bounds were classified as severely degraded. These findings point out the necessity for restoration and sustainable landscape management to safeguard the catchment's ecological functions.

Keywords Human induced vegetation loss · Climate variability · Land use transitions · Intensity analysis · Ecosystem services

1 Introduction

The Anthropocene epoch is characterized by pervasive environmental challenges, with vegetation degradation among the most pressing issues (Arfasa et al. 2024; Prăvălie 2021). Catchment systems, such as the greater uMngeni Catchment in South Africa, epitomize this interplay, experiencing acute vegetation loss due to human-induced land use transitions and natural climatic perturbations. Vegetation degradation, whether driven by deforestation, urbanization, agricultural expansion, or climate change diminishes ecosystem resilience and functionality, thereby impairing essential services including carbon sequestration, water regulation, and soil stabilization (Bhungeni et al. 2024; Mumtaz et al. 2023). In support, Pandey et al. (2019) iterated that a key concern at the global stage is the impact of LULC change on the ecosystems and the services they provide. Disentangling the climate and anthropogenic contribution on vegetation growth trends is crucial in identifying areas undergoing degradation solely due to anthropogenic stressors (Hentze et al. 2017; Sun et al. 2022). This differentiation enables targeted interventions that address human impacts while bolstering ecosystem resilience. Furthermore, spatially precise information on the scale and severity of vegetation degradation guides catchment management planners in prioritizing areas for restoration, rehabilitation, or degradation control strategies (Abel et al. 2020; Li et al. 2017).

The efficacy of earth observation methods in the analysis of LULC change at diverse geographic and time-based scales is emphasized (Bhungeni et al. 2024; Namugize et al. 2018; Tilahun et al. 2024). Remote sensing technologies offer unparalleled opportunities to investigate vegetation dynamics (Mullapudi et al. 2023; Nyatuame et al. 2023). These technologies have been successfully employed to detect vegetation loss and degradation, as demonstrated in research, for example, Mullapudi et al. (2023) in India and Nyatuame et al. (2023) in Ghana. However, Akinyemi et al. (2016) purported that the most commonly used traditional change detection technique has become mundane since it misses out in highlighting the connections between the underlying variations.

Intensity analysis technique has been widely used to unveil insights into quantification and distribution of land use transitions (Akinyemi et al. 2016; Kilama Luwa et al. 2024). While the technique effectively identifies land cover changes over time, it lacks depth in distinguishing climate-driven degradation from human-induced impacts (Montfort et al. 2020; Sun et al. 2022). On the other hand, the residual trend (RESTREND) approach isolates human-induced degradation by accounting for climatic influences, such as rainfall and soil moisture (Ibrahim et al. 2015; Matin et al. 2019) from human-induced degradation, using vegetation indices like NDVI (Wessels et al. 2004). This approach has been executed across broad regions, including the African Sahel (Ibrahim et al. 2015) and South African rangelands (Wessels et al. 2012). However, most of these studies focused on broad regions, overlooking fine-scale dynamics at the catchment level.

Studies integrating RESTREND and intensity analysis demonstrate improved degradation assessments (Akodéwou et al. 2020; Ibrahim et al. 2015; Mumtaz et al. 2023). However, these efforts are geographically skewed towards regions like Europe (Mumtaz et al. 2023), China (Chen et al. 2018, 2023), West Africa (Zhao et al. 2011) and sub-Saharan Africa (Ibrahim et al. 2015). Southern Africa, particularly South Africa, remains underrepresented in such analyses despite its unique ecological and socio-economic conditions. Whilst Wessels et al. (2012) applied RESTREND in South Africa, the study did not incorporate intensity analysis in addition to being on a broader scale, stretching into Botswana. It can be inferred that most of the aforementioned studies quantified degradation trends at broad scales, either national or regional levels (Ibrahim et al. 2015; Zhao et al. 2011). Detailed spatial mapping of degradation severity at the local or catchment level is lacking.

The greater uMngeni Catchment, an essential water resource for South Africa's KwaZulu-Natal region, faces significant challenges from land use changes and vegetation degradation, which exacerbate soil erosion, thus, compromising water quality (Makaya et al. 2019; Namugize et al. 2018; Rangeti and Dzwairo 2021). Soil erosion and subsequent sediment deposition in aquatic ecosystems have been regarded as manifestations of deforested landscape (Hughes et al. 2018; Moodley et al. 2023; Namugize et al. 2018). According to Hughes et al. (2018), nearly 48% of the natural land cover has been lost, severely impacting the catchment's ecological functions, particularly its hydrological responses. The removal of riparian vegetation, exacerbated by agricultural activities (Moodley et al. 2023; Rangeti and Dzwairo 2021), together with wetland and natural vegetation degradation (Hughes et al. 2018) have compromised water supply and quality in the catchment. Some research has been done, inclined towards erosion risk assessment (Makaya et al. 2019), though, either in terms of modelling surface runoff (Schütte and Schulze 2017), or sediment yield (Hughes et al. 2018), prompting numerous catchment development programs on soil erosion control (Mabuda et al. 2024). According to Seutloali and Beckedahl (2015), sparse vegetation cover increases soil vulnerability to erosional processes, particularly due to rainfall impact and surface runoff. Advancing the argument, Mander et al. (2017), in their study of assessing possible hydrological yields from investment in natural capital in the uMngeni catchment, considered investing in vegetation cover monitoring as key in modelling evaluation of investments in ecological investment.

While studies have been done in the uMngeni catchment, indicating LULC changes as major drivers (Namugize et al. 2018), comprehension of variations in land use change patterns over time is limited. Manandhar et al. (2010) propounded that thorough analysis of LULC patterns affords scientists opportunity to concentrate on the most important indicators of landscape transitions, that will ultimately be tied to driving forces of the change. In addition, most studies examining degradation in the region did so without explicitly separating human-induced from climate-induced influences. More importantly, there is hardly any documentation that categorized levels of degradation in the catchment. Whilst Endale et al. (2023) stressed the cruciality of defining the degree, extent, and severity of degradation to guide rehabilitation efforts effectively, Veach et al. (2017) enhanced the assertion by citing dearth of data on the distribution pattern of degradation threats as critical limitation in the identification of spatial priorities for conservation actions.

Owing to these gaps, the study seeks to give a more nuanced assessment of degradation risk within the Greater uMngeni catchment, using intensity analysis and residual trend analysis. This research will not only identify degradation hotspots but also categorize these areas based on the severity of degradation. To the extent of our knowledge, this is the first study of its kind that quantitatively assesses the severity of human-induced vegetation loss in the study region. Thus, the research aims to investigate the vegetation degradation pattern within the region and map hotspots of human-induced degradation. The findings will provide a decision-making reference for the management and protection of vegetated spaces and the determination of areas of priority for ecological investment.

The main objectives of the study are:

1. Quantify and map the severity of vegetation degradation within the Greater uMngeni Catchment using intensity analysis and residual trend analysis (RESTREND).
2. Separate anthropogenic influences from climate variability on vegetation trends within the catchment, using RESTREND.
3. Identify hotspots of significant degradation as priority areas for potential investment in restoration activities.

2 Materials and Methods

2.1 Study Area

The greater uMngeni catchment (Fig. 1), situated in KwaZulu-Natal province of South Africa, spans an area of about 14,330 km^2. The region lies between longitudes 29°78′ to 30°42′ East and latitudes 29°34′ to 29°63′ South (Fig. 1). The greater uMngeni River catchment encompasses upper Mooi and Mvoti catchments supporting more than five million inhabitants, amounting to about 42.5% of the total KwaZulu-Natal (KZN) population (WWF and Booklet 2015). Umngeni catchment is significant for contributing about a fifth of South Africa's gross national product (Schulze et al. 2004). Numerous tributaries and water storage dams feed into the catchment and these include Nagle, Albert Falls, Midmar, and Inanda dams (Namugize et al. 2018), which regulate the water flow and provide water for domestic, industrial, and agricultural use. The catchment experiences a subtropical climate

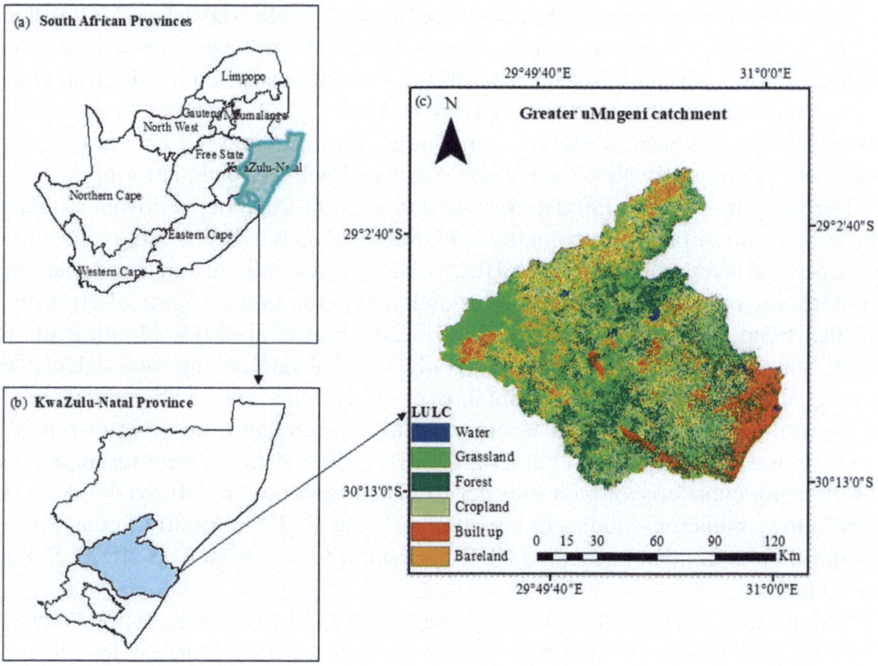

Fig. 1 Study area map layout for the Greater uMngeni catchment

with significant seasonal variations in rainfall averaging 1000 mm. The topography of the region can be described as ranging from coastal plains to mountainous regions, contributing to varied vegetation such farmlands, wetlands and grasslands (Schulze et al. 2004). Although the region is depicted as a hotspot of ecological diversity, the catchment landscape has been largely modified and fragmented by human unsustainable actions and is under increasing pressure to meet the growing water demands of the region (Moodley et al. 2023). The catchment faces numerous challenges, including water shortages, poor water quality, and sediment yield issues (Awuah et al. 2023).

2.2 *Acquisition of Data and Processing*

The data used in the present study were retrieved from the United States Geological Survey Global Visualization Viewer (USGS) (www.earthexplorer.usgs.gov). Landsat 4-5 TM and Landsat 8 OLI, acquired during 2000 and 2003 were employed in the present study. Four scenes were downloaded that corresponded with path 163 and row 52, path 164, row 52, path 163, row 53, and path 164, row 53.

The NDVI maps were generated using the 16-day MODIS NDVI data (MOD13Q1 level-3 gridded product, version 6), accessible at 250 m spatial resolution. The MODIS product was preferred because of its time series consistency (Montfort et al. 2021; Wingate et al. 2019). Yearly maximum MODIS NDVI images, covering the 2000–2023 period, were retrieved from Google Earth Engine (GEE) and for each pixel, maximum yearly NDVI value was calculated within cloud platform.

The study utilized rainfall data for the period 2000–2023 derived from Climate Hazards Group InfraRed Precipitation with Station data (CHIRPS) using GEE platform. Several investigators utilized CHIRPS data to assess rainfall trends and examine relationships between climate variables and vegetation indices, particularly using residual trend approach (Hentze et al. 2017; Ibrahim et al. 2015; Montfort et al. 2020; Sun et al. 2022). In the current study, optimal rainfall sum was calculated at pixel level, and as cumulated rainfall over the growing season (November (n − 1) to April (n)) each year to take into account the time lag from onset of rainfall to vegetation response (Abel et al. 2020). All CHIRPS datasets were resampled to 250 m using cubic convolution in order to match with the MODIS NDVI dataset. This follows numerous studies in literature (Abel et al. 2018; Montfort et al. 2021; Solano et al. 2022) that resampled CHIRPS dataset to match MODIS NDVI 250 m resolution.

Soil moisture data was downloaded from NASA POWER website (https://power.larc.nasa.gov/data-access-viewer/). Root zone soil wetness (Surface to 100 cm below) was utilized. The soil data was resampled to the MODIS NDVI resolution to ensure similarity and consistency among all the datasets.

2.3 *Methodology*

2.3.1 Maximum Likelihood Classification

The greater Umngeni was categorized into six distinct land cover classes, basing on insights gathered from prior researches on uMngeni catchment. These are built up, cropland, forest, grassland, water, and bareland. Three hundred training polygons were established in Google earth Pro, 50 for each of the six LU/LC classes, following guidelines provided by Congalton and Green (2019). Maximum likelihood classifier, performed in ArcMap version 10.7, was used for classification. The data used for validation consisted of 390 samples that were randomly distributed among the LULC classes and independent from the training data.

2.3.2 Intensity Analysis

Land use/land cover transitions over the 24-year period were obtained using post classification change detection considering two time points, in ArcGis Pro. Transition matrices were generated to compute the area changes between different land use categories. The land use transitions' intensity was calculated between the observed versus expected changes, following the methods outlined by Pontius et al. (2004). Graphs showing the categorical and transitional intensities were produced.

2.3.3 Residual Trend Analysis (RESTREND)

The RESTREND technique was implemented to separate the impact of climate variations and anthropogenic influences from the NDVI trend. The study executed pixel-based linear regression analysis using the yearly maximum NDVI, and time series of rainfall and moisture, following some studies (Ibrahim et al. 2015; Matin et al. 2019). According to Sehler et al. (2019), stronger correlations exist between soil moisture and precipitation particularly in arid regions and areas characterized by limited vegetation. Ibrahim et al. (2015) posited that accounting for rainfall and moisture in the RESTREND approach in such areas allows for a nuanced assessment of vegetation health, critical for managing ecosystems under climate and anthropogenic pressures. Initially, the approach involved computation of a linear regression between yearly maximum NDVI, a proxy for vegetation vigour (Chen et al. 2023; Sun et al. 2022) and precipitation or moisture. Further, the regression models were used to capture the residual NDVI (disparity between observed NDVI and predicted rainfall-derived NDVI) for the respective models, which when regressed against time show areas of either improvement or degradation (Feng et al. 2019; Zhao et al. 2021, 2011). Following studies by Ibrahim et al. (2015) and Montfort et al. (2021), significance of trends was assessed using p-values at 95% confidence interval in order to map only significant trends. Further, positive significant trends were masked and only significant negative trends were considered in further analyses. In addition, the Mann–Kendall coefficient τ, applied to annual data, was computed to infer monotonicity of residual trends from 2000 to 2023. The R statistical software was used for the RESTREND analyses. The spatial analysis of the residual trend was executed to establish regions characterized by significant declining or increasing trends of NDVI residuals.

2.3.4 Accuracy Assessment

To validate the RESTREND model, a deforestation map was produced from change detection map, where deforestation encompassed transitions where either forests or grasslands were converted to either bareland, cropland or built-up areas. That is, LULC changes that involved loss of vegetation. All the other changes were lumped into the class "other changes". The Jaccard Index, sometimes referred to as the Jaccard Similarity Coefficient, was used as a metric to establish similarity and accuracy between two raster maps. This approach was used by Chung et al. (2019) in quantitative comparison of presence-absence data in evaluation of species co-existences as well as by Kulkarni and Vijaya (2021) in validating land cover classification using spectral indices. Jaccard Index involves measuring of the degree of spatial overlap between observed (reference) data and predicted (classified) data (Chung et al. 2019).

Jaccard Index Formula

$$J(A, B) = (|A \cap B|)/(|AUB|), \text{ where,}$$

$|A \cap B|$ is the number of records (or area) in the intersection of the two rasters, that is, the pixels classified as degraded in both reference and classified maps. And, $|A \cup B|$ is the number of records (or area) in the union of the two sets, that is, all pixels classified as degraded in either reference or classified maps.

3 Results

3.1 Land Use/land Cover Change Dynamics

The spatial distribution of LULC in 2000 and 2023 is represented in Fig. 2. The obtained classification accuracies were 87.1, and 92.6% for the period 2000, and 2023, respectively. In other respects, intensity analysis results showed that the naturally vegetated area (forest and grassland) was estimated at 44.2 and 33.2% of the study area for 2000 and 2023). Importantly, the analysis revealed substantial shifts in land use patterns, with natural vegetation cover (forests and grassland) decreasing by about 16 and 27% (Table 1), respectively, over the period from 2000 to 2023. Whilst cropland and built-up areas increased by 300 km^2 and 678 km^2 respectively, forest areas shrank by about 437 km^2, signalling deforestation, while grassland saw a 926 km^2 reduction, potentially reflecting increased land use pressure and conversion for agriculture or urbanization. At the same time, built up and water classes showed increasing trends during the period with bareland declining by about 42% (Fig. 3).

3.2 Transition Between Natural Vegetation Cover, Built up and Cropland

The graphical representations in Fig. 4 highlight land-use changes from 2000 to 2023, focusing on transitions involving forest and grassland. Intensity analysis reveals that built-up and cropland are the actively expanding categories, as their gain intensities exceed the uniform line, whereas bareland is primarily losing, indicated by its intensity lying behind the line. In Fig. 4, d, bareland exhibits the highest transition intensities into both built-up and cropland, confirming its susceptibility to transformation. Forest encroachment is more pronounced for cropland gain than for built-up, with transition intensities slightly above the 10% threshold. Grassland also experiences notable transitions, primarily towards cropland. Figure 4a shows that land conversion

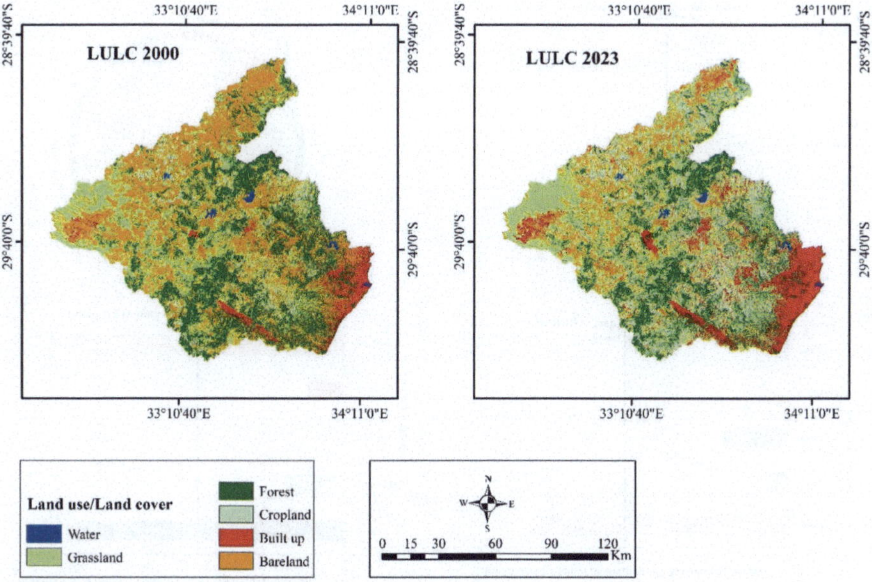

Fig. 2 Maps showing LULC for the years 2000 and 2023

Table 1 Table showing magnitudes of LULC changes from 2000 to 2023 in km^2, as well as percentage area changes

LULC category	2000	2023	Net change (%)
Water	63.6	78.0	22.6
Forest	2790.7	2353.6	−15.7
Built up	931.0	1608.8	72.8
Cropland	2995.3	3294.9	10.0
Grassland	3383.2	2457.3	−27.4
Bareland	5455.7	3141.5	−42.4

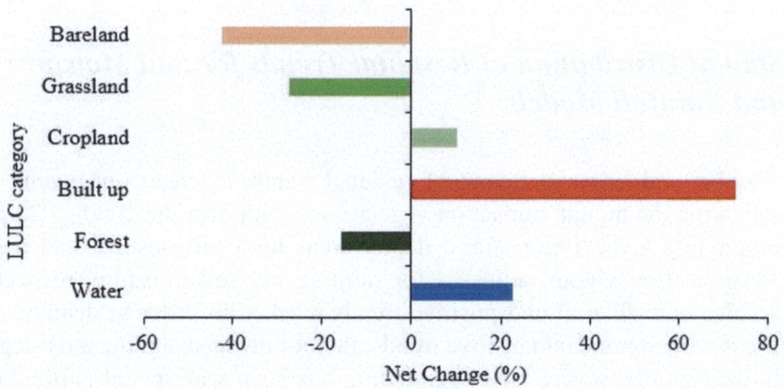

Fig. 3 Bar graph showing net change in area of land use/land cover categories from 2000 to 2023

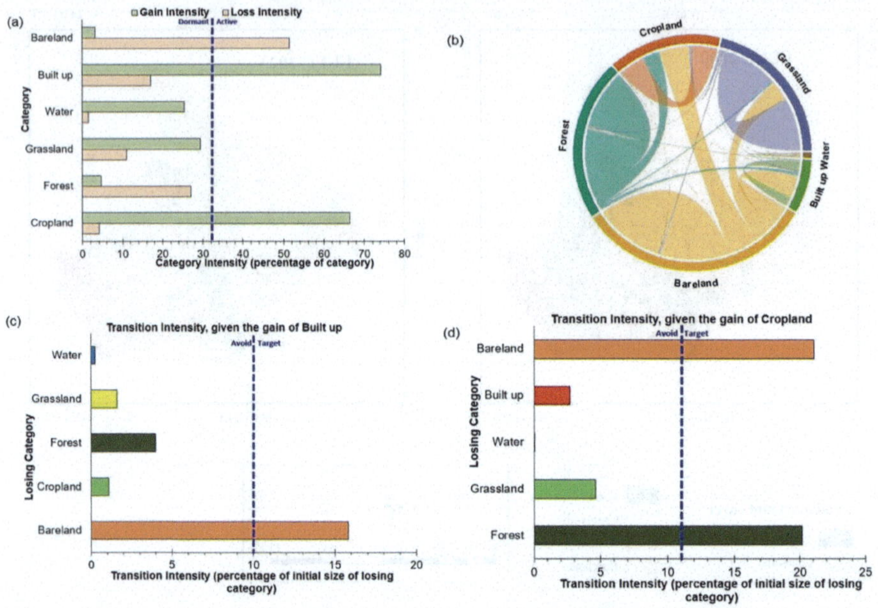

Fig. 4 Intensity analysis for category-level changes (**a**), interconnection between the changes (sankey plot) (**b**), observed transitions given the gross gain of built up (**c**), and gross gain of cropland (**d**)

into built-up areas mainly targets bareland, with transition intensities exceeding 15%. Forest and grassland follow as the next targeted categories. The Sankey diagram in Fig. 4b further illustrates significant flows from forest to cropland and built-up areas, and from grassland to cropland, signalling substantial losses of these natural land cover types to agriculture and urban expansion.

3.3 Spatial Distribution of Residual Trends for Soil Moisture and Rainfall Models

Figure 5a shows the spatial layout of residual trends in greater uMngeni catchment, indicating the human impact on vegetation health over the 2000–2023 period at a municipality level. Figure 5a, c depict areas with full positive and negative trends in vegetation vigour, adjusted for rainfall and soil moisture, respectively. These results were filtered and positive trends masked in order to display, exclusively, areas with significant negative trends, thereof deemed significantly degraded (Fig. 5b, d). Notably, severe land degradation has been widespread particularly in the Durban Metro for both rainfall-controlled and moisture controlled RESTREND,

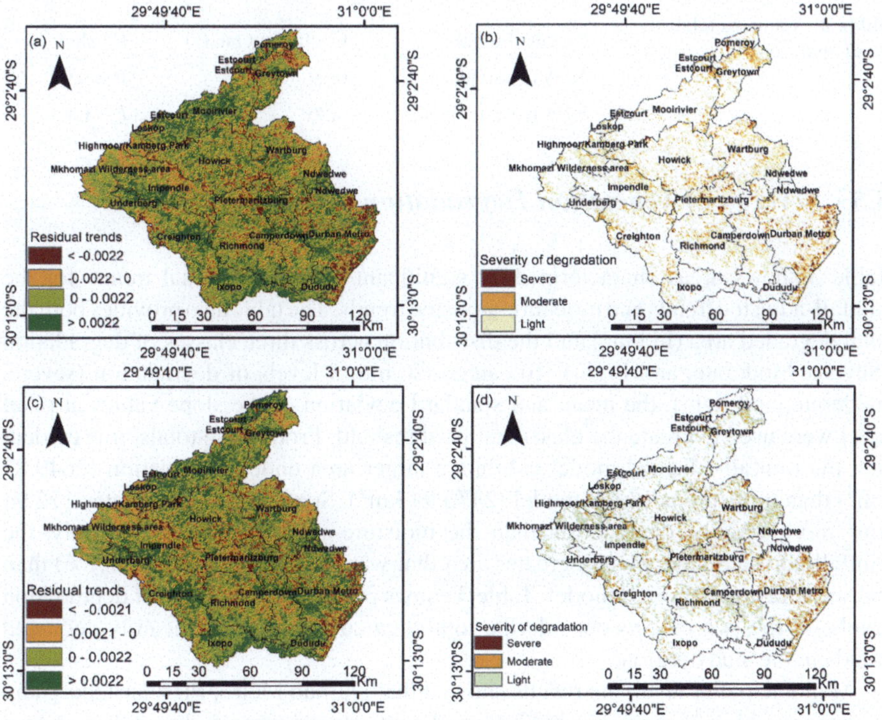

Fig. 5 Spatial layout of slopes of NDVI residuals from NDVI-rainfall regression (**a**) areas with significant negative trends at 95% confidence interval (**b**), and slopes of NDVI residuals from NDVI-moisture regression (**c**) and areas with significant negative trends at 95% confidence interval (**d**)

covering 32.7 and 31.3% of the area covered by significant negative trends, respectively. Clearly though, it is more pronounced for rainfall controlled RESTREND than moisture controlled RESTREND.

3.4 Consistency Analysis Using Mann–Kendall Approach

The two models show weakly positive trends in NDVI. For both models, the positive trend is statistically significant, ($\tau = 0.29, P < 0.05$) and ($\tau = 0.26, P < 0.05$) (Table 2). This implies that vegetation has been gradually improving in the study area over the years, even after accounting for the two climatic variables. The results suggest that the increasing NDVI trend is unlikely due to random chance.

Table 2 Mann–Kendall trend analysis

Residual trend	Coefficient tau (τ)	P value
NDVI-rainfall	0.29	$P < 0.05$
NDVI-soil moisture	0.26	$P < 0.05$

3.5 Severity of Vegetation Degradation

Table 3 shows areas characterized by significant negative residual trends for the rainfall-adjusted trends and moisture-adjusted trends. The table also provides both the total degraded area (in km^2) and the distribution across three classes of degradation (Severe, Moderate, and Light). To categorize into 3 levels of degradation (severe, moderate, and light), the mean and standard deviation of the slope values at pixel level were used to create the classification threshold. From these totals, it is evident that the rainfall-adjusted model exhibits a larger area under degradation (2649.87 km^2) than moisture-adjusted model (2476.94 km^2). Specifically, it indicates 172.94 km^2 more extreme degradation than the moisture-adjusted model. Similarly, the rainfall-adjusted model had a greater area that was severely degraded (17.1%) than the soil moisture adjusted model. Table 3 shows percentage coverage of degradation levels for the two models considering total area covered by significantly degraded trends in the study region.

At municipality level, the results showed that Durban Metro, Piermaritzburg and Pomeroy had larger areas covered by significant severe human-induced degradation over the study period. As indicated in Table 4, severely degraded patches covered 32.7, 29, and 22% of those areas, respectively. Highmoor/Kamberg park (1.8%) had the smallest areas categorized as severely degraded among all class categories.

The results of analysis indicate that during years when NDVI was higher than expected (positive residuals), the residuals exhibited more variability when adjusted for soil moisture than for rainfall. Rainfall adjusted residuals exhibited more consistency and predictable influence on NDVI. A scrutiny of the distribution pattern of positive and negative trends of the NDVI residuals from rainfall and moisture models portray that the rainfall-adjusted model (Fig. 6a) had a more pronounced negative trend as indicated by a larger spread, reaching lower values than the soil moisture

Table 3 Percentages of significant residual trend values for rainfall adjusted model and moisture adjusted model

Class	Rainfall-adjusted trends		Moisture adjusted-trends	
	Area/km^2	Percentage (%)	Area/km^2	Percentage (%)
Severe	451.75	17.1	402.63	16.2
Moderate	350.88	13.2	320.88	13.0
Light	1847.25	69.7	1753.44	70.8
Total	2649.88	100	2476.94	100

Table 4 Table presenting the extent of human-driven degradation across all municipalities

Municipality	Area (km^2)	Severity of degradation		
		Severe (%)	Moderate (%)	Light (%)
Richmond	183.9	10.8	13.0	76.2
Highmoor/Kamberg park	42	1.8	7.7	90.5
Wartburg	290.56	14.3	10.9	74.8
Durban Metro	523.69	32.7	15.0	52.3
Greytown	161.38	15.1	18.1	66.8
Loskop	18.2	4.5	12.0	83.5
Impendle	177.88	5.6	12.5	81.9
Escourt	10	18.1	18.8	63.1
Pietermaritzburg	146.56	29.0	12.1	58.9
Creighton	209.8	13.2	14.8	72.0
Howick	298.63	14.9	12.1	73.0
Mkhomazi wilderness area	110.88	1.3	7.4	91.3
Camperdown	132.75	11.5	13.9	74.6
Pomeroy	52.8	22.0	12.2	65.8
Mooi River	196.56	15.4	14.2	70.4
Underberg	25.4	5.9	13.5	80.5
Ixopo	28.9	8.2	10.4	81.4

model. This suggests that vegetation health, when adjusted for rainfall, experienced more considerable and variable declines compared to the model adjusted for soil moisture. The results also indicate that while vegetation improves after adjusting for either rainfall or soil moisture, soil adjusted model exhibited highly variable positive growth outcomes (larger IQR) which could reflect the complexities of water availability in different soil and ecological conditions. The rainfall adjusted NDVI residuals, in contrast, show a stronger positive trend (Fig. 6b) which were more consistent.

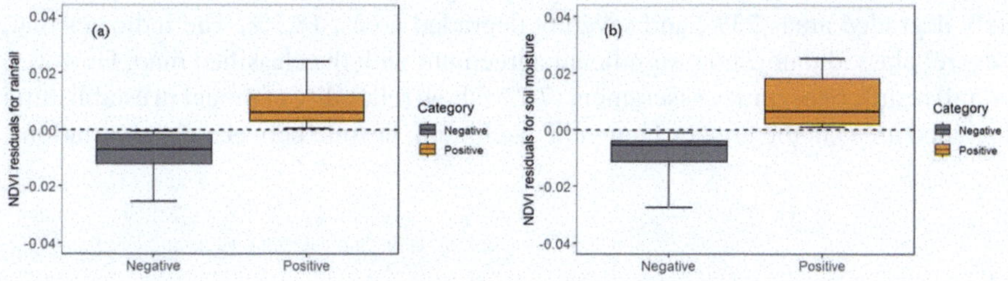

Fig. 6 Adjusted trends of NDVI residuals for **a** rainfall and **b** soil moisture models, categorized into negative and positive groups. The box plots show the median (central line), and interquartile range (box length), representing the distribution and variability of NDVI residuals

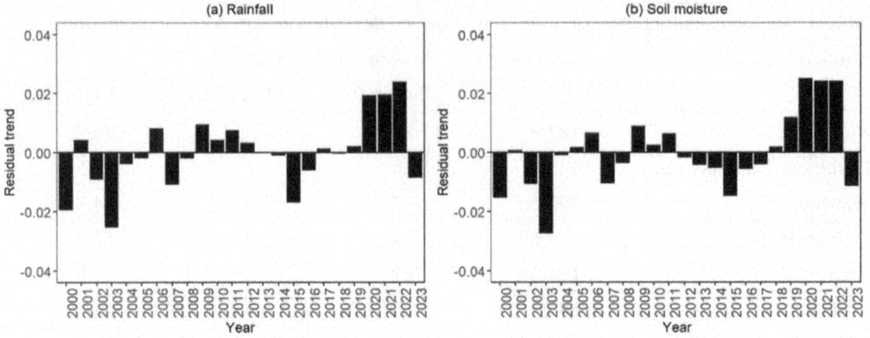

Fig. 7 Mean annual NDVI residuals for 2000 to 2023. Values were adjusted for **a** rainfall and **b** soil moisture

3.6 Temporal Variation of Annual NDVI Residual Trends

Figure 7 shows the temporal trend of the mean annual residuals spread across the study area. These graphs help identify years that experienced negative residual trends (land degradation) or positive residual NDVI (improvement). Throughout the study period, temporal analysis results revealed that the years 2000, 2003, and 2015 exhibited the most pronounced negative trends. The soil moisture model results (Fig. 7b) depicted a much clearer negative extent compared to the rainfall model from 2012 to 2017. Beyond 2017, contrarily, the land degradation trend reversed, with positive residuals becoming dominant in the time series. Period 2009–2011 as well as 2019–2022 showed pronounced positive residual trends for both models.

3.7 Accuracy Assessment

Results of Jaccard correlation coefficient revealed that the level of overlap between classified deforested area and degraded areas from RESTREND model (Fig. 8) was 74%. Of the points of overlap, the lightly degraded areas covered 58.7%, moderately degraded areas 23%, and severely degraded areas, 18.3%. The indicates that, 'severe' class, did not show significant agreement with the classified map. However, overall results of accuracy assessment (74%) show reliability of model in establishing degraded areas in the greater Umngeni catchment, despite between class variations.

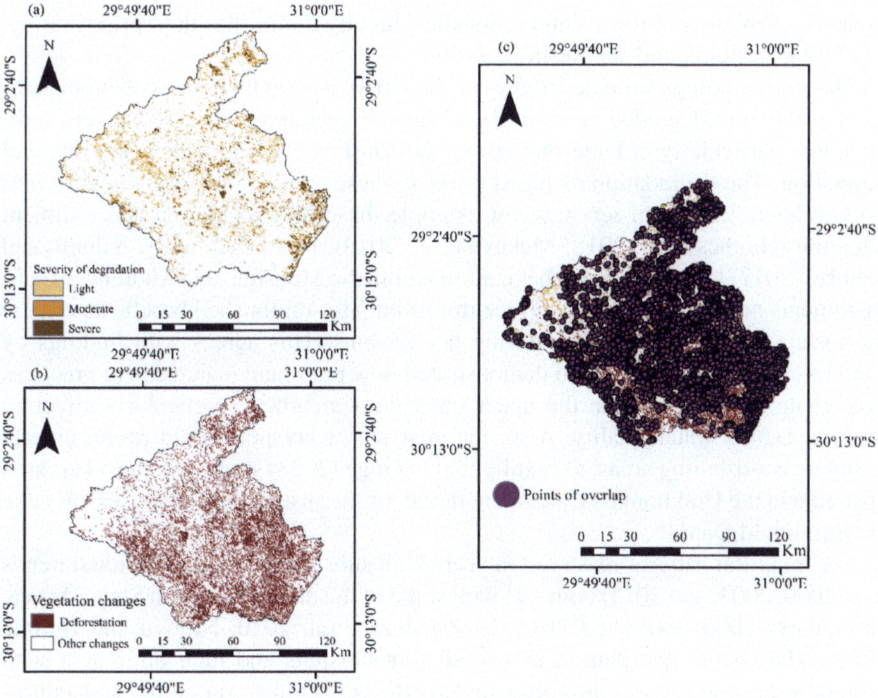

Fig. 8 Spatial distribution of areas of overlap (**c**) between degraded areas from RESTREND model (**a**) and deforested areas from LULC transition map (**b**)

4 Discussion

Vegetation degradation within the greater uMngeni Catchment emerges as a multifaceted challenge, intricately linked to land use transitions and anthropogenic activities. The study revealed areas of human induced degradation risk within greater uMngeni catchment through integration of land cover changes as well as residual trends. These degraded areas have profound implications for ecosystem services, which are integral to the ecological and socio-economic well-being of the region.

The incorporation of intensity analysis and RESTREND techniques provided robust intuitions into the spatiality of degradation patterns. Intensity analysis revealed that urban expansion and agricultural intensification were the predominant drivers of vegetation loss. The RESTREND results, on the other hand, revealed that approximately 17.1 and 16.2% of the catchment area were identified as severely degraded in rainfall- and moisture-adjusted models, respectively, with Durban Metro and Pietermaritzburg municipalities being the most degraded. The spatial distribution of degradation hotspots, identified through RESTREND analysis, reveals implications beyond vegetation loss, highlighting potential declines in ecosystem service capacity, as the loss of forests and grasslands directly diminishes their provisioning, regulating, cultural, and supporting services.

The study findings validate Hughes et al. (2018), who in their assessment of sediment yield from degraded areas in the uMngeni catchment, observed severe sediment loss particularly in Pietermaritzburg and Durban, due to the demise of natural vegetation. The degradation of forest cover in these areas significantly undermines water-related ecosystem services, for example, base flow regulation and sediment retention (Hughes et al. 2018; Makaya et al. 2019). In another study, Schutte and Schulze (2017) identified that urbanization within the Mpushini and Mkhondeni sub-catchments not only increased surface runoff but also diminished baseflows critical for sustained water availability during dry seasons. This agrees with findings by van Deventer et al. (2021), who demonstrated how poor land management practices exacerbate nutrient loads in the upper uMngeni Catchment, particularly affecting Midmar Dam's water quality. Also, the analysis of hotspots could reveal critical sediment contributing areas, as highlighted in Goge (2023) study, where soil erosion upstream in the Umkhomazi Catchment threatens the sustainability of reservoirs like the Smithfield Dam.

The temporal patterns observed in years with substantial negative residual trends (e.g. 2000, 2003, and 2015) coincide with some of the drought years in South Africa, particularly 2000–2005, and 2015–2018 (Meza et al. 2020; Ndlovu and Demlie 2020). The spatial distribution of degradation hotspots and their alignment with drought years underscores the compounded effects of climate variability and anthropogenic pressures. Since residuals in RESTREND analysis represent deviations from expected vegetation productivity based on rainfall, a negative residual during a drought reflects even poorer performance exacerbated by water unavailability emphasizing the compounded effects of climate variability and anthropogenic stressors on ecosystem service decline. The increasing frequency and severity of droughts under climate change, combined with the unrelenting degradation of natural ecosystems, exacerbate the loss of critical water-related services (Evans 2023).

Addressing vegetation degradation requires strategic EI investments in priority areas, particularly these degraded hotspots. Efforts such as those proposed by Hughes et al. (2018) and Gokool and Jewitt (2019), including sediment control and water retention interventions, can mitigate the adverse effects of fragmentation of vegetation loss ecosystem services. Additionally, the study by van Deventer et al. (2021) highlights the importance of land use planning that prioritizes ecological infrastructure protection to sustain water quality and ecological health.

Integrated catchment management approaches are essential to addressing these challenges. Investments in ecological infrastructure, such as riparian buffers and wetland rehabilitation, can enhance resilience to both anthropogenic and climatic pressures. As Hughes et al. (2018) emphasize, protecting these ecosystems sustains water-related services and safeguards the livelihoods and health of dependent communities. Proactive measures, such as those modeled by Schutte and Schulze (2017), demonstrate the importance of strategic urban planning in mitigating the hydrological impacts of land use transitions.

5 Conclusion

This study presented an investigation of vegetation degradation in the greater uMngeni catchment by combining land-use transitions and residual trend analysis over the past 24 years. The results revealed significant shifts in land use patterns, primarily driven by anthropogenic activities. This study has demonstrated that natural vegetation, particularly forests and grasslands, has declined substantially owing to increased urbanization and agricultural intensification. Severely degraded areas were evident mainly in Durban and Pietermaritzburg where 32.7 and 29% of the area were classified as severely degraded, respectively. This study highlights probable areas of intersection of degraded areas and ecosystem service provision within the greater uMngeni Catchment. By leveraging advanced remote sensing techniques, it provides actionable insights into the drivers and impacts of degradation, offering a roadmap for sustainable landscape management. The study showcases the critical areas for either rehabilitation or restoration of ecological infrastructure within the catchment. Implementing targeted interventions, such as erosion control, afforestation, and sustainable agricultural practices, is essential to mitigate further degradation and enhance the catchment's ecological functions.

Acknowledgements This research was supported by NRF Chair in Land use Planning and Management (Grant no. 84157).

Author Contributions Dadirai Matarira: Conceptualization, Data curation, Formal analysis, Methodology, Writing—original draft, Writing—review and editing. Onisimo Mutanga: Conceptualization, Methodology, Supervision, Funding acquisition, Writing—review and editing. Marco Vizzari: Conceptualization, Data curation, Writing—review and editing.

Funding This research was financially supported by the NRF Chair in Landuse Planning and Management (Grant no. 84157).

Data Availability Statement Data are available from the authors upon reasonable request.

Competing Interests The authors declare no conflict of interest.

References

Abel C, Brandt M, Tagesson T, Fensholt R (2018) Improved characterization of dryland degradation using trends in vegetation/rainfall sequential linear regression (SERGS-TREND). In: IGARSS 2018–2018 IEEE international geoscience and remote sensing symposium

Abel C, Horion S, Tagesson T, De Keersmaecker W, Seddon AWR, Abdi AM, Fensholt R (2020) The human–environment nexus and vegetation–rainfall sensitivity in tropical drylands. Nat Sustain 4(1):25–32. https://doi.org/10.1038/s41893-020-00597-z

Akinyemi FO, Pontius RG, Braimoh AK (2016) Land change dynamics: insights from Intensity Analysis applied to an African emerging city. J Spatial Sci 1–15. https://doi.org/10.1080/14498596.2016.1196624

Akodéwou A, Oszwald J, Saïdi S, Gazull L, Akpavi S, Akpagana K, Gond V (2020) Land use and land cover dynamics analysis of the Togodo protected area and its surroundings in Southeastern Togo, West Africa. Sustainability 12(13):https://doi.org/10.3390/su12135439

Arfasa GF, Owusu-Sekyere E, Doke DA (2024) Spatio-Temporal analysis of land use/land cover change and the implications on sustainable development goals in the Vea catchment of Ghana. Rangel Ecol Manage 94:83–94. https://doi.org/10.1016/j.rama.2024.02.005

Awuah AK, Stuart-Hill S, Trotter-Boardman D (2023) Pandora's box: assessing the current trends and challenges of IWRM in the uMngeni catchment. Water Policy 25(1):23–37. https://doi.org/10.2166/wp.2023.183

Bhungeni O, Ramjatan A, Gebreslasie M (2024) Evaluating machine-learning algorithms for mapping LULC of the uMngeni catchment area, KwaZulu-Natal. Remote Sens 16(12) https://doi.org/10.3390/rs16122219

Chen H, Liu X, Ding C, Huang F (2018) Phenology-based residual trend analysis of MODIS-NDVI time series for assessing human-induced land degradation. Sensors (Basel) 18(11). https://doi.org/10.3390/s18113676

Chen Z, Zhang X, Jiao Y, Cheng Y, Zhu Z, Wang S, Zhang H (2023) Investigating the spatio-temporal pattern evolution characteristics of vegetation change in Shendong coal mining area based on kNDVI and intensity analysis. Front Ecol Evol 11:1344664. https://doi.org/10.3389/fevo.2023.1344664

Chung NC, Miasojedow B, Startek M, Gambin A (2019) Jaccard/Tanimoto similarity test and estimation methods for biological presence-absence data. BMC Bioinf 20(Suppl 15):644. https://doi.org/10.1186/s12859-019-3118-5

Congalton RG, Green K (2019) Assessing the accuracy of remotely sensed data: principles and practices. CRC Press

Endale T, Diels J, Tsegaye D, Kassaye A, Belayneh L, Verdoodt A (2023) Farmer-science-based soil degradation metrics guide prioritization of catchment-tailored control measures. Environ Dev 45:100783. https://doi.org/10.1016/j.envdev.2022.100783

Evans W (2023) Mapping of Woody invasive alien plants in the uMngeni catchment

Feng D, Wang J, Fu M, Liu G, Zhang M, Tang R (2019) Spatiotemporal variation and influencing factors of vegetation cover in the ecologically fragile areas of China from 2000 to 2015: a case study in Shaanxi Province. Environ Sci Pollut Res Int 26(28):28977–28992. https://doi.org/10.1007/s11356-019-06096-9

Goge M (2023) Identifying potentially significant sediment contributing areas and assessment of interventions to limit sediment yield into the proposed smithfield dam on the umkhomazi river. University of KwaZulu-Natal

Gokool S, Jewitt GPW (2019) Identifying hotspots for investment in ecological infrastructure within the uMngeni catchment, South Africa. Phys Chem Earth Parts a/b/c 114:102807. https://doi.org/10.1016/j.pce.2019.11.003

Hentze K, Thonfeld F, Menz G (2017) Beyond trend analysis: how a modified breakpoint analysis enhances knowledge of agricultural production after Zimbabwe's fast track land reform. Int J Appl Earth Obs Geoinformation 62:78–87. https://doi.org/10.1016/j.jag.2017.05.007

Hughes CJ, De Winnaar G, Schulze RE, Mander M, Jewitt GPW (2018) Mapping of water-related ecosystem services in the uMngeni catchment using a daily time-step hydrological model for prioritisation of ecological infrastructure investment—Part 2: outputs. Water SA 44(4 October). https://doi.org/10.4314/wsa.v44i4.08

Ibrahim Y, Balzter H, Kaduk J, Tucker C (2015) Land degradation assessment using residual trend analysis of GIMMS NDVI3g, soil moisture and rainfall in Sub-Saharan West Africa from 1982 to 2012. Remote Sens 7(5):5471–5494. https://doi.org/10.3390/rs70505471

Kilama Luwa J, Majaliwa JGM, Bamutaze Y, Wasswa D, Pilesjö P, Espoir B (2024) Intensity analysis applied to land use and land cover change and transitions in a fragile tropical mountain environment: a case of sironko catchment on Mt. Elgon, Eastern Uganda. Afr Geogr Rev 1–21. https://doi.org/10.1080/19376812.2024.2331493

Kulkarni K, Vijaya PA (2021) Multi-index validation mechanisms for the land cover classification of multispectral images: a case study of Kabini Reservoir. In: Data management, analytics and innovation: Proceedings of ICDMAI, vol 1. Springer, pp 75–89

Li J, Peng S, Li Z (2017) Detecting and attributing vegetation changes on China's Loess Plateau. Agric for Meteorol 247:260–270. https://doi.org/10.1016/j.agrformet.2017.08.005

Mabuda MO, Shoko C, Dube T, Mazvimavi D (2024) An analysis of the effects of changes in land use and land cover on runoff in the Luvuvhu catchment, South Africa. Remote Sens Appl: Soc Environ 33:101144. https://doi.org/10.1016/j.rsase.2024.101144

Makaya N, Dube T, Seutloali K, Shoko C, Mutanga O, Masocha M (2019) Geospatial assessment of soil erosion vulnerability in the upper uMgeni catchment in KwaZulu Natal, South Africa. Phys Chem Earth Parts a/b/c 112:50–57. https://doi.org/10.1016/j.pce.2019.02.012

Manandhar R, Odeh IOA, Pontius RG (2010) Analysis of twenty years of categorical land transitions in the Lower Hunter of New South Wales, Australia. Agr Ecosyst Environ 135(4):336–346. https://doi.org/10.1016/j.agee.2009.10.016

Mander M, Jewitt G, Dini J, Glenday J, Blignaut J, Hughes C, Marais C, Maze K, van der Waal B, Mills A (2017) Modelling potential hydrological returns from investing in ecological infrastructure: case studies from the Baviaanskloof-Tsitsikamma and uMngeni catchments, South Africa. Ecosyst Serv 27:261–271. https://doi.org/10.1016/j.ecoser.2017.03.003

Matin S, Ghosh S, Behera MD (2019) Assessing land transformation and associated degradation of the west part of Ganga River Basin using forest cover land use mapping and residual trend analysis. J Arid Land 11:29–42

Meza I, Eyshi Rezaei E, Siebert S, Ghazaryan G, Nouri H, Dubovyk O, Gerdener H, Herbert C, Kusche J, Popat E, Rhyner J, Jordaan A, Walz Y, Hagenlocher M (2020) Drought risk for agricultural systems in South Africa: Drivers, spatia l patterns, and implications for drought risk management. Sci Total Environ 799:149505. https://doi.org/10.1016/j.scitotenv.2021.149505

Montfort F, Begue A, Leroux L, Blanc L, Gond V, Cambule AH, Remane IA, Grinand C (2021) From land productivity trends to land degradation assessment in Mozambique: effects of climate, human activities and stakeholder definitions. Land Degrad Dev 32(1):49–65

Montfort F, Bégué A, Leroux L, Blanc L, Gond V, Cambule AH, Remane IAD, Grinand C (2020) From land productivity trends to land degradation assessment in Mozambique: Effects of climate, human activities and stakeholder definitions. Land Degrad Dev 32(1):49–65. https://doi.org/10.1002/ldr.3704

Moodley K, Toucher ML, Lottering RT (2023) Simulating future land-use within the uThukela and uMngeni catchments in KwaZulu-Natal. Sci Afr 20:e01666. https://doi.org/10.1016/j.sciaf.2023.e01666

Mullapudi A, Vibhute AD, Mali S, Patil CH (2023) Spatial and seasonal change detection in vegetation cover using time-series landsat satellite images and machine learning methods. SN Comput Sci 4(3):254. https://doi.org/10.1007/s42979-023-01710-7

Mumtaz F, Li J, Liu Q, Arshad A, Dong Y, Liu C, Zhao J, Bashir B, Gu C, Wang X, Zhang H (2023) Spatio-temporal dynamics of land use transitions associated with human activities over Eurasian Steppe: Evidence from improved residual analysis. Sci Total Environ 905:166940. https://doi.org/10.1016/j.scitotenv.2023.166940

Namugize JN, Jewitt G, Graham M (2018) Effects of land use and land cover changes on water quality in the uMngeni river catchment, South Africa. Phys Chem Earth Parts a/b/c 105:247–264. https://doi.org/10.1016/j.pce.2018.03.013

Ndlovu MS, Demlie M (2020) Assessment of meteorological drought and wet conditions using two drought indices Across KwaZulu-Natal Province, South Africa. Atmosphere 11(6):623. https://doi.org/10.3390/atmos11060623

Nyatuame M, Agodzo S, Amekudzi LK, Brako B (2023) Assessment of past and future land use/cover change over Tordzie watershed in Ghana. Front Environ Sci 11:1139264. https://doi.org/10.3389/fenvs.2023.1139264

Pandey PC, Koutsias N, Petropoulos GP, Srivastava PK, Ben Dor E (2019) Land use/land cover in view of earth observation: data sources, input dimensions, and classifiers—A review of the state of the art. Geocarto Int 36(9):957–988. https://doi.org/10.1080/10106049.2019.1629647

Prăvălie R (2021) Exploring the multiple land degradation pathways across the planet. Earth Sci Rev 220:103689. https://doi.org/10.1016/j.earscirev.2021.103689

Rangeti I, Dzwairo BR (2021) uMngeni Basin water quality trend analysis for river health and treatability fitness. In: River basin management-sustainability issues and planning strategies
Schulze RE, Lorentz S, Kienzle S, Perks L (2004) Case study 3: modelling the impacts of land-use and climate change on hydrological responses in the mixed underdeveloped/developed Mgeni catchment, South Africa. In: Vegetation, water, humans and the climate: a new perspective on an interactive system. Springer, Berlin, Heidelberg, pp 441–453
Schutte S, Schulze RE (2017) Projected impacts of urbanisation on hydrological resource flows: a case study within the uMngeni Catchment, South Africa. J Environ Manage 196:527–543. https://doi.org/10.1016/j.jenvman.2017.03.028
Schütte S, Schulze RE (2017) Projected impacts of urbanisation on hydrological resource flows: a case study within the uMngeni Catchment, South Africa. J Environ Manage 196:527–543. https://doi.org/10.1016/j.jenvman.2017.03.028
Sehler R, Li J, Reager J, Ye H (2019) Investigating relationship between soil moisture and precipitation globally using remote sensing observations. J Contemp Water Res Educ 168(1):106–118
Seutloali KE, Beckedahl HR (2015) Understanding the factors influencing rill erosion on roadcuts in the south eastern region of South Africa. Solid Earth 6(2):633–641. https://doi.org/10.5194/se-6-633-2015
Solano F, Praticò S, Ripa MN, Modica G (2022) Assessing differences in land productivity trends to climatic data in arid and semi-arid zones: a study case in Northern Mozambique. In: Conference of the Italian Society of Agricultural Engineering
Sun R, Chen S, Su H (2022) Trend analysis and driving factors of vegetation dynamics in Northern China from 1982 to 2015. Remote Sens 14(23). https://doi.org/10.3390/rs14236163
Tilahun ZA, Bizuneh YK, Mekonnen AG (2024) A spatio-temporal analysis of the magnitude and trend of land use/land cover changes in Gilgel Gibe Catchment, Southwest Ethiopia. Heliyon 10(2):e24416. https://doi.org/10.1016/j.heliyon.2024.e24416
van Deventer R, Morris CD, Hill TR, Rivers-Moore NA (2021) Use of biological and water quality indices to evaluate conditions of the Upper uMngeni Catchment, KwaZulu-Natal, South Africa. Afr J Aquat Sci 47(1):11–22. https://doi.org/10.2989/16085914.2021.1941743
Veach V, Moilanen A, Di Minin E (2017) Threats from urban expansion, agricultural transformation and forest loss on global conservation priority areas. PLoS ONE 12(11):e0188397. https://doi.org/10.1371/journal.pone.0188397
Wessels KJ, Prince S, Frost P, Van Zyl D (2004) Assessing the effects of human-induced land degradation in the former homelands of northern South Africa with a 1 km AVHRR NDVI time-series. Remote Sens Environ 91(1):47–67
Wessels KJ, Van Den Bergh F, Scholes R (2012) Limits to detectability of land degradation by trend analysis of vegetation index data. Remote Sens Environ 125:10–22
Wingate VR, Phinn SR, Kuhn N (2019) Mapping precipitation-corrected NDVI trends across Namibia. Sci Total Environ 684:96–112. https://doi.org/10.1016/j.scitotenv.2019.05.158
WWF & Booklet (2015) Water StewardShip in the Greater UMngeni River catchment
Zhao C, Yan Y, Ma W, Shang X, Chen J, Rong Y, Xie T, Quan Y (2021) RESTREND-based assessment of factors affecting vegetation dynamics on the Mongolian Plateau. Ecol Model 440:109415
Zhao Y, He C, Zhang Q (2011) Monitoring vegetation dynamics by coupling linear trend analysis with change vector analysis: a case study in the Xilingol steppe in northern China. Int J Remote Sens 33(1):287–308. https://doi.org/10.1080/01431161.2011.594102

Assessment of Soil Organic Carbon Across a Woody Encroached Bisley Nature Reserve Using PlanetScope Data

Sfundo Mthiyane, Trylee Nyasha Matongera, and John Odindi

Abstract Globally, grasslands constitute one of the largest land ecosystems that sequestrate substantial amounts of Soil Organic Carbon (SOC) and provide various ecological services such as regulating the climate, improving water quality, and supporting biodiversity. Woody encroachment has pronounced impacts on grassland ecosystems, exerting substantial influence on their structure, ecological dynamics, and SOC distribution. To the best of our knowledge, concerns and discrepancies about the influence of woody proliferation on SOC within protected grasslands remain largely underexplored. Remote sensing offers cost-effective, time-efficient, and environmentally friendly means to quantify SOC distribution and variability in grasslands. Consequently, this chapter sought to quantify the spatial variability of SOC across a pristine and woody encroached grassland using PlanetScope image data. Employing a stratified sampling technique, 254 samples were collected, and the Loss-on-Ignition procedure adopted to derive SOC. The PlanetScope spectral bands and derived vegetation indices were utilized to quantify SOC variability between woody encroached and pristine grassland. A tenfold cross validation technique was adopted to train the Deep Neural Network (DNN) model using 21 input variables including SOC values. To improve model accuracy, hyper-parameter tuning was employed. The findings indicate a higher concentration of SOC sequestration across landscapes affected by proliferation of woody plants compared to pristine grasslands. The DNN model produced acceptable accuracies with Root Mean Square Error of 1.90 t/ha and R^2 value of 0.64. The research establishes a framework for continuous assessment of SOC stock in protected grasslands.

S. Mthiyane (✉) · J. Odindi
Discipline of Geography and Environmental Science, School of Agricultural Earth and Environmental Sciences, University of KwaZulu-Natal, Scottsville, Pietermaritzburg, South Africa
e-mail: 218030355@stu.ukzn.ac.za

J. Odindi
e-mail: odindi@ukzn.ac.za

T. N. Matongera
School of Biosciences, University of Nottingham Malaysia, Semenyih, Malaysia
e-mail: Trylee.Matongera@nottingham.edu.my

O. Mutanga et al. (eds.), *Revealing Ecosystem Services Through Geospatial Technologies*, Springer Remote Sensing/Photogrammetry,
https://doi.org/10.1007/978-3-031-98048-0_18

Keywords Grassland; Woody encroachment; Soil organic carbon · PlanetScope · Deep neural network

1 Introduction

Globally, approximately 40% of the Earth's terrestrial ecosystems are covered by grasslands (Wang and Fang 2009). As one of the predominant ecosystems (Loveland and Webb 2003; Merino et al. 2004), grasslands are essential for providing unique and significant ecological services, such as reducing soil erosion, preserving biodiversity, and enhancing food production (Bugalho and Abreu 2008; Carlier et al. 2009). Furthermore, grasslands constitute one of the most significant repositories of Soil Organic Carbon (SOC), crucial for providing various ecological services such as regulating the climate, improving water quality, and supporting biodiversity (Sommer and Bossio 2014; Vågen and Winowiecki 2013). In grasslands, SOC helps to improve soil structure, retain water, and provide plant nutrients (Conant and Paustian 2002; Gerke 2022); characteristics that are critical for increasing the resilience of soils towards droughts and floods. In this regard, a decline in SOC can trigger a loss of soil quality and productivity. The accumulation and loss of SOC in grasslands are primarily influenced by land cover characteristics and change (Ratajczak et al. 2012). Woody proliferation is a natural phenomenon that has been dominant in several grassland ecosystems across the globe for the past 200 years (Ding and Eldridge 2024; Ratajczak et al. 2012). According to Honda and Durigan (2016), the proliferation of woody plants is a key phenomenon causing landscape degradation and significantly affect the ecology, land surface, and soil biophysical characteristics of grasslands.

Understanding the trade-offs and adverse influence of woody encroachment is important because various natural grasslands across South Africa are degrading due to woody proliferation (Grellier et al. 2013). According to Mureva et al. (2018), woody proliferation affects about 15 million hectares of grasslands in South Africa. Additionally, Gartzia et al. (2014) established that the invasion of woody plants amplifies water and nutrient balance changes at various spatial and temporal extents, which are primary resources for numerous animal species thriving in grasslands. Ratajczak et al. (2012) notes that woody proliferation has an adverse influence on the abundance of species (grazers) in grassland ecosystems, particularly in areas receiving high precipitation. Their study observed a decrease of species abundance and richness in grasslands experiencing woody proliferation. According to Mochi et al. (2022), woody encroachment minimizes food source availability for numerous grazers within the grassland. Although the productivity of herbivore biomass can be maintained in areas influenced by woody proliferation, the herbivore assemblage re-structuring causes intense changes in their average body size (Smit and Prins 2015). Moreover, woody proliferation reduces the amount of water available in the soil and promotes surface runoff, which may potentially result in seasonal drought, consequently influencing grassland productivity (Schreiner-McGraw et al. 2020).

Literature reveals that a switch from a grassland to a woody dominated ecosystem may potentially impact SOC sequestration (Alberti et al. 2011; Archer et al. 2017; Blaser et al. 2014). Jackson et al. (2002) established that the dominance of woody plants substantially amplifies SOC sink in grassland ecosystems. However, Li et al. (2016) notes that there are still contradictions and debates on the potential influence of proliferation of woody plants on SOC an ecosystem level. Whereas Alberti et al. (2011) suggest that woody encroachment increases the concentration of SOC stocks in grasslands, others (Mureva et al. 2018; Zhou et al. 2017) observed a reduction of SOC due to the encroachment of woody plants in localized grasslands. These inconsistencies can be linked to various environmental factors that include historical land-use patterns and soil biophysical attributes (Bahri et al. 2022). Furthermore, the proliferation of woody vegetation could amplify the spatial heterogeneity of soil characteristics (Liu et al. 2011), complicating SOC monitoring in grasslands. Consequently, there is a necessity for further evaluation of the relationship between SOC stocks and woody proliferation at an ecosystem level. Additionally, there is a need to create a reliable approach for measuring SOC stocks in localized grasslands. This is useful to environmental managers in formulating effective environmental management strategies and policies.

Historically, SOC monitoring relied primarily on conventional methods that includes wet soil analysis, laboratory procedures, and field measurements (Odebiri et al. 2022). Although conventional methods offer good accuracy, drawbacks that include time-consuming processes, intensive labour, and high costs make their adoption challenging across substantial geographical extents (Bartholomeus et al. 2011; Xu and Zhai 2022). Recognizing these shortcomings, there is a growing interest to formulate rapid, reliable and lucrative strategies for quantifying SOC within heterogeneous landscapes (Eggleston et al. 2006; Odebiri et al. 2020). Hence, remote sensing technology has become a reliable tool offering cost-effective, time-efficient, and environmentally friendly approaches to assess SOC variability at various spatial extents (Bahri et al. 2022; Odebiri et al. 2021). Notably, over the past few years, the utilization of new generation commercial sensors, including WorldView series, for SOC monitoring in different ecological systems have extensively increased (Broderick et al. 2015; Thaler et al. 2019). These sensors consist of distinct band configurations strategically positioned inside the red-edge portions of the electromagnetic spectrum, which are essential for enhancing the spectral sensitivity to vegetation (Rapinel et al. 2014). Unfortunately, despite their robustness in quantifying SOC, these sensors are unsuitable for measuring SOC stocks in less developed regions such as Southern African region (Mngadi et al. 2021). This is because these sensors are associated with small swath width, high acquisition cost and are not readily available (Matongera et al. 2017). Consequently, SOC monitoring is still not clearly understood in many heterogeneous landscapes, particularly in Sub-Saharan grasslands. These limitations have necessitated the adoption of freely accessible multispectral sensors that offer distinct spatial and spectral resolution attributes that are critical for accurately quantifying SOC (Bhunia et al. 2019; Mutanga et al. 2016). These advancements promise to revolutionize SOC monitoring, offering vital insights for sustainable land management practices.

Global interest in remote sensing methods for SOC research has been sparked by new innovations in reliable, open-source sensor technology, like the PlanetScope Multispectral sensor (Andreatta et al. 2022; Koparan 2019; Tan et al. 2023). The sensor has gained popularity for mapping and assessing SOC stocks due to its cutting-edge sensing properties (Andreatta et al. 2022; Koparan et al. 2022). It consists of eight spectral bands, including the red-edge regions, that are particularly responsive to biophysical characteristics of green vegetation (Frazier and Hemingway 2021; Roy et al. 2021). According to Odebiri et al. (2020), green biomass is a key indicator of SOC stocks as its accumulation is highly affected by its litter and decomposition rate. Studies have utilized red-edge wavebands to assess and predict SOC stocks at various landscapes (Castaldi et al. 2019; Koparan et al. 2022; Xie et al. 2022). Furthermore, PlanetScope satellite is characterized by an exceptionally fine spatial resolution of 3 m (Francini et al. 2020). The sensor's finer pixel size is valuable for capturing finer details, which is crucial for mapping the variability of SOC, particularly at small spatial extents. While the characteristics of PlanetScope data to SOC mapping are crucial, its heterogeneity and complexity call for a fast and reliable analytical technique that can successfully quantify SOC stocks to various extents are adopted. To date, Deep Learning (DL) techniques have gained popularity in numerous remote sensing studies attributed to their capability to read big and complex data (Abdar et al. 2021; Zhou et al. 2017). DL models are capable of automatically extracting abstract and invariant components from remote sensing dataset, hence have been widely adopted for quantification of complex non-linear data, including SOC stocks (Zhang et al. 2019).

Although studies evaluating the influence of woody proliferation on SOC at a national and regional scale have employed diverse remote sensing techniques, information about the influence of this phenomenon on SOC at an ecosystem level is still lacking, especially in protected areas (Jackson et al. 2002; Odebiri et al. 2021). Moreover, literature reveals the success of spectral vegetation indices, acquired from PlanetScope bands, in quantifying SOC stocks in complex landscapes. Previous studies suggest that sensors with explicit spatial and spectral resolution may offer enhanced information, essential for regular monitoring of SOC in grasslands affected by woody proliferation (Tan et al. 2023). Therefore, the improved spatial and spectral characteristics including red-edge region of freely available PlanetScope are critical for SOC quantification within woody encroached landscapes with limited spatial extent. Furthermore, current uncertainties that exist in the literature about the influence of woody proliferation on SOC pool advocates for further assessment of the relationship that exist between SOC and woody encroachment. This can be achieved by assessing SOC stocks variability between pure grasses and encroached grasslands. Therefore, this chapter sought to quantify the spatial variability of SOC stocks between a pristine and woody encroached grassland using PlanetScope spectral bands and vegetation indices.

2 Materials and Methods

2.1 Description of Study Site

The research was performed at the Bisley Valley Nature Reserve (29° 39′ 23″ S; 30° 23′ 44″ E), approximately 7.8 km Southwest of Pietermaritzburg Central Business District (Fig. 1). The reserve is approximately 3.5 km^2 and is covered by Thornveld and Ngongoni Veld of the sub-escarpment of the KwaZulu-Natal bioregion (Kraai et al. 2023). The reserve experiences hot summers, with average temperatures around 26 °C, and mild winters, with average temperatures of 17 °C. The area receives an average annual precipitation of 694 mm. The study area is characterized by a variety of soil types, namely Mispah, Dundee, Glenrosa, Clovely, and Avalon (Jaganyi 1998). Among these, Mispah soils predominate, covering most of the landscape. Mispah soils are typically shallow and rocky, often forming over hard rock or partially weathered material, which limits their effective depth (Otto 2019). This shallow depth contributes to low organic matter content, as the restricted rooting zone constrains the accumulation and retention of organic residues (Rossouw 2016; Van Huyssteen et al. 2021). Furthermore, these soils exhibit moderate vulnerability to erosion, particularly under improper land management or when protective vegetation cover is reduced (Bühmann et al. 2006). In contrast, the Dundee, Glenrosa, Clovely, and Avalon soils, although present, are less extensive. They vary in their mineral composition, texture, and organic matter content, potentially creating small-scale heterogeneity across the landscape (Gaston et al. 1996; Moberly and Meyer 1984). For instance, Glenrosa and Clovely soils often exhibit slightly higher organic content and better drainage characteristics than Mispah soils, while Avalon soils tend to be deeper and can support more robust vegetation under favourable conditions (Zablotowicz et al. 2000). However, due to the prevalence of Mispah soils, the overall capacity of the area to store and maintain SOC is limited, making SOC an important indicator of land degradation or improvement in local land management. In the recent past, several control practices have been adopted to minimize the dominance of woody vegetation in the reserve. Hence, whereas most landscapes are dominated by woody vegetation, there are several pristine grassland patches across the reserve.

2.2 Field Data Sampling

Samples were gathered using a stratified random sampling approach from September 21st and 23rd, 2023 (during the spring season). Prior knowledge of the site facilitated the delineation of two homogeneous strata, namely pristine grassland and woody-encroached grassland (Fig. 2). Within the woody-encroached stratum, 15 transects were established, each measuring 110 m in length. Along each transect, eight sampling plots (5 m^2 each) were marked at 10 m intervals. Consequently, a total

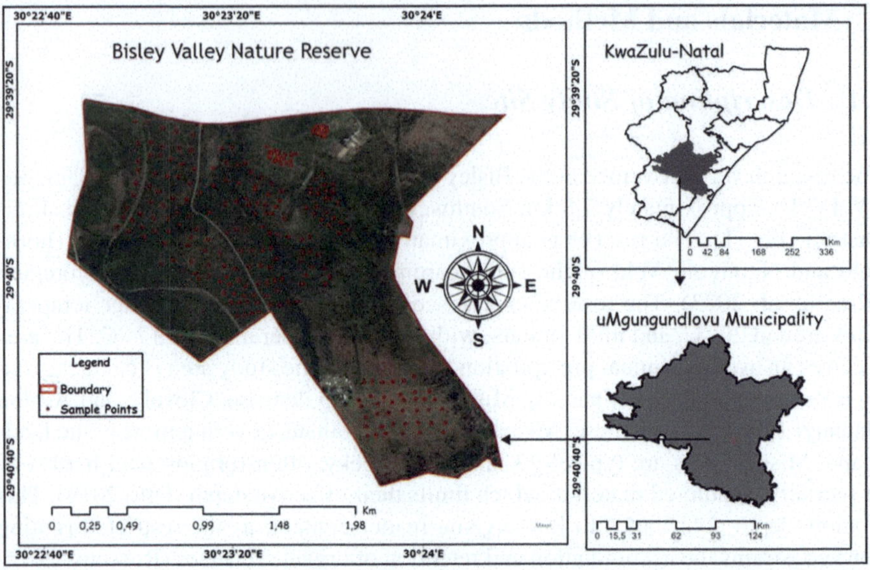

Fig. 1 Location of Bisley Valley Nature Reserve in KwaZulu-Natal Province of South Africa

of 120 sampling plots were established in the stratum. At each plot, a 30 cm deep soil auger was utilized to collect soil samples, following the procedure recommended by Vågen and Winowiecki (2013). The same sampling protocol was applied in the pristine grassland stratum, ensuring consistency in data collection. A Trimble Handheld GPS with sub-meter accuracy was used to record the coordinates of each sampling point for subsequent spatial analysis. In total, 240 soil samples were collected during field sampling.

Furthermore, to clearly visualize encroached and pristine landscapes within Bisley Nature Reserve, a supervised landcover classification was executed through Google Earth Engine (GEE) platform. The study site was separated into three landcover classes namely, woody encroached grassland, pristine grassland and other landcover classes that include buildings, roads, water bodies and bare soils. A total of 65 presence locations were collected for each landcover class using the Trimble Handheld GPS with sub-meter accuracy, 7–3 ratio of the collected points were used for training and testing, respectively.

2.3 Laboratory Analysis

To eliminate excess water, soil samples were dried outside for three days and put through a 2-mm sieve. Subsequently, at the temperature of 105 °C, the samples were subjected to oven-drying for 24 h (Konen et al. 2002). Then, the soil samples were

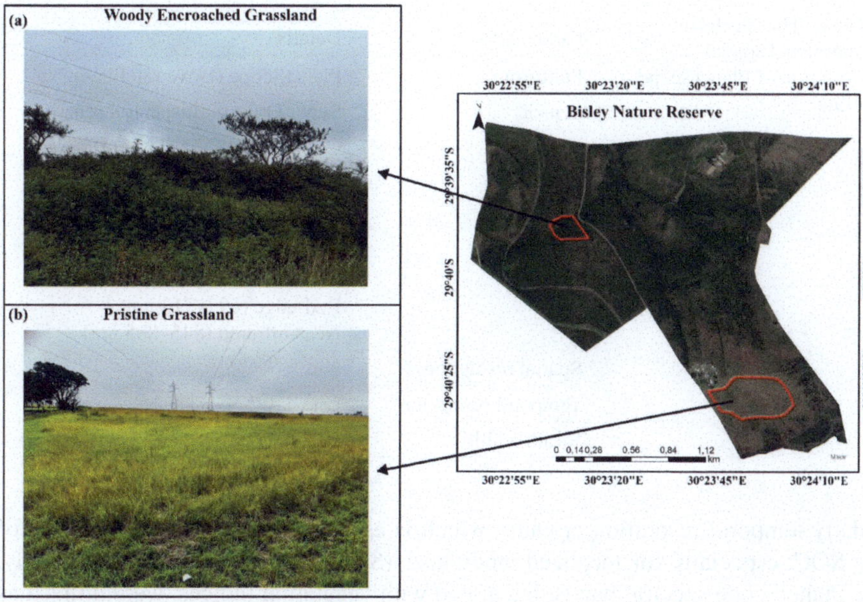

Fig. 2 Imagery of a woody encroached (**a**) and pristine (**b**) grasslands in Bisley Nature Reserve

analysed to determine the SOC concentration. The Loss-On-Ignition (LOI) technique was adopted to derive SOC values as described by Schulte and Hopkins (1996). A muffle furnace (Model Exation 1200-30 L) was used to combust soil samples at the temperature of 360 ∘C for two hours. An estimation of the SOC percentage from the LOI method (SOCLOI) was calculated using Eq. 1 (Schulte and Hopkins 1996).

$$SOCLOI = \left[\frac{soil\ weight\ after\ ignition - soil\ weight\ before\ ignition}{soil\ weight\ before\ ignition} \right] * 100 \quad (1)$$

2.4 Image Data Acquisition

The PlanetScope imagery at level 3B with less than 10% cloud cover was acquired on the 1st of September 2023 from the Planet Labs Explorer (https://www.planet.com). No pre-processing was employed since Planet Labs provide product level 3B PlanetScope images that are already atmospherically and radiometrically corrected (Žížala et al. 2019). Moreover, Planet offers two primary radiometry options for PlanetScope scenes, i.e., visual, and analytic products. PlanetScope imagery incorporate eight spectral bands (Table 1), including the Red-Edge region that is paramount for quantifying SOC sequestration across woody invaded grassland ecosystems (Andreatta et al. 2022). Moreover, PlanetScope is characterized by

Table 1 The spectral, temporal and spatial information of PlanetScope satellite

Parameter	Details
Platform	PlanetScope (Dove satellites)
Sensor	Super Dove multispectral sensor
Bands	Coastal blue (432–452 nm) Blue (465–515 nm) Green I (513–546 nm) Green (547–583 nm) Yellow (600–620 nm) Red (650–680 nm) Red-edge (697–713 nm) Near-infrared (845–885 nm)
Spatial resolution	3 m
Temporal resolution	One day
Swath width	24 km at nadir

a daily temporal resolution at nadir, which is essential for frequent monitoring of the SOC, especially for localized landscapes (Shimizu et al. 2020). Subsequently, the PlanetScope spectral bands, integrated with vegetation indices, were utilized as predictor variables.

2.5 Image Classification

Furthermore, PlanetScope image was utilized for conducting landcover classification. The satellite images were classified using Random Forest (RF) classier. The RF is a joint learning technique that quantifies the output variable by combining multiple decision trees. According to (Odebiri et al. 2020), the RF classifier is time efficient and requires less manual intervention when compared to other classifiers such as Support Vector Machine.

2.6 Spectral Vegetation Indices

This research employed various spectral band combinations derived from PlanetScope imagery to generate widely used spectral vegetation indices for the quantification of SOC. These indices include Modified Soil Adjusted Vegetation Index (MSAVI), Normalized Difference Vegetation Index (NDVI), Enhance Vegetation Index (EVI), Green Normalized Difference Vegetation Index (GNDVI), Coloration Index (CI), Photosynthetic Vigour Ratio (PVR), Greenness Index (GI), Anthocyanin Reflectance Index (ARI), Transformed Vegetation Index (TVI), Renormalized Difference Vegetation Index (RDVI), Enhanced Vegetation Index 2 (EVI_2), Chlorophyll Index Green (CIG), Simple Ratio Red/NIR Ratio Vegetation Index

(SRRE), Blue Normalized Difference Vegetation Index (BNDVI), Plant Pigment Ratio (PPR), Infrared Percentage Vegetation Index (IPVI), Log Ratio (LogR), Modified Normalized Difference Vegetation Index (MNDVI), Normalized Difference Red-Edge (NDRE) and Adjusted Vegetation Index (SAVI) (Table 2). Evidence from prior literature supports the utility of vegetation indices for enhancing model performance in quantifying the accumulation of SOC across diverse landscapes (Odebiri et al. 2020, 2022; Wang et al. 2021). Consequently, the derived vegetation indices were integrated with corresponding spectral bands to develop an optimized model for the accurate quantification of SOC.

Table 2 Spectral indices derived from PlanetScope and their formulae

Indices	Formulae	References
NDVI	(NIR − RED) / (NIR + RED)	Xie et al. (2007)
EVI	2.5*((NIR − RED) / NIR + 6 * RED − 7.5BLUE + 1))	Wu et al. (2007)
MSAVI	$2 * (\mathrm{NTR} + \sqrt{(2 * NIR+)^2 - 8(NIR - RED)})/2$	Qi et al. (1994)
GNDVI	(NIR − GREEN) / (NIR + GREEN)	Huang et al. (2010)
CI	(RED − BLUE) / RED	Escadafal et al. (1994)
PVR	(GREEN − RED) / (GREEN + RED)	Metternicht (2003)
GI	GREEN / RED	Smith and Yeaton (1998)
ARI	(1 / GREEN) − (1 / RED_EDGE)	Gitelson et al. (2001)
TVI	$\sqrt{(NDVI)} + 0.5$	Bannari et al. (1995)
RDVI	$(\mathrm{NIR} - \mathrm{RED}) / (\mathrm{NIR} + \mathrm{RED})^{0.5}$	Ehammer et al. (2010)
EVI_2	2.4 * (NIR – RED) / (NIR + RED + 1)	Miura et al. (2008)
CIG	(NIR / GREEN) − 1	Ahamed et al. (2011)
SRRE	RED / NIR	Bannari et al. (1995)
BNDVI	(NIR − BLUE) / (NIR + BLUE)	Hancock and Dougherty (2007)
PPR	(GREEN – BLUE) / (GREEN + BLUE)	Gobron et al. (2000)
IPVI	[((NIR / NIR + RED) / 2) * (NDVI + 1)]	Crippen (1990)
LogR	Log (NIR/RED)	Henrich et al. (2011)
MNDVI	(NIR − Red) / (NIR + Red − 2 ∗ Blue)	Huete et al. (1997)
NDRE	(NIR – REDEDGE) / (NIR + REDEDGE)	Nichol et al. (2000)
SAVI	(NIR − Red) / (NIR + Red + L) ∗ (1 + L)	Zarco-Tejada and Sepulcre-Cantó (2007)

2.7 Deep Neural Networks (DNN)

Recently, DNN has gained popularity in modelling and predicting SOC distribution in various ecosystems ascribed to the ability of the model to explain complex relationships that exist between explanatory and target variables (García-Martínez et al. 2020; Odebiri et al. 2020). DNN model is a collection of multiple neurons placed in a specific sequence of numerous layers (Montavon et al. 2018; Odebiri et al. 2022). Neurons from a single layer are connected to the subsequent layer up to the final estimated neuron. In essence, neurons receive activations from the previous layer (Larochelle et al. 2009). The layers that contain a network of neurons produce a complex nonlinear mapping from the input to the output using backpropagation (Montavon et al. 2018). Additionally, to avoid model overfitting, the dropout regularization method can be adopted to hidden layers. An example of the DNN structural design is illustrate in Fig. 3.

The python programming language was used to build the DNN model. Fourteen vegetation indices, mixed with eight PlanetScope spectral bands, were used as input variables to build the model. The input variables were separated into ten even parts and used for training and testing the model. To yield the best output for the model, the hyperparameter tuning technique, where hyperparameters were randomly optimized repeatedly until the best results were achieved, was adopted.

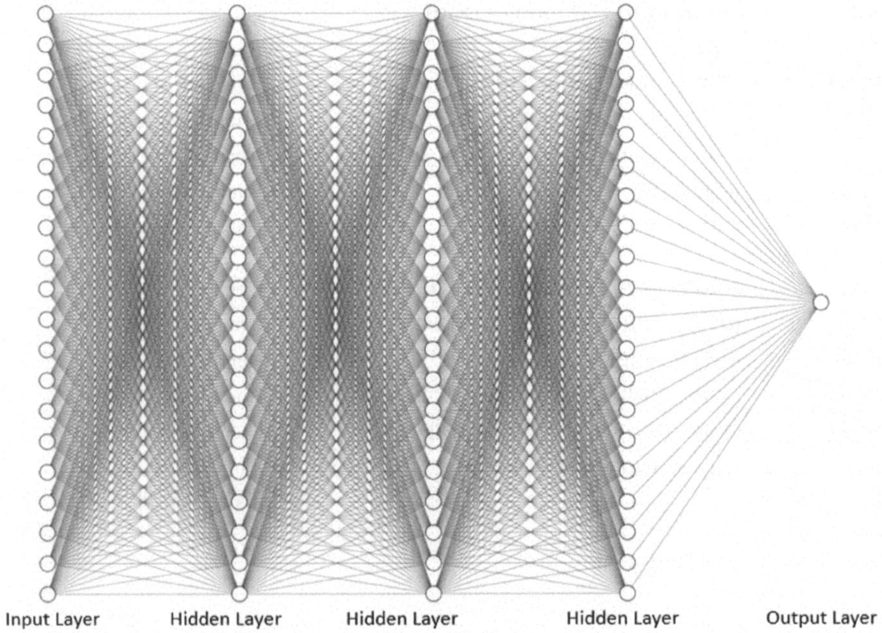

Fig. 3 A graphical example of Deep Neural Network (DNN) architecture

2.8 Feature Selection

Generally, regression analysis is challenged by multicollinearity, which is a statistical phenomenon in which two or more predictors are moderately or highly correlated. Hence, it is advisable that the model is developed using the least correlated predictors (Ismail and Mutanga 2010; Odebiri et al. 2020). To counteract the challenge of multicollinearity, a Pearson correlation test was executed. Therefore, one predictor between two highly correlated predictor was removed from the input dataset using the method suggested by Behnamian et al. (2017). The predictor residuals that have proven in the past research to be the best predictors for SOC stocks, particularly in grasslands, were selected for predicting SOC distribution.

2.9 Model Evaluation Metrics

Three commonly used metrics namely, Lin's Concordance Correlation Coefficient (LCCC), coefficient of determination (R^2), and Root Mean Squared Error (RMSE), were reported to evaluate model. RMSE demonstrates the variance between the observed and estimated SOC values. The coefficient of determination measures the robustness of the algorithm in quantifying SOC and ranges between 0 and 100%, while LCCC reports on how well bivariate pairs of the observed SOC conform to predicted SOC values (King and Chinchilli 2001; Odebiri et al. 2022). These metrics can be expressed as;

$$RMSE = \sqrt{\frac{\sum_{i=1}^{n}\left(X_{O,i} - X_{p,i}\right)^2}{n}} \tag{2}$$

$$R^2 = \left[\frac{\sum_{i=1}^{n}(X_0 - P)^2}{\sum_{i=1}^{n}(X_o - O')^2}\right] \tag{3}$$

$$LCCC = \frac{2r\sigma_0\sigma_p}{\sigma_p^2 + \sigma_p^2 + \left[O' - p'\right]^2} \tag{4}$$

where X_o and X_p represent the measured and predicted SOC, while n signify the total number of observations. Furthermore, σ_0 and σ_p are the respective variance of the observed and predicted value, while o' signifies the averages of the observed and predicted SOC.

Furthermore, tenfold cross-validation was executed to ensure that each dataset was used at least once to avoid bias for DNN algorithms. The best-fitted model normally produces a higher R^2 and LCCC, and a lower RMSE. Hence, the performance of the model was identified using the three-accuracy metrics. Furthermore, variable importance ranking was executed using the relative variable importance and

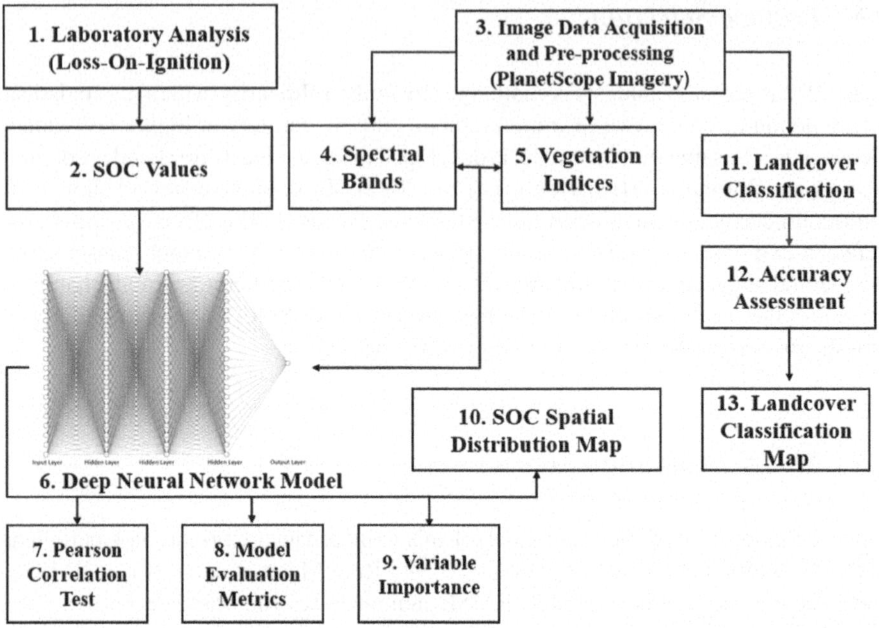

Fig. 4 Summary of procedures followed in quantifying SOC

SHapely Additive exPlanations (SHAP), respectively, to identify the most significant predictors of SOC. A digital SOC map was assembled using the best predictor variables derived from the DNN model.

Additionally, the accuracy assessment for landcover classification was conducted within the GEE and the confusion matrix technique adopted. The model employed the overall accuracy, producer's accuracy, and user's accuracy to evaluate RF classifier. Accuracy assessment was executed by comparing training dataset with testing dataset as suggested by Nasiri et al. (2022) and (Mahdianpari et al. 2020) (Fig. 4).

2.10 *Statistical Analysis*

The Statistical Package for the Social Science (SPSS version 28.0) software was used to execute a paired t-test to investigate the level of significance of SOC distribution between pristine and woody encroached grassland. The measures of central tendency, particularly mode, mean, maximum, minimum, standard deviation and standard error, were executed using SPSS. Microsoft excel was utilized to investigate the Coefficient of Variation (CV), skewness, and kurtosis for woody and pristine grassland. Also, a box and whisker plot was executed on Microsoft excel.

3 Results

3.1 Descriptive Statistics

Table 3 summarize the descriptive statistics for both pristine and woody encroached grasslands. The paired t-test between woody and pristine grassland ($p = 0.05$) indicate that there is a significant difference in SOC distribution between the two sites. In addition, for pristine grassland, data showed strong negative skewness (– 0.75) and kurtosis (1.34), satisfying the normality rule (Hair et al. 2021), while woody encroached grassland SOC data indicated a strong positive skewness (0.75) and kurtosis (1.33). Moreover, the CV for pristine and woody encroached grassland was 11.89%. The CV value suggests that there was a variation between pristine and encroached grasslands (Fig. 5).

Table 3 Descriptive statistics of the dataset

Landcover	Minimum value (%)	Maximum value (%)	Mean value (%)	Standard deviation (%)	Standard error (%)
Pristine grassland	3.6	9.4	6.6	1.46	0.19
Woody encroached grassland	2.4	13.1	7.56	2.76	0.31

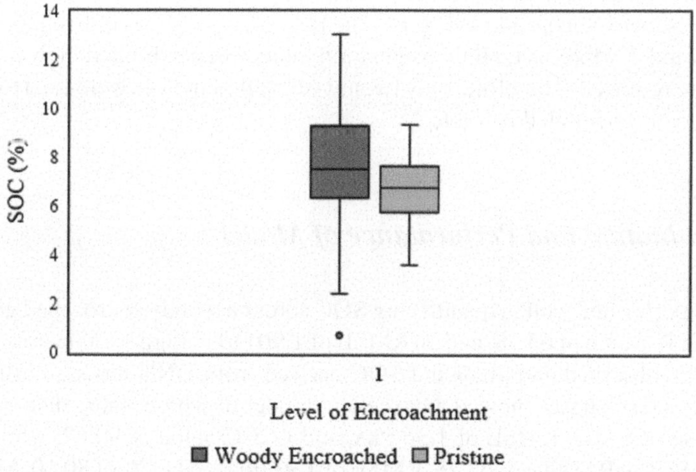

Fig. 5 The distribution of SOC between a pristine and woody encroached grassland

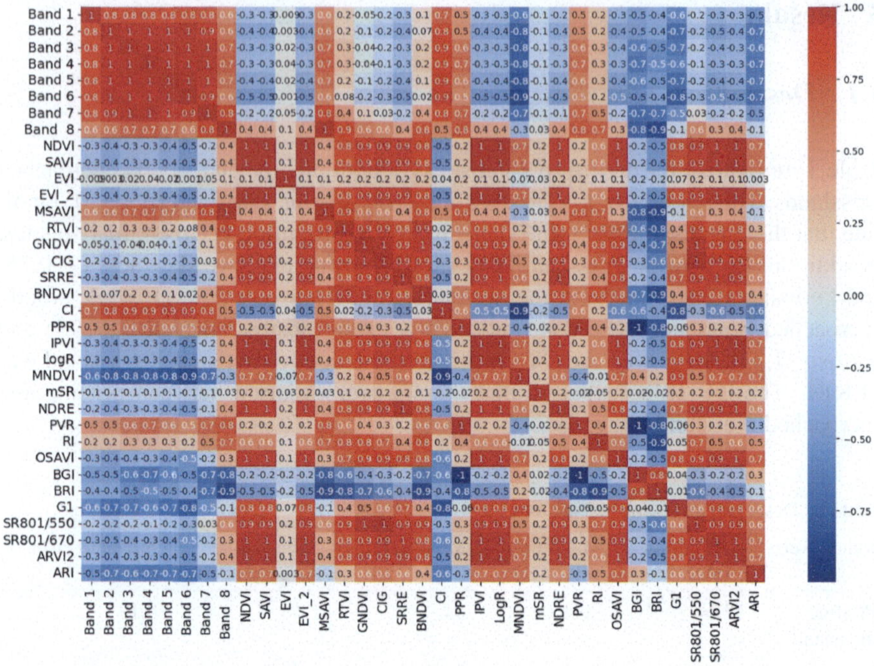

Fig. 6 Pearson correlation test of predictor variables

According to the Pearson correlation test, all spectral bands exhibited high correlation values, between 0.9 and 1 (Fig. 6). The Red-edge band was utilized for building the DNN model. Moreover, all the vegetation indices that exhibited high correlation values were removed. Therefore, only twenty vegetation indices were used to develop the best model as detailed in Table 2.

3.2 *Evaluation and Performance of Model*

The DNN performed well in quantifying SOC across a woody encroached grassland, yielding an R^2 value of 63.78, and an RMSE of 1.901 t/ha. Figure 7 illustrates the link between the observed and predicted SOC derived from DNN model. Furthermore, the train and test dataset showed an excellent model fit with training data exhibiting an R^2 value of 67,02, RMSE of 1.36 t/ha, and LCCC value of 90.67, while testing data exhibited an R^2 value of 63.78, RMSE of 1.91 t/ha, and LCC of 80.60. Moreover, Table 4 explains the performance of RF classifier.

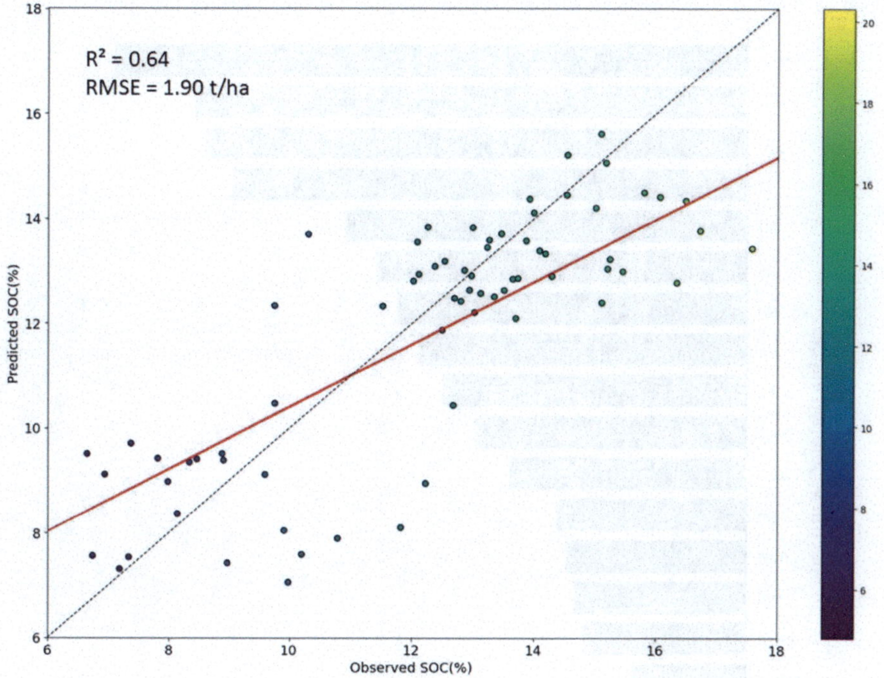

Fig. 7 Relationship between predicted and measured SOC using Deep Neural Network (DNN)

Table 4 The accuracy assessment of RF classifier

Classifier	Overall accuracy (%)	Producer accuracy (%)	Kappa accuracy (%)	Consumer accuracy (%)
RF	0.95	0.88	0.92	0.96

3.3 Variable Importance Assessment of the Model

Figure 8 illustrates the importance and contribution of each predictor variable using SHapely Additive exPlanations (SHAP) for DNN. The best three explanatory variables for the DNN model were NDVI, MSAVI and TVI, while LogR, EVI and PPR were the least important variables. Figure 8 further indicates that NDRE and MNDVI exhibited similar contributions towards quantifying SOC.

3.4 Spatial Distribution of SOC

Figure 9a depicts the distribution of SOC in both woody encroached and pristine landscapes, respectively, within the Bisley Nature Reserve, while Fig. 9b explains

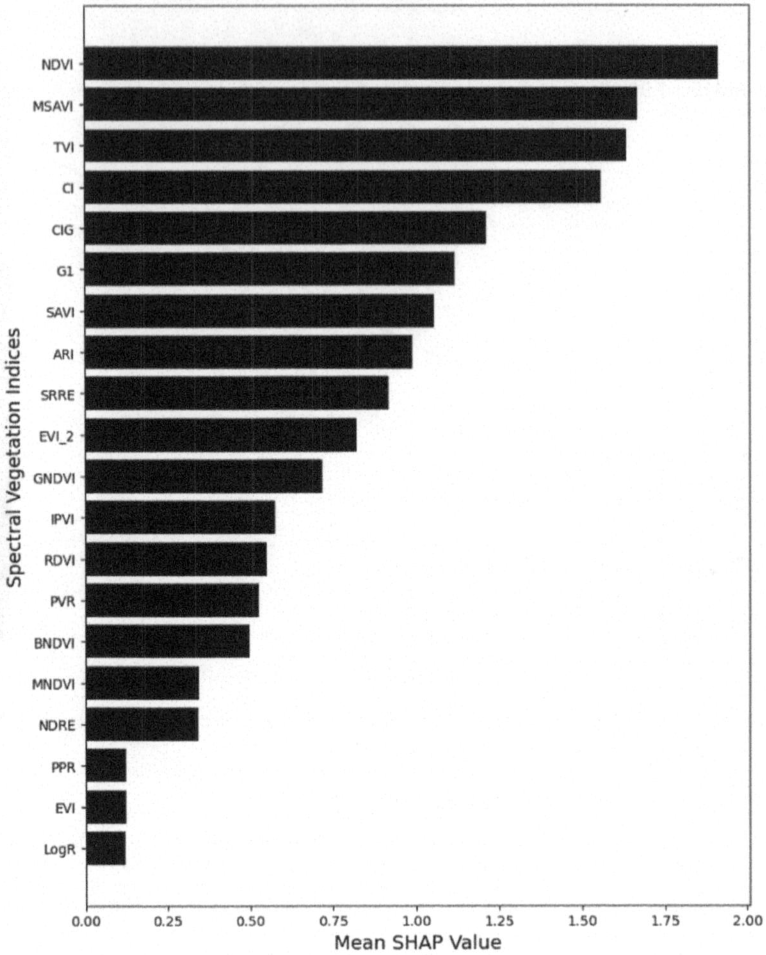

Fig. 8 Importance ranking of variables used for quantifying SOC within Bisley Nature Reserve

different land cover classes within the study area. According to Fig. 9a, woody encroached landscapes are dominated by medium and high concentrations of SOC, while low concentrations of SOC stocks were dominant in pristine grasslands. Furthermore, SOC concentration was higher along the river channels located across the central parts of the reserve. The map further reveals that woody encroached landscapes located central, West and South of the reserve are dominated by high concentration of SOC. The SOC spatial distribution maps reveal that there is a noticeable SOC variability between pristine and woody encroached landscapes. For instance, the distribution of SOC is generally higher across woody encroached grasslands in comparison to pristine grasslands, which exhibited low SOC concentrations.

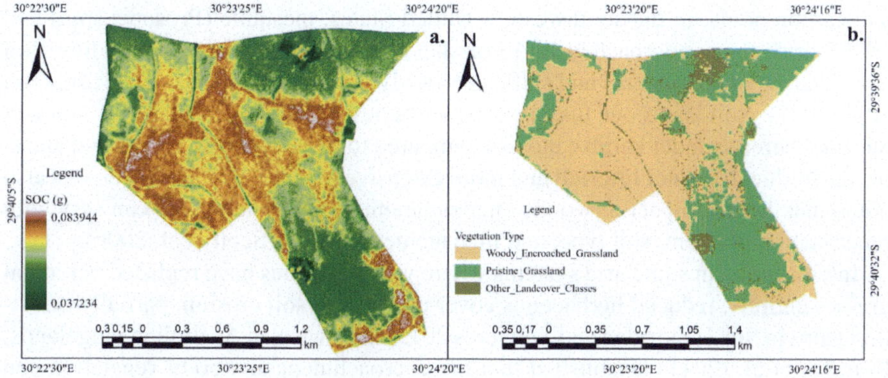

Fig. 9 The spatial distribution of SOC for Deep Neural Network (DNN) (**a**) and landcover classification map (**b**) using PlanetScope MSI data

4 Discussion

This chapter quantified SOC variability between a pristine and woody encroached grassland using PlanetScope's spectral bands and vegetation indices. The chapter examined the influence of woody vegetation proliferation on SOC storage in grasslands using the DNN model. The findings show that woody encroached areas have a high concentration of SOC stock compared to pristine grasslands (Fig. 9). A two-sample t-test confirmed this difference ($p < 0.05$), indicating a statistically significant increase in SOC stock under woody encroachment. Fantappiè et al. (2010) and Rasse et al. (2005) established that SOC accumulation is highly influenced by the surface litter and root biomass. Therefore, areas with high litter biomass are expected to exhibit high SOC concentrations. The study by Yang et al. (2021) established SOC to be concentrated in areas with high shrub encroachment compared to pure grasslands and suggested that woody encroached landscapes are associated with high mean litter and root biomass. As a result, SOC distribution is expected to be more in grasslands with a proliferation of woody vegetation compared to pure grasslands (Fig. 5). Similarly, Odebiri et al. (2020) established that SOC stocks are commonly dominant in landscapes with higher vegetation biomass compared to areas with less vegetation cover.

Our findings further shows that woody encroached grassland has the highest maximum percentage of SOC (13.1%) when compared to pristine grassland (9.4%) (Fig. 5). Several regional studies corroborate our observations that woody encroachment often increases SOC. For instance, African savannas, Sankaran et al. (2004) and Barger et al. (2011) noted that shrub invasion can elevate carbon sequestration by increasing both above- and belowground biomass, particularly under intermediate rainfall regimes where woody species thrive without outcompeting herbaceous vegetation entirely. This is consistent with the notion that the additional litter and root inputs from woody plants augment organic matter accumulation, enhancing soil carbon pools. In the southwestern United States, mesquite (Prosopis spp.) and other shrubs have encroached into grasslands over the past century, leading to a restructuring of ecosystems and shifts in SOC dynamics (Barger et al. 2011; Jackson et al. 2002). Similar to our findings, these studies demonstrated that SOC content tends to increase under shrub canopies compared to adjacent grass-dominated areas, primarily due to greater litterfall and more extensive root systems. However, it is also noted that the net impact of woody encroachment on ecosystem SOC can vary with precipitation gradient, soil type, and management history (Scott et al. 2006).

Interestingly, in some arid systems where woody species have replaced perennial grasses entirely, reduced herbaceous cover can lead to soil erosion, partially offsetting gains in SOC (Eldridge and Soliveres 2015). Likewise, in Australian rangelands, Eldridge et al. (2011) established that the encroachment of woody vegetation can immensely increase SOC in the topsoil layer, although the magnitude of this impact depends on grazing intensity, fire regimes, and other land-use pressures. Where woody encroachment is managed or occurs alongside moderate grazing, the combination of woody litter input and reduced soil disturbance often promotes organic matter accumulation. Furthermore, Li et al. (2016) reported that in semi-arid and humid regions, leguminous shrubs encroachment in grasslands have greatly amplified SOC pool, while non-legumes have shown to have less impact on SOC stocks, particularly in topsoil. The overall consensus across various regional studies, including those from African savannas, the southwestern United States, and Australian rangelands, is that woody proliferation can enhance SOC accumulation, at least in the upper soil layers, by adding substantial organic inputs from litterfall and belowground biomass.

Woody encroached areas are also associated with high soil metabolism due to high litter fall and dead matter accumulation, which increases the concentration of SOC (Mayer et al. 2020; Odebiri et al. 2020). Consequently, forested regions possess elevated levels of SOC in comparison to pristine grasslands. Furthermore, areas with high soil moisture exhibit high vegetation biomass (Lozano-Parra et al. 2018; Ruiz-Sinoga et al. 2011). Hence, areas with high soil moisture are associated with high SOC compared to dry areas. Figure 8 reveals that areas along the river channels have high concentrations of SOC. This aligns with the discoveries by Evans et al. (2011), which indicate a robust positive relationship between SOC and soil moisture. However, the study by Kerr and Ochsner (2020) noted that the correlation between SOC and soil moisture can vary depending on other soil characteristics including soil temperature, inter-site vegetation variability, and soil physical attributes. According to Chen et al. (2016), soil moisture promotes vegetation growth, productivity, and decomposition, which enhance SOC sequestration. The predominant soil types in this study area are Mispah, Dundee, Glenrosa, Clovely, and Avalon soils. These soils were dry and shallow, thus explaining low SOC distribution across the grassland. However, the results presented by the current study may not be conclusive because all the soil samples were obtained at a depth of 30 cm. According to Zhou et al. (2017), there is a significant different between SOC located in topsoil and SOC in deeper layers of the soil profile. Therefore, it is recommended that further investigations about the spatial variability of SOC in woody encroached grasslands be done, considering the depth.

The DNN model produced acceptable accuracy ($R^2 = 0.64$) for quantifying SOC stocks. This is consistent with Odebiri et al. (2022) who sought to compare the performance of the DL approach with commonly used classical models in SOC stock mapping. Their study established that the DNN model outcompete classical models when estimating SOC stocks. According to Wang et al. (2021), DL approaches commonly outperform classical models when dealing with large and complicated datasets. Unlike other classical models, the DL models can read and understand intricate relationships between many ecological parameters and utilize feature representations that are solely acquired from the data (Odebiri et al. 2022). Additionally, the DNN model is characterized by many interconnected hidden layers and neurons that enable the model to extract numerous representative characteristics from the input dataset. The results of the current study were influenced by the attributes of the nature reserve and non-linearity of the dataset. The field data was collected from a localized encroached grassland and the dataset used to build the DNN model was complex. As a result, the DNN was the best model for quantifying SOC stocks across a woody encroached grassland.

Among the PlanetScope-derived vegetation indices, the most important index was NDVI. According to Odebiri et al. (2020), there is an elevated correlation between SOC and NDVI, hence the index is the most important variable when estimating SOC, particularly within vegetated landscape. The NDVI can better minimize the influence of topography and possess a good classification capability when dealing with heterogenous landscapes (Barati et al. 2011; Zhang et al. 2019). The study by Yang et al. (2020) observed that NDVI can also be utilized as a primary determinant of vegetation health, which will automatically inform the spatial distribution of SOC in grassland ecosystems. Spectral indices including MSAVI, TVI and CI were also ranked high by the model when quantifying SOC. Yet, PPR, EVI and LogR had less significant contribution to the DNN model. The spectral bands of PlanetScope were combined with derived spectral vegetation indices to develop the best model for quantifying SOC stocks. Previous literature suggests that red-edge spectral bands are the most valuable for quantifying SOC stocks (Andreatta et al. 2022; Mngadi et al. 2021; Odebiri et al. 2020). The red-edge wavelength regions of PlanetScope imagery provide important reflectance information, essential for accurate SOC mapping. These regions, being highly sensitive to vegetation characteristics, offer reliable means to assess SOC distribution. The sensor's remarkable spatial resolution of three meters contributes significantly to the wealth of information available for SOC quantification at an ecosystem level.

While few studies have endeavoured to quantify SOC using the spectral information of PlanetScope, especially in the Sub-Saharan African, noteworthy contributions have been made. For instance, Koparan et al. (2022) explored the capability of PlanetScope imagery in estimating and mapping SOC stocks. Their findings concluded that models built from PlanetScope-derived data exhibited excellent accuracies compared to models from other multispectral sensors. Additionally, it is advisable to conduct further investigations into the utilization of PlanetScope imagery bands rather than red-edge regions, particularly in the context of SOC mapping studies. The outcomes of the current study hold practical significance for the uMsunduzi Municipality on insights into SOC distribution within Bisley Nature Reserve. Additionally, nature reserve managers and indeed similar invaded landscapes can benefit from this study's insights for continuous monitoring of SOC stock.

5 Conclusion

The proliferation of woody plants in grassland ecosystems alters the SOC stock cycle. This study employed a remote sensing algorithm (DNN) and PlanetScope MSI data to quantify the spatial variability of SOC within a woody encroached and pristine grassland. The chapter established that areas exposed to woody encroachment exhibit high SOC concentration. The DNN model produced acceptable accuracy when quantifying SOC variability within woody encroached grasslands at a localized scale. Utilizing the SHAP technique, the study identified the most crucial predictor variables for SOC estimation. Notably, NDVI, MSAVI, TVI, and CI emerged as the most significant variables. The developed model is an inexpensive and time-efficient tool for quantifying SOC in both pristine and encroached grasslands. It is recommended that future studies further explore the distribution of SOC in areas experiencing woody proliferation. Furthermore, the impact of bioclimatic, soil depth, and topographic residuals on SOC should be investigated using remotely derived information, particularly within woody encroached grasslands.

References

Abdar M, Pourpanah F, Hussain S, Rezazadegan D, Liu L, Ghavamzadeh M, Fieguth P, Cao X, Khosravi A, Acharya UR, Makarenkov V, Nahavandi S (2021) A review of uncertainty quantification in deep learning: techniques, applications and challenges. Inf Fusion 76:243–297. https://doi.org/10.1016/j.inffus.2021.05.008

Alberti G, Leronni V, Piazzi M, Petrella F, Mairota P, Peressotti A, Piussi P, Valentini R, Gristina L, La Mantia T, Novara A, Rühl J (2011) Impact of woody encroachment on soil organic carbon and nitrogen in abandoned agricultural lands along a rainfall gradient in Italy. Reg Environ Change 11(4):917–924. https://doi.org/10.1007/s10113-011-0229-6

Andreatta D, Gianelle D, Scotton M, Dalponte M (2022) Estimating grassland vegetation cover with remote sensing: A comparison between Landsat-8, Sentinel-2 and PlanetScope imagery. Ecol Ind 141:109102. https://doi.org/10.1016/j.ecolind.2022.109102

Archer SR, Andersen EM, Predick KI, Schwinning S, Steidl RJ, Woods SR (2017) Woody plant encroachment: causes and consequences. In: Rangeland systems: Processes, management and challenges, pp 25–84

Bahri H, Raclot D, Barbouchi M, Lagacherie P, Annabi M (2022) Mapping soil organic carbon stocks in Tunisian topsoils. Geoderma Reg 30:e00561. https://doi.org/10.1016/j.geodrs.2022.e00561

Bannari A, Morin D, Bonn F, Huete A (1995) A review of vegetation indices. Remote Sens Rev 13(1–2):95–120

Barati S, Rayegani B, Saati M, Sharifi A, Nasri M (2011) Comparison the accuracies of different spectral indices for estimation of vegetation cover fraction in sparse vegetated areas. Egypt J Remote Sens Space Sci 14(1):49–56. https://doi.org/10.1016/j.ejrs.2011.06.001

Barger NN, Archer SR, Campbell JL, Huang CY, Morton JA, Knapp AK (2011) Woody plant proliferation in North American drylands: a synthesis of impacts on ecosystem carbon balance. J Geophys Res: Biogeosci 116(G4)

Bartholomeus H, Kooistra L, Stevens A, van Leeuwen M, van Wesemael B, Ben-Dor E, Tychon B (2011) Soil organic carbon mapping of partially vegetated agricultural fields with imaging spectroscopy. Int J Appl Earth Obs Geoinf 13(1):81–88

Behnamian A, Millard K, Banks SN, White L, Richardson M, Pasher J (2017) A systematic approach for variable selection with random forests: achieving stable variable importance values. IEEE Geosci Remote Sens Lett 14(11):1988–1992

Bhunia GS, Kumar Shit P, Pourghasemi HR (2019) Soil organic carbon mapping using remote sensing techniques and multivariate regression model. Geocarto Int 34(2):215–226

Blaser WJ, Shanungu GK, Edwards PJ, Olde Venterink H (2014) Woody encroachment reduces nutrient limitation and promotes soil carbon sequestration. Ecol Evol 4(8):1423–1438

Broderick DE, Frey KE, Rogan J, Alexander HD, Zimov NS (2015) Estimating upper soil horizon carbon stocks in a permafrost watershed of Northeast Siberia by integrating field measurements with Landsat-5 TM and WorldView-2 satellite data. Gisci Remote Sens 52(2):131–157

Bugalho MN, Abreu J (2008). The multifunctional role of grasslands. In: Sustainable Mediterranean grasslands and their multifunctions, pp 25–30.

Bühmann C, Beukes DJ, Turner DP (2006) Clay mineral associations in soils of the Lusikisiki area, Eastern Cape Province, and their agricultural significance. South Afr J Plant and Soil 23(2):78–86. https://doi.org/10.1080/02571862.2006.10634735

Carlier L, Rotar I, Vlahova M, Vidican R (2009) Importance and functions of grasslands. Notulae Botanicae Horti Agrobotanici Cluj-Napoca 37(1):25–30

Castaldi F, Chabrillat S, Don A, van Wesemael B (2019) Soil organic carbon mapping using LUCAS topsoil database and Sentinel-2 data: an approach to reduce soil moisture and crop residue effects. Remote Sensing 11(18):2121

Chen X, Zhang D, Liang G, Qiu Q, Liu J, Zhou G, Liu S, Chu G, Yan J (2016) Effects of precipitation on soil organic carbon fractions in three subtropical forests in southern China. Journal of Plant Ecology 9(1):10–19

Conant RT, Paustian K (2002) Spatial variability of soil organic carbon in grasslands: implications for detecting change at different scales. Environ Pollut 116:S127–S135. https://doi.org/10.1016/S0269-7491(01)00265-2

Crippen RE (1990) Calculating the vegetation index faster. Remote Sens Environ 1(34):71–73. https://doi.org/10.1016/0034-4257(90)90085-z

Ding J, Eldridge DJ (2024) Woody encroachment: social–ecological impacts and sustainable management. Biol Rev 99(6):1909–1926. https://doi.org/10.1111/brv.13104

Eggleston H, Buendia L, Miwa K, Ngara T, Tanabe K (2006) 2006 IPCC guidelines for national greenhouse gas inventories

Eldridge DJ, Bowker MA, Maestre FT, Roger E, Reynolds JF, Whitford WG (2011) Impacts of shrub encroachment on ecosystem structure and functioning: towards a global synthesis. Ecol Lett 14(7):709–722

Eldridge DJ, Soliveres S (2015) Are shrubs really a sign of declining ecosystem function? Disentangling the myths and truths of woody encroachment in Australia. Aust J Bot 62(7):594–608

Escadafal R, Belghith A, Ben Moussa H (1994) Indices spectraux pour la télédétection de la dégradation des milieux naturels en Tunisie aride. In: 6Ã¨me symposium on international mesures physiques et signatures en TÃ©lÃ©dÃ©tection, ISPRS-CNES, Val dâ™lsÃ¨re, France

Evans SE, Burke IC, Lauenroth WK (2011) Controls on soil organic carbon and nitrogen in Inner Mongolia, China: a cross-continental comparison of temperate grasslands. Glob Biogeochem Cycles 25(3)

Fantappiè M, L'Abate G, Costantini E (2010) Factors influencing soil organic carbon stock variations in Italy during the last three decades. In: Land degradation and desertification: assessment, mitigation and remediation, pp 435–465

Francini S, McRoberts RE, Giannetti F, Mencucci M, Marchetti M, Scarascia Mugnozza G, Chirici G (2020) Near-real time forest change detection using PlanetScope imagery. Eur J Remote Sens 53(1):233–244

Frazier AE, Hemingway BL (2021) A technical review of planet smallsat data: practical considerations for processing and using planetscope imagery. Remote Sens 13(19):3930

García-Martínez H, Flores-Magdaleno H, Ascencio-Hernández R, Khalil-Gardezi A, Tijerina-Chávez L, Mancilla-Villa OR, Vázquez-Peña MA (2020. Corn grain yield estimation from vegetation indices, canopy cover, plant density, and a neural network using multispectral and RGB images acquired with unmanned aerial vehicles. Agriculture 10(7):277. https://www.mdpi.com/2077-0472/10/7/277

Gartzia M, Alados CL, Pérez-Cabello F (2014) Assessment of the effects of biophysical and anthropogenic factors on woody plant encroachment in dense and sparse mountain grasslands based on remote sensing data. Progr Phys Geogr: Earth Environ 38(2):201–217. https://doi.org/10.1177/0309133314524429

Gaston L, Locke M, Zablotowicz R (1996) Sorption and degradation of bentazon in conventional-and no-till Dundee soil, pp 0047–2425

Gerke J (2022) The central role of soil organic matter in soil fertility and carbon storage. Soil Syst 6(2):33

Gobron N, Pinty B, Verstraete MM, Widlowski J-L (2000) Advanced vegetation indices optimized for up-coming sensors: Design, performance, and applications. IEEE Trans Geosci Remote Sens 38(6):2489–2505

Grellier S, Ward D, Janeau J-L, Podwojewski P, Lorentz S, Abbadie L, Valentin C, Barot S (2013) Positive versus negative environmental impacts of tree encroachment in South Africa. Acta Oecologica 53:1–10

Honda EA, Durigan G (2016) Woody encroachment and its consequences on hydrological processes in the savannah. Philos Trans Roy Soc B: Biol Sci 371(1703):20150313

Huang Y, Lan Y, Ge Y, Hoffmann W, Thomson S (2010) Spatial modeling and variability analysis for modeling and prediction of soil and crop canopy coverage using multispectral imagery from an airborne remote sensing system. Trans ASABE 53(4):1321–1329

Ismail R, Mutanga O (2010) A comparison of regression tree ensembles: predicting Sirex noctilio induced water stress in Pinus patula forests of KwaZulu-Natal, South Africa. Int J Appl Earth Obs Geoinf 12:S45–S51

Jackson RB, Banner JL, Jobbágy EG, Pockman WT, Wall DH (2002) Ecosystem carbon loss with woody plant invasion of grasslands. Nature 418(6898):623–626

Jaganyi JNU (1998) Response of carabid and cicindelid beetles to various types of landscape disturbances

Kerr DD, Ochsner TE (2020) Soil organic carbon more strongly related to soil moisture than soil temperature in temperate grasslands. Soil Sci Soc Am J 84(2):587–596. https://doi.org/10.1002/saj2.20018

King TS, Chinchilli VM (2001) Robust estimators of the concordance correlation coefficient. J Biopharm Stat 11(3):83–105

Konen ME, Jacobs PM, Burras CL, Talaga BJ, Mason JA (2002) Equations for predicting soil organic carbon using loss-on-ignition for north central US soils. Soil Sci Soc Am J 66(6):1878–1881

Koparan, Rekabdarkolaee H, Sood K, Westhoff S, Reese C, Malo D (2022) Estimating soil organic carbon levels in cultivated soils from satellite image using parametric and data-driven methods. Int J Remote Sens 43(9):3429–3449

Koparan MH (2019) Estimating Soil organic carbon in cultivated soils using soil test data, remote sensing imagery from satellites (Landsat 8 and Planetscope), and web soil survey data. South Dakota State University

Kraai UM, Kraai M, Tsvuura Z, Mkhize NR, Tjelele TJ (2023) The impacts of Cape porcupines on woody plant mortality. Austral Ecol

Larochelle H, Bengio Y, Louradour J, Lamblin P (2009) Exploring strategies for training deep neural networks. J Mach Learn Res 10(1)

Li H, Shen H, Chen L, Liu T, Hu H, Zhao X, Zhou L, Zhang P, Fang J (2016) Effects of shrub encroachment on soil organic carbon in global grasslands. Sci Rep 6(1):28974. https://doi.org/10.1038/srep28974

Liu F, Wu XB, Bai E, Boutton TW, Archer SR (2011) Quantifying soil organic carbon in complex landscapes: an example of grassland undergoing encroachment of woody plants. Glob Change Biol 17(2):1119–1129

Loveland P, Webb J (2003) Is there a critical level of organic matter in the agricultural soils of temperate regions: a review. Soil Tillage Res 70(1):1–18

Lozano-Parra J, Schnabel S, Pulido M, Gómez-Gutiérrez Á, Lavado-Contador F (2018) Effects of soil moisture and vegetation cover on biomass growth in water-limited environments. Land Degrad Dev 29(12):4405–4414

Mahdianpari M, Jafarzadeh H, Granger JE, Mohammadimanesh F, Brisco B, Salehi B, Homayouni S, Weng Q (2020) A large-scale change monitoring of wetlands using time series Landsat imagery on Google Earth Engine: a case study in Newfoundland. Gisci Remote Sens 57(8):1102–1124

Matongera TN, Mutanga O, Dube T, Sibanda M (2017) Detection and mapping the spatial distribution of bracken fern weeds using the Landsat 8 OLI new generation sensor. Int J Appl Earth Obs Geoinf 57:93–103

Mayer M, Prescott CE, Abaker WE, Augusto L, Cécillon L, Ferreira GW, James J, Jandl R, Katzensteiner K, Laclau J-P (2020) Tamm review: influence of forest management activities on soil organic carbon stocks: a knowledge synthesis. For Ecol Manage 466:118127

Merino AN, Pérez-Batallón P, Macías F (2004) Responses of soil organic matter and greenhouse gas fluxes to soil management and land use changes in a humid temperate region of southern Europe. Soil Biol Biochem 36(6):917–925

Metternicht G (2003) Vegetation indices derived from high-resolution airborne videography for precision crop management. Int J Remote Sens 24(14):2855–2877

Mngadi M, Odindi J, Mutanga O (2021) The utility of Sentinel-2 spectral data in quantifying aboveground carbon stock in an urban reforested landscape. Remote Sens 13(21):4281. https://www.mdpi.com/2072-4292/13/21/4281

Moberly P, Meyer J (1984) Soils: a management factor in sugarcane production in the South African sugar industry. In: Proceedings of the South African Sugar Technologists' Association—June 1993

Mochi LS, Mazía N, Biganzoli F, Aguiar MR (2022) Contrasting effects of grazing on the early stages of woody encroachment in a Neotropical savanna. Basic Appl Ecol 60:13–24. https://doi.org/10.1016/j.baae.2022.01.007

Montavon G, Samek W, Müller K-R (2018) Methods for interpreting and understanding deep neural networks. Digital Signal Process 73:1–15. https://doi.org/10.1016/j.dsp.2017.10.011

Mureva A, Ward D, Pillay T, Chivenge P, Cramer M (2018) Soil organic carbon increases in semi-arid regions while it decreases in humid regions due to woody-plant encroachment of grasslands in South Africa. Sci Rep 8(1):15506

Mutanga O, Rouget M, Hlanguza N, Odindi J (2016) Mapping alien and indigenous vegetation in the KwaZulu-Natal Sandstone Sourveld using remotely sensed data. Bothalia-Afr Biodivers Conserv 46(2):1–9

Nasiri V, Deljouei A, Moradi F, Sadeghi SM, Borz SA (2022) Land use and land cover mapping using Sentinel-2, Landsat-8 satellite images, and google earth engine: a comparison of two composition methods. Remote Sens 14(9)

Nichol CJ, Huemmrich KF, Black TA, Jarvis PG, Walthall CL, Grace J, Hall FG (2000) Remote sensing of photosynthetic-light-use efficiency of boreal forest. Agric for Meteorol 101(2–3):131–142

Odebiri O, Mutanga O, Odindi J (2022) Deep learning-based national scale soil organic carbon mapping with Sentinel-3 data. Geoderma 411:115695. https://doi.org/10.1016/j.geoderma.2022.115695

Odebiri O, Mutanga O, Odindi J, Naicker R, Masemola C, Sibanda M (2021) Deep learning approaches in remote sensing of soil organic carbon: a review of utility, challenges, and prospects. Environ Monit Assess 193:1–18

Odebiri O, Mutanga O, Odindi J, Peerbhay K, Dovey S (2020) Predicting soil organic carbon stocks under commercial forest plantations in KwaZulu-Natal province, South Africa using remotely sensed data. Gisci Remote Sens 57(4):450–463. https://doi.org/10.1080/15481603.2020.1731108

Otto D (2019) Musina-Makhado special economic zone development, Vhembe District Municipality, Limpopo province soils and land capability

Rapinel S, Clément B, Magnanon S, Sellin V, Hubert-Moy L (2014) Identification and mapping of natural vegetation on a coastal site using a Worldview-2 satellite image. J Environ Manage 144:236–246

Rasse DP, Rumpel C, Dignac M-F (2005) Is soil carbon mostly root carbon? Mechanisms for a specific stabilisation. Plant Soil 269(1–2):341–356

Ratajczak Z, Nippert JB, Collins SL (2012) Woody encroachment decreases diversity across North American grasslands and savannas. Ecology 93(4):697–703. https://doi.org/10.1890/11-1199.1

Rossouw P (2016) Soil, agricultural potential, land capability and land use study: impact of a proposed tailings pipeline on the farms Frischgewaagd and Mimosa, near the town of Ledig, North West Province

Roy DP, Huang H, Houborg R, Martins VS (2021) A global analysis of the temporal availability of PlanetScope high spatial resolution multi-spectral imagery. Remote Sens Environ 264:112586

Ruiz-Sinoga J, Martínez-Murillo J, Gabarrón-Galeote M, García-Marín R (2011) The effects of soil moisture variability on the vegetation pattern in Mediterranean abandoned fields (Southern Spain). CATENA 85(1):1–11

Sankaran M, Ratnam J, Hanan NP (2004) Tree–grass coexistence in savannas revisited–insights from an examination of assumptions and mechanisms invoked in existing models. Ecol Lett 7(6):480–490

Schreiner-McGraw AP, Vivoni ER, Ajami H, Sala OE, Throop HL, Peters DP (2020) Woody Plant encroachment has a larger impact than climate change on Dryland water budgets. Sci Rep 10(1):8112

Schulte E, Hopkins B (1996) Estimation of soil organic matter by weight loss-on-ignition. Soil Organ Matter: Anal Interpret 46:21–31

Scott RL, Huxman TE, Williams DG, Goodrich DC (2006) Ecohydrological impacts of woody-plant encroachment: Seasonal patterns of water and carbon dioxide exchange within a semiarid riparian environment. Glob Change Biol 12(2):311–324

Shimizu K, Ota T, Mizoue N, Saito H (2020) Comparison of multi-temporal PlanetScope data with Landsat 8 and Sentinel-2 data for estimating Airborne LiDAR Derived Canopy Height in Temperate forests. Remote Sens 12(11):1876

Smit IP, Prins HH (2015) Predicting the effects of woody encroachment on mammal communities, grazing biomass and fire frequency in African savannas. PLoS ONE 10(9):e0137857

Smith F, Yeaton R (1998) Disturbance by the mound-building termite, shape Trinervitermes trinervoides, and vegetation patch dynamics in a semi-arid, southern African grassland. Plant Ecol 137:41–53

Sommer R, Bossio D (2014) Dynamics and climate change mitigation potential of soil organic carbon sequestration. J Environ Manage 144:83–87

Tan J, Ding J, Han L, Ge X, Wang X, Wang J, Wang R, Qin S, Zhang Z, Li Y (2023) Exploring PlanetScope satellite capabilities for soil salinity estimation and mapping in arid regions oases. Remote Sensing 15(4):1066

Thaler EA, Larsen IJ, Yu Q (2019) A new index for remote sensing of soil organic carbon based solely on visible wavelengths. Soil Sci Soc Am J 83(5):1443–1450

Vågen T-G, Winowiecki LA (2013) Mapping of soil organic carbon stocks for spatially explicit assessments of climate change mitigation potential. Environ Res Lett 8(1):015011

Van Huyssteen C, Rantoa N, Du Preez C (2021) Organic carbon content in the diagnostic horizons and materials of South African soil forms. South Afr J Plant Soil 38(2):140–151

Wang, Qi Y, Guo W, Zhang J, Chang Q (2021) Retrieval and mapping of soil organic carbon using Sentinel-2A spectral images from bare cropland in autumn. Remote Sens 13(6):1072

Wang W, Fang J (2009) Soil respiration and human effects on global grasslands. Global Planet Change 67(1–2):20–28

Wu J, Wang D, Bauer ME (2007) Assessing broadband vegetation indices and QuickBird data in estimating leaf area index of corn and potato canopies. Field Crop Res 102(1):33–42. https://doi.org/10.1016/j.fcr.2007.01.003

Xie B, Ding J, Ge X, Li X, Han L, Wang Z (2022) Estimation of soil organic carbon content in the Ebinur Lake wetland, Xinjiang, China, based on multisource remote sensing data and ensemble learning algorithms. Sensors 22(7):2685

Xie H, Tian Y, Granillo J, Keller G (2007) Suitable remote sensing method and data for mapping and measuring active crop fields. Int J Remote Sens 28(2):395–411

Xu X, Zhai X (2022) Mapping soil organic matter content during the bare soil period by using satellite data and an improved deep learning network. Sustainability 15(1):323

Yang L, Cai Y, Zhang L, Guo M, Li A, Zhou C (2021) A deep learning method to predict soil organic carbon content at a regional scale using satellite-based phenology variables. Int J Appl Earth Obs Geoinf 102:102428

Yang L, He X, Shen F, Zhou C, Zhu A-X, Gao B, Chen Z, Li M (2020) Improving prediction of soil organic carbon content in croplands using phenological parameters extracted from NDVI time series data. Soil Tillage Res 196:104465

Zablotowicz R, Locke M, Gaston L, Bryson C (2000) Interactions of tillage and soil depth on fluometuron degradation in a Dundee silt loam soil. Soil Tillage Res 57(1–2):61–68

Zarco-Tejada P, Sepulcre-Cantó G (2007) Remote sensing of vegetation biophysical parameters for detecting stress condition and land cover changes. Estudios de la Zona no Saturada del Suelo 8(11):37–44

Zhang L, Shao Z, Liu J, Cheng Q (2019) Deep learning based retrieval of forest aboveground biomass from combined LiDAR and landsat 8 data. Remote Sens 11(12):1459

Zhou Y, Boutton TW, Wu XB (2017) Soil carbon response to woody plant encroachment: importance of spatial heterogeneity and deep soil storage. J Ecol 105(6):1738–1749. https://doi.org/10.1111/1365-2745.12770

Žížala D, Minařík R, Zádorová T (2019) Soil organic carbon mapping using multispectral remote sensing data: prediction ability of data with different spatial and spectral resolutions. Remote Sensing 11(24):2947

Challenges and Opportunities to Sustainable Future

Holistic Approaches to Ecosystem Services: A Key to Sustainable Future

Sanjana Chauhan, Anjali Padhi, Siddhi Koli, Prince Amos Gbatoe, and Ashutosh K. Pandey

Abstract Ecosystem services are crucial to human wellbeing, contributing to basic benefits including clean water, food security, and combating the ill effects of climate changes. Amidst the intensification of global environmental challenges, it is crucial to understand and manage these services using integrated, holistic approaches that pave the way to sustainable development. This research focuses on ecosystem service management comparing local practices with global strategies. By implementing a holistic strategy comprising responsible resource management, sustainable land practices, and conservation of biodiversity, we can therefore enhance ecosystem vitality while addressing societal demands. The formulation of effective policies and governance frameworks is important in regulating interactions between people and nature while directing sustainability practices. Such policies should be based on evidence and stakeholder involvement to make them more resilient, flexible, and competent in inducing long-term ecosystem resilience and sustainability.

Keywords Ecosystem services · Sustainable development · Biodiversity conservation · Policy formulation · Ecosystem resilience

1 Introduction

Ecosystems provide an array of services essential for sustaining human life, ranging from provisioning and regulating services to cultural and supporting functions. However, in an era marked by rapid environmental change and increasing anthropogenic pressures, conventional approaches to ecosystem management often fall short of addressing the complexity and interconnectedness of ecosystem processes.

S. Chauhan · A. Padhi · S. Koli · P. A. Gbatoe · A. K. Pandey (✉)
Department of Life Sciences, Sharda School of Basic Sciences and Research, Food Science and Technology, Sharda University, Greater Noida, India
e-mail: Aashu.p20@gmail.com

O. Mutanga et al. (eds.), *Revealing Ecosystem Services Through Geospatial Technologies*, Springer Remote Sensing/Photogrammetry,
https://doi.org/10.1007/978-3-031-98048-0_19

A holistic approach to ecosystem services (ES) emphasizes the integration of ecological, social, and economic dimensions to ensure sustainable management and equitable benefit-sharing. This paradigm is increasingly gaining traction as a pathway to balance human needs with ecological integrity.

Ecosystem services encompass the myriad benefits that humans derive from ecosystems, including provisioning services such as food and water, regulating services like climate and disease control, cultural services that provide aesthetic and recreational value, and support services essential for the production of all other ecosystem services (Daily 1997; Fisher et al. 2009). As we grapple with pressing global challenges, from food security to biodiversity loss, adopting holistic approaches to ecosystem services is not just beneficial but essential for sustainable development. The urgency for a holistic perspective on ecosystem services is underscored by the complex interdependencies that characterize ecological systems. Traditional management approaches often compartmentalize ecosystem functions, leading to unintended consequences such as resource depletion and habitat destruction. For instance, agricultural practices that prioritize short-term yield often neglect the long-term health of soil ecosystems, which are vital for sustaining agricultural productivity (Kremen et al. 1994). A holistic approach recognizes these interconnections and advocates for integrated management strategies that consider ecological, social, and economic dimensions simultaneously.

In Asia, where rapid urbanization and industrialization have intensified pressures on natural resources, the need for holistic ecosystem management is particularly pronounced. The region is home to diverse ecosystems that provide critical services to millions of people. However, these ecosystems are increasingly threatened by unsustainable practices and climate change (Chopra et al. 2023). By embracing holistic approaches, we can better access and manage these vital resources to ensure their sustainability for future generations. Understanding the complex relationships between human activities and natural environments calls for sustaining development while reducing adverse impacts to ecosystems and its services. A means of achieving this ends is by examining compromise among different ecosystem services, as well as assessing the prospects of regime shifts—sudden transformations in the conditions that may impede ecosystems from delivering vital services. Sustainability can only be attained with use and preservation of natural resources in a well-balanced manner, in order to consequently reduce human exploitation and adverse outcomes.

Today, when the Earth's population has reached almost eight billion, it is becoming increasingly important to balance human needs with ecological systems. Ecosystem various ecosystem services which play a crucial role for human sustenance. These services include tangible resources, like food and water, as well as intangible benefits, such as, climate resilience, preservation of biodiversity, and water conservation. However, human activities, such as population explosion, agriculture intensification, industrialization, urbanization, and resource extraction, are rapidly degrading these systems, thus threatening the continued availability of services. These challenges require an interdisciplinary and global effort to achieve a sustainable balance between humans and nature.

In an era marked by rapid environmental degradation and climate change, the concept of ecosystem services has emerged as a critical framework for understanding the intricate relationships between human well-being and the natural world. This review delves into the current state of holistic approaches to ecosystem services, with a particular focus on their valuation, management, and application across diverse geographical contexts, including Asia.

1.1 Ecological Services Provide by Ecosystems

Ecosystem provides four major ecological services; the four primary categories include provisioning, regulating, cultural, and supporting services. Provisioning services involve the essential supply of food, water and fuel. Regulating services involve natural processes such as carbon sequestration, water filtration, or climate resilience. From a cultural service perspective, recreational, spiritual and aesthetic experiences are provided, the supporting services comprise nutrient cycling and habitat formation, crucial to the functioning of all other services (MEA 2005). Overall, they form the underpinning of human well-being, economic stability, and environmental resilience. Human sustenance have been dependent since history on ecosystem services, and modern life remains an integral part of a healthy natural environment. Soil fertility, requisite amount of water supply, and the service of pollinations are only some of them for food production. Climate is regulated and extreme weather impacts are reduced by forests and oceans. The presence of natural landscapes enhances mental health and cultural values.

Notwithstanding this significant dependence, such services are frequently regarded as "free," resulting in minimal understanding and motivation for its preservation policies. This approach towards the ecosystem services may results in resource exploitation and quality deterioration. It is also evidenced by species extinction, increased environmental pollution, and diminishing soil fertility, ultimately adversely affecting economic output and the overall quality of human life. Anthropogenic activities, such as, intense human activities like deforestation, urbanization, pollution, and climate change. What people have been calling the "lungs of the planet," that is, forests, have been seriously, being destroyed because of agricultural expansion and logging, biodiversity loss, and the release of carbon emissions. These disturbances highlight the vital role ecosystem services play and the challenges facing regenerative restoration of degraded environments.

In trade-off decisions, people tend to focus more on provision services like food, fiber, and water supply, while the regulating services are often under-estimated. Mainly it is because regulating services rely on the gradual modification of ecosystem components like biodiversity and soil health. This undervaluation would eventually diminish the sustainability of provision services, as well as the overall integrity of ecosystems. More precisely, these slowly changing factors are typically overlooked in policy-making decisions, unknowingly encouraging areas of vulnerabilities that ultimately undermine sustainable resource use.

Ecosystem resilience- recovery abilities—remain integral in the course of maintaining the flow of ecosystem services, factors such as biodiversity conservation and restoration enhance, while crossing critical thresholds may lead to irreversible changes. On the other hand, overgrazing could be the cause for some grassland-poor quality landscapes, and global warming could disrupt ocean currents. Therefore, identification and management through sustainable practices such as soil conservation are important for ecosystem stability. Incremental changes in ecosystems often trigger unexpected "regime shifts"—substantial and possibly irreversible changes that are difficult or impossible to reverse. Such regime shifts can produce significant consequences, such as economic declines in fisheries, eutrophication, or the conversion of grasslands to shrublands (Department of the Environment, Water, Heritage and the Arts 2009).

These changes often occur when ecosystems pass critical thresholds, which are driven both by subtle declines in resilience and by external factors like climate fluctuations or human impacts. For example, the accumulation of phosphorus in soils can make lakes susceptible to toxic algal blooms and fish kills, and steady increases in overfishing can turn coral reefs ever more vulnerable to storms and diseases. Similarly, decreased grass cover in regions like the Sahel triggers vegetation shifts, reducing forage and intensifying drought feedback. These cascading changes can cause major shifts in ecosystem services, which in turn alter social dynamics in ways that amplify ecosystem degradation. Therefore, recognizing and managing ecosystem variability, resilience, and thresholds is crucial for sustainable development. If these factors are incorporated into decision-making processes, societies will ensure environmental, economic, and social stability to guarantee long-term human well-being and planetary health.

1.2 The Role of Valuation in Holistic Approaches

Valuation of ecosystem services is a fundamental component of holistic approaches. Understanding the economic value of ecosystem services can inform decision-making processes and encourage sustainable practices. For instance, Krzemień et al. (2023) highlight how valuing carbon allowances in coal mining areas can facilitate optimal recovery strategies that balance economic development with environmental conservation. Similarly, studies in India have shown that systematic valuation of ecosystem services can guide policy decisions towards more sustainable land use practices (Chopra et al. 2023).

However, valuation must be approached with caution. The complexity of ecosystem functions often defies simple economic quantification. Ignoring non-market values—such as cultural significance or intrinsic ecological worth—can lead to misguided policies that prioritize short-term economic gains over long-term sustainability (Farber et al. 2002). Therefore, a pluralistic approach that incorporates diverse values is essential for effective ecosystem service valuation.

1.3 The Urgency of a Holistic Perspective

The urgency for a holistic perspective on ecosystem services is underscored by the complex interdependencies that characterize ecological systems. Traditional management approaches often compartmentalize ecosystem functions, leading to unintended consequences such as resource depletion and habitat destruction. For instance, agricultural practices that prioritize short-term yield often neglect the long-term health of soil ecosystems, which are vital for sustaining agricultural productivity (Kremen et al. 1994). A holistic approach recognizes these interconnections and advocates for integrated management strategies that consider ecological, social, and economic dimensions simultaneously.

In Asia, where rapid urbanization and industrialization have intensified pressures on natural resources, the need for holistic ecosystem management is particularly pronounced. The region is home to diverse ecosystems that provide critical services to millions of people. However, these ecosystems are increasingly threatened by unsustainable practices and climate change (Chopra et al. 2023). By embracing holistic approaches, we can better access and manage these vital resources to ensure their sustainability for future generations.

1.4 Defining Holistic Approaches to Ecosystem Services

Holistic approaches to ecosystem services involve integrating various disciplines and stakeholder perspectives to create comprehensive management frameworks. Such approaches emphasize collaboration among scientists, policymakers, local communities, and businesses to develop strategies that enhance ecosystem resilience while meeting human needs (Meraj et al. 2022). This integration is crucial in addressing the multifaceted challenges posed by environmental change and socio-economic pressures.

One exemplary framework is the Ecosystem-Based Management (EBM) approach, which seeks to manage natural resources in a way that maintains ecosystem health while providing for human needs (Rosenberg and McLeod 2011). EBM recognizes the interconnectedness of ecosystems and human activities, promoting sustainable practices that enhance both ecological integrity and socio-economic benefits. This approach has gained traction in various contexts across Asia, where local communities are increasingly involved in managing their natural resources.

1.5 The Role of Valuation in Holistic Approaches

Valuation of ecosystem services is a fundamental component of holistic approaches. Understanding the economic value of ecosystem services can inform decision-making processes and encourage sustainable practices. For instance, Krzemień et al. (2023) highlight how valuing carbon allowances in coal mining areas can facilitate optimal recovery strategies that balance economic development with environmental conservation. Similarly, studies in India have shown that systematic valuation of ecosystem services can guide policy decisions towards more sustainable land use practices (Chopra et al. 2023).

However, valuation must be approached with caution. The complexity of ecosystem functions often defies simple economic quantification. Ignoring non-market values—such as cultural significance or intrinsic ecological worth—can lead to misguided policies that prioritize short-term economic gains over long-term sustainability (Farber et al. 2002). Therefore, a pluralistic approach that incorporates diverse values is essential for effective ecosystem service valuation.

1.6 Challenges and Opportunities in Implementing Holistic Approaches

Despite the clear benefits of holistic approaches to ecosystem services, several challenges hinder their widespread adoption. One major obstacle is the lack of interdisciplinary collaboration among stakeholders. Effective management requires input from various fields—ecology, economics, sociology—yet traditional academic structures often promote siloed research efforts (Berkes 2009). Bridging these gaps through collaborative research initiatives can foster innovative solutions that address complex environmental issues.

Moreover, there is a pressing need for robust assessment tools that can evaluate multiple ecosystem services simultaneously. Current assessment frameworks often fall short in capturing the synergies and trade-offs between different services (Müller et al. 2021). Developing integrated assessment models that account for these interactions will be crucial in guiding effective management strategies.

1.7 Case Studies: Successful Holistic Approaches in Asia

Several case studies illustrate the potential of holistic approaches to enhance ecosystem service management in Asia. In China's Heihe River Basin, researchers have identified societal drivers influencing the evolution of ecosystem services over millennia (Wang et al. 2022a; b). By integrating historical data with contemporary assessments, this study highlights the importance of linking societal values with environmental outcomes.

In India, integrated farming systems have been developed as a holistic strategy to enhance food security while conserving ecosystem services (Kumar et al. 2024). These systems combine various agricultural practices—such as crop rotation and agroforestry—to optimize resource use and improve resilience against climate variability.

2 Man-Used Resources from Ecosystem

Humans depend on ecosystems for many essential resources-from food and water to building materials, clothing, and fuel. Particularly these are the most often exploited at non-renewing rates that could lead to ecological imbalance and degradation of availability for future generations. Growing human populations and economies, which also lead to increased resource extraction and consumption, are the main drivers of the debate over overexploitation, environmental degradation, and biodiversity loss (Foley et al. 2005; Newman and Cragg 2016).

2.1 Food and Energy Resources

Nutritional assets ecosystems provide widespread food supply in the form of crops, livestock, fish, and indigenous plants. Farming occupies about 40% of the Earth's land mass and has variable impacts on the environment; these include clearance of forests, soil erosion, and habitat destruction (Foley et al. 2005). While fishing and aquaculture are crucial components in human diets, the trend of overfishing has reduced the populations of fish and interfered with marine ecosystems. Sustainable agriculture and responsible fishing practices are needed to ensure these resources remain available while minimizing ecological harm.

Humans abstract different forms of energy from ecosystems, from fossil fuels to wood and biomass. Fossil fuels, which feed much of the modern economy, are made up of ancient biological material and are a leading source of greenhouse gas emissions, driving climate change. Wood is still a major source of energy, especially in developing regions, where it is used for heating and cooking. Following the recent trends, wind, solar, and hydropower renewable energy sources are increasingly favored due to reduced ecological impacts which reduce reliance on finite resources and alleviate the impacts of environmental degradation (International Renewable Energy Agency 2021).

2.2 *Wood and Forestry Resources*

Deforestation for timber and agricultural expansion have great influences on biodiversity and carbon reservoirs, and among these, tropical rainforests are most affected (Laurance et al. 2014). Conserving these ecosystems simultaneously with the fulfillment of human needs through the use of renewable resources and reforestation as well as effective forest management strategies can combat against the climate change impact.

2.3 *Pharmaceutical Resources*

Many plants, animals, and microorganisms are good sources of bioactive compounds which are utilized in traditional and modern medicine. Today, pharmaceutical discovery depends significantly on biodiversity which serves as a source of new treatments in the future. Therefore, conservation of biodiversity-rich ecosystems is critical to enable continued use of these medicinal resources, which are being destroyed through habitat destruction and loss of species, thus making the idea of new medicines to be discovered a far cry and necessitating ecosystem preservation (Newman and Cragg 2016).

3 Ecological Service Dynamics: Drivers and Trade-Offs

The classification of ecosystem service and its interactions with different communities can be broadly classified into two primary mechanisms: the influence of shared drivers on various ecosystem services and the direct interactions among these services. This framework makes it possible to delve deeper into the complexities involved in ecosystem management by emphasizing the trade-offs and synergies that arise from human activities. (Bennett et al. 2009).

- Common Factors' Impacts on Ecosystem Services:

  Many ecosystem services may be interactively changed by such anthropogenic activities as change in land use, infrastructure development, and fertilization sometimes at the expense of others-trade-offs-and sometimes in mutualistic ways-synergies (Bennett et al. 2009). For example, infrastructural development for visitors to view maple syrup production could improve recreation but not influence the syrup. Conversely, excessive fertilizer application to enhance crop yields may deteriorate water quality through nutrient runoff, so there is a trade-off. Synergies are found when a single service enhancement benefits multiple services: for example, wetland restoration enhances simultaneously water quality and flood control (Bennett et al. 2009).
- Direct Interactions Among Ecosystem Services:

Ecosystem services can also interact directly in either a unidirectional or a bidirectional manner. In unidirectional interactions, one service influences another with no feedback. For instance, forest patches can provide enhanced pollination that elevates the yields of coffee while coffee production itself doesn't impact pollination. Conversely, bidirectional interactions involve a feedback loop (Bennett et al. 2009). For example, in dryland agriculture, cultivation practices that reduce erosion increase crop yields, thereby triggering new land with improved soil health and agricultural productivity (Enfors and Gordon 2007).

In addition, ecosystem services often share common drivers that bring synergies in their relationship. For instance, marine protected areas save fish populations that in turn support tourism and maintain an important ecological role, such as algal grazing. These interactions highlight how one management action can simultaneously support multiple services. Beyond direct interactions and common drivers, several indirect drivers also have an effect on ecosystem services and its outcomes. One of the factors is demographic drivers like global population growth, especially within Sub-Saharan Africa and South Asia, influences resource consumption and environmental impact. Economic growth, urbanization, and globalization increase resource use unless balanced with sustainable policies to the environment. Sociopolitical drivers, including government decisions and local participation, also shape resource management. Similarly, technological innovations, particularly in agriculture, have increased food production but also caused environmental degradation. Direct drivers include climate change, conversion, nutrient application, and biological invasions. Actually, these drivers further engross the complications of ecosystem service dynamics. Human activities consist of land-use changes and fertilizer use that have caused immense environmental damage, ranging from biodiversity loss, eutrophication, and changed climate patterns. Maintenance of ecosystem resilience amidst these drivers necessitates integrated approaches that consider both direct and indirect influences on ecosystem services (Nelson et al. 2006).

4 Human Impact on the Ecosystem

4.1 Demand and Production for Food

Food constitutes one of the principal services of terrestrial and aquatic ecosystems, essential for human health and lifestyle. Increasing food demand across the world is primarily driven by population growth, improvements in the economic situation, and changes in diets patterns. The United Nations further predicts that by 2050, the world population will have reached 9.8 billion. To meet this demand, food production needs to be increased compared with current levels. As incomes rise, people tend to consume more resource-intense foods such as meat, which puts further pressure on ecosystems (Rodríguez et al. 2006). Such increased production mounts huge pressures on these systems that lead to clearing forests, biodiversity losses, and depletion of both water and arable land.

Agriculture is a vital point of intersection for critical trade-offs among ecosystem services. For example, clearing forest land for agricultural fields might increase food supply but will generally lead to decreased biodiversity, reduced water purifying, and weakened climate regulation previously supported by the forest. A balancing of this environmental conundrum therefore necessarily reconciles growing demands for food production with efforts to minimize environmental degradation. The shift in dietary preferences, especially towards resource-intensive foods, underlines the importance of innovations in food production, improved resource management, and promoting sustainable consumption patterns.

For food supply and demand pattern analysis and prediction, models like IMPACT are increasingly used (Rosegrant et al. 2001). This tool examines the regional need for food, the regional production capacity, and the role played by international trade in determining the food prices. Through the simulation of diverse future scenarios, these models assist policymakers and resource managers in the development of sustainable food systems. For instance, the "Global Orchestration" scenario anticipates swift economic advancement and a rise in the consumption of premium foods such as meat, thereby heightening the demand for livestock feed. In contrast, the "TechnoGarden" scenario encourages plant-based dietary practices, resulting in decreased meat consumption, mitigating ecosystem pressures, and diminishing the need for feed resources (Alcamo et al. 2005).

Additionally, regional disparities further influence the worldwide food landscape. Asia - driven by rising incomes and urbanization - will experience a significant growth in food demand, particularly in meat and dairy products. This will lift demand for grains used as animal feed and impact cereal prices and availability. Sub-Saharan Africa, however, with generally slower economic growth, will experience modestly higher growth in demand for food, reflecting ongoing challenges to economic growth and food security. By incorporating regional knowledge and predictive models into strategic planning, societies can move toward resilient food systems that balance growing demands with ecosystem health and sustainability (Alcamo et al. 2005).

4.2 Land Use Patterns

Land is one of the most critical resources for food production, and food production significantly affects other ecosystem services. Conversion of natural grasslands to arable land can reduce benefits of biodiversity and water regulation (Schröter et al. 2005). On the other hand, in wealthier regions where agriculture could potentially be abandoned, the landscape may be restored as a native habitat, such as a grassland or forest, and enhance ecosystem services and promote environmental balance. Balancing these competing interests requires careful discussion because each food production and land use decision creates a long-lasting impact on ecosystems and to the suite of services they provide.

King et al. (2015) analyzed the trade-offs and suggested a conceptual framework for the proposition of land use policies with the goal of maximizing synergies and minimizing tradeoffs. Yang et al. (2018) showed that, the Yanhe watershed in China and its influence of land-use changes on the ecosystem-service trade-offs within its ecological restoration zone by using the InVEST model to calculate the tradeoffs of carbon sequestration, habitat quality, nutrient retention, sediment retention, and seasonal water yield services. The findings indicate that ecosystem services had both positive and negative impacts due to land use changes; for example, afforestation and grassland expansion improved carbon sequestration and habitat quality but diminished water yield, while expansion of cropland boosted food production at the cost of declining nutrient and sediment retention.

4.3 *Fresh Water Resource*

Freshwater ecosystems are also vital for human life and well-being, because freshwater can be used for drinking, agriculture, industry, and recreation purposes, while freshwater fisheries support significant portions of the world's food supply. The WaterGAP model has been used to estimate water availability in rivers and basins, relating these to data on rainfall, evaporation, groundwater, and river runoff. It is also possible to estimate future water usages based on changes in income, population, and energy demand with this model (Alcamo et al. 2003).

Overuse of water also creates stress in freshwater ecosystems and may even result in scarcity. Already, "severe water stress," where demand outpaces availability, affects around 18% of the world's river basins and serves some 2.3 billion people. By 2050, this may stress up to 4.9 to 5.5 billion people, mainly in water-scarce regions like Africa and parts of Asia. Many basins in the developing world could experience increasing water stress because of growing demand (Alcamo et al. 2005).

Increased water consumption tends to cause more pollution, as wastewater "return flows" find their way back into rivers and lakes. These "return flows," or amounts of wastewater, are predicted to sharply rise by 2050-most severely in Latin America and Sub-Saharan Africa, potentially doubling or tripling. Untreated wastewater will degrade the quality of the water, jeopardize ecosystems, and make water less safe for human consumption (Alcamo et al. 2005).

In general, the heightened demand for freshwater services does not ensure sustainability, since numerous regions will encounter competition over water resources, elevated pollution risks, and a greater frequency of water shortages. Tackling these issues will necessitate meticulous management, particularly in rapidly developing areas.

4.4 *Fish Production*

Fish are a critical component of global protein sources; therefore, fish supply is an essential service extracted from aquatic ecosystems. Predictions of future consumption patterns for fish often apply models like IMPACT to evaluate worldwide demand and use EcoSim/EcoPath to model fish production in specific regions (Alcamo et al. 2005). The study concluded that the depletion of fish stocks is mainly due to the interaction of overfishing, the natural life histories of the fishes, and climate change.

When fish are overfished in the short term, it leads to sudden drops in their numbers. However, if overfishing continues over a longer period, it leads to gradual depletion of the population (Pinsky and Byler 2015). Also, natural life-history traits of the fish, such as how long they live and how fast they reproduce, also determine their sensitivity to overfishing. Finally, climate changes can alter the population of fish and the productivity of marine systems, raising further risks.

4.5 *Biodiversity Loss*

Biodiversity-the staggering diversity of life on Earth—provides priceless services, from food and medicine to cultural enrichment. Yet human activities are driving an unprecedented loss of biodiversity. Habitat loss, pollution, overfishing, and climate change have pushed many species to the edge of extinction, while changing food webs, destabilizing ecosystems, and reducing the ability of nature to respond to environmental changes. The decline of pollinators such as bees is of great concern to global food security, as many crops rely on insects for pollination.

4.5.1 Case Study

The sudden decline of Gyps vultures in eastern India is a good example of how the loss of one species can disrupt many ecosystem services and result in unintended socio-ecological consequences.

Vultures play an important role in controlling cattle carcasses in areas where beef consumption is prohibited. They also facilitate the Parsi community in Amritsar in the removal of human remains in traditional burial grounds. Serious ramifications of the plummeting population of vultures have been reported. Since there are fewer vultures to clean carcasses, members of the Parsi community must store bodies, thereby exposing themselves to health issues. Reduced competition for carcasses means a boom in feral dog populations; this increase has heightened the risk of rabies and heightened public health concerns. The decline in vulture populations has been associated with the use of diclofenac, an anti-inflammatory medication administered to cattle that poses a threat to vultures. This instance highlights the manner in which a measure intended to promote the health of livestock can cascade into multiple consequences for numerous ecosystem services, including human health, thus underlining the interdependence among species and the necessity for careful assessment of ecological implications in human interventions (Rodríguez et al. 2006).

Coral Reefs and Marine Protection: Coral reefs are vital habitats that support a rich diversity of marine life and shield coastal regions from erosion. However, climate change and pollution have caused extensive coral bleaching. Marine protected areas help safeguard coral ecosystems, while initiatives to reduce pollution and overfishing aid in reef restoration (Department of the Environment, Water, Heritage and the Arts 2009).

4.6 Climate Change and Ecosystem Alterations

Anthropogenic climate change, caused mostly by burning fossil fuels, is widely transforming ecosystems across the world, exerting immense strain on natural systems like forests, wetlands, and coral reefs. Rising temperatures, melting of the ice caps, chemical changes in the water of the oceans, and increased extreme weather events threaten biodiversity even as these threaten vital ecosystem services, such as carbon sequestration, water filtration, and flood management, that are essential to human life (Alcamo et al. 2005; Pinsky and Byler 2015). The growing impact of climate change, caused by human activities, poses a significant danger to ecosystems and humankind. The biosphere, defined as the global sum of all living organisms on Earth, is integral to climate regulation. It contributes by both absorbing and emitting CO_2, reflecting solar radiation, and controlling water movement. For example, plants take in CO_2 during photosynthesis and subsequently release it via respiration, thereby affecting the atmospheric CO_2 equilibrium. Nevertheless, the rapidly intensifying effects of climate change are progressively threatening these essential natural processes (Alcamo et al. 2005).

4.7 Soil Erosion

Erosion takes place when soil and land surfaces are ground away, leading to a greater sediment load within rivers and streams. It can hinder navigation and availability of water while reducing the suitability of agricultural land. Soil erosion is a major concern in agriculture because the degradation of the land may lower crop production by up to 12.7% over time. In this regard, the ability of soil to resist erosion has been considered a fundamental ecosystem service (Alcamo et al. 2005).

5 Methods for Measuring and Evaluating Ecosystem Services

Integrating qualitative and quantitative analysis can be useful at all stages of a broad ecosystem services assessment. In general, quantitative ecosystems services assessments are more technical in nature, as they require the collection and analysis of biophysical data on ecosystem services or the ability to undertake spatial analyses using GIS-based modeling software.

An ecosystem's services measurement can be employed to determine a baseline measure for monitoring changes over time, or to allow evaluation of the effects of management actions or policy changes on ecosystems services provision. Some reasons to measure ecosystems services are as follows; (1) Public/policy support, (2) Site management, (3) Human well-being, (4) Planning, (5) Private sector management, (6) Fund investment and, (7) Knowledge generation (Table 1).

Among the diverse methods followed, research attention have been mostly given to the approach of ecological flow followed by actual use accounting, another method of flow, and the spatial connection method.

Table 1 Methods for measuring and evaluating ecosystem services

Method of classification	Specific method	Principles
Actual use accounting method	Fast-track method	Measures ES actual use with easily available data (e.g., statistical data, LULC)
	Spatial separation method	Differentiates ES actual use from potential ES using spatial relationships between ES potential and demand
Spatial connection method	Overlay flow and benefitting area	Considers flow as a spatial connection between providing and benefiting areas
Ecological flow method	Service flow effect (intraregional)	Reflects ES flow in the final relationship between ES supply and demand
	ES delivery chain (intraregional)	Tracks flow quantity delivered along paths from service-providing to benefiting areas
	ARIES modeling (intraregional)	Uses service path attribution networks to map ES flow between source, sink, and use locations
Interregional flow method	Interregional flow method	Transfers ES from surplus areas (positive balance) to demand areas (negative balance) using energy or human input

1. Actual use accounting methods actually define ESF as the total amount of services consumed. Fast-track is a method often used for provisioning services, which employs statistical data, participatory datasets, or land use and land cover information. Although the methodology is fairly straightforward, the derived outputs are marred by biases in expert perception, low resolution in space and errors that have been put at the national level (Bukvareva et al. 2017; Huang et al. 2021). In contrast, the spatial separation approach, which is widely used for services regulating service, integrates potential ecosystem services with demand variables, creating a mix of, for instance, population density and infrastructure, including reservoirs. This approach allows for multiscalar assessment and integrates ecological and societal data; it, however, tends to oversimplify the complexity that exists between ecosystems and human interactions (Schirpke et al. 2019).
2. Spatial connectivity conceptualizes ESF as the spatial connection between SPA and SBA. ESF is calculated according to the overlay percentage of flow areas with SBA. In this regard, this method is most efficient for services characterized by flows, such as food provision and climatic regulation, according to Serna-Chavez et al. (2014). However, reliance on simple overlay analyses may lead to inaccuracies in the mapping, valuation, and trade-offs of ecosystem services because such analyses often fail to capture specific flow pathways (Bagstad et al. 2014).
3. Ecological flow process methods give a more dynamic view because these consider the flow processes between supply and demand regions. The impacts of service flow are quantified in ESF based on the relationship of the supply and demand of ecosystem services. This makes it particularly useful for flow-dependent services such as water supply and local climate regulation (Goldenberg et al. 2017). Similarly, the methodology associated with the delivery chain of ecosystem services includes aspects of flow paths, such as distance and route specifications. However, although this approach is more holistic, it fails to realistically model the dynamic processes that accompany ecosystem service flows (Vigl et al. 2017). More complex models such as ARIES (Artificial Intelligence for Ecosystem Services) provide a conceptual framework for the quantification of ESF with service path attribution networks, enabling holistic assessments but requiring further technological refinement (Bagstad et al. 2013).
4. Interregional Flow methods take a broader perspective and describe the transfer of ecosystem services from surplus to deficit regions. For instance, food production and carbon sequestration often flow from regions of surplus to those with higher demand (Xie et al. 2019). These methods incorporate spatial and ecological dynamics and provide a holistic understanding of ESF while solving the challenges of balancing complexity with practical application.

6 Challenges in Estimating Economic Value

Monetary valuation of ecosystem services can sometimes be in conflict with conservation objectives if the economic values associated with conservation are not as high as alternative land uses in the short term (Schröter et al. 2014). This does not mean that the site should be converted, but that the conservation value needs to be assessed from a non-monetary perspective, such as globally significant biodiversity values, irreplaceable cultural values or relational values (Chan et al. 2016). Also, certain ES (such as cultural heritage) are difficult to assess in monetary terms and may be better evaluated using non-monetary measures.

7 Tools and Frameworks for Integrated Assessments

Ecosystem assessments apply the judgement of experts to existing knowledge generated from the scientific community (and other forms of knowledge) to provide credible answers to policy-relevant questions. And therefore, ecosystem assessments are a tool that can support the development of an evidence base that meets the needs of different sectors and encourages integration (Berghöfer and Brown 2016; Ash et al. 2010). By taking into account a range of biophysical, social and economic parameters, integrated ecosystem assessments can be a useful tool for characterising and communicating the true societal value of ecosystem services. The level of integration within existing ecosystem assessments varies, but usually includes: i) combining; ii) interpreting; and, iii) communicating knowledge from diverse disciplines.

Integrated risk assessments, which apply both social and ecological perspectives, are critical to informing on development trajectories of these vulnerable landscapes (Hagenlocher et al. 2018).

Early engagement of a diversity of stakeholders has been emphasised by a number of respondents as being a key element within an integrated ecosystem assessment. Developing a good contextual understanding and potential complexity involved in engaging certain stakeholders will help mitigate problems and potential conflict, and should form an important part of the scoping stage. Good communication is emphasized as a priority throughout the assessment process. Important stages within which to involve stakeholders are during the framing and decision-making stages.

It is essential when designing an ecosystem assessment to consider how and where the concepts of integration will be considered. The integrated assessment framework for ESMERALDA, highlights the importance of the design phase of an assessment as well as the many tasks which can take place within an ecosystem assessment. It is important to note that the integrated assessment framework presented here represents an ideal situation, and should be adapted to suit the national situation.

However, while integrated ecosystem assessment processes have not been well documented or evaluated, the evidence that is available suggests that integration through the governance structure (inclusion of stakeholders), combining of different data sources and the use of tools allows for greater impact of the ecosystem assessment report within decision making. What also has emerged, is that the key contribution that the notion of integrated assessment provides is the ability to consider the synergies and trade-offs of a range of ecosystem services associated with one or more ecosystems. Overcoming the barriers and limitations of 'siloed thinking' is perhaps the main feature of the integrated frameworks in general and especially of the one proposed here. Without such cross-sectoral thinking it is difficult to see how a proper account of biodiversity and ecosystem services can be fully taken into account by decision makers.

8 Management of Ecosystem Services

The management of ecosystem services requires a comprehensive understanding of how ecological, social, economic, and technological factors interact (MEA 2005). These interactions shape the trade-offs and synergies between services, influencing how resources are managed, distributed, and utilised. Economic drivers, such as market demand, pricing, and subsidies, can either promote sustainable practices or exacerbate ecosystem degradation (Nelson et al. 2006). For example, economic incentives in agriculture can boost food production but may contribute to deforestation and biodiversity loss. On the other hand, policies like payments for ecosystem services align economic benefits with ecological conservation, encouraging sustainable land management practices.

Technological advancements, such as precision agriculture, renewable energy, and wastewater treatment, offer significant opportunities to reduce trade-offs and enhance synergies among different ecosystem services. For example, innovations in irrigation technology can improve water use efficiency, benefiting both agricultural productivity and water conservation. Similarly, advancements in aquaculture can mitigate overfishing while meeting the rising demand for fish, although challenges such as reliance on fishmeal remain. However, technology cannot fully replace natural processes—desalination, for instance, may substitute the natural hydrological cycle, but it is costly, and its long-term effects remain uncertain. Technological solutions must, therefore, be carefully monitored to prevent unintended environmental harm Schirpke et al. (2023).

In the context of ecosystem governance, technology plays a pivotal role in enhancing monitoring and management capabilities. Technologies like tracking systems for offshore fisheries can improve resource management and reinforce property rights. Legal frameworks also play a critical role, balancing economic and ecological priorities through zoning laws that restrict land-use changes harmful to ecosystems and international agreements like the Paris Climate Accord, which promote sustainable practices to mitigate climate-related trade-offs Porras (2021).

Optimal land management strategies require integrating scenario analysis and optimization methods. Scenario analysis projects socio-economic and environmental indicators, supporting global and sub-global assessments to quantify trade-offs and synergies between land use intensity, biodiversity, and ecosystem services. However, these relationships are often non-linear, where maximising one service can compromise another. Optimization methods identify Pareto-frontiers, providing actionable insights that balance competing objectives like economic gains and environmental conservation Salmani et al. (2021).

By combining scenario-based projections with optimization-based insights, researchers can navigate constraints and identify sustainable pathways. These methods offer tools for decision-making, such as trade-off gradients, thresholds, and risk assessment maps, enabling stakeholders to make informed choices. Computational advancements in modeling and algorithms further support the systematic exploration of land management configurations, offering a global-scale understanding that incorporates place-based studies and broader analyses. This integration provides actionable, data-driven solutions that balance human and ecological needs, paving the way for sustainable land management practices Zhang et al. (2020).

Ultimately, technology can significantly enhance ecosystem service management and human well-being, but its development and application must be responsible. When integrated into appropriate legal and economic frameworks, technology can foster interactions that sustain and nurture vital ecosystem services, transforming humanity's relationship with the biosphere Millennium Ecosystem Assessment (2003).

9 Conclusions

As we confront escalating environmental challenges globally, embracing holistic approaches to ecosystem services is imperative. These strategies offer pathways towards sustainable development by recognizing the interconnectedness of ecological health and human well-being. By fostering interdisciplinary collaboration, enhancing valuation methods, and learning from successful case studies across Asia and beyond, we can create resilient ecosystems that support both current and future generations.

A holistic approach to ecosystem services is not merely a theoretical construct but a pragmatic necessity in addressing the complex challenges of sustainable development. By integrating ecological, economic, and socio-cultural dimensions, such approaches promise to foster resilience and equity in ecosystem management. The insights from this review underscore the transformative potential of holistic strategies, urging researchers, policymakers, and practitioners to embrace this paradigm for the benefit of both nature and society.

References

Agrawal A, Gibson CC (1999) Enchantment and disenchantment: the role of community in natural resource conservation. World Dev 27(4):629–649

Alcamo J, Döll P, Henrichs T, Kaspar F, Lehner B, Rösch T, Siebert S (2003) Development and testing of the WaterGAP 2 global model of water use and availability. Hydrol Sci J 48(3):317–337

Alcamo J, van Vuuren D, Ringler C, Cramer W, Masui T, Alder J, Schulze K (2005) Changes in nature's balance sheet: Model-based estimates of future worldwide ecosystem services. Ecol Soc 10(2)

Arkema KK, Abramson SC, Dewsbury BM (2015) Marine ecosystem-based management: from characterization to implementation. Front Ecol Environ 4(10):525–532

Ash N, Blanco H, Brown C, García K, Henrichs T, Lucas N, Raudsepp Hearne C, Simpson RD, Scholes R, Tomich TP, Vira B, Zurek M (2010) Ecosystems and human well being: a manual for assessment practitioners. Island Press

Bagstad KJ, Johnson GW, Voigt B, Villa F (2013) Spatial dynamics of ecosystem service flows: a comprehensive approach to quantifying actual services. Ecosyst Serv 4:117–125

Bagstad KJ, Villa F, Batker D, Harrison-Cox J, Voigt B, Johnson GW (2014) From theoretical to actual ecosystem services: mapping beneficiaries and spatial flows in ecosystem service assessments. Ecol Soc 19(2):64

Baskent EZ, Borges JG, Vacik H, Reynolds KM, Rodriguez LCE (2022) Management of multiple ecosystem services under climate change, bioeconomy, and participation. Sustainability 14(5):1–15

Bennett EM, Peterson GD, Gordon LJ (2009) Understanding relationships among multiple ecosystem services. Ecol Lett 12(12):1394–1404

Berghöfer A, Brown C (2016, April) Increasing the policy impact of ecosystem service assessments and valuations: insights from practice. GIZ ValuES, Deutsche Gesellschaft für Internationale Zusammenarbeit (GIZ)

Berkes F (2009) Evolution of co-management: role of knowledge generation, bridging organizations and social learning. J Environ Manage 90(5):1692–1702

Brander LM, Schägner JP, de Groot R (2023) On the potential use of the ecosystem services valuation database for valuation in the system of environmental economic accounting. Ecol Econ 203(1):1–15

Bukvareva E, Zamolodchikov D, Kraev G, Grunewald K, Narykov A (2017) Supplied, demanded and consumed ecosystem services: prospects for national assessment in Russia. Ecol Indic 78:351–360

Carpenter SR, Bennett EM, Peterson GD (2006) Scenarios for ecosystem services: an overview. Ecol Soc 11(1)

Chan KMA, Balvanera P, Benessaiah K, Chapman M, Díaz S, Gómez Baggethun E, Hannahs N, Jax K, Klain S, Luck GW, Martín López B, Muraca B, Norton B, Ott K, Pascual U, Satterfield T, Tadaki M, Taggart J, Turner N (2016) Opinion: why protect nature? Rethinking values and the environment. Proc Nat Acad Sci 113(6):1462–1465

Chopra B, Khuman YSC, Dhyani S (2023) Advances in ecosystem services valuation studies in India: learnings from a systematic review. Ecol Econ 203(1):1–10

Dade MC, Mitchell MGE, McAlpine CA, Rhodes JR (2018) Assessing ecosystem service trade-offs and synergies: the need for a more mechanistic approach. Ambio 47(1):1–12

Daily GC (1997) Nature's services: societal dependence on natural ecosystems. Island Press, Washington, DC

Department of the Environment, Water, Heritage and the Arts (2009) Ecosystem services: key concepts and applications, occasional Paper No. 1. Canberra

Dwyer J, Short C, Berriet-Solliec M, Gael-Lataste F, Pham HV, Affleck M, Courtney P, Déprès C (2012) Refocusing on nature: holistic assessment of ecosystem services. Integr Environ Assess Manag 8(4):501–510

Enfors EI, Gordon LJ (2007) Analysing resilience in dryland agro-ecosystems: a case study of the Makanya catchment in Tanzania over the past 50 years. Land Degrad Dev 18(6):680–696
Farber SC, Costanza R, Wilson MA (2002) Economic and ecological concepts for valuing ecosystem services. Ecol Econ 41(3):375–392
Fisher B, Turner RK, Morling P (2009) Defining and classifying ecosystem services for decision making. Ecol Econ 68(3):643–653
Foley JA, DeFries R, Asner GP et al (2005) Global consequences of land use. Science 309(5734):570–574
Food and Agriculture Organization of the United Nations (FAO) (2017) The future of food and agriculture—Trends and challenges
Food and Agriculture Organization of the United Nations (FAO) (2020) Water use in agriculture. Retrieved from https://www.fao.org/water/en/
Goldenberg R, Kalantari Z, Cvetkovic V, Mörtberg U, Deal B, Destouni G (2017) Distinction, quantification and mapping of potential and realized supply-demand of flow-dependent ecosystem services. Sci Total Environ 593–594:599–609
Hagenlocher M, Cramer L, López-Carr D, Ruiz J (2018) A framework for integrating ecosystem services indicators into vulnerability and risk assessments of deltaic social–ecological systems. Reg Environ Change 18(3):735–749
Huang Z, Zhang Y, Wang F, Cao W (2021) Advancing a novel large-scale assessment integrating ecosystem service flows and real human needs: a comparison between China and the United States. J Cleaner Prod 314:128022
Ignatyeva M, Yurak V, Dushin A (2023) Valuating natural resources and ecosystem services: systematic review of methods in use. Environ Sci Policy 132(1):1–12
International Renewable Energy Agency (IRENA) (2021) World energy transitions outlook
IPCC (2021) Climate Change 2021: The Physical Science Basis. Contribution of Working Group I to the Sixth Assessment Report of the Intergovernmental Panel on Climate Change. Cambridge University Press.
King E, Cavender-Bares J, Balvanera P, Mwampamba TH, Polasky S (2015) Trade-offs in ecosystem services and varying stakeholder preferences: evaluating conflicts, obstacles, and opportunities. Ecol Soc 20(3)
Kremen C (2005) Managing ecosystem services: what do we need to know about their ecology? Ecol Lett 8(5):468–479
Kremen C, Merenlender AM, Murphy DD (1994) Ecological monitoring: a vital need for integrated conservation and development programs in the tropics. Conserv Biol 8(2):388–397
Krzemień A, Álvarez Fernández JJ, Riesgo Fernández P, Fidalgo Valverde G, Garcia-Cortes S (2023) Valuation of ecosystem services based on EU carbon allowances—Optimal recovery for a coal mining area. Environ Sci Policy 132(1):1–12
Kumar S, Madegowda M, Singh RK, Kaushal R … et al (2024) Changes in the value of ecosystem services due to watershed development in India's Eastern Ghats and incentives for better stewardship. Ecosyst Serv 65(2):Article 101580
LaCanne CE, Lundgren JG (2018) Regenerative agriculture: merging farming and natural resource conservation profitably. PeerJ 6:e4428
Laurance WF, Sayer J, Cassman KG (2014) Agricultural expansion and its impacts on tropical nature. Trends Ecol Evol 29(2):107–116
Lavorel S, Locatelli B, Colloff MJ, Bruley E (2020) Co-producing ecosystem services for adapting to climate change. Philos Trans R Soc B 375(1794):20190119
Meraj G, Kanga S, Kranjčić N, Đurin B, Singh SK (2022) Role of natural capital economics for sustainable management of earth resources. Environ Sci Policy 132(1):1–12
Millennium Ecosystem Assessment (MA) (2003) Ecosystems and human well-being. A framework for assessment. Island Press, Washington, DC, USA
Millennium Ecosystem Assessment (2005) Ecosystems and human well-being: synthesis. Island Press

Montgomery DR (2007) Soil erosion and agricultural sustainability. Proc Natl Acad Sci 104(33):13268–13272

Müller F, de Groot R, Willemen L (2021) Ecosystem services at the landscape scale: The need for integrative approaches. Ecol Ind 133:1–13

Nelson E, Mendoza G, Regetz J, Polasky S, Tallis H, Cameron D, Chan KM, Daily GC, Goldstein J, Kareiva PM, Lonsdorf E (2009) Modeling multiple ecosystem services, biodiversity conservation, commodity production, and tradeoffs at landscape scales. Front Ecol Environ 7(1):4–11

Nelson GC, Bennett E, Berhe AA, Cassman K, DeFries R, Dietz T, Dobermann A, Dobson A, Janetos A, Levy M, Marco D (2006) Anthropogenic drivers of ecosystem change: an overview. Ecol Soc 11(2)

Nepstad D, Stickler CM, Soares-Filho B, Merry F (2008) Interactions among Amazon land use, forests, and climate: prospects for a near-term forest tipping point. Philos Trans Roy Soc B: Biol Sci 363(1498):1737–1746

Newman DJ, Cragg GM (2016) Natural products as sources of new drugs from 1981 to 2014. J Nat Prod 79(3):629–661

Pinsky ML, Byler D (2015) Fishing, fast growth, and climate variability increase the risk of collapse. Proc Roy Soc B: Biol Sci 282(1813):20151053

Porras–Agüera JA, Moreno–García J, García–Martínez T, Moreno J, Mauricio JC (2021) Impact of CO_2overpressure on yeast mitochondrial associated proteome during the "prise de mousse" of sparkling wine production. Int J Food Microbiol 348:109226

Revell D, King P, Giliam J, Calil J, Jenkins S, Helmer C, Nakagawa J, Snyder A, Ellis J, Jamieson M (2022) A holistic framework for evaluating adaptation approaches to coastal hazards and sea level rise: a case study from Imperial Beach, California. Coastal Manage 50(6):1–15

Rockström J, Steffen W, Noone K et al (2009) Planetary boundaries: exploring the safe operating space for humanity. Ecol Soc 14(2):32–50

Rodríguez JP, Beard Jr TD, Bennett EM, Cumming GS, Cork SJ, Agard J, Dobson AP, Peterson GD (2006) Trade-offs across space, time, and ecosystem services. Ecol Soc 11(1)

Rodríguez JP, Beard T, Agard J, Bennett E, Cork S, Cumming G, Deane D, Dobson A, Lodge D, Mutale M, Nelson G, Peterson G, Ribeiro T (2005) Interactions among ecosystem services. In: Carpenter S, Pingali P, Bennett E, Zurek M (eds) Scenarios Millennium Ecosyst Assessment (12) (Island Press)

Rosegrant MW, Paisner MS, Meijer S, Witcover J (2001) Global food projections to 2020: emerging trends and alternative futures. 2020 Vision Food Policy Report. International Food Policy Research Institute

Ruotsalainen P, Blobel B (2022) Future pHealth ecosystem-holistic view on privacy and trust. Future pHealth 1(1):1–10

Salman A, Jaafar M, Mohamad D, Malik S (2021) Ecotourism development in Penang Hill: a multistakeholder perspective towards achieving environmental sustainability. Environ Sci Pollut Res 28:42945–42958

Schirpke U, Candiago S, Egarter Vigl L, Jäger H, Labadini A, Marsoner T, Meisch C, Tasser E, Tappeiner U (2019) Integrating supply, flow and demand to enhance the understanding of interactions among multiple ecosystem services. Sci Total Environ 651:928–941

Schirpke U, Ghermandi A, Sinclair M, Van Berkel D, Fox N, Vargas L, Willemen L (2023) Emerging technologies for assessing ecosystem services: a synthesis of opportunities and challenges. Ecosyst Serv 63:Article 101558

Schröter D, Cramer W, Leemans R, Prentice IC, Araújo MB, Arnell NW, Bondeau A, Bugmann H, Carter TR, Gracia CA, de la Vega-Leinert AC, Erhard M, Ewert F, Glendining M, House JI, Kankaanpää S, Klein RJT, Lavorel S, Lindner M ... Zierl B (2005) Ecosystem service supply and vulnerability to global change in Europe. Sci 310(5752):1333–1337

Schröter M, van der Zanden EH, van Oudenhoven APE, Remme RP, Serna Chavez HM, de Groot RS, Opdam P (2014) Ecosystem services as a contested concept: a synthesis of critique and counter arguments. Conserv Lett 7(6):514–523

Searchinger T et al (2018) Creating a sustainable food future: a menu of solutions to feed nearly 10 billion people by 2050. *World Resources Institute.*
Seppelt R, Lautenbach S, Volk M (2013) Identifying trade-offs between ecosystem services, land use, and biodiversity: a plea for combining scenario analysis and optimization on different spatial scales. Curr Opin Environ Sustain 5(5):458–463
Serna-Chavez HM, Schulp CJE, van Bodegom PM, Bouten W, Verburg PH, Davidson MD (2014) A quantitative framework for assessing spatial flows of ecosystem services. Ecol Ind 39:24–33
Souther S, Colombo S, Lyndon NN (2023) Integrating traditional ecological knowledge into US public land management: knowledge gaps and research priorities. Ecol Appl 33(4):e2519
Tallis H, Mooney H, Andelman S, Balvanera P, Cramer W, Karp D, Polasky S, Reyers B, Ricketts T, Running S, Thonicke K (2012) A global system for monitoring ecosystem service change. Bioscience 62(11):977–986
Trégarot E, Failler P (2023) Adequacy of ecosystem services assessment tools and approaches to current policy needs and gaps in the European Union overseas entities. Environ Manage 61(2):245–258
Vallecillo S, La Notte A, Ferrini S, Maes J (2019a) How ecosystem services are changing: an accounting application at the EU level. Ecosyst Serv 40:101044
Vallecillo S, La Notte A, Zulian G, Ferrini S, Maes J (2019b) Ecosystem services accounts: valuing the actual flow of nature-based recreation from ecosystems to people. Ecol Model 392:196–211
Vigl LE, Depellegrin D, Pereira P, de Groot R, Tappeiner U (2017) Mapping the ecosystem service delivery chain: capacity, flow, and demand pertaining to aesthetic experiences in mountain landscapes. Sci Total Environ 574:422–436
Wang T, Chen X, Zheng X, Lu Y, Han F, Yang Z (2022a) Identification of priority conservation areas for natural heritage sites integrating landscape ecological risks and ecosystem services: a case study in the Bogda, China. Landscape Ecol 37(4):1237–1250
Wang X, Peng J, Luo Y (2022b) Exploring social-ecological impacts on trade-offs and synergies among ecosystem services. Ecol Econ 197:107369
Xie G, Liu J, Xu J, Xiao Y, Zhen L, Zhang C ... Jiang Y (2019) A spatio-temporal delineation of transboundary ecosystem service flows from Inner Mongolia. Environ Res Lett 14(6):065002
Yang S, Zhao W, Liu Y, Wang S, Wang J, Zhai R (2018) Influence of land use change on the ecosystem service trade-offs in the ecological restoration area: dynamics and scenarios in the Yanhe watershed, China. Sci Total Environ 644:556–566
Zhang Z, Liu Y, Wang Y (2020) What factors affect the synergy and trade-off between ecosystem services, and how, from a geospatial perspective? J Clean Prod 257:120536
Zheng H, Wang L, Wu T (2019) Coordinating ecosystem service trade-offs to achieve win-win outcomes: a review of the approaches. J Environ Sci 82:26–34

The manufacturer's authorised representative in the EU is Springer Nature Customer Service Centre GmbH, Europaplatz 3, 69115 Heidelberg, Germany. If you have any concerns regarding our products, please contact ProductSafety@springernature.com

Printed and bound by CPI Group (UK) Ltd, Croydon, CR0 4YY
30/09/2026
02225567-0018